Principles of Biochemistry

Geoffrey L. Zubay

Columbia University

William W. Parson

University of Washington

Dennis E. Vance

University of Alberta

WCB **Wm. C. Brown Publishers**
Dubuque, Iowa•Melbourne, Australia•Oxford, England

Book Team

Editor *Elizabeth M. Sievers*
Developmental Editor *Robin P. Steffek*
Publishing Services Coordinator *Julie Avery Kennedy*
Photo Editor *Lori Hancock*
Permissions Coordinator *Karen L. Storlie*

Wm. C. Brown Publishers
A Division of Wm. C. Brown Communications, Inc.

Vice President and General Manager *Beverly Kolz*
Vice President, Publisher *Kevin Kane*
Vice President, Director of Sales and Marketing *Virginia S. Moffat*
Vice President, Director of Production *Colleen A. Yonda*
National Sales Manager *Douglas J. DiNardo*
Marketing Manager *Patrick E. Reidy*
Advertising Manager *Janelle Keeffer*
Production Editorial Manager *Renée Menne*
Publishing Services Manager *Karen J. Slaght*
Royalty/Permissions Manager *Connie Allendorf*

Wm. C. Brown Communications, Inc.

President and Chief Executive Officer *G. Franklin Lewis*
Senior Vice President, Operations *James H. Higby*
Corporate Senior Vice President, President of WCB Manufacturing *Roger Meyer*
Corporate Senior Vice President and Chief Financial Officer *Robert Chesterman*

Cover Images:
   Main Cover and Volume 1: Image copyright 1994 by the Scripps Research Institute/
   Molecular Graphics Images by Michael Pique using software by Yng Chen, Michael
   Connolly, Michael Carson, Alex Shah, and AVS, Inc. Endonuclease III by Kuo,
   McRee, Fisher, O'Handley, Cunningham, and Tainer

   Volume II: Courtesy of Dr. Klaus Piontek

   Volume III: Courtesy of Dr. John Kuriyan

Freelance Permissions Editor Karen Dorman
Copyediting and Production by York Production Services
Design by York Production Services
Composition by York Graphic Services

The credits section for this book begins on page C-1 and is considered
an extension of the copyright page.

Contributors to the End-of-Chapter Problems
   Chapters 2–24
      Hugh Akers
      Lamar University

   Chapters 25–31
      Caroline Breitenberger
      Ohio State University

Copyright © 1995 by Wm. C. Brown Communications, Inc.
All rights reserved

A Times Mirror Company

Library of Congress Catalog Card Number: 94–70034 Complete
                                    94–70098 Volume One, Two, and Three

ISBN 0–697–24169–6—Volume One
     0–697–24170–X—Volume Two
     0–697–24171–8—Volume Three
     0–697–14275–2—Complete

Printed in the United States of America by Wm. C. Brown Communications, Inc.,
2460 Kerper Boulevard, Dubuque, IA 52001

10 9 8 7 6 5 4 3 2 1

This text is dedicated to:

Bongsoon
Polly, and
Jean

Publisher's Note

Principles of Biochemistry, first edition, is available as a single, full-length casebound text or in three paperback volumes, which may be used separately or together in any combination that best suits your course needs.

Binding Option	Description	ISBN
Principles of Biochemistry, 1/E (Casebound)	The full-length text, Chapters 1–31, plus four Supplements.	0-697-14275-2
Principles of Biochemistry, 1/E Volume 1: Energy, Proteins, and Catalysis (Paperback)	Chapters 1–10	0-697-24169-6
Principles of Biochemistry, 1/E Volume 2: Metabolism (Paperback)	Chapters 11–24, plus Supplements 1 and 2.	0-697-24170-X
Principles of Biochemistry, 1/E Volume 3: Molecular Genetics (Paperback)	Chapters 25–31, plus Supplements 3 and 4.	0-697-24171-8

Brief Contents

Volume One

Energy, Proteins, and Catalysis

Part 1

An Overview of Biochemical Structures and Reactions That Occur in Living Systems 1

Chapter 1 Cells, Biomolecules, and Water 3
Chapter 2 Thermodynamics in Biochemistry 29

Chapter 5 Functional Diversity of Proteins 101
Chapter 6 Methods for Characterization and Purification of Proteins 118

Part 2

Protein Structure and Function 47

Chapter 3 The Building Blocks of Proteins: Amino Acids, Peptides, and Polypeptides 49
Chapter 4 The Three-Dimensional Structures of Proteins 77

Part 3

Catalysis 133

Chapter 7 Enzyme Kinetics 135
Chapter 8 How Enzymes Work 154
Chapter 9 Regulation of Enzyme Activities 175
Chapter 10 Vitamins and Coenzymes 198

Volume Two

Metabolism

Part 4

Metabolism of Carbohydrates 225

Chapter 11 Metabolic Strategies 227
Chapter 12 Glycolysis, Gluconeogenesis, and the Pentose Phosphate Pathway 242

Chapter 13 The Tricarboxylic Acid Cycle 282
Chapter 14 Electron Transport and Oxidative Phosphorylation 305
Chapter 15 Photosynthesis 330
Chapter 16 Structures and Metabolism of Oligosaccharides and Polysaccharides 356

Part 5

Metabolism of Lipids 379

Chapter 17 Structure and Functions of Biological
Membranes 381
Chapter 18 Metabolism of Fatty Acids 411
Chapter 19 Biosynthesis of Membrane Lipids 436
Chapter 20 Metabolism of Cholesterol 459

Part 6

Metabolism of Nitrogen-Containing Compounds 485

Chapter 21 Amino Acid Biosynthesis and Nitrogen
Fixation in Plants and Microorganisms 436
Chapter 22 Amino Acid Metabolism in Vertebrates 511
Chapter 23 Nucleotides 533
Chapter 24 Integration of Metabolism and Hormone
Action 562
Supplement 1 Principles of Physiology and
Biochemistry: Neurotransmission 602
Supplement 2 Principles of Physiology and
Biochemistry: Vision 614

Volume Three

Molecular Genetics

Part 7

Storage and Utilization of Genetic Information 625

Chapter 25 Structures of Nucleic Acids and
Nucleoproteins 627
Chapter 26 DNA Replication, Repair, and
Recombination 650
Chapter 27 DNA Manipulation and Its Applications 678
Chapter 28 RNA Synthesis and Processing 700
Chapter 29 Protein Synthesis, Targeting, and
Turnover 730

Chapter 30 Regulation of Gene Expression in
Prokaryotes 768
Chapter 31 Regulation of Gene Expression in
Eukaryotes 800
Supplement 3 Principles of Physiology and
Biochemistry: Immunobiology 830
Supplement 4 Principles of Physiology and
Biochemistry: Carcinogenesis and
Oncogenes 848

Extended Contents

List of Supplementary Text Elements xx
Preface xxi

Part 1

An Overview of Biochemical Structures and Reactions That Occur in Living Systems 1

Chapter 1

Cells, Biomolecules, and Water 3

The Cell Is the Fundamental Unit of Life 4
Cells Are Composed of Small Molecules, Macromolecules, and Organelles 7
Macromolecules Fold into Complex Three-Dimensional Structures 9
Water Is a Primary Factor in Determining the Type of Structures That Form 9
Biochemical Reactions Are a Subset of Ordinary Chemical Reactions 16
Biochemical Reactions Take Place on the Catalytic Surfaces of Enzymes 18
Many Biochemical Reactions Require Energy 19
Biochemical Reactions Are Localized in the Cell 20
Biochemical Reactions Are Organized into Pathways 20
Biochemical Reactions Are Regulated 21

Organisms Are Biochemically Dependent on One Another 21
Information for the Synthesis of Proteins Is Carried by the DNA 22
Biochemical Systems Have Been Evolving for Almost Four Billion Years 22
All Living Systems Are Related through a Common Evolution 25

Chapter 2

Thermodynamics in Biochemistry 29

Thermodynamic Quantities 30
The First Law of Thermodynamics: Any Change in the Energy of a System Requires an Equal and Opposite Change in the Surroundings 30
The Second Law of Thermodynamics: In Any Spontaneous Process the Total Entropy of the System and the Surroundings Increases 31
Free Energy Provides the Most Useful Criterion for Spontaneity 35
Applications of the Free Energy Function 35
Values of Free Energy Are Known for Many Compounds 36
The Standard Free Energy Change in a Reaction Is Related Logarithmically to the Equilibrium Constant 36
Free Energy Is the Maximum Energy Available for Useful Work 38
Biological Systems Perform Various Kinds of Work 38
Favorable Reactions Can Drive Unfavorable Reactions 38
ATP as the Main Carrier of Free Energy in Biochemical Systems 40
The Hydrolysis of ATP Yields a Large Amount of Free Energy 40

Part 2

Protein Structure and Function 47

Chapter 3

The Building Blocks of Proteins: Amino Acids, Peptides, and Polypeptides 49

Amino Acids 50
    Amino Acids Have Both Acid and Base Properties 52
    Aromatic Amino Acids Absorb Light in the Near-Ultraviolet 55
    All Amino Acids except Glycine Show Asymmetry 56
Peptides and Polypeptides 56
Determination of Amino Acid Composition of Proteins 58
Determination of Amino Acid Sequence of Proteins 61
Chemical Synthesis of Peptides and Polypeptides 66

Chapter 4

The Three-Dimensional Structures of Proteins 77

Pauling and Corey Provided the Foundation for Our Understanding of Fibrous Protein Structures 73
    The Structure of the α-Keratins Was Determined with the Help of Molecular Models 75
    The β-Keratins Form Sheetlike Structures with Extended Polypeptide Chains 77
    Collagen Forms a Unique Triple-Stranded Structure 79
Globular Protein Structures Are Extremely Varied and Require a More Sophisticated Form of Analysis 80
Folding of Globular Proteins Reveals a Hierarchy of Structural Organization 82
    Visualizing Folded Protein Structures 82
    Primary Structure Determines Tertiary Structure 82
    Secondary Valence Forces Are the Glue That Holds Polypeptide Chains Together 86
    Domains Are Functional Units of Tertiary Structure 88
    Predicting Protein Tertiary Structure 90
    Quaternary Structure Involves the Interaction of Two or More Proteins 91

Chapter 5

Functional Diversity of Proteins 101

Hemoglobin Is an Allosteric Oxygen-Binding Protein 101
    The Binding of Certain Factors to Hemoglobin Influences Oxygen Binding in a Negative Way 103
    X-ray Diffraction Studies Reveal Two Conformations for Hemoglobin 104
    Changes in Conformation Are Initiated by Oxygen Binding 107
    Alternative Theories on How Hemoglobins and Other Allosteric Proteins Work 109
Muscle Is an Aggregate of Proteins Involved in Contraction 110

Chapter 6

Methods for Characterization and Purification of Proteins 118

Methods of Protein Fractionation and Characterization 119
    Differential Centrifugation Divides a Sample into Two Fractions 119
    Differential Precipitation Is Based on Solubility Differences 119
    Column Procedures Are the Most Versatile and Productive Purification Methods 120
    Electrophoresis Is Used for Resolving Mixtures 122
    Sedimentation and Diffusion Are Used for Size and Shape Determination 123
Protein Purification Procedures 124
    Purification of an Enzyme with Two Catalytic Activities 125
    Purification of a Membrane-Bound Protein 127

Part 3

Catalysis 133

Chapter 7

Enzyme Kinetics 135

The Discovery of Enzymes 136
Enzyme Terminology 136
Basic Aspects of Chemical Kinetics 137
    A Critical Amount of Energy Is Needed for the Reactants to Reach the Transition State 138
    Catalysts Speed up Reactions by Lowering the Free Energy of Activation 139
Kinetics of Enzyme-Catalyzed Reactions 140
    Kinetic Parameters Are Determined by Measuring the Initial Reaction Velocity as a Function of the Substrate Concentration 140

The Henri-Michaelis-Menten Treatment Assumes That the Enzyme–Substrate Complex Is in Equilibrium with Free Enzyme and Substrate 140

Steady-State Kinetic Analysis Assumes That the Concentration of the Enzyme–Substrate Complex Remains Nearly Constant 141

Kinetics of Enzymatic Reactions Involving Two Substrates 144

Effects of Temperature and pH on Enzymatic Activity 146

Enzyme Inhibition 146

Competitive Inhibitors Bind at the Active Site 147

Noncompetitive and Uncompetitive Inhibitors Do Not Compete Directly with Substrate Binding 149

Irreversible Inhibitors Permanently Alter the Enzyme Structure 149

Chapter 8

How Enzymes Work 154

General Themes in Enzymatic Mechanisms 154

The Proximity Effect: Enzymes Bring Reacting Species Close Together 155

General-Base and General-Acid Catalysis Avoids the Need for Extremely High or Low pH 155

Electrostatic Interactions Can Promote the Formation of the Transition State 156

Enzymatic Functional Groups Provide Nucleophilic and Electrophilic Catalysis 157

Structural Flexibility Can Increase the Specificity of Enzymes 158

Detailed Mechanisms of Enzyme Catalysis 159

Serine Proteases: Enzymes That Use a Serine Residue for Nucleophilic Catalysis 159

Ribonuclease A: An Example of Concerted Acid–Base Catalysis 165

Triosephosphate Isomerase Has Approached Evolutionary Perfection 169

Chapter 9

Regulation of Enzyme Activities 175

Partial Proteolysis Results in Irreversible Covalent Modifications 176

Phosphorylation, Adenylylation, and Disulfide Reduction Lead to Reversible Covalent Modifications 177

Allosteric Regulation Allows an Enzyme to Be Controlled Rapidly by Materials That Are Structurally Unrelated to the Substrate 180

Allosteric Enzymes Typically Exhibit a Sigmoidal Dependence on Substrate Concentration 180

The Symmetry Model Provides a Useful Framework for Relating Conformational Transitions to Allosteric Activation or Inhibition 182

Phosphofructokinase: Allosteric Control of Glycolysis Is Consistent with the Symmetry Model 183

Aspartate Carbamoyl Transferase: Allosteric Control of Pyrimidine Biosynthesis 187

Glycogen Phosphorylase: Combined Control by Allosteric Effectors and Phosphorylation 191

Chapter 10

Vitamins and Coenzymes 198

Water-Soluble Vitamins and Their Coenzymes 199

Thiamine Pyrophosphate Is Involved in C—C and C—X Bond Cleavage 199

Pyridoxal-5'-Phosphate Is Required for a Variety of Reactions with α-Amino Acids 200

Nicotinamide Coenzymes Are Used in Reactions Involving Hydride Transfers 203

Flavins Are Used in Reactions Involving One or Two Electron Transfers 207

Reactions Requiring Acyl Activation Frequently Use Phosphopantetheine Coenzymes 210

α-Lipoic Acid Is the Coenzyme of Choice for Reactions Requiring Acyl-Group Transfers Linked to Oxidation–Reduction 212

Biotin Mediates Carboxylations 213

Folate Coenzymes Are Used in Reactions for One-Carbon Transfers 215

Ascorbic Acid Is Required to Maintain the Enzyme That Forms Hydroxyproline Residues in Collagen 216

Vitamin B_{12} Coenzymes Are Associated with Rearrangements on Adjacent Carbon Atoms 216

Iron-Containing Coenzymes Are Frequently Involved in Redox Reactions 217

Metal Cofactors 220

Lipid-Soluble Vitamins 220

Part 4

Metabolism of Carbohydrates 225

Chapter 11

Metabolic Strategies 227

Living Cells Require a Steady Supply of Starting Materials and Energy 227

Organisms Differ in Sources of Energy, Reducing Power, and Starting Materials for Biosynthesis 228
Reactions Are Organized into Sequences or Pathways 229
Sequentially Related Enzymes Are Frequently Clustered 229
Pathways Show Functional Coupling 231
The ATP–ADP System Mediates Conversions in Both Directions 232
Conversions Are Kinetically Regulated 233
Pathways Are Regulated by Controlling Amounts and Activities of Enzymes 234
   Enzyme Activity Is Regulated by Interaction with Regulatory Factors 234
   Regulatory Enzymes Occupy Key Positions in Pathways 234
   Regulatory Enzymes Often Show Cooperative Behavior 235
   Both Anabolic and Catabolic Pathways Are Regulated by the Energy Status of the Cell 235
Strategies for Pathway Analysis 236
   Analysis of Single-Step Pathways 237
   Analysis of Multistep Pathways 237
   Radiolabeled Compounds Facilitate Pathway Analysis 238
   Pathways Are Usually Studied Both in Vitro *and in* Vivo 239

Chapter 12

Glycolysis, Gluconeogenesis, and the Pentose Phosphate Pathway 242

Carbohydrates Important in Energy Metabolism 243
   Monosaccharides and Related Compounds 243
   In Disaccharides the Monosaccharides Are Linked by Glycosidic Bonds 245
   Starch and Glycogen Are Major Energy-Storage Polysaccharides 248
   Cellulose Serves a Structural Function Despite Its Similarity in Composition to the Energy-Storage Polysaccharides 248
   The Configurations of Glycogen and Cellulose Dictate Their Roles 249
The Synthesis and Breakdown of Sugars 249
Overview of Glycolysis 249
   Three Hexose Phosphates Constitute the First Metabolic Pool 251
   Phosphorylase Converts Storage Carbohydrates to Glucose Phosphate 251

   Hexokinase and Glucokinase Convert Free Sugars to Hexose Phosphates 253
   Phosphoglucomutase Interconverts Glucose-1-phosphate and Glucose-6-phosphate 254
   Phosphohexoseisomerase Interconverts Glucose-6-phosphate and Fructose-6-phosphate 254
   Formation of Fructose-1,6-bisphosphate Signals a Commitment to Glycolysis 256
   Fructose-1,6-bisphosphate and the Two Triose Phosphates Constitute the Second Metabolic Pool in Glycolysis 256
   Aldolase Cleaves Fructose-1,6-bisphosphate 257
   Triose Phosphate Isomerase Interconverts the Two Trioses 257
   The Conversion of Triose Phosphates to Phosphoglycerates Occurs in Two Steps 257
   The Three-Carbon Phosphorylated Acids Constitute a Third Metabolic Pool 259
   Conversion of Phosphoenolpyruvate to Pyruvate Generates Another ATP 259
   The NAD^+ Reduced in Glycolysis Must Be Regenerated 259
   Summary of Glycolysis 261
Gluconeogenesis 262
   Gluconeogenesis Consumes ATP 263
   Conversion of Pyruvate to Phosphoenolpyruvate Requires Two High Energy Phosphates 263
   Conversion of Phosphoenolpyruvate to Fructose-1,6-bisphosphate Uses the Same Enzymes as Glycolysis 264
   Fructose-bisphosphate Phosphatase Converts Fructose-1,6-bisphosphate to Fructose-6-phosphate 264
   Hexose Phosphates Can Be Converted to Storage Polysaccharides 264
   Summary of Gluconeogenesis 266
Regulation of Glycolysis and Gluconeogenesis 266
   How Do Intracellular Signals Regulate Energy Metabolism? 267
   Hormonal Controls Can Override Intracellular Controls 267
   Hormonal Effects of Glucagon Are Mediated by Cyclic AMP 268
   The Hormone Epinephrine Stimulates Glycolysis in Both Liver Cells and Muscle Cells 270
   The Hormonal Regulation of the Flux between Fructose-6-phosphate and Fructose-1,6-bisphosphate Is Mediated by Fructose-2,6-bisphosphate 270
   Summary of the Regulation of Glycolysis and Gluconeogenesis 270
The Pentose Phosphate Pathway 272
   Two NADPH Molecules Are Generated by the Pentose Phosphate Pathway 272
   Transaldolase and Transketolase Catalyze the Interconversion of Many Phosphorylated Sugars 273
   Production of Ribose-5-phosphate and Xylulose-5-phosphate 274

Chapter 13

The Tricarboxylic Acid Cycle 282

Discovery of the TCA Cycle 283
Steps in the TCA Cycle 285
  The Oxidative Decarboxylation of Pyruvate Leads to Acetyl-CoA 287
  Citrate Synthase Is the Gateway to the TCA Cycle 289
  Aconitase Catalyzes the Isomerization of Citrate to Isocitrate 289
  Isocitrate Dehydrogenase Catalyzes the First Oxidation in the TCA Cycle 289
  α-Ketoglutarate Dehydrogenase Catalyzes the Decarboxylation of α-Ketoglutarate to Succinyl-CoA 290
  Succinate Thiokinase Couples the Conversion of Succinyl-CoA to Succinate with the Synthesis of GTP 291
  Succinate Dehydrogenase Catalyzes the Oxidation of Succinate to Fumarate 291
  Fumarase Catalyzes the Addition of Water to Fumarate to Form Malate 292
  Malate Dehydrogenase Catalyzes the Oxidation of Malate to Oxaloacetate 292
Stereochemical Aspects of TCA Cycle Reactions 293
ATP Stoichiometry of the TCA Cycle 293
Thermodynamics of the TCA Cycle 294
The Amphibolic Nature of the TCA Cycle 295
The Glyoxylate Cycle Permits Growth on a Two-Carbon Source 295
Oxidation of Other Substrates by the TCA Cycle 295
The TCA Cycle Activity Is Regulated at Metabolic Branchpoints 299
  The Pyruvate Branchpoint Partitions Pyruvate between Acetyl-CoA and Oxaloacetate 299
  Citrate Synthase Is Negatively Regulated by NADH and the Energy Charge 300
  Isocitrate Dehydrogenase Is Regulated by the NADH-to-NAD$^+$ Ratio and the Energy Charge 300
  α-Ketoglutarate Dehydrogenase Is Negatively Regulated by NADH 301

Chapter 14

Electron Transport and Oxidative Phosphorylation 305

The Mitochondrial Electron-Transport Chain 307
  A Bucket Brigade of Molecules Carries Electrons from the TCA Cycle to O$_2$ 307
  The Sequence of Electron Carriers Was Deduced from Kinetic Measurements 309
  Redox Potentials Give a Measure of Oxidizing and Reducing Strengths 310
  Most of the Electron Carriers Occur in Large Complexes 312
  Reconstitution Experiments Demonstrate the Need for Carriers to Mediate Electron Transfer between Complexes 316
Oxidative Phosphorylation 316
  Electron Transfer Is Coupled to ATP Formation at Three Sites 316
  Uncouplers Release Electron Transport from Phosphorylation 317
  The Chemiosmotic Theory Proposes That Phosphorylation Is Driven by Proton Movements 318
  Electron Transport Creates an Electrochemical Potential Gradient for Protons across the Inner Membrane 318
  How Do the Electron-Transfer Reactions Pump Protons across the Membrane? 321
  Flow of Protons Back into the Matrix Drives the Formation of ATP 321
  Reconstitution Experiments Demonstrate the Components of the Proton-Conducting ATP-Synthase 322
Transport of Substrates, P$_i$, ADP, and ATP into and out of Mitochondria 324
  Uptake of P$_i$ and Oxidizable Substrates Is Coupled to the Release of Other Compounds 324
  Export of ATP Is Coupled to ADP Uptake 324
  Electrons from Cytosolic NADH Are Imported by Shuttle Systems 325
  Complete Oxidation of Glucose Yields about 30 Molecules of ATP 325

Chapter 15

Photosynthesis 330

The Photochemical Reactions of Photosynthesis Take Place in Membranes 332
Photosynthesis Depends on the Photochemical Reactivity of Chlorophyll 332
Light Is Composed of Photons 333
Photons of Light Interact with Electrons in Molecules 335
Photooxidation of Chlorophyll Generates a Cationic Free Radical 336
The Reactive Chlorophyll Is Bound to Proteins in Reaction Centers 337
In Purple Bacterial Reaction Centers, Electrons Move from P870 to Bacteriopheophytin and Then to Quinones 338
A Cyclic Electron-Transport Chain Moves Protons Outward across the Membrane That Drive the Formation of ATP 339

An Antenna System Transfers Energy to the Reaction
   Centers 340
Chloroplasts Have Two Photosystems Linked in
   Series 342
Photosystem I Reduces NADP$^+$ by Way of Iron–Sulfur
   Proteins 345
O_2 Evolution Requires the Accumulation of Four
   Oxidizing Equivalents in the Reaction Center of
   Photosystem II 345
Flow of Electrons from H_2O to NADP$^+$ Drives Proton
   Transport into the Thylakoid Lumen; Protons Return
   to the Stroma through an ATP-Synthase 346
Carbon Fixation Utilizes the Reductive Pentose
   Cycle 348
Ribulose-Bisphosphate Carboxylase-Oxygenase:
   Photorespiration and the C-4 Cycle 350

Chapter 16

Structures and Metabolism of Oligosaccharides and
Polysaccharides 356

Many Types of Monosaccharides Become Integrated into
   Polysaccharides 357
   *Monosaccharides Are Related through a Network of
      Metabolic Interconversions 357*
   *Galactosemia Is a Genetically Inherited Disease That
      Results from the Inability to Convert Galactose into
      Glucose 358*
Structural Polysaccharides Include Homopolymers and
   Heteropolymers 358
Oligosaccharides Are Found in Glycoconjugates 359
   *Oligosaccharides Are Synthesized by Specific
      Glycosyltransferases 362*
   *N-Linked Oligosaccharides Utilize a Lipid Carrier in the
      Early Stages of Synthesis 362*
   *Carbohydrate Modification Is Crucial to Targeting
      Lysosomal Enzymes 365*
   *Biosynthesis of O-Linked Oligosaccharides Begins in the
      Cis-Golgi 367*
   *O-Linked Oligosaccharides Are Responsible for Different
      Blood Group Types 368*
Specific Inhibitors and Mutants Are Used to Explore the
   Roles of Glycoprotein Carbohydrates 368
Bacterial Cell Wall Synthesis 370
   *Synthesis of Activated Monomers Occurs in the
      Cytoplasm 370*
   *Formation of Linear Polymers Is Membrane-
      Associated 371*
   *Cross-Linking of Linear Polymers Occurs Outside the
      Plasma Membrane 372*
   Penicillin Inhibits the Transpeptidation Reaction 374

Part 5

Metabolism of Lipids 379

Chapter 17

Structure and Functions of Biological Membranes 381

The Structure of Biological Membranes 383
   Membranes Contain Complex Mixtures of Lipids 383
   *Phospholipids Spontaneously Form Ordered Structures in
      Water 386*
   *Membranes Have Both Integral and Peripheral
      Proteins 388*
   *Integral Membrane Proteins Contain Transmembrane α
      Helices 389*
   *Proteins and Lipids Can Move around within
      Membranes 390*
   Biological Membranes Are Asymmetrical 393
   *Membrane Fluidity Is Sensitive to Temperature and Lipid
      Composition 395*
   *Some Proteins of Eukaryotic Plasma Membranes Are
      Connected to the Cytoskeleton 396*
Transport of Materials across Membranes 398
   Most Solutes Are Transported by Specific Carriers 398
   *Some Transporters Facilitate Diffusion of a Solute down an
      Electrochemical Potential Gradient 400*
   *Active Transport against an Electrochemical Potential
      Gradient Requires Energy 401*
   *Isotopes, Substrate Analogs, Membrane Vesicles, and
      Bacterial Mutants Are Used to Study Transport 402*
   Molecular Models of Transport Mechanisms 403
   *The Catalytic Cycle of the Na$^+$–K$^+$ Pump Includes Two
      Phosphorylated Forms of the Enzyme 404*
   Some Membranes Have Relatively Large Pores 406
   *Other Specific Interactions Mediated by Membrane
      Proteins 407*

Chapter 18

Metabolism of Fatty Acids 411

Fatty Acid Degradation 412
   *Fatty Acids Originate from Three Sources: Diet, Adipocytes,
      and de novo Synthesis 412*
   *Fatty Acid Breakdown Occurs in Blocks of Two Carbon
      Atoms 414*
   *The Oxidation of Saturated Fatty Acids Occurs in
      Mitochondria 414*
   Fatty Acid Oxidation Yields Large Amounts of ATP 414

Additional Enzymes Are Required for Oxidation of Unsaturated Fatty Acids in Mitochondria 416

Ketone Bodies Formed in the Liver Are Used for Energy in Other Tissues 418

Summary of Fatty Acid Degradation 419

Biosynthesis of Saturated Fatty Acids 419

The First Step in Fatty Acid Synthesis Is Catalyzed by Acetyl-CoA Carboxylase 420

Seven Reactions Are Catalyzed by the Fatty Acid Synthase 421

The Organization of the Fatty Acid Synthase Is Different in E. coli and Animals 424

Biosynthesis of Monounsaturated Fatty Acids Follows Distinct Routes in E. coli and Animal Cells 424

Biosynthesis of Polyunsaturated Fatty Acids Occurs Mainly in Eukaryotes 426

Summary of Pathways for Synthesis and Degradation 427

Regulation of Fatty Acid Metabolism 427

The Release of Fatty Acids from Adipose Tissue Is Regulated 427

Transport of Fatty Acids into Mitochondria Is Regulated 429

Fatty Acid Biosynthesis Is Limited by Substrate Supply 430

Fatty Acid Synthesis Is Regulated by the First Step in the Pathway 430

The Controls for Fatty Acid Metabolism Discourage Simultaneous Synthesis and Breakdown 432

Long-Term Dietary Changes Lead to Adjustments in the Levels of Enzymes 432

Chapter 19

Biosynthesis of Membrane Lipids 436

Phospholipids 438

In E. coli, Phospholipid Synthesis Generates Phosphatidylethanolamine, Phosphatidylglycerol, and Diphosphatidylglycerol 438

Phospholipid Synthesis in Eukaryotes Is More Complex 438

Diacylglycerol Is the Key Intermediate in the Biosynthesis of Phosphatidylcholine and Phosphatidylethanolamine 441

Fatty Acid Substituents at SN-1 and SN-2 Positions Are Replaceable 441

Phosphatidylinositol-4,5-Bisphosphate, a Precursor of Second Messengers, Is Synthesized via CDP-Diacylglycerol 441

The Metabolism of Phosphatidylserine and Phosphatidylethanolamine Is Closely Linked 443

The Final Reactions for Phospholipid Biosynthesis Occur on the Cytosolic Surface of the Endoplasmic Reticulum 445

In the Liver, Regulation Gives Priority to Formation of Structural Lipids Over Energy-Storage Lipids 445

Phospholipases Degrade Phospholipids 447

Sphingolipids 447

Sphingomyelin Is Formed Directly from Ceramide 447

Glycosphingolipid Synthesis Also Starts from Ceramide 448

Glycosphingolipids Function as Structural Components and as Specific Cell Receptors 450

Defects in Sphingolipid Catabolism Are Associated with Metabolic Diseases 450

Eicosanoids Are Hormones Derived from Arachidonic Acid 452

Eicosanoid Biosynthesis 453

Eicosanoids Exert Their Action Locally 454

Chapter 20

Metabolism of Cholesterol 459

Biosynthesis of Cholesterol 461

Mevalonate Is a Key Intermediate in Cholesterol Biosynthesis 461

The Rate of Mevalonate Synthesis Determines the Rate of Cholesterol Biosynthesis 462

Six Mevalonates and 10 Steps Are Required to Make Lanosterol, the First Tetracyclic Intermediate 463

From Lanosterol to Cholesterol Takes Approximately 20 Steps 464

Summary of Cholesterol Biosynthesis 464

Lipoprotein Metabolism 465

There Are Five Types of Lipoproteins in Human Plasma 465

Lipoproteins Are Made in the Endoplasmic Reticulum of the Liver and Intestine 469

Chylomicrons and Very-Low-Density Lipoproteins (VLDLs) Transport Cholesterol and Triacylglycerol to Other Tissues 470

Low-Density Lipoproteins (LDLs) Are Removed from the Plasma by the Liver, Adrenals, and Adipose Tissue 471

Serious Diseases Result from Cholesterol Deposits 472

High-Density Lipoproteins (HDLs) May Reduce Cholesterol Deposits 472

Bile Acid Metabolism 473

Metabolism of Steroid Hormones 475

Overview of Mammalian Cholesterol Metabolism 477

Part 6

Metabolism of Nitrogen-Containing Compounds 485

Chapter 21

Amino Acid Biosynthesis and Nitrogen Fixation in Plants and Microorganisms 436

The Pathways to Amino Acids Arise as Branchpoints from the Central Metabolic Pathways 488
Our Understanding of Amino Acid Biosynthesis Has Resulted from Genetic and Biochemical Investigations 489
The Number of Proteins Participating in a Pathway Is Known through Genetic Complementation Analysis 489
Biochemists Use the Auxotrophs Isolated by Geneticists 489
The Glutamate Family of Amino Acids and Nitrogen Fixation 489
The Direct Amination of α-Ketoglutarate Leads to Glutamate 491
Amidation of Glutamate Is an Elaborately Regulated Process 491
The Nitrogen Cycle Encompasses a Series of Reactions in Which Nitrogen Passes through Many Forms 493
The Serine Family and Sulfur Fixation 495
Cysteine Biosynthesis Occurs by Sulfhydryl Transfer to Activated Serine 495
Sulfate Must Be Reduced to Sulfide before Incorporation into Amino Acids 497
The Aspartate and Pyruvate Families Both Make Contributions to the Synthesis of Isoleucine 497
Isoleucine and Valine Biosynthesis Share Four Enzymes 497
Amino Acid Pathways Absent in Mammals Offer Targets for Safe Herbicides 499
Chorismate Is a Common Precursor of the Aromatic Amino Acid Family 499
Tryptophan Is Synthesized in Five Steps from Chorismate 499
Amino Acids Act as Negative Regulators of Their Own Synthesis 502
Histidine Constitutes a Family of One 502
Nonprotein Amino Acids Are Derived from Protein Amino Acids 502
A Wide Variety of D-Amino Acids Are Found in Microbes 503

Chapter 22

Amino Acid Metabolism in Vertebrates 511

Humans and Rodents Synthesize Less Than Half of the Amino Acids They Need for Protein Synthesis 512
Many Amino Acids Are Required in the Diet for Good Nutrition 513
Essential Amino Acids Must Be Obtained by Degradation of Ingested Proteins 514
Amino Acids May Be Reutilized or They May Be Degraded When Present in Excess 515
Transamination Is the Most Widespread Form of Nitrogen Transfer 515
Net Deamination via Transamination Requires Oxidative Deamination 515
In Many Vertebrates Ammonia Resulting from Deamination Must Be Detoxified prior to Elimination 516
Urea Formation Is a Complex and Costly Mode of Ammonia Detoxification 517
The Urea Cycle and the TCA Cycle Are Linked by the Krebs Bicycle 520
More Than One Carrier Exists for Transporting Ammonia from the Muscle to the Liver 520
Amino Acid Catabolism Can Serve as a Major Source of Carbon Skeletons and Energy 521
For Many Genetic Diseases the Defect Is in Amino Acid Catabolism 523
Most Human Genetic Diseases Associated with Amino Acid Metabolism Are Due to Defects in Their Catabolism 523
Amino Acids Serve as the Precursors for Compounds Other Than Proteins 526
Porphyrin Biosynthesis Starts with the Condensation of Glycine and Succinyl-CoA 526
Glutathione Is γ-Glutamylcysteinylglycine 526

Chapter 23

Nucleotides 533

Nucleotide Components: A Phosphoryl Group, a Pentose, and a Base 535
Overview of Nucleotide Metabolism 538
Synthesis of Purine Ribonucleotides *de Novo* 538
Inosine Monophosphate (IMP) Is the First Purine Nucleotide Formed 538
IMP Is Converted into AMP and GMP 543

Synthesis of Pyrimidine Ribonucleotides *de Novo* 543
  *UMP Is a Precursor of Other Pyrimidine
    Mononucleotides* 544
  CTP Is Formed from UTP 545
Biosynthesis of Deoxyribonucleotides 545
  Thymidylate Is Formed from dUMP 546
Formation of Nucleotides from Bases and Nucleosides
  (Salvage Pathways) 548
  *Purine Phosphoribosyltransferases Convert Purines to
    Nucleotides* 548
  *Conversion of Nucleoside Monophosphates to Triphosphates
    Goes through Diphosphates* 549
Inhibitors of Nucleotide Synthesis 549
Catabolism of Nucleotides 553
  *Purines Are Catabolized to Uric Acid and Then to Other
    Products* 555
  *Pyrimidines Are Catabolized to β-Alanine, NH_3, and
    CO_2* 555
Regulation of Nucleotide Metabolism 556
  Purine Biosynthesis Is Regulated at Two Levels 556
  *Pyrimidine Biosynthesis Is Regulated at the Level of
    Carbamoyl Aspartate Formation* 558
  *Deoxyribonucleotide Synthesis Is Regulated by Both
    Activators and Inhibitors* 559
  *Enzyme Synthesis Also Contributes to Regulation of
    Deoxyribonucleotides during the Cell Cycle* 559
  *Intracellular Concentrations of Nucleotides Vary According
    to the Physiological State of the Cell* 559

Chapter 24

Integration of Metabolism and Hormone Action 562

Tissues Store Biochemical Energy in Three Major
  Forms 563
Each Tissue Makes Characteristic Demands and
  Contributions to the Energy Pool 563
  *Brain Tissue Makes No Contributions to the Fuel Needs of
    the Organism* 563
  *Heart Muscle Utilizes Fatty Acids in Preference to Glucose
    to Fulfill Its Energy Needs* 564
  *Skeletal Muscle Can Function Aerobically or
    Anaerobically* 565
  *Adipose Tissue Maintains Vast Fuel Reserves in the Form
    of Triacylglycerols* 566
  *The Liver Is the Central Clearing House for All Energy-
    Related Metabolism* 566
  *Pancreatic Hormones Play a Major Role in Maintaining
    Blood Glucose Levels* 567
Hormones Are Major Vehicles for Intercellular
  Communication 570

Hormones Are Synthesized and Secreted by Specialized
  Endocrine Glands 570
  *Polypeptide Hormones Are Stored in Secretory Granules
    after Synthesis* 570
  *Thyroid Hormones and Epinephrine Are Amino Acid
    Derivatives* 574
  Steroid Hormones Are Derived from Cholesterol 574
The Circulating Hormone Concentration Is
  Regulated 578
Hormone Action Is Mediated by Receptors 578
  *Many Plasma Membrane Receptors Generate a Diffusible
    Intracellular Signal* 580
  *The Adenylate Cyclase Pathway Is Triggered by a
    Membrane-Bound Receptor* 580
  *The G Protein Cycle Is a Target for Certain Bacterial
    Toxins* 583
  *Multicomponent Hormonal Systems Facilitate a Great
    Variety of Responses* 583
  The Guanylate Cyclase Pathway 583
  Calcium and the Inositol Trisphosphate Pathway 584
  Steroid Receptors Modulate the Rate of Transcription 586
  Hormones Are Organized into a Hierarchy 586
Diseases Associated with the Endocrine System 589
  *Overproduction of Hormones Is Commonly Caused by
    Tumor Formation* 589
  Underproduction of Hormones Has Multiple Causes 590
  *Target-Cell Insensitivity Results from a Lack of Functional
    Receptors* 590
Growth Factors 591
Plant Hormones 592

Supplement 1

Principles of Physiology and Biochemistry:
Neurotransmission 602

Pumping and Diffusion of Ions Creates an Electric
  Potential Difference across the Plasma
  Membrane 603
An Action Potential Is a Wave of Local Depolarization
  and Repolarization of the Plasma Membrane 604
Action Potentials Are Mediated by Voltage-Gated Ion
  Channels 605
Synaptic Transmission Is Mediated by Ligand-Gated Ion
  Channels 609

Supplement 2

Principles of Physiology and Biochemistry: Vision 614

The Visual Pigments Are Found in Rod and Cone
  Cells 614

Rhodopsin Consists of 11-*cis*-Retinal Bound to the
   Protein, Opsin 615
   *Light Isomerizes the Retinal of Rhodopsin to All-*trans 616
   Transformations of Rhodopsin Can Be Detected by Changes
      in Its Absorption Spectrum 616
   Isomerization of the Retinal Causes Other Structural
      Changes in the Protein 618
The Conductivity Change That Results from Absorption
   of a Photon 619
The Effect of Light Is Mediated by Guanine
   Nucleotides 621

Part 7

Storage and Utilization of Genetic Information 625

Chapter 25

Structures of Nucleic Acids and Nucleoproteins 627

The Genetic Significance of Nucleic Acids 628
   Transformation Is DNA-Mediated 628
Structural Properties of DNA 628
   The Polynucleotide Chain Contains Mononucleotides Linked
      by Phosphodiester Bonds 630
   Most DNAs Exist as Double-Helix (Duplex) Structures 631
   Hydrogen Bonds and Stacking Forces Stabilize the Double
      Helix 633
   Conformational Variants of the Double-Helix Structure 635
   Duplex Structures Can Form Supercoils 636
   DNA Denaturation Involves Separation of Complementary
      Strands 638
   DNA Renaturation Involves Duplex Formation from Single
      Strands 639
Chromosome Structure 641
   Physical Structure of the Bacterial Chromosome 641
   The Genetic Map of Escherichia coli 642
   Eukaryotic DNA Is Complexed with Histones 642
   Organization of Genes within Eukaryotic
      Chromosomes 643

Chapter 26

DNA Replication, Repair, and Recombination 650

The Universality of Semiconservative Replication 651
Overview of DNA Replication in Bacteria 652

Growth during Replication Is Bidirectional 652
Growth at the Replication Forks Is Discontinuous 653
Proteins Involved in DNA Replication 654
Characterization of DNA Polymerase I in Vitro 656
Crystallography Combined with Genetics to Produce a
   Detailed Picture of DNA PolI Function 656
Establishing the Normal Roles of DNA Polymerases I and
   III 658
Other Proteins Required for DNA Synthesis in Escherichia
   coli 659
Replication of the Escherichia coli *Chromosome 660*
Initiation and Termination of Escherichia coli *Chromosomal*
   Replication 660
DNA Replication in Eukaryotic Cells 661
   Eukaryotic Chromosomal DNA 662
   SV40 Is Similar to Its Host in Its Mode of Replication 663
Several Systems Exist for DNA Repair 664
   Synthesis of Repair Proteins Is Regulated 665
DNA Recombination 666
   Enzymes Have Been Found in Escherichia coli *That*
      Mediate the Recombination Process 668
   Other Types of Recombination 671
RNA-Directed DNA Polymerases 671
   Retroviruses Are RNA Viruses That Replicate through a
      DNA Intermediate 671
   Hepatitis B Virus Is a DNA Virus That Replicates through
      an RNA Intermediate 671
   Some Transposable Genetic Elements Encode a Reverse
      Transcriptase That Is Crucial to the Transposition
      Process 671
   Bacterial Reverse Transcriptase Catalyzes Synthesis of a
      DNA–RNA Molecule 673
   Telomerase Facilitates Replication at the Ends of
      Eukaryotic Chromosomes 673
Other Enzymes That Act on DNA 673

Chapter 27

DNA Manipulation and Its Applications 678

Sequencing DNA 679
Methods for Amplification of Select Segments of
   DNA 679
Amplification by the Polymerase Chain Reaction 679
DNA Cloning 682
   Restriction Enzymes Are Used to Cut DNA into Well-
      Defined Fragments 682
   Plasmids Are Used to Clone Small Pieces of DNA 683
   Bacteriophage λ Vectors Are Useful for Cloning Larger
      DNA Segments 685
   Cosmids Are Used to Clone the Largest Segments of DNA
      686
   Shuttle Vectors Can Be Cloned into Cells of Different
      Species 686

Constructing a ''Library'' 686
   A Genomic DNA Library Contains Clones with Different
      Genomic Fragments 686
   A cDNA Library Contains Clones Reflecting the mRNA
      Sequences 687
   Numerous Approaches Can Be Used to Pick the Correct
      Clone from a Library 688
   Cloning in Systems Other than Escherichia coli 688
Site-Directed Mutagenesis Permits the Restructuring of
   Existing Genes 689
Recombinant DNA Techniques Were Used to
   Characterize the Globin Gene Family 689
   DNA Sequence Differences Were Used to Detect Defective
      Hemoglobin Genes 690
   The β-Globin cDNA Probe Was Used to Characterize the
      Normal β-Globin Gene 694
   Chromosome Walking Permitted Identification and Isolation
      of the Regions around the Adult β-Globin Genes 694
   Walking and Jumping Were Both Used to Map the Cystic
      Fibrosis Gene 694

Chapter 28

RNA Synthesis and Processing 700

The First RNA Polymerase to Be Discovered Did Not
   Require a DNA Template 701
DNA–RNA Hybrid Duplexes Suggest That RNA Carries
   the DNA Sequences 701
There Are Three Major Classes of RNA 702
   Messenger RNA Carries the Information for Polypeptide
      Synthesis 702
   Transfer RNA Carries Amino Acids to the Template for
      Protein Synthesis 703
   Ribosomal RNA Is an Integral Part of the Ribosome 705
   The Fine Structure of the Ribosome Is Beginning to
      Emerge 705
Overview of the Transcription Process 705
   Bacterial RNA Polymerase Contains Five Subunits 706
   Binding at Promoters 708
   Initiation at Promoters 709
   Alternative Sigma Factors Trigger Initiation of
      Transcription at Different Promoters 709
   Elongation of the Transcript 709
   Termination of Transcription 710
   Comparison of Escherichia coli RNA Polymerase with DNA
      PolI and PolIII 710
Important Differences Exist between Eukaryotic and
   Prokaryotic Transcription 711
   Eukaryotes Have Three Nuclear RNA Polymerases 712
   Eukaryotic RNA Polymerases Are Not Fully Functional by
      Themselves 712
   Recent Evidence Suggests That the TATA-Binding Protein
      May Be Required by All Three Polymerases 713
   In Eukaryotes Promoter Elements Are Located at a
      Considerable Distance from the Polymerase Binding
      Site 715
Many Viruses Encode Their Own RNA
   Polymerases 715
   RNA-Dependent RNA Polymerases of RNA Viruses 715
Other Types of RNA Synthesis 716
Posttranscriptional Alterations of Transcripts 717
   Processing and Modification of tRNA Requires Several
      Enzymes 717
   Processing of Ribosomal Precursor Leads to Three
      RNAs 719
   Eukaryotic Pre-mRNA Undergoes Extensive
      Processing 719
Some RNAs Function Like Enzymes 722
   Some RNAs Are Self-Splicing 722
   Some Ribonucleases Are RNAs 722
   Ribosomal RNA Catalyzes Peptide Bond Formation 723
Catalytic RNA May Have Evolutionary Significance 723
Inhibitors of RNA Metabolism 725
   Some Inhibitors Act by Binding to DNA 725
   Some Inhibitors Bind to RNA Polymerase 725
   Some Inhibitors Are Incorporated into the Growing RNA
      Chain 725

Chapter 29

Protein Synthesis, Targeting, and Turnover 730

The Cellular Machinery of Protein Synthesis 731
   Messenger RNA Is the Template for Protein Synthesis 731
   Transfer RNAs Order Activated Amino Acids on the mRNA
      Template 731
   Ribosomes Are the Site of Protein Synthesis 735
The Genetic Code 736
   The Code Was Deciphered with the Help of Synthetic
      Messengers 736
   The Code Is Highly Degenerate 737
   Wobble Introduces Ambiguity into Codon–Anticodon
      Interactions 738
   The Code Is Not Quite Universal 740
   The Rules Regarding Codon–Anticodon Pairing Are Species-
      Specific 741
The Steps in Translation 742
   Synthases Attach Amino Acids to tRNAs 742
   Each Synthase Recognizes a Specific Amino Acid and
      Specific Regions on Its Cognate tRNA 743
   Aminoacyl-tRNA Synthases Can Correct Acylation
      Errors 744
   A Unique tRNA Initiates Protein Synthesis 745
   Translation Begins with the Binding of mRNA to the
      Ribosome 746
   Dissociable Protein Factors Play Key Roles at the Different
      Stages in Protein Synthesis on the Ribosome 746

Protein Factors Aid Initiation 747
Three Elongation Reactions Are Repeated with the
*   Incorporation of Each Amino Acid 748*
Two GTPs Are Required for Each Step in Elongation 750
Termination of Translation Requires Release Factors and
*   Termination Codons 754*
Ribosomes Can Change Reading Frame during
*   Translation 755*

Targeting and Posttranslational Modification of
    Proteins 757
Proteins Are Targeted to Their Destination by Signal
*   Sequences 757*
Some Mitochondrial Proteins Are Transported after
*   Translation 757*
Eukaryotic Proteins Targeted for Secretion Are Synthesized
*   in the Endoplasmic Reticulum 758*
Proteins That Pass through the Golgi Apparatus Become
*   Glycosylated 760*
Processing of Collagen Does Not End with Secretion 760
Bacterial Protein Transport Frequently Occurs during
*   Translation 760*

Protein Turnover 760
The Lifetimes of Proteins Differ 761
Structural Features Can Determine Protein Half-Lives 761
Abnormal Proteins Are Selectively Degraded 762
Proteolytic Hydrolysis Occurs in Mammalian
*   Lysosomes 763*
Ubiquitin Tags Proteins for Proteolysis 763

Chapter 30

Regulation of Gene Expression in Prokaryotes 768

Control of Transcription Is the Dominant Mode of
    Regulation in *Escherichia coli* 769
The Initiation Point for Transcription Is a Major Site for
    Regulating Gene Expression 769
Regulation of the Three-Gene Cluster Known as the *Lac*
    Operon Occurs at the Transcription Level 770
β-Galactosidase Synthesis Is Augmented by a Small-
*   Molecule Inducer 771*
A Gene Was Discovered That Leads to Repression of
*   Synthesis in the Absence of Inducer 772*
A Locus Adjacent to the Operon Is Found to Be Required
*   for Repressor Action 774*
Genetic Studies on the Repressor Gene and the Operator
*   Locus Lead to a Model for Repressor Action 774*
Biochemical Investigations Verify the Operon
*   Hypothesis 774*
An Activator Protein Is Discovered That Augments Operon
*   Expression 775*
Enzymes That Catalyze Amino Acid Biosynthesis Are
*   Regulated at the Level of Transcription Initiation 777*
The trp *Operon Is Also Regulated after the Initiation Point*
*   for Transcription 777*

Genes for Ribosomes Are Coordinately Regulated 780
Control of rRNA and tRNA Synthesis by the rel *Gene 781*
Translational Control of Ribosomal Protein Synthesis 783
Regulation of Gene Expression in Bacterial Viruses 783
λ Metabolism Is Directed by Six Regulatory Proteins 784
The Dormant Prophage State of λ Is Maintained by a
*   Phage-Encoded Repressor 784*
Events That Follow Infection of Escherichia coli *by*
*   Bacteriophage λ Can Lead to Lysis or Lysogeny 784*
The N Protein Is an Antiterminator That Results in
*   Extension of Early Transcripts 785*
Another Antiterminator, the Q Protein Is the Key to Late
*   Transcription 785*
Cro Protein Prevents Buildup of cI Protein during the Lytic
*   Cycle 785*
Late Expression along the Lysogenic Pathway Requires a
*   Rapid Buildup of the cII Regulatory Protein 786*

Interaction between DNA and DNA-Binding
    Proteins 789
Recognizing Specific Regions in the DNA Duplex 789
The Helix-Turn-Helix Is the Most Common Motif Found in
*   Prokaryotic Regulatory Proteins 789*
Helix-Turn-Helix Regulatory Proteins Are Symmetrical 790
DNA–Protein Cocrystals Reveal the Specific Contacts
*   between Base Pairs and Amino Acid Side Chains 790*
Some Regulatory Proteins Use the β-Sheet Motif 791
Involvement of Small Molecules in Regulatory Protein
*   Interaction 791*

Chapter 31

Regulation of Gene Expression in Eukaryotes 800

Gene Regulation in Yeast: A Unicellular Eukaryote 801
Galactose Metabolism Is Regulated by Specific Positive and
*   Negative Control Factors in Yeast 801*
The GAL4 Protein Is Separated into Domains with Different
*   Functions 804*
Mating Type Is Determined by Transposable Elements in
*   Yeast 804*
Gene Regulation in Multicellular Eukaryotes 807
Nuclear Differentiation Starts in Early Development 807
Chromosome Structure Varies with Gene Activity 809
Giant Chromosomes Permit Direct Visualization of Active
*   Genes 809*
In Some Cases Entire Chromosomes Are
*   Heterochromatic 809*
Biochemical Differences between Active and Inactive
*   Chromatin 810*
DNA-Binding Proteins That Regulate Transcription in
    Eukaryotes Are Often Asymmetrical 812
The Homeodomain 813
Zinc Finger 814
Leucine Zipper 815

Helix-Loop-Helix 815

*Transcription Activation Domains of Transcription
Factors 815*

Alternative Modes of mRNA Splicing Present a Potent
Mechanism for Posttranscriptional Regulation 817

Gene Expression Is Also Regulated at the Levels of
Translation and Polypeptide Processing 817

Patterns of Regulation Associated with Developmental
Processes 819

During Embryonic Development in the Amphibian,
Specific Gene Products Are Required in Large
Amounts 819

*Ribosomal RNA in Frog Eggs Is Elevated by DNA
Amplification 819*

*5S rRNA Synthesis in Frogs Requires a Regulatory
Protein 820*

Early Development in *Drosophila* Leads to a Segmented
Structure That Is Preserved to Adulthood 820

Early Development in Drosophila *Involves a Cascade of
Regulatory Proteins 822*

*Three Types of Regulatory Genes Are Involved in Early
Segmentation Development in* Drosophila 822

Analysis of Genes That Control the Early Events of
Drosophila *Embryogenesis 823*

*Maternal-Effect Gene Products for Oocytes Are Frequently
Made in Helper Cells 823*

*Gap Genes Are the First Segmentation Genes to Become
Active 824*

Supplement 3

Principles of Physiology and Biochemistry:
Immunobiology 830

Overview of the Immune System 830

The Humoral Response: B Cells and T Cells Working
Together 831

*Immunoglobulins Are Extremely Varied in Their
Specificities 832*

*Antibody Diversity Is Augmented by Unique Genetic
Mechanisms 834*

*Interaction of B Cells and T Cells Is Required for Antibody
Formation 838*

*The Complement System Facilitates Removal of
Microorganisms and Antigen–Antibody
Complexes 841*

The Cell-Mediated Response: A Separate Response by
T Cells 841

*Tolerance Prevents the Immune System from Attacking Self-
Antigens 842*

T Cells Recognize a Combination of Self and Nonself 843

MHC Molecules Account for Graft Rejection 843

*There Are Two Major Types of MHC Proteins: Class I and
Class II 843*

*T-Cell Receptors Resemble Membrane-Bound
Antibodies 844*

*Additional Cell Adhesion Proteins Are Required to Mediate
the Immune Response 844*

*Immune Recognition Molecules Are Evolutionarily
Related 845*

Supplement 4

Principles of Physiology and Biochemistry:
Carcinogenesis and Oncogenes 848

Cancers: Cells Out of Control 849

*Transformed Cells Are Closely Related to Cancer
Cells 850*

*Environmental Factors Influence the Incidence of
Cancers 850*

Cancerous Cells Are Genetically Abnormal 851

*Some Tumors Arise from Genetically Recessive
Mutations 852*

*Many Tumors Arise by Mutational Events in Cellular
Protooncogenes 852*

*Oncogenes Are Frequently Associated with Tumor-Causing
Viruses 853*

*The Role of DNA Viral Genes in Transformation Reflects
Their Role in the Permissive Infectious Cycle 854*

*p53 Is the Most Common Gene Associated with Human
Cancers 855*

Retroviral-Associated Oncogenes That Are Involved in
Growth Regulation 856

The src Gene Product 856

The sis Gene Product 856

The erbB Gene Product 858

The ras Gene Product 859

The myc Gene Product 860

The jun and fos Gene Products 861

The Transition from Protooncogene to Oncogene 861

Carcinogenesis Is a Multistep Process 861

Comparative Sizes of Biomolecules, Viruses, and
Cells A-2

Common Abbreviations in Biochemistry A-4

Organic Chemistry and Its Relationship to
Biochemistry A-6

Appendix A: Some Landmark Discoveries in
Biochemistry A-19

Appendix B: A Guide to Career Paths in the Biological
Sciences A-23

Appendix C: Answers to Selected Problems A-25

Glossary G-1

Credits C-1

Index I-1

List of Supplementary Text Elements

Boxes

20A   Cholesterol Metabolism and Heart Disease   474

24A   The Lactate Dehydrogenase of Heart Muscle   566

29A   Site-Specific Variation in Translation Elongation   739

29B   Mechanisms of Damage Produced by Certain Toxins   752

29C   Antibiotics Inhibit by Binding to Specific Sites on the Ribosome   756

30A   Genetic Concepts and Genetic Notation   773

S3A   The Role of Interleukins   842

Supplemental Chapters
Principles of Physiology and Biochemistry

S1   Neurotransmission   602

S2   Vision   614

S3   Immunobiology   830

S4   Carcinogenesis   848

Notes on Nutrition   598

Methods of Biochemical Analysis*

* Please Note: Most of the material on **Methods of Biochemical Analysis** and **Biochemical Experiments** presented in this textbook is **integrated** into the main part of the text. This material is **"flagged"** throughout the text by the **lab door icon. Supplemental biochemical methods material which can be identified within the text by the blue-tabbed pages is listed below.**

3A   Measurement of Ultraviolet Absorption in Solution   70

3B   The Dansyl Chloride Method for N-Terminal Amino Acid Determination   71

4A   X-Ray Diffraction: Applications to Fibrous Proteins   96

4B   Three-Dimensional Protein Structure Can Be Analyzed by X-Ray Diffraction of Protein Crystals   97

4C   Radiation Techniques for Examining Protein Structure   99

12A   Methods for Structural Analysis: Polarized Light and Polarimetry   281

21A   Demonstration That Indole Is Not a True Intermediate in the Tryptophan Biosynthetic Pathway   509

25A   X-Ray Diffraction of DNA   649

28A   Use of the Foot-Printing Technique to Determine the Binding Site of a DNA Binding Protein   729

Preface

As the subject of biochemistry has grown, the need for texts that are appropriate for different types of courses has also grown. There is a need for texts at three distinct levels: the "comprehensive" level that includes a great deal of detail, the "middle" level, and the "short course" level. Most up-to-date texts are written at the "comprehensive" level despite the fact that most students would benefit more from a less detailed, "middle" level text that emphasizes *principles*. Currently, there are very few up-to-date "middle" level texts available. The main goal of our text, as indicated by the title, is to serve the needs of the *principles-oriented*, "middle" level audience.

In addition to presenting material that we think is important for the "middle" level audience, we have tried to write this text with the following goals in mind:

- to give a balanced presentation that explains the biochemistry of both prokaryotes and eukaryotes;

- to indicate the biomedical significance of defects in metabolism;

- to describe metabolism with special emphasis on how pathways are organized and how they are regulated;

- to give the student a feel for the process of scientific research by describing how some major biochemical discoveries were made; and finally,

- to present the techniques used by biochemists during the course of their research.

Unique Features of *Principles of Biochemistry*

Some of the unique features of our text can be classified as *general features* in that they apply to the entire text, while other *content-specific features* apply to an individual chapter or group of chapters.

General Features

1. **The text is available as a single, full-length (31 Chapters; plus 4 Supplements) casebound text or in three paperback volumes (Volume One: Chapters 1–10, Volume Two: Chapters 11–24, plus Supplements 1 & 2; and Volume Three: Chapters 25–31, plus Supplements 3 & 4).** It is generally useful for students to bring their texts to class, but students of biochemistry are not inclined to do this because of the large size of the texts. The three volume binding format makes it convenient for students, and professors alike, to carry only one third of the textbook around at any one time. An additional advantage of the three volume binding format is that each of the volumes may be used separately or together in any combination that best suits the needs of the biochemistry course. The three volumes therefore allow greater flexibility for professors and greater manageability and affordability for students.

2. *Eight icons, graphic images that are intended to help the textbook reader remember and mentally cross-reference* 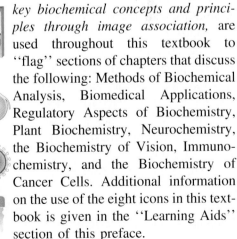 *key biochemical concepts and principles through image association,* are used throughout this textbook to "flag" sections of chapters that discuss the following: Methods of Biochemical Analysis, Biomedical Applications, Regulatory Aspects of Biochemistry, Plant Biochemistry, Neurochemistry, the Biochemistry of Vision, Immunochemistry, and the Biochemistry of Cancer Cells. Additional information on the use of the eight icons in this textbook is given in the "Learning Aids" section of this preface.

Content-Specific Features

1. Chapter 6: Detailed step-by-step descriptions of the purification schemes for two proteins are presented to give the student an understanding of how different purification procedures may be effectively combined.

2. Chapters 7–9: Emphasis on basic enzymology—enzyme kinetics, mechanisms and regulation.

3. Chapter 10: An exclusive chapter devoted to vitamins and coenzymes.

4. Chapter 11: A comprehensive description of basic metabolic strategies that govern substrate and energy flow in metabolic pathways.

5. Chapter 12: A single cohesive chapter on all major aspects of glucose metabolism—glycolysis, gluconeogenesis, and regulation.

6. Chapters 14 and 18: Correct values for ATP production resulting from catabolism. Other texts do not take into account the energy required for ATP, ADP, and P_i transport across the inner mitochondrial membrane. This makes a substantial difference in the calculated energy yields.

7. Chapter 20: A chapter exclusively devoted to cholesterol metabolism.

8. Chapter 22: A chapter exclusively devoted to amino acid metabolism in vertebrates.

9. Chapters 12–16, 17–23: Organization of carbohydrate metabolism, lipid metabolism, and the metabolism of nitrogen-containing compounds, respectively, into self-contained units.

10. Chapters 30 and 31: Separate chapters devoted to the regulation of gene expression in prokaryotes and eukaryotes, respectively.

Learning Aids for the Student

In order for students to succeed in their study of biochemistry, they must be able to understand the material presented, utilize the text as a tool for learning, and enjoy reading the text. Therefore, we have included many aids to make the study of biochemistry efficient and enjoyable.

The text chapters contain the following:

Chapter Opening Outline and Overview   The chapter opening outline and overview were written to help students preview how the chapter is organized and what major concepts are to be covered in the chapter.

Declarative Statement Headings   Clear, informative headings help introduce students to each important topic.

Underlined Terms and Key Concepts   Underlined important terms and key concepts are easy for students to locate.

> The second law of thermodynamics is that the universe inevitably proceeds from states that are more ordered to states that are more disordered. This phenomenon is measured by a thermodynamic function called entropy, which is denoted

Numbered Equations   Important equations within a chapter are numbered so that they can be easily located and referenced by students.

Summary Tables   Summary tables are designed to help students more easily understand and utilize important facts or characteristics presented for a specific topic.

Eight Concept and Applications Icons   Throughout the text, the student will find text sections "flagged" by eight different graphic icons or images. These icons are intended

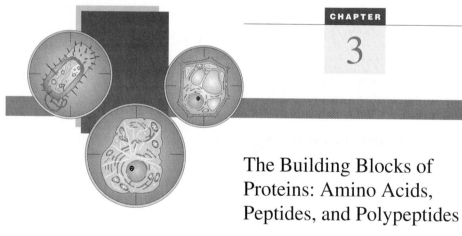

CHAPTER

3

The Building Blocks of Proteins: Amino Acids, Peptides, and Polypeptides

Chapter Outline

Amino Acids
  Amino Acids Have Both Acid and Base Properties
  Aromatic Amino Acids Absorb Light in the Near-Ultraviolet
  All Amino Acids except Glycine Show Asymmetry
Peptides and Polypeptides
Determination of Amino Acid Composition of Proteins
Determination of Amino Acid Sequence of Proteins
Chemical Synthesis of Peptides and Polypeptides

Amino acids have common features that permit them to be linked together into polypeptide chains and uncommon features that give each polypeptide chain its unique character.

In the middle of the nineteenth century, the Dutch chemist Gerardus Mulder extracted a substance common to animal tissues and the juices of plants, which he believed to be "without doubt the most important of all substances of the organic kingdom, and without it life on our planet would probably not exist." At the suggestion of the famous Swedish chemist Jons Jakob Berzelius, Mulder named this substance protein (from the Greek *proteios*, meaning "of first importance") and assigned to it a specific chemical formula to help the student remember and mentally cross-reference the following concepts and applications:

 Methods of Biochemical Analysis

  Biomedical Applications

 Regulatory Aspects of Biochemistry

 Plant Biochemistry

 Neurochemistry

 Biochemistry of Vision

 Immunochemistry

Biochemistry of Cancer Cells

Boxed Readings   Thought-provoking boxed readings on relevant topics are featured in various chapters—for example, Box 20A, Metabolism and Heart Disease.

BOX
20A   Cholesterol Metabolism and Heart Disease

Molecular Graphics and Illustration Program Molecular graphics images of key molecules enable students to "see" and interpret three-dimensional structures.

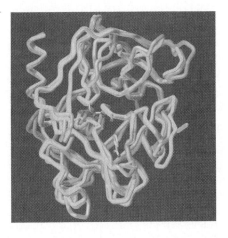

Figure 8.6a

Trypsin, Chymotrypsin, Elastase

End-of-Chapter Enumerated Summary This concise summary of chapter concepts is designed to serve as a guide for chapter study.

End-of-Chapter Selected Readings Each chapter concludes with carefully selected references that contain further information on the topics covered in that chapter.

End-of-Chapter Problems These problems (over 400) are designed to help students access their understanding of the chapter's basic concepts. Brief solutions for the odd-numbered problems are found in the back of the text.

End-of-Chapter Methods of Biochemical Analysis Supplemental information on experimental techniques and procedures are presented at the end of some chapters. The "Methods of Biochemical Analysis" can be easily located throughout the text as their text pages are tabbed with a blue band. The lab door icon is used to cross-reference the

Summary

In this chapter we discussed the ways in which biochemistry parallels ordinary chemistry and those in which it is quite different. The chief points to remember are the following.

1. The basic unit of life is the cell, which is a membrane-enclosed, microscopically visible object.
2. Cells are composed of small molecules, macromolecules, and organelles. The most prominent small molecule is water, which constitutes 70% of the cell by weight. Other small molecules are present only in quite small amounts; they are precursors or breakdown products of macromolecules or coenzymes. There are four types of macromolecules: lipids, carbohydrates, proteins, and nucleic acids.

3. Noncovalent intermolecular forces largely determine the folded structure adopted by a macromolecule, particularly the relative affinity of different groupings on the macromolecule for water. In general, hydrophobic groupings are buried within the folded macromolecular structure, whereas hydrophilic groupings are located on the surface, where they can interact with water.
4. Biochemical reactions utilize a limited number of elements, most prominently carbon, hydrogen, oxygen, nitrogen, sulfur, and phosphorus. Many biochemical reactions are simple organic reactions.

"Methods of Biochemical Analysis" to pertinent text sections within a chapter.

Methods of Biochemical Analysis **3 A**

Measurement of Ultraviolet Absorption in Solution

The general quantitative relationship that governs all absorption processes is called the Beer-Lambert law:

$$I = I_0 10^{-\epsilon c d}$$

where I_0 is the intensity of the incident radiation, I is the intensity of the radiation transmitted through a cell of thickness d (in centimeters) that contains a solution of concentration c (expressed either in moles per liter or in grams per 100 ml), and ϵ is the extinction coefficient, a characteristic of the substance being investigated (see fig. 3.7).

Light absorption is measured by a spectrophotometer as shown in the illustration. The spectrophotometer usually is capable of directly recording the absorbance A, which is related to I and I_0 by the equation

$$A = \log_{10}\left(\frac{I_0}{I}\right)$$

Hence $A = \epsilon c d$, and A is a direct measure of concentration. We can see from figure 3.7 that the ϵ values are largest for tryptophan and smallest for phenylalanine.

Because protein absorption maxima in the near-ultraviolet (240–300 nm) are determined by the content of the

Figure 1

Schematic diagram of a spectrophotometer for measuring light absorption. Laboratory instruments for making measurements are much more complex than this, but they all contain the same basic components: a light source, a monochromator, a sample, and a detector. λ is the wavelength of the light, I_0 and I are the incident light intensity and the transmitted light intensity, respectively, and d is the thickness of the absorbing solution.

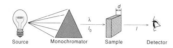

aromatic amino acids and their respective values, most proteins have absorption maxima in the 280-nm region. By contrast, absorption in the far-ultraviolet (around 190 nm) is shown by all polypeptides regardless of their aromatic amino acid content. The reason is that absorption in this region is due primarily to the peptide linkage.

Common Abbreviations in Biochemistry   These abbreviations can be found at the end of the text.

Organic Chemistry for Biochemistry   This appendix, which describes the similarities and the differences between organic chemistry and biochemistry, is also located at the end of this text.

Appendix A: Some Major Discoveries in Biochemistry   Summarizes major research discoveries from the past and present.

Appendix B: Career Paths in the Biological Sciences   A useful starting point for career information for biochemistry students.

Appendix C: Answers to Selected Problems   Brief solutions are provided to help the student determine if they are on the right track.

Glossary   Over 700 important biochemical terms are defined in this end-of-book glossary.

Index   An easy-to-use, comprehensive index is provided.

Endsheets with Reference Material   The endsheets of the text contain useful easily accessible reference information.

Organization of the Text

This text is divided into seven parts: part 1, An Overview of Biochemical Structures and Reactions That Occur in Living Systems; part 2, Protein Structure and Function; part 3, Catalysis; part 4, Metabolism of Carbohydrates; part 5, Metabolism of Lipids; part 6, Metabolism of Nitrogen-Containing Compounds; and part 7, Storage and Utilization of Genetic Information.

The subjects presented in each of the seven parts, the level of the discussion, the organization of the parts, and the organization within each of the chapters are all designed to satisfy the teaching needs of introductory biochemistry courses.

We may think of biochemistry as being divided into three main areas: biomolecular structure, intermediary me-

In addition to these chapter aids, this text also features the following:

Four "Principles of Physiology and Biochemistry" Supplements   These supplements cover the topics of **neurotransmission, vision, immunobiology, and carcinogenesis.** The Neurotransmission and Vision supplements, Supplements 1 and 2, are located at the end of Part 6 of the text. The Immunobiology and Carcinogenesis supplements, Supplements 3 and 4, are located at the end of Part 7 of the text. Physiological discussions within the text that relate to these topics are also "flagged" by the corresponding concept icon.

Page-Referenced Listing of Supplementary Learning Aids   This helpful ready reference tool is located in the frontmatter of this text.

Illustrated Guide to the "Comparative Sizes of Biomolecules, Viruses, and Cells"   This highly visual guide is located at the end of the text.

tabolism, and nucleic acid and protein metabolism. It would be possible to segregate the coverage of these topics in a biochemistry text and present them as three consecutive blocks. However, we did not take this approach in the development of this text because we felt it would not be good pedagogy for two reasons: (1) students get bored if they sit through a third of a year where they only hear about structure; and (2) students often forget what was said about the structure of a biomolecule by the time they read the section on its metabolism and therefore have to review it before going on to the metabolism material.

Our general approach in writing *Principles of Biochemistry* was to integrate the presentation of biomolecular structure and metabolism throughout the text. We were successful in applying this integrative approach to presenting biomolecular structure and metabolism in five of the seven parts of the text. Specifically, amino acid structure and protein structure are discussed in part 2, well before considering their metabolism in parts 6 and 7, respectively. This deviation from our general approach is essential because one must understand amino acid structure before one can understand the function of proteins and enzymes, which is discussed in early chapters (chapters 5, 7, 8, and 9 of parts 2 and 3). An additional advantage to discussing protein structure early in the text is that most students find it exciting. Notwithstanding the early discussion of basic protein structure in chapter 4, additional discussions pertaining to protein structure are included within the text where they are most appropriate to furthering the student's understanding protein function. Such discussions can be found in chapter 17 where membrane proteins are discussed and in chapters 30 and 31 where gene regulatory proteins are discussed.

The structures of sugars and polysaccharides are covered in the appropriate chapters within part 4 just prior to discussing their metabolism. Similarly, the structures of lipids are presented in the lipid metabolism chapters found in part 5. Nucleotide structures are addressed in chapter 23 before considering their metabolism. Finally, nucleic acid and nucleoprotein structures are examined in the first chapter (chapter 25) of part 7 prior to the discussion of the roles these molecules play in nucleic acid and protein metabolism in the six subsequent chapters.

Metabolism is regulated at the level of the gene and at the level of the enzyme. Regulation is emphasized in this text both because of its importance and because it reflects some of the most unique characteristics of biochemical reactions. Of necessity, our treatment of the subject of regulation at the gene level is confined primarily to part 7, as this is where the metabolism of nucleic acids is discussed. We cover metabolic regulation at the level of the enzyme throughout this text. Specifically, the regulation of the oxygen binding capacity of hemoglobin and the process of mus-

cle contraction is discussed in chapter 5. An entire chapter, chapter 9, is devoted to regulatory enzymes. With respect to the metabolism of carbohydrates (chapter 12) and fatty acids (chapter 18), the regulation of the catabolic and anabolic pathways of these biomolecules is presented together within the respective chapters. We feel that this approach simplifies the discussion of the regulation of these opposing pathways.

Detailed Chapter-by-Chapter Description of the Text

Having indicated some aspects of the overall organization of this text, we now turn to a more detailed chapter-by-chapter description of the text. Part 1, An Overview of Biochemical Structures and Reactions That Occur in Living Systems, contains only two chapters. Chapter 1, Cells, Biomolecules and Water, gives an introductory presentation of the structure of cells and the molecules from which cells are composed. The central role of water in determining the structures that are formed is emphasized. A general description of the principles that govern cellular organization and determine the course of these reactions is presented. An introductory survey of the types of molecules that make up the cell is given. Finally, the overall strategy of living systems and the evolution of living systems is briefly discussed.

The main function of chapter 1 is to present an overview of the principles and facts that are presented in greater depth in various parts of the text. Most of the material in this chapter is presented in a good introductory biology course. However, there are a number of students coming from a chemistry background that may not have had the benefits of such a course. Reference should also be made at this point to the appendix ''Organic Chemistry for Biochemistry'' (located in the endmatter of the text), which gives a brief review of the similarities and differences between organic chemistry and biochemistry. It is assumed that all students reading this text will have had a year of general chemistry and one term of organic chemistry. Or, at the very least, that they will be taking organic chemistry concurrently with biochemistry.

In chapter 2, thermodynamics is presented in a way that is most relevant to the consideration of biochemical phenomena. We chose to present the rather difficult topic of thermodynamics in the beginning of the text because thermodynamics is an important consideration in all aspects of biochemistry. The central role of ATP as the main carrier of free energy in biochemical systems is emphasized.

The discussion of thermodynamics presented in chapter 2 gives a perspective of bioenergetic considerations, which are discussed at many points throughout the text.

Thermodynamic applications that relate to biomolecular structure and biochemical reactions are also elaborated on in specific sections of subsequent chapters.

Part 2, Protein Structure and Function, contains four chapters that relate to the structures and functions of proteins. In chapter 3, The Building Blocks of Proteins: Amino Acids, Peptides, and Polypeptides, we discuss basic structural and chemical properties of amino acids, peptides and polypeptides. In chapter 4, The Three-Dimensional Structures of Proteins, we describe how and why polypeptide chains fold into long fibrous molecules in some cases, or into compact globular molecules in other cases. In chapter 5, Functional Diversity of Proteins, we turn to the question of how protein structure relates to protein function. To explore this question, two protein systems, hemoglobin and the actin-myosin complex are examined in detail. In chapter 6, Methods for Purification and Characterization of Proteins, the primary goal is to acquaint the reader with the techniques used for protein purification. The first part of chapter 6 presents methods for protein fractionation. In the second part of this chapter, purification procedures for two proteins, UMP synthase and lactose carrier protein, are presented so that the student can see how different purification steps are combined for maximum effectiveness.

Part 3, Catalysis, is divided into four chapters. The first three chapters in part 3 deal with the properties and functions of enzymes. The last chapter explains the essential properties of coenzymes and cofactors.

Enzyme kinetics is studied for two reasons: (1) it is a practical concern to determine the activity of the enzyme under different conditions; (2) frequently the analysis of enzyme kinetics gives information about the mechanism of enzyme action. Chapter 7, Enzyme Kinetics, begins with an introductory section on the discovery of enzymes, basic enzyme terminology and a description of the six main classes of enzymes and the reactions they catalyze. The remainder of the chapter deals with basic aspects of chemical kinetics, enzyme-catalyzed reactions and various factors that affect the kinetics.

Chapter 8, How Enzymes Work, starts with a description of the basic chemical mechanisms that are exploited by enzymes. The latter half of this chapter presents a detailed description of how three enzymes—chymotrypsin, RNase, and triose phosphate isomerase—exploit these basic mechanisms of enzyme catalysis.

Most enzymes spontaneously process substrates when present, as long as inhibitory factors do not prevent this from happening. A few enzymes, known as regulatory enzymes, do not react spontaneously with their substrates unless signaled to do so by overiding metabolic conditions. Chapter 9, Regulation of Enzyme Activities, describes a wide range of mechanisms that are used to control the activity of regulatory enzymes. This chapter concludes with a detailed description of how the activity of three enzymes— phosphofructokinase, aspartate transcarbamylase and glycogen phosphorylase—are regulated.

Frequently enzymes act in concert with small molecules, coenzymes or cofactors, which are essential to the function of the amino acid side chains of the enzyme. Coenzymes or cofactors are distinguished from substrates by the fact that they function as catalysts. They are also distinguishable from inhibitors or activators in that they participate directly in the catalyzed reaction. Chapter 10, Vitamins and Coenzymes, starts with a description of the relationship of water-soluble vitamins to their coenzymes. Next, the functions and mechanisms of action of coenzymes are explained. In the concluding sections of this chapter, the roles of metal cofactors and lipid-soluble vitamins in enzymatic catalysis are briefly discussed.

Part 4, Metabolism of Carbohydrates, begins with a general overview in chapter 11, Metabolic Strategies. This chapter also relates strongly to parts 5 and 6. Chapter 11 starts with an explanation of how biochemical reactions are organized into energy-generating and energy requiring pathways and a handful of principles that explain the design of metabolic pathways. The next five chapters describe the metabolism of carbohydrates insofar as they relate to energy-generating catabolism and energy-consuming biosynthesis. Chapter 12, Glycolysis, Gluconeogenesis, and the Pentose Phosphate Pathway, deals with all aspects of glucose metabolism. After a brief section describing structures, glycolysis, the pathway for the breakdown of glucose, is examined. This is followed by a shorter section on gluconeogenesis, the synthesis of glucose from three carbon compounds. While glycolysis produces energy, gluconeogenesis consumes energy. These two processes are discussed side-by-side so that we can consider the closely related question of the regulation of these pathways. This chapter concludes with a short section on an oxidative route for glucose catabolism, the pentose phosphate pathway which serves multiple purposes.

Under anaerobic conditions the breakdown of glucose stops at the three carbon compound stage. Further catabolism requires oxygen as described in chapter 13, The Tricarboxylic Acid Cycle. The tricarboxylic acid cycle is an energy-producing process although this is not immediately obvious because the major products of the cycle outside of CO_2—H_2O and a single ATP molecule—are the reduced forms of the coenzymes NAD^+ and FAD. These coenzymes contain the potential chemical energy for ATP production, but an elaborate process of membrane-associated electron transport and proton transport must precede the synthesis of ATP. This process is described in chapter 14, Electron Transport and Oxidative Phosphorylation.

Most of the energy that is used to drive biochemical processes originates from the sun. The way in which solar energy is harnessed to produce chemical energy and to fix atmospheric carbon dioxide into reduced organic compounds is described in chapter 15, Photosynthesis.

Up to this point in part 4, the focus has been on the roles that sugars and carbohydrates play in energy metabolism. Polymeric carbohydrates are major structural components in plant and bacterial cell walls and in the extracellular matrix of vertebrates. Branched-chain oligosaccharides covalently linked to proteins are used to give proteins unique signatures that guide them to their final destination within a cell and facilitate specific interactions between free proteins (ligands) and proteins attached to cells (receptors). In chapter 16, Structures and Metabolism of Oligosaccharides and Polysaccharides, the structures and functions of some of these polymeric carbohydrates are discussed.

Part 5, Metabolism of Lipids, comprises four chapters that deal with the structure and metabolism of lipids. In chapter 17, Structure and Function of Biological Membranes, we start by examining the constituents of membranes with the aim of developing a general model for membrane structure. We then turn to the question of how cells transport materials across membranes.

In chapter 18, Metabolism of Fatty Acids, we discuss the synthesis and breakdown of fatty acids. The chapter starts with a discussion of fatty acid breakdown. A second section covers the pathway for fatty acid biosynthesis. Finally, we consider the regulatory mechanisms that determine the conditions under which each of these processes occurs. As in the case of glucose metabolism, it is convenient to discuss the synthesis and breakdown in the same chapter so that the closely related topic of regulation can be considered alongside.

Thus far we have been concerned with the metabolism of fatty acids in relationship to the storage and release of energy. In chapter 19, Biosynthesis of Membrane Lipids, we focus on the metabolism of lipids that serve other roles. Many types of lipids are essential membrane components. A number of lipids also function as metabolic signals in response to hormonal signals. These lipid molecules are known as second messengers.

The final chapter in part 5, chapter 20, Metabolism of Cholesterol, deals with the synthesis of cholesterol and some of its derivatives, the steroid hormones and the bile acids. This chapter considers the structure, function and metabolism of these molecules. Also, the health-related concerns associated with cholesterol excess are addressed.

Part 6, Metabolism of Nitrogen-Containing Compounds, is concerned mostly with the metabolism of amino acids and nucleotides. Chapter 24, the last chapter in this part, deals with the integration of metabolism.

Amino acid metabolism is divided into two chapters. Amino acids are best known as the building blocks of proteins. In addition to this role, amino acids serve as precursors to many important low molecular weight compounds including nucleotides, porphyrins, parts of lipid molecules and precursors for several coenzymes. Amino acids also serve as the "vehicles" for converting inorganic forms of nitrogen and sulfur to organic forms. As an alternative energy source amino acid catabolism can be coupled to the regeneration of ATP from ADP of AMP. In chapter 21, Amino Acid Biosynthesis and Nitrogen Fixation in Plants and Microorganisms, we focus on the biosynthesis of amino acids and their role in bringing inorganic nitrogen and sulfur into the biological world. In addition, nonprotein amino acids are discussed briefly.

Amino acid metabolism in vertebrates contrasts sharply with amino acid metabolism in plants and microorganisms. Most striking is the fact that plants and microorganisms can synthesize all twenty amino acids required for protein synthesis whereas vertebrates can only synthesize about half this number. This leads to complex nutritional needs for vertebrates, which are discussed in chapter 22, Amino Acid Metabolism in Vertebrates. Vertebrate amino acid degradation pathways are also discussed in chapter 22 along with the existence of many pathological states that result from enzyme deficiencies in the degradative pathways.

Chapter 23, Nucleotides, deals with the biosynthesis of ribonucleotides, deoxyribonucleotides, the roles of these biomolecules in metabolic processes, and the pathways for their degradation. Medically related topics such as nucleotide metabolism deficiencies or the use of nucleotide analogs in chemotherapy are also considered.

Chapter 24, Integration of Metabolism and Hormone Action, explains the organization strategies used to integrate metabolic processes in a multicellular organism. Like the first chapter in part 4, the content of chapter 24 relates to all of the chapters on metabolism (chapters 11–24). This chapter emphasizes the fact that hormones and closely related growth factors play a dominant role in regulating metabolic activities in different tissues.

Part 6 concludes with two brief, informative supplements that integrate physiological and biochemical principles as they apply to the "nonmetabolic" functions of amino acids and lipids: Supplement 1—Principles of Physiology and Biochemistry: Neurotransmission; and Supplement 2—Principles of Physiology and Biochemistry: Vision.

Part 7, Storage and Utilization of Genetic Information, is composed of seven chapters, plus two supplements. In this part, we examine the means by which genetic information is replicated and expressed in the cell. This discussion includes a description of the relevant structures, their function, and their metabolism. In addition, this part includes a chapter that explains the procedures by which DNA is experimentally manipulated to build new genes or new combinations of genes. In chapter 25, Structures of Nucleic Acids and Nucleoproteins, we begin by considering the key experiments that revealed the genetic significance of DNA. At every stage in this chapter and the subsequent chapters of this section, explanations of genetic observations essential to the understanding of biochemical phenomena are given.

From the complementary duplex structure of DNA described in chapter 25, it is a short intuitive hop to a model for replication that satisfies the requirement for one round of DNA duplication for every cell division. In chapter 26, DNA Replication, Repair, and Recombination, key experiments demonstrating the semiconservative mode of replication in vivo are presented. This is followed by a detailed examination of the enzymology of replication, first for how it occurs in bacteria and then for how it occurs in animal cells. Also included in this chapter are select aspects of the metabolism of DNA repair and recombination. The novel process of DNA synthesis using RNA-directed DNA polymerases is also considered. First discovered as part of the mechanisms for the replication of nucleic acids in certain RNA viruses, this mode of DNA synthesis is now recognized as occurring in the cell for certain movable genetic segments and as the means whereby the ends of linear chromosomes in eukaryotes are synthesized.

Before going into the processes for the utilization of genetic information in the cell, DNA manipulation and some of its applications are considered. In chapter 27, DNA Manipulation and Its Applications, DNA sequencing is the first subject to be explained. Different approaches for amplifying and isolating specific genes are also explored. Following this, methods for restructuring existing DNA sequences are described. Finally, some applications of new technologies are considered. These applications focus on the mapping of the human globin gene family and the gene responsible for the genetically inherited disease, cystic fibrosis.

In chapter 28, RNA Synthesis and Processing, the DNA-directed synthesis of RNA is considered. As with the other chapters in part 7, a balanced presentation of how these processes occur in bacteria and eukaryotes is given. First, the structures of different classes of RNA are described. The RNA classes include messenger RNA, transfer RNA, and ribosomal RNA. A single enzyme is responsible for the transcription of the RNAs in bacteria. The initial transcripts for transfer RNA and ribosomal RNA undergo extensive modification after synthesis, while the messenger RNA is used without modification. In eukaryotes, even the messenger RNA undergoes extensive modifications before

it can be utilized. Also, in eukaryotes the process of transcription is much more complicated, involving several RNA polymerases and many more protein subunits which are associated with the polymerases. Every attempt is made to treat these complications without presenting an overwhelming amount of detail. RNA synthesis in certain viruses is also discussed and the fascinating subject of RNA enzymes is briefly considered. Lastly, selective inhibitors of RNA polymerases that have diagnostic value or medical value in chemotherapy are discussed.

In chapter 29, Protein Synthesis, Targeting, and Turnover, the processes of protein synthesis and transport are described. First the process whereby amino acids are ordered and polymerized into polypeptide chains is described. Next, posttranslational alterations of newly synthesized polypeptides is considered. This is followed by a discussion of the targeting processes whereby proteins migrate from their site of synthesis to their target sites of function. Finally, proteolytic reactions that result in the return of proteins to their starting materials, the amino acids, are considered.

Regulation of gene expression is on the cutting edge of research in molecular biology and biochemistry. This topic is so expansive that we chose to divide our coverage of it into two chapters: one that considers bacteria and another that considers eukaryotes. In chapter 30, Regulation of Gene Expression in Prokaryotes, the classical systems of the *lac* and *trp* operons and regulation of bacteriophage lambda are considered, with particular emphasis on the nature of gene regulatory proteins and how they interact with DNA. While most forms of regulation of gene expression that are understood involve regulation at the transcriptional level, one excellent example of translational control, the regulation of ribosomal protein synthesis, is also presented in chapter 30. Chapter 30 ends with a comprehensive summary section on the different types of regulatory proteins that are used to regulate transcription in prokaryotes.

The subject of regulation of gene expression in eukaryotes is complex and diffuse because so many different types of systems have been studied and the level of research effort in this field has reached unprecedented heights. Studies on this subject are truly at the cutting edge of modern biological investigations. In chapter 31, Regulation of Gene Expression in Eukaryotes, regulation of gene expression is first examined in yeast, a unicellular organism. We then examine regulatory mechanisms prevalent in multicellular eukaryotes. A section on the types of regulatory proteins most frequently found in eukaryotic systems is presented next. Finally, modes of regulation specific to developmental processes are considered.

Part 7 concludes with two brief, informative supplements that integrate key physiological and biochemical principles as they apply to two specialized processes that utilize genetic information: Supplement 3—Principles of Physiology and Biochemistry: Immunobiology and Supplement 4—Principles of Physiology and Biochemistry: Carcinogenesis and Oncogenes.

Ancillary Materials
For the Instructor

An *Instructor's Manual with Test Item File* contains suggestions on how to utilize the text in different course situations and detailed, worked-out solutions for the even-numbered problems found in the text chapters. These answers are **not** included in the *Student's Solutions Manual* that accompanies the text. In addition, this manual offers an average of 30 objective test questions for each chapter which can be used to generate exams. (ISBN 14276)

Classroom Testing Software is offered free upon request to adopters of this text. The software provides a database of questions for preparing exams. No programming experience is required. The software is available in IBM and Macintosh formats: IBM DOS 3.5 (ISBN 14278), IBM DOS 5.25 (ISBN 14279), WINDOWS 3.5 (ISBN 14280), and MAC 3.5 (ISBN 14282).

A set of *150 full-color acetate transparencies* is available free to adopters. These acetates feature key illustrations that can be used to enhance your classroom lectures. (ISBN 14277)

A set of *150 full-color projection slides* derived from the transparency illustrations is also available free to adopters. (ISBN 22871)

Electronic acetates, computerized image files for a majority of the text illustrations, will also be available free to adopters upon request. The electronic acetates will be available in a Mac/Windows (ISBN 26204). These electronic acetates can be clearly projected on large lecture hall screens using a LCD projection system.

A set of *175 transparency masters* is available free to adopters upon request. These black-and-white versions of in-text tables and illustrations can be used to prepare course handouts or additional transparency acetates. (ISBN 22872)

For the Student

A *Student's Solutions Manual* by Hugh Akers, Lamar University, and Caroline Breitenberger, Ohio State University, provides detailed, worked-out solutions for the odd-numbered problems found in the text. This solutions manual can help students to better understand how to solve the problems in the text and prepare for exams. (ISBN 22870)

A *Student Study Art Notebook,* a lecture companion containing the illustrations from the text that correspond to the acetate transparency images, is designed to help students spend more of their lecture time listening to the professor and less time copying down art from the overhead transparencies. A copy of this student study art notebook is packaged FREE with each new text. (ARTPAK ISBN 27016)

Computer Software and CD-ROM for Instructors and Students

 NOTE: PLEASE ALSO REFER TO THE CD-SAMPLER THAT WAS PACKAGED WITH YOUR COPY OF *PRINCIPLES OF BIOCHEMISTRY*

Gene Game Software, by Bill Sofer of The State University of New Jersey–Rutgers, is an interactive software game that tests students' critical-thinking skills and knowledge of the scientific method as they attempt to work through "dry" lab protocols to clone a fictitious "Fountain of Youth" gene. The software provides direct feedback and hints to the student as protocols are completed. Protocols used and the results obtained are automatically recorded in the program's "lab notebook." Contact your bookstore, or call Wm. C. Brown Publishers at 1-800-338-5578 to place an order or request more information on this challenging, interactive software game. (ISBN 24893)

Biochemical Pathways Software, also created by Bill Sofer, is an easy-to-use tutorial review software program for Macintosh that provides quizzes/memory exercises that test the students' knowledge of Glycolysis and the TCA Cycle. Contact your bookstore, or call Wm. C. Brown Publishers at 1-800-338-5578 to place an order or request more information on this software. (ISBN 25100)

Molecules of Life CD-ROM, developed by a talented team of professionals from Purdue University, is a four CD-ROM set that allows the user to visualize key biomolecular structures through *interactive* animations, simulations, drills and tutorials. The set includes material on (1) Amino acids and Proteins, (2) Carbohydrates and Lipids, (3) Nucleic acids, and (4) Metabolism and Photosynthesis. The four CD-ROMs are available individually, or as a set. Contact your bookstore, or call Wm. C. Brown Publishers at 1-800-338-5578 to place an order or request more information on this software (Four CD set/MAC ISBN 27235) (Four CD set/WINDOWS ISBN 27264).

Acknowledgments

We wish to thank our biochemistry colleagues and contributing authors to the third edition of *Biochemistry;* Dr. Raymond L. Blakley, Dr. James W. Bodley, Dr. Ann Baker Burgess, Dr. Richard Burgess, Dr. Perry Frey, Irving Geis, Dr. Lloyd L. Ingram, Dr. Gary R. Jacobson, Dr. Julius Marmur, Dr. Richard Palmiter, Dr. Milton H. Saier, Jr., Dr. Pamela Stanley and Dr. H. Edwin Umbarger, for providing us with a wealth of comprehensive information from which to draw from in the development of *Principles of Biochemistry.* Special recognition is also due to Dr. Raymond L. Blakley for his work in writing chapter 23, and to Dr. Perry A. Frey, Dr. Gary R. Jacobson, Dr. Pamela Stanley, and Dr. H. Edwin Umbarger for their invaluable feedback and proofing efforts during the many phases of writing and publishing this text. In addition, we thank Hugh Akers, Lamar University, and Caroline Breitenberger, Ohio State University, for their insightful contributions to the end-of-chapter problems in this text. We are also indebted to Michael Pique, The Scripps Research Institute, and Holly Miller, Wake Forest University Medical Center, for their special contributions to the development and generation of many of the molecular graphics images in this text.

My fellow authors and I would also like to extend a special thank you to our many colleagues across the country for reviewing the text manuscript and making many helpful suggestions. The reviewers included:

Hugh Akers
Lamar University

Richard M. Amasino
UW–Madison

Dean R. Appling
The University of Texas at Austin

John N. Aronson
University of Arizona

Paul Austin
Hanover College

Derek Baisted
Oregon State University

Terry A. Barnett, Ph.D.
Southwestern College

E. J. Behrman
The Ohio State University

Paul Arthur Berkman
Ohio State University

Frank O. Brady, Ph.D.
University of South Dakota School of Medicine

Caroline A. Breitenberger
Ohio State University

Ronald W. Brosemer
Washington State University

Oscar P. Chilson
Washington University

David P. Chitharanjan
University of Wisconsin, Stevens Point

Alan D. Cooper
Worcester State College

Rick H. Cote
University of New Hampshire

Mukul C. Datta
Tuskegee University

Lawrence C. Davis
Kansas State University

Dr. Paul H. Demchick
Barton College

Michael W. Dennis
Montana State University-Billings

Kathleen A. Donnelly, Ph.D.
Russell Sage College

Lawrence K. Duffy
University of Alaska Fairbanks

John R. Edwards
Villanova University

Alfred T. Ericson
Emporia State University

Robert J. Evans
Illinois College

David Fahrney
Colorado State University

H. Richard Fevold
University of Montana

Christopher Francklyn
University of Vermont College of Medicine

Edward A. Funkhouser
Texas A & M University

Edwin J. Geels
Dordt College

Darrel Goll
University of Arizona

Dr. Eugene Gooch
Elon College

Milton Gordon
University of Washington, Seattle

Joan M. Griffiths
Cornell University

Lonnie J. Guralnick
Western Oregon State College

James H. Hageman
New Mexico State University

B. A. Hamkalo
University of California, Irvine

Kenneth D. Hapner
Montana State University

Gerald W. Hart
University of Alabama at Birmingham

Terry L. Helser
S.U.N.Y College at Oneonta

Pui Shing Ho
Oregon State University

Joel Hockensmith
University of Virginia

Daniel Holderbaum
Case Western Reserve University

Charles F. Hosler, Jr.
University of Wisconsin-La Crosse

Larry Jackson
Montana State University

Ralph A. Jacobson
CAL POLY

John R. Jefferson
Luther College

Colleen B. Jonssen
New Mexico State
University

Floyd W. Kelly
Casper College

Mary B. Kennedy
California Institute of
Technology

R. L. Khandelwal
University of Saskatchewan

Nazir A. Khatri
Franklin College of Indiana

Ramaswamy
Krishnamoorhi
Kansas State University

James I. Lankford
St. Andrews Presbyterian
College

Daniel J. Lavoie
Saint Anselm College

Franklin R. Leach
Oklahoma State University

Carol Leslie
Union University

Michael Leung
State University of New
York/Old Westbury

Randolph V. Lewis
University of Wyoming

Albert Light
Purdue University,
West Lafayette

Donald R. Lueking
Michigan Technological
University

Dr. Celia L. Marshak
University of San Diego
1994
(Emeritus, San Diego State
University)

Lynn M. Mason
Lubbock Christian
University

Harry R. Matthews
University of California at
Davis

Martha McBride
Norwich University

William L. Meyer
University of Vermont

Holly Miller
Wake Forest University
Medical Center

Michael J. Minch
University of the Pacific

Debra M. Moriarity
University of Alabama in
Huntsville

Mary E. Morton
College of the Holy Cross

Melvyn W. Mosher
Missouri Southern State
College

Stephen H. Munroe
Marquette University

Richard M. Niles
Marshall University School
of Medicine

Jennifer K. Nyborg
Colorado State University

William R. Oliver
Northern Kentucky
University

Dr. Richard Steven Pappas
Georgia State University

Raymond Earl Poore,
Ph.D.
Jacksonville State
University

William T. Potter
The University of Tulsa

Gary L. Powell
Clemson University

Michael Eugene Pugh
Bloomsburg University of
Pennsylvania

Paul D. Ray
University of North Dakota

Philip Reyes
University of New Mexico
School of Medicine

John M. Risley
The University of North
Carolina at Charlotte

H. Alan Rowe
Norfolk State University

John E. Robbins
Montana State University

Norman G. Sansing
University of Georgia

Roy A. Scott, III
The Ohio State University

Steven E. Seifried
John A. Burns School
of Medicine
University of Hawaii

Ralph Shaw
Southeastern Louisiana
University

J. M. Shively
Clemson University

Roger D. Sloboda
Dartmouth College

Deborah Kay Smith
Meredith College

Thomas Sneider
Colorado State University

Wesley E. Stites
University of Arkansas

Eric R. Taylor
University of Southwestern
Louisiana

Martin Teintze
Montana State University

Arrel D. Toews
University of North
Carolina at Chapel Hill

H. Edwin Umbarger
Purdue University

Harry van Keulen
Cleveland State University

Robert J. Van Lanen
Saint Xavier University

Charles Vigue
University of New Haven

William H. Voige
James Madison University

Raymond E. Waldner
Palm Beach Atlantic
College

Arthur C. Washington
Tennessee State University

Daniel Weeks
University of Iowa

Steven M. Wietstock,
Ph.D.
Alma College

Steven Woeste
Scholl College of Podiatric
Medicine

Robert Zand
University of Michigan

We are grateful for the assistance of the editorial staff at Wm. C. Brown Publishers, especially Kevin Kane, publisher, Liz Sievers, our editor, Robin Steffek, developmental editor, Julie Kennedy, in-house production services coordinator, and Lori Hancock, photo editor. In taking this text from the raw manuscript stage to a production-ready stage, the authors have received tremendous assistance from Robin Steffek. Her input covers a wide range of activities from the meticulous checking of the manuscript for accuracy to numerous suggestions for modifications and special learning aids. She has pursued this project with great enthusiasm, skill, and dedication.

Once the project was ready for the production stage, we knew that a highly technical multicolor text of this magnitude would require an outstanding production team and a leader to see that every element in the final product would attain the highest quality. Laura Skinger directed the York Production Services team with confidence and determination to turn out the best possible final product. Whenever we the authors complained that something was not quite the way we wanted it, Laura would most willingly see that the concern was properly addressed without hesitation. In addition, she went way beyond our concerns to produce a text in which we could take great pride.

We hope very much that this text will be interesting and educational for students and a help to their instructors. We would appreciate any comments and suggestions from our readers. If you should find errors, please notify us so that we can make corrections in the second printing.

Geoffrey L. Zubay   William W. Parson   Dennis E. Vance

An Overview of Biochemical Structures and Reactions that Occur in Living Systems

Chapter 1   Cells, Biomolecules, and Water   3
Chapter 2   Thermodynamics in Biochemistry   29

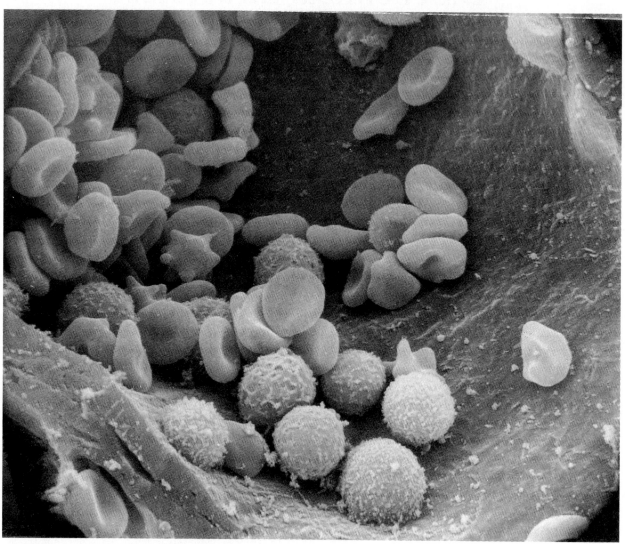

Scanning electron micrograph of erythrocytes and leucocytes in a small blood vessel.
Erythrocytes are the biconcave cells with a relatively smooth outer membrane structure. Most
of the protein in erythrocytes is hemoglobin, which transports oxygen from the lungs to other
tissues. Leucocytes are the rounded cells that possess filamentous protrusions. They are
involved in the immune response. The complex surface structures are required for interaction
with other cells. Despite the dramatic difference in composition and function these cells are
both derived from the same pluripotential hematopoietic stem cells. (From R. G. Kessel and
R. H. Kardon, *Tissues and Organs*, W.H. Freeman, Copyright © 1979.)

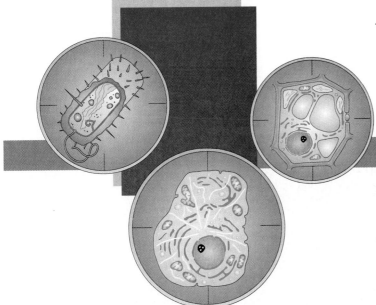

Cells, Biomolecules, and Water

Chapter Outline

The Cell Is the Fundamental Unit of Life

Cells Are Composed of Small Molecules, Macromolecules, and Organelles

Macromolecules Fold into Complex Three-Dimensional Structures

  Water Is a Primary Factor in Determining the Type of Structures that Form

Biochemical Reactions Are a Subset of Ordinary Chemical Reactions

Biochemical Reactions Take Place on the Catalytic Surfaces of Enzymes

Many Biochemical Reactions Require Energy

Biochemical Reactions Are Localized in the Cell

Biochemical Reactions Are Organized into Pathways

Biochemical Reactions Are Regulated

Organisms Are Biochemically Dependent on One Another

Information for the Synthesis of Proteins Is Carried by the DNA

Biochemical Systems Have Been Evolving for Almost Four Billion Years

  All Living Systems Are Related through a Common Evolution

The most unique feature of the living cell is the way in which so many reactions are organized to serve a single purpose.

Aside from a number of striking geological features, the most prominent and unique aspect of the Earth's surface is that it is virtually covered with living organisms. In general biology we are introduced to the extraordinary diversity of living organisms, a diversity so great that it is generally acknowledged that there are more species than could ever be classified. Life seems incredibly complicated. If we limit our inspection of living things to the gross organismic level or what we can see with the naked eye, it is very easy to be overwhelmed by this complexity. Furthermore we will not find explanations for why organisms are constructed the way they are at this level of inquiry. If we probe more deeply into organismic structure, we find that all organisms are composed of much smaller units called cells. If we probe even further we find that cells are composed of a limited number of small molecules and macromolecules. We also find that the macromolecules, which make up the bulk of the solid matter of cells, are constructed from a small number of building blocks that are closely related in structure. It is at the molecular level that we find the ultimate explanations for organismic structure and behavior. We may revel in the discovery that life is not as complicated as we first thought.

Thanks to the investigations of many biochemists over the past half century we are on the verge of a thorough understanding of life at the molecular level. This is a most

Figure 1.1

Generalized representations of the internal structures of animal and plant cells (eukaryotic cells). Cells are the fundamental units in all living systems, and they vary tremendously in size and shape. All cells are functionally separated from their environment by the plasma membrane that encloses the cytoplasm. Plant cells have two structures not found in animal cells: a cellulose cell wall, exterior to the plasma membrane, and chloroplasts. The many different types of bacteria (prokaryotes) are all smaller than most plant and animal cells. Bacteria, like plant cells, have an exterior cell wall, but it differs greatly in chemical composition and structure from the cell wall in plants. Like all other cells, bacteria have a plasma membrane that functionally separates them from their environment. Some bacteria also have a second membrane, the outer membrane, which is exterior to the cell wall.

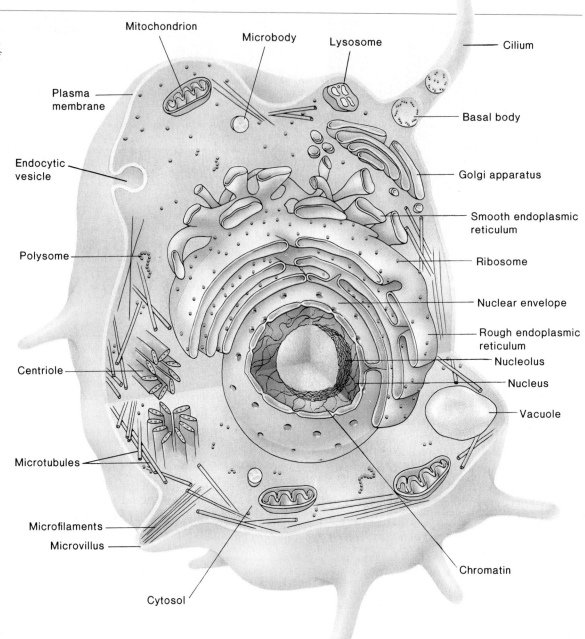

satisfying and exciting time for biochemists; their investigations and accomplishments place biochemistry in the forefront of the biological sciences. The object of this text is to acquaint the beginning student of biochemistry with the basic facts and principles of this subject.

In the introduction to this part of the text, we said that certain common principles underlie all of biochemistry. Let's look briefly now at some of these principles.

The Cell Is the Fundamental Unit of Life

Microscopic examination of any organism reveals that it is composed of membrane-enclosed structures called cells. The enclosing membrane is called the cell membrane, or the plasma membrane. Cells vary enormously in size and shape, but even the largest cells would have to be much larger to be visible to the naked eye. Within this tiny object thousands of chemical reactions are taking place, all regulated, all designed to serve a specific function. Collectively these reactions serve the function of maintaining the cell and permitting it to replicate when the time is right. Perhaps the most amazing thing about the living cell is that so much organized activity takes place in such a small space. Figure 1.1 shows prototypical animal and plant cells, along with some common shapes and sizes of bacteria. Bacteria are single-

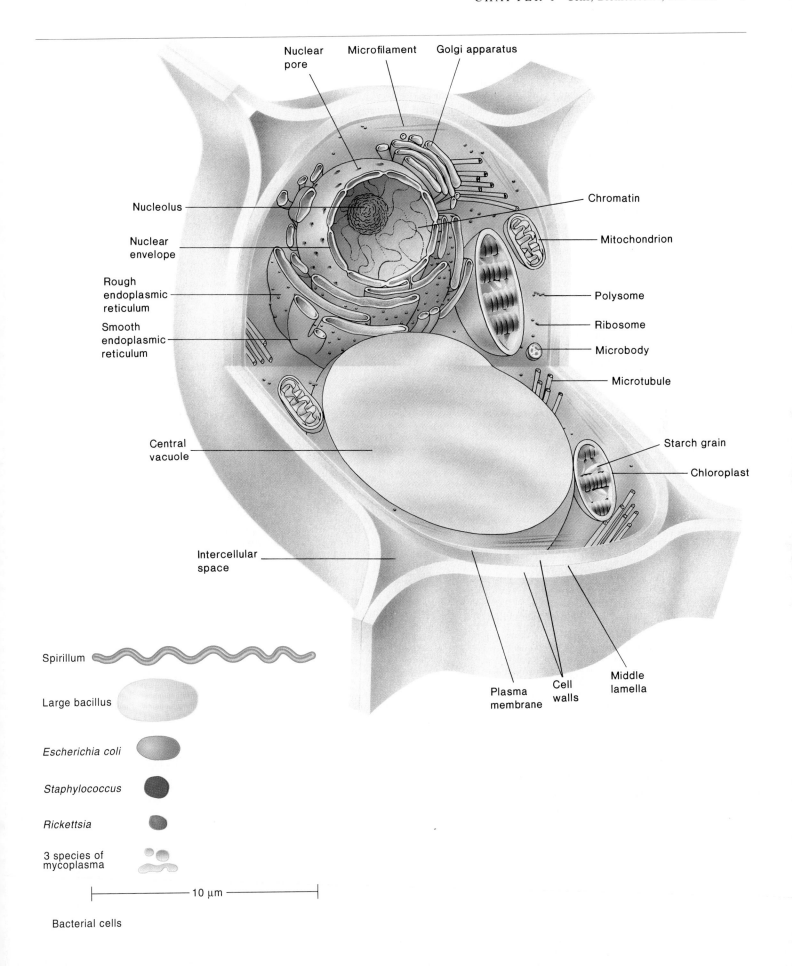

Nuclear pore

Microfilament

Golgi apparatus

Nucleolus

Nuclear envelope

Rough endoplasmic reticulum

Smooth endoplasmic reticulum

Central vacuole

Intercellular space

Chromatin

Mitochondrion

Polysome

Ribosome

Microbody

Microtubule

Starch grain

Chloroplast

Plasma membrane

Cell walls

Middle lamella

Spirillum

Large bacillus

Escherichia coli

Staphylococcus

Rickettsia

3 species of mycoplasma

10 μm

Bacterial cells

Figure 1.2

Specialized cell types found in the human. Although all cells in a multicellular organism have common constituents and functions, specialized cell types have unique chemical compositions, structures, and biochemical reactions that establish and maintain their specialized functions. Such cells arise during embryonic development by the complex processes of cell proliferation and cell differentiation. Except for the sex (germ) cells, all cell types contain the same genetic information, which is faithfully replicated and partitioned to daughter cells. Cell differentiation is the process whereby some of this genetic information is activated in some cells, resulting in the synthesis of certain proteins and not other proteins. Thus, specialized cells come to have different complements of enzymes and metabolic capacities.

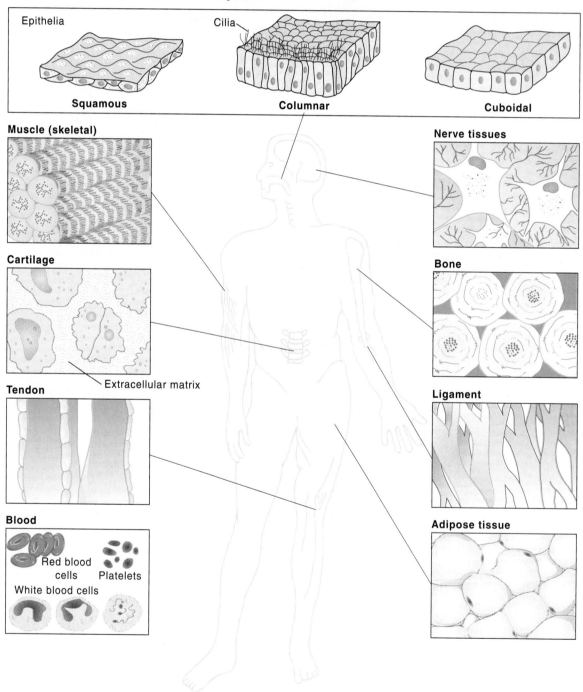

celled organisms, but sometimes the cells are connected into long chains. In multicellular organisms the cells associate to form specialized tissues.

The plasma membrane is a delicate, semipermeable, sheetlike covering for the entire cell. Forming an enclosure prevents gross loss of the intracellular contents; the semipermeable character of the membrane permits the selective absorption of nutrients and the selective removal of metabolic waste products. In many plant and bacterial (but not animal) cells, a cell wall encompasses the plasma membrane. The cell wall is a more porous structure than the plasma membrane, but it is mechanically stronger because it is constructed of a covalently cross-linked, three-dimensional network. The cell wall maintains a cell's three-dimensional form when it is under stress.

The contents enclosed by the plasma membrane constitute the cytoplasm. The purely liquid portion of the cytoplasm is called the cytosol. Within the cytoplasm are a number of macromolecules and larger structures, many of which can be seen by high-power light microscopy or by electron microscopy. Some of the structures are membranous and are called organelles. Organelles commonly found in plant and animal cells include the nucleus, the mitochondria, the endoplasmic reticulum, the Golgi apparatus, and the lysosomes (see fig. 1.1). Chloroplasts are an important class of organelles found in many plant cells but never in animal cells. Each type of organelle is a specialized biochemical factory in which certain biochemical products are synthesized. In addition to organelles, animal and plant cells contain a collection of filamentous structures termed the cytoskeleton, which is important in maintaining the three-dimensional integrity of the cell.

As we will see, the evolutionary tree is bisected into a lower prokaryotic domain and an upper eukaryotic domain. The terms prokaryote and eukaryote refer to the most basic division between cell types. The fundamental difference is that eukaryotic cells contain a membrane-bounded nucleus, whereas prokaryotes do not. The cells of prokaryotes usually lack most of the other membrane-bounded organelles as well. Plants, fungi, and animals are eukaryotes, and bacteria are prokaryotes. The biochemical functions associated with organelles are frequently present in bacteria, but they are usually located on the inner plasma membrane.

Cells are organized in a variety of ways in different living forms. Prokaryotes of a given type produce cells that are very similar in appearance. A bacterial cell replicates by a process in which two identical daughter cells arise from an identical parent cell. Simple eukaryotes can also exist as single nonassociating cells. Eukaryotes of increasing complexity can contain many cells with specialized structures and functions. For example, humans contain about 10^{14}

Table 1.1

The Approximate Chemical Composition of a Bacterial Cell

	Percent of Total Cell Weight	Number of Types of Each Molecule
Water	70	1
Inorganic ions	1	20
Sugars and precursors	3	200
Amino acids and precursors	0.4	100
Nucleotides and precursors	0.4	200
Lipids and precursors	2	50
Other small molecules	0.2	≈200
Macromolecules (proteins, nucleic acids, and polysaccharides)	22	≈5,000

cells of more than a hundred different types. Specialized cells make up the skin, connective tissue, nerve tissue, muscles, blood, sensory functions, and reproductive organs (fig. 1.2). In such a complex organism, the capacity of different cells for replication is limited. When a skin cell or a muscle cell precursor replicates, it makes more cells of the same type. The only cells in a complex eukaryote capable of reproducing an entire organism are the germ cells, that is, the sperm and the egg.

Cells Are Composed of Small Molecules, Macromolecules, and Organelles

Of the many different types of molecules in the various organelles and the cytosol that constitute the living cell, water is by far the most abundant, constituting about 70% by weight of most living matter (table 1.1). As a result, most other components exist essentially in an aqueous environment.

Except for water, most of the molecules found in the cell are lipids or macromolecules, which can be classified into four different categories: lipids, carbohydrates, proteins, and nucleic acids. Each type of macromolecule possesses distinct chemical properties that suit it for the functions it serves in the cell.

Figure 1.3

The structures of common lipids. (*a*) The structures of saturated and unsaturated fatty acids, represented here by stearic acid and oleic acid. (*b*) Three fatty acids covalently linked to glycerol by ester bonds form a triacylglycerol. (*c*) The general structure for a phospholipid consists of two fatty acids esterified to glycerol, which is linked through phosphate to a polar head group. The polar head group may be any one of several different compounds—for example, choline, serine, or ethanolamine.

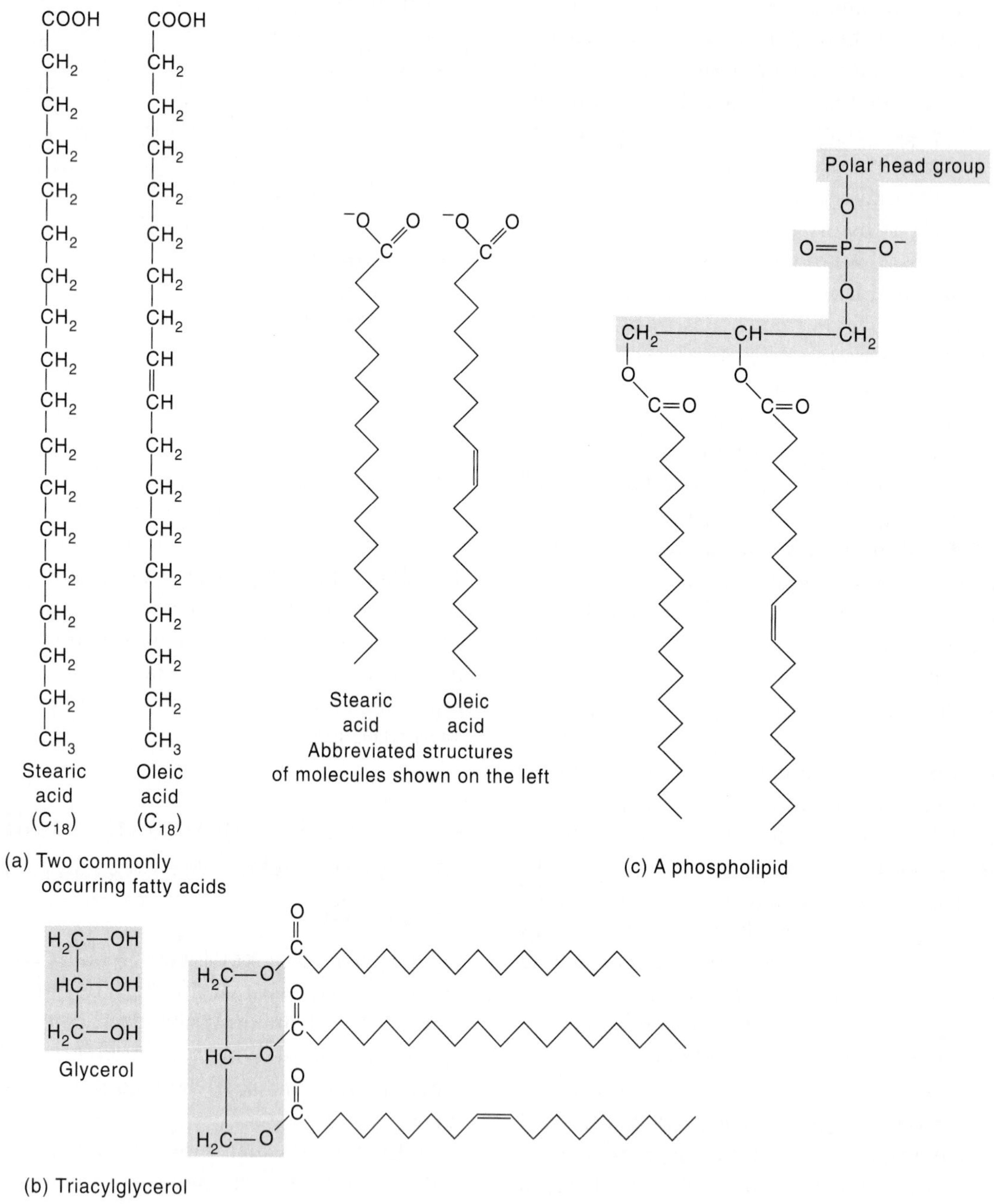

(a) Two commonly occurring fatty acids

(b) Triacylglycerol

(c) A phospholipid

Lipids are primarily hydrocarbon structures (fig. 1.3). They tend to be poorly soluble in water and are therefore particularly well suited to serve as a major component of the various membrane structures found in cells. Lipids also serve as a compact means of storing chemical energy to drive the metabolism of the cell.

Carbohydrates, like lipids, contain a carbon backbone, but they also contain many polar hydroxyl (—OH) groups and are therefore very soluble in water. Large carbohydrate molecules called polysaccharides consist of many small, ringlike sugar molecules; these sugar monomers are attached to one another by glycosidic bonds in a linear or branched array to form the sugar polymer (fig. 1.4). In the cell, such polysaccharides often form storage granules that may be readily broken down into their component sugars. With further chemical breakdown these sugars release chemical energy and may also provide the carbon skeletons for the synthesis of a variety of other molecules. Important structural functions are also served by polysaccharides. Linear polysaccharides form a major component of plant cell walls, and bacterial cell walls are composed of linear polysaccharides that are cross-linked by short polypeptide chains.

Proteins are the most complex macromolecules found in the cell. They are composed of linear polymers called polypeptides, which contain amino acids connected by peptide bonds (fig. 1.5). Each amino acid contains a central carbon atom attached to four substituents: (1) a carboxyl group, (2) an amino group, (3) a hydrogen atom, and (4) an R group. The R group gives each amino acid its unique characteristics. Twenty different amino acids occur in proteins. Some R groups are charged, some are neutral but still polar, and some are apolar.

The linear polypeptide chains of a protein fold in a highly specific way that is determined by the sequence of amino acids in the chains. Many proteins are composed of two or more polypeptides. Certain proteins function in structural roles. Some structural proteins interact with lipids in membrane structures. Others aggregate to form part of the cytoskeleton that helps to give the cell its shape. Still others are the chief components of muscle or connective tissue. Enzymes constitute yet another major class of proteins, which function as catalysts that accelerate and direct biochemical reactions.

Nucleic acids are the largest macromolecules in the cell. They are very long, linear polymers, called polynucleotides, composed of many nucleotides. A nucleotide contains (1) a five-carbon sugar molecule, (2) one or more phosphate groups, and (3) a nitrogenous base. It is the nitrogenous base that gives each nucleotide a distinct character (fig. 1.6). Five different types of nitrogenous bases are found in the two main types of nucleic acids, deoxyribonucleic acid (DNA) and ribonucleic acid (RNA). DNA contains the genetic information that is inherited when cells divide and organisms reproduce. This genetic information is used in the cell to make ribonucleic acids and proteins.

In addition to water and the macromolecules and organelles, the cytosol contains a large variety of small molecules that differ greatly in both structure and function. These never make up more than a small fraction of the total cell mass despite their great variety (see table 1.1). One class of small molecules consists of the monomer precursors of the different types of macromolecules. These monomers are derived by a series of chemical modifications from the nutrients absorbed through the cell membrane. The intermediate molecules between nutrients and monomers are present in small concentrations in the cytosol. Another class of molecules found in the cytosol includes molecules formed as side products in important synthetic reactions and as degradation products of the macromolecules. Finally, the cytosol contains small bioorganic molecules known as coenzymes, which act in concert with the enzymes in a highly specific manner to catalyze a wide variety of reactions.

Macromolecules Fold into Complex Three-Dimensional Structures

The complex folding of biomacromolecules rarely entails making or breaking covalent linkages. Rather the folding process is dictated by the primary structure and the way in which different elements of the macromolecule interact with each other and with water. The forces that determine folding are noncovalent in character. As a rule, specific interactions amount to only a fraction of the interaction energy that occurs when a covalent bond is made or broken, but because so many of these interactions occur and their effects are additive, the energies involved can be quite large.

Water Is a Primary Factor in Determining the Type of Structures that Form

Water, as we have seen, is the major component of living systems, and it interacts with many biomolecules. Some molecules are water-loving, or hydrophilic, others are water-abhorring, or hydrophobic, and still others are amphipathic, or in between. What properties of a molecule make it hydrophilic or hydrophobic? First, consider the molecular properties of water and how water interacts with itself.

An individual water molecule has a significant dipole that is due to the greater electronegativity of the oxygen

F i g u r e   1 . 4

Monomers and polymers of carbohydrates. (*a*) The most common carbohydrates are the simple six-carbon (hexose) and five-carbon (pentose) sugars. In aqueous solution, these sugar monomers form ring structures. (*b*) Polysaccharides are usually composed of hexose monosaccharides covalently linked together by glycosidic bonds to form long straight-chain or branched-chain structures.

(a) Two common monosaccharides that circularize in aqueous solution

(b) Polysaccharides composed of covalently linked monosaccharides

Figure 1.5

Amino acids and the structure of the polypeptide chain. Polypeptides are composed of L-amino acids covalently linked together in a sequential manner to form linear chains. (*a*) The generalized structure of the amino acid. The zwitterion form, in which the amino group and the carboxyl group are ionized, is strongly favored. (*b*) Structures of some of the R groups found for different amino acids. (*c*) Two amino acids become covalently linked by a peptide bond, and water is lost. (d) Repeated peptide bond formation generates a polypeptide chain, which is the major component of all proteins.

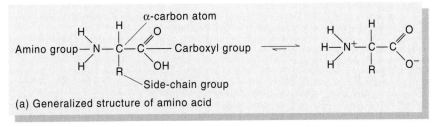

Figure 1.6

The structural components of nucleic acids. Nucleic acids are long linear polymers of nucleotides, called polynucleotides. (*a*) The nucleotide consists of a five-carbon sugar (ribose in RNA or deoxyribose in DNA) covalently linked at the 5′ carbon to a phosphate, and at the 1′ carbon to a nitrogenous base. (*b*) Nucleotides are distinguished by the types of bases they contain. These are either of the two-ring purine type or of the one-ring pyrimidine type. (*c*) When two nucleotides become linked they form a dinucleotide, which contains one phosphodiester bond. Repetition of this process produces a polynucleotide.

(a) Generalized structure of a nucleotide

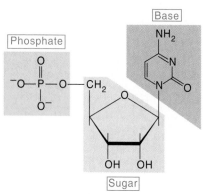

(b) Different bases found in nucleotides

(c) Two nucleotides reacting to form a dinucleotide

Figure 1.7

The structure of water and the interaction of water with other water molecules.

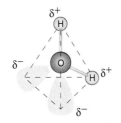

(a) Single water molecule

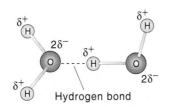

(b) Two interacting water molecules

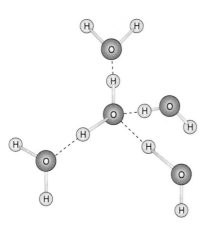

(c) Cluster of interacting water molecules

Figure 1.8

The arrangement of molecules in an ice crystal. Water molecules are oriented so that one proton along each oxygen–oxygen axis is closer to one or the other of the two oxygen atoms.

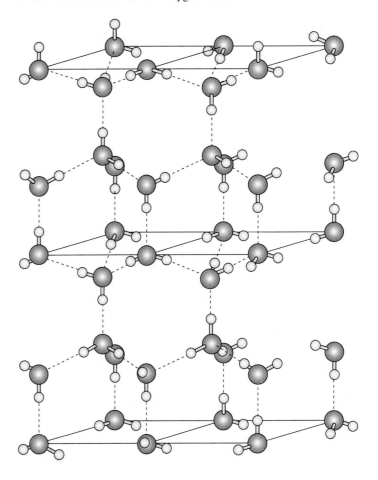

atom over the hydrogen atoms (fig. 1.7). This dipole leads to strong interactions between water molecules, in the form of hydrogen bonds. A hydrogen bond is a noncovalent interaction between polar molecules, one of which is an unshielded proton. In solid water, or ice, the polar forces hold the individual molecules together in a regular three-dimensional lattice (fig. 1.8). Most of the hydrogen bonds present in ice are also present in liquid water. Hence water is a highly hydrogen-bonded structure, not too different from ice, but with a somewhat less regular structure in which the individual molecules have greater mobility.

The dipolar properties of water molecules affect the interaction between water and other molecules that dissolve in water. For example, a favorable interaction accounts for the high solubility of sodium chloride in water (fig. 1.9). The kinds of ion–dipole interactions that take place between water and simple ions such as Na^+ and Cl^- are also important in the interactions between the charged, or polar, groups on biomolecules and water. Thus biomolecules that contain charged residues, hydrogen-bond-forming substituents, or other kinds of polar groups are hydrophilic. In the form of small molecules such groups tend to be very soluble in water. When attached to biopolymers they determine which parts of the molecule will be oriented on the exposed surface, where they can make contact with water.

Apolar groups such as neutral hydrocarbon side chains do not contain significant dipoles or the capacity for forming hydrogen bonds. Consequently, they have nothing to gain by interacting with water, as evidenced by their poor

Figure 1.9

The water molecule is composed of two hydrogen atoms covalently bonded to an oxygen atom with tetrahedral (sp^3) electron orbital hybridization. As a result, two lobes of the oxygen sp^3 orbital contain pairs of unshared electrons, giving rise to a dipole in the molecule as a whole. The presence of an electric dipole in the water molecule allows it to solvate charged ions because the water dipoles can orient to form energetically favorable electrostatic interactions with charged ions.

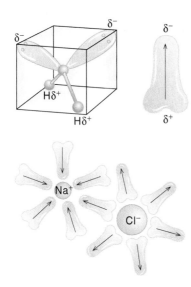

Figure 1.10

Clathrate structures are ordered cages of water molecules around hydrocarbon chains. A portion of the cage structure of $(nC_4H_9)_3S^+F^- \cdot 23\ H_2O$ is shown. The trialkyl sulfur ion nests within the hydrogen-bonded framework of water molecules. In the intact framework, each oxygen is tetrahedrally coordinated to four others. One such oxygen atom and its associated hydrogens are shown by the arrow. (Illustration copyright by Irving Geis. Reprinted by permission.)

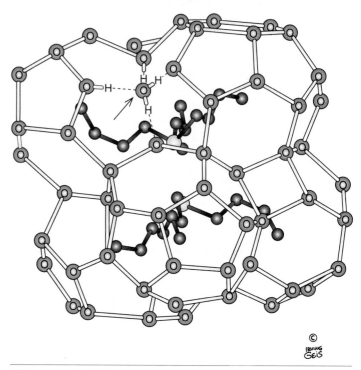

solubility in water. When such hydrophobic molecules are present in water, the water forms a rigid clathrate (cagelike) structure around them (fig. 1.10). Apolar groups in biopolymers tend to bury themselves within the structure of the biopolymer, where they are in the proximity of other apolar groups and avoid contact with water.

Some structures illustrating these principles for macromolecular interaction are shown in figures 1.11 through 1.13. Phospholipids (see fig. 1.3c), which have a hydrophilic polar group on one end and long hydrophobic side chains attached to it, produce multimolecular aggregates in an aqueous environment (fig. 1.11). These phospholipid aggregates form monomolecular layers at the air–water interface or bilayer vesicles within the water. In all of these structures, the polar head groups of the lipid are in contact with water, whereas the apolar side chains are excluded from the solvent structure.

As another example of polarity effects on macromolecular structure, consider polypeptide chains, which usually contain a mixture of amino acids with hydrophilic and hydrophobic side chains. Enzymes fold into complex three-dimensional globular structures with hydrophobic residues located on the inside of the structure and hydrophilic residues located on the surface, where they can interact with water (fig. 1.12).

Figure 1.11

Structures formed by phospholipids in aqueous solution. Phospholipids may form a monomolecular layer at the air–water interface, or they may form spherical aggregations surrounded by water. A vesicle consists of a double molecular layer of phospholipids surrounding an internal compartment of water.

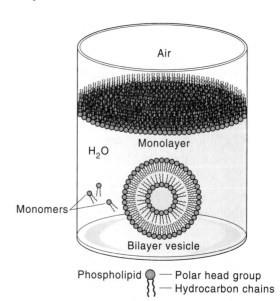

Figure 1.12

A graphic representation of a three-dimensional model of the protein, cytochrome *c*. Amino acids with nonpolar, hydrophobic side chains (color) are found in the interior of the molecule, where they interact with one another. Polar, hydrophilic amino acid side chains (gray) are on the exterior of the molecule, where they interact with the polar aqueous solvent. (Illustration copyright by Irving Geis. Reprinted by permission.)

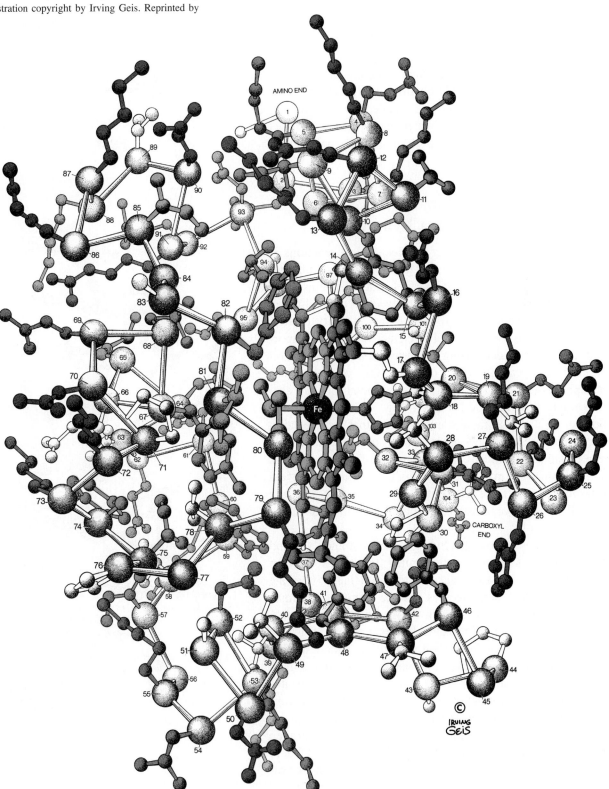

Figure 1.13

The right-handed helical structure of DNA. DNA normally exists as a two-chain structure held together by hydrogen bonds (dashed lines) formed between the bases in the two chains. Along the chain the planar surfaces of these bases interact and, together with the hydrogen bonds, contribute to the stability of the two-chain structure. The negatively charged phosphate groups are on the outside of the structure, where they interact with water, ions, or charged molecules. (Illustration copyright by Irving Geis. Reprinted by permission.)

DNA forms a complementary structure of two helically oriented polynucleotide chains (fig. 1.13). The polar sugar and phosphate groups are situated on the surface, where they can interact with water; the nitrogenous bases from the two chains form intermolecular hydrogen bonds in the core of the structure.

Biochemical Reactions Are a Subset of Ordinary Chemical Reactions

Even though the total number of biochemical reactions is very large, it is still much smaller than the potential number of reactions that occur in ordinary chemical systems. This simplification results partly from the fact that only a limited number of elements account for the vast majority of substances found in living cells. The elements of major importance, in order of decreasing numerical abundance, are hydrogen (H), carbon (C), oxygen (O), nitrogen (N), phosphorus (P), and sulfur (S). Certain metal ions are also important; these include Na^+, K^+, Mg^{2+}, Ca^{2+}, Zn^{2+}, and Fe^{2+} or Fe^{3+}. Other metals and elements that are needed in very small amounts are iodine, cobalt, molybdenum, selenium, vanadium, nickel, chromium, tin, fluorine, silicon, and arsenic. In some cases we don't know the biological roles of these ''trace elements'' but only that they are needed by some organisms for normal growth or development.

The types of covalent linkages most commonly found in biomolecules are also quite limited (table 1.2). Only 16 different types of linkages account for more than 95% of the linkages found in biomolecules. All the elements can form single or double bonds, except for hydrogen, which only makes single bonds; all the elements exist primarily in one valence state, except for carbon and sulfur, which are frequently found in more than one valence state (table 1.3). Despite this overall simplicity, many other valence states can be found in unusual cases, and some of these are very important. For example, the biochemistry of nitrogen involves consideration of all the valence states of nitrogen from +5 to 0 to −3. A major source of nitrogen available to biosystems is gaseous nitrogen found in the atmosphere (valence state 0). Biochemical reactions convert gaseous nitro-

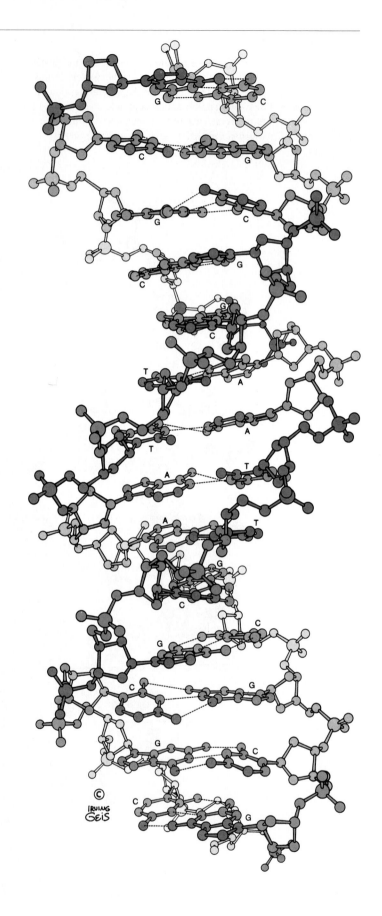

Table 1.2

Types of Covalent Linkages Most Commonly Found in Biomolecules

	H	C	O	N	P	S
H						
C	—C—H	—C—C— C=C				
O	—O—H	—C—O— C=O				
N	N—H	—C—N= C=N—	—			
P	—	—	P—O P=O	—		
S	—S—H	—C—S—	S—O— S=O	—	—	—S—S—

Table 1.3

Most Common Valences Displayed by Atoms in Covalent Linkages

Element	Valence
H	+1
C	−4 to +4
O	−2
P	+5
N	−3
S	+6, −2, −1

Figure 1.14

Different functional groups found in biomolecules. This figure includes the major functional groups. Other functional groups are found in minor amounts.

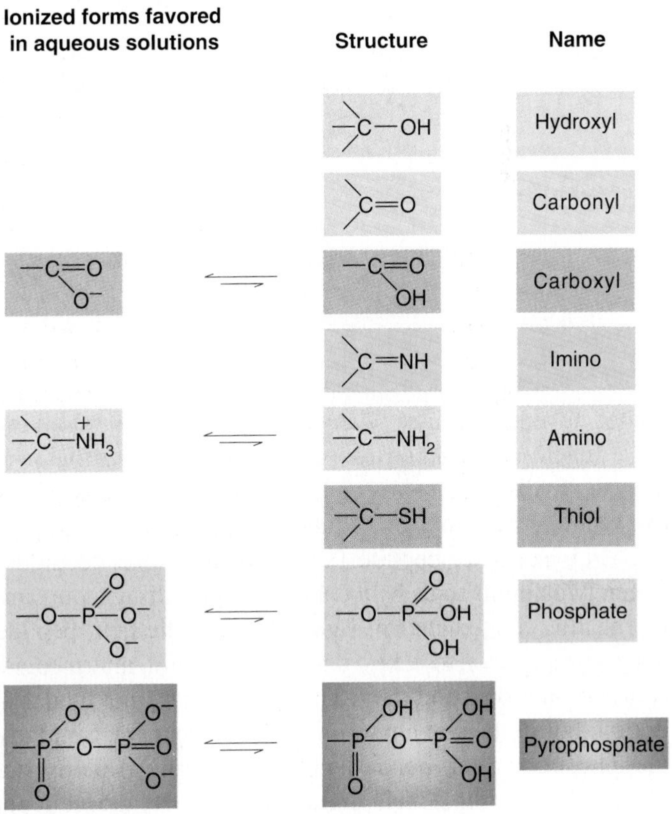

gen into other forms of nitrogen by reactions which occur uniquely in a select group of microorganisms.

Biochemical reactions involving the different classes of substances use a limited number of functional groups, some of which are illustrated in figure 1.14. All of the functional groups depicted are electrostatically neutral in organic solvents. In water or the cell cytosol, however, many of these functional groups either lose or gain protons to become charged species (as shown on the left side of fig. 1.14). Most of the reactive groups in biomolecules contain one or more of these functional groups or ones closely

Figure 1.15

The structure of the complex formed between the enzyme lysozyme and its substrate. The crevice that forms the site for substrate binding (the active site) runs horizontally across the enzyme molecule. The individual hexose sugars of the hexasaccharide substrate are shown in a darker color and labeled A–F. (Coordinates courtesy of D. C. Philips, Oxford, England.) (Illustration copyright by Irving Geis. Reprinted by permission.)

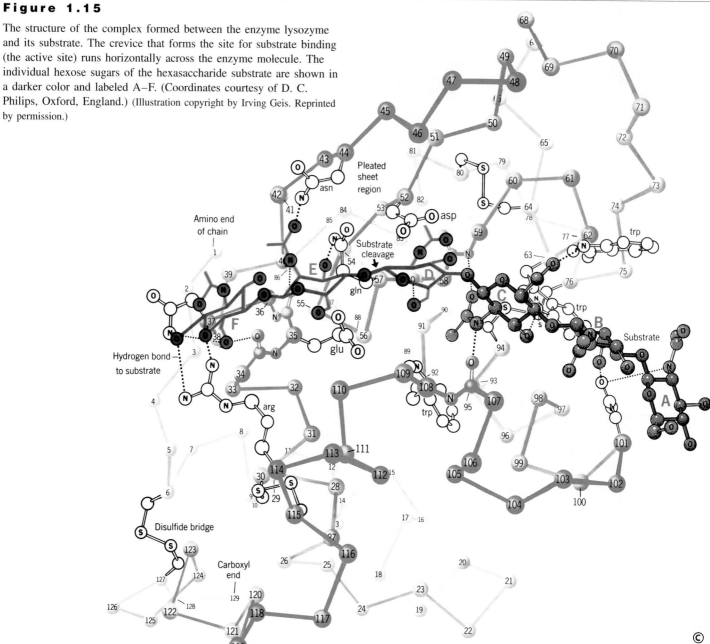

related to these groups. Many cellular reactions involving these functional groups closely resemble reactions that take place in nonliving systems under different conditions. These extracellular reactions are studied in organic chemistry.

For example, peptide bond formation can occur between two amino acids by a dehydration resulting from simple heating as depicted in figure 1.5c. In the cell, peptide bond formation also takes place, but several intermediate steps are involved and the reaction takes place not by dehydrating but in the wet environment of the cytosol. Similarly the phosphodiester bond depicted in figure 1.6c is not formed by a simple dehydration reaction in the cell but

rather when one nucleotide in the triphosphate form loses a pyrophosphate group as it becomes linked to the hydroxyl group of another nucleotide.

Biochemical Reactions Take Place on the Catalytic Surfaces of Enzymes

Although biochemical reactions resemble ordinary chemical reactions, they differ in some important ways. Chemical reactions are frequently carried out in nonaqueous solvents, using elevated temperatures and pressures, acids or bases, or

Figure 1.16

The different fates of an amino acid. Depending on which enzymes are present and active and on the needs of the organism, an amino acid can be metabolized in different ways. Each of these conversions involves one or more steps, and usually each step requires a specific enzyme.

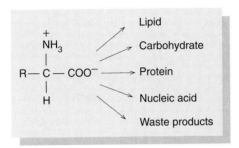

Figure 1.17

Flow of energy in the biosphere. The sun's rays are the ultimate source of energy. These rays are absorbed and converted into chemical energy (ATP) in the chloroplasts. The chemical energy is used to make carbohydrates from carbon dioxide and water. The energy stored in the carbohydrates is then used, directly or indirectly, to drive all the energy-requiring processes in the biosphere.

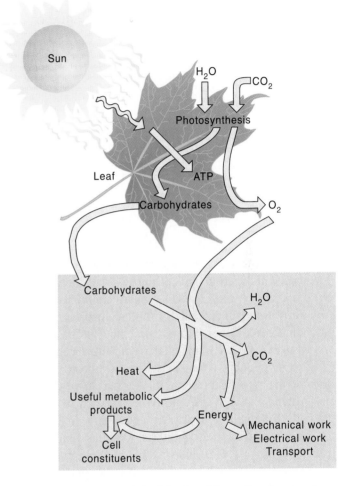

other harsh reagents—conditions that would destroy the functional organization of a living cell. Biochemical reactions usually take place under very mild conditions in aqueous solution. However, many chemical reactions do not proceed at reasonable rates under such conditions. Biochemical reactions proceed at substantially faster rates because of the very special nature of the enzyme catalysts that accelerate them.

Enzymes are structurally complex, highly specific catalysts; each enzyme usually catalyzes only one type of reaction. The enzyme surface binds the interacting molecules, or substrates, so that they are favorably disposed to react with one another (fig. 1.15). The specificity of enzyme catalysis also has a selective effect, so that only one of several potential reactions takes place. For example, a simple amino acid can be used in the synthesis of any of the four major classes of macromolecules or can simply be secreted as waste product (fig. 1.16). The fate of the amino acid is determined as much by the presence of specific enzymes as by its reactive functional groups.

Many Biochemical Reactions Require Energy

An appreciable amount of energy is needed to build a cell. Even maintaining a cell in a steady nongrowing state requires energy input. Chemical energy is needed to drive many biochemical reactions, to do mechanical work, and for transport of substances across the plasma membrane. The ultimate source of energy that drives a cell's reactions is sunlight (fig. 1.17). Light energy is converted into chemical energy in the chloroplasts of plant cells or in the photosynthetic structures of certain microorganisms. The main form

of chemical energy produced in the chloroplast is a nucleotide containing three phosphoric acid groups attached in sequence, adenosine triphosphate, or ATP (fig 1.18). Organisms that cannot harness the light rays of the sun themselves to make ATP are able to make ATP from the breakdown of organic nutrients originating from plants or other organisms.

Most biochemical reactions fall into one of two classes: degradative or synthetic. Degradative, or catabolic, reactions result in the breakdown of organic compounds to simpler substances. Synthetic, or anabolic, reactions lead to the assembly of biomolecules from simpler molecules. Anabolic processes require energy to drive them. This energy is usually supplied by coupling the energy-requiring biosynthetic reactions to energy-releasing catabolic reactions.

Figure 1.18

The structures of ATP and ADP and their interconversion. The two compounds differ by a single phosphate group.

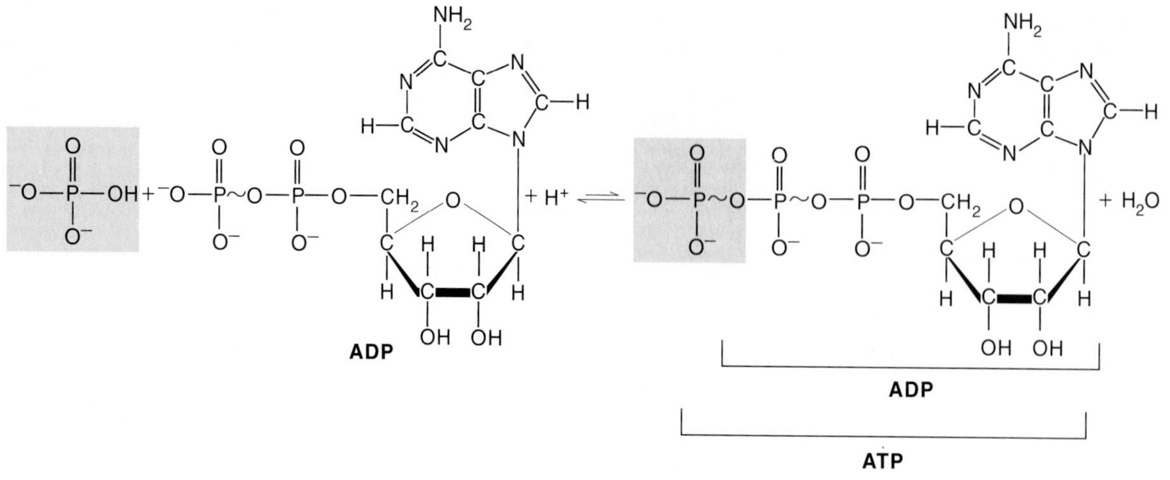

Biochemical Reactions Are Localized in the Cell

Biochemical reactions are organized so that different reactions occur in different parts of the cell. This organization is most apparent in eukaryotes, where membrane-bounded structures are visible proof for the localization of different biochemical processes. For example, the synthesis of DNA and RNA takes place in the nucleus of a eukaryotic cell. The RNA is subsequently transported across the nuclear membrane to the cytoplasm, where it takes part in protein synthesis. Proteins made in the cytoplasm are used in all parts of the cell. A limited amount of protein synthesis also occurs in chloroplasts and mitochondria. Proteins made in these organelles are used exclusively in organelle-related functions. Most ATP synthesis occurs in chloroplasts and mitochondria. A host of reactions that transport nutrients and metabolites occur in the plasma membrane and the membranes of various organelles. The localization of functionally related reactions in different parts of the cell concentrates reactants and products at sites where they can be most efficiently utilized.

Biochemical Reactions Are Organized into Pathways

Most biochemical reactions are integrated into multistep pathways using several enzymes. For example, the breakdown of glucose into CO_2 and H_2O involves a series of reactions that begins in the cytosol and continues to completion in the mitochondrion. A complex series of reactions like this is referred to as a biochemical pathway (fig. 1.19).

Figure 1.19

Summary diagram of the breakdown of glucose to carbon dioxide and water in a eukaryotic cell. As depicted here, the process starts with the absorption of glucose at the plasma membrane and its conversion into glucose-6-phosphate. In the cytosol, this six-carbon compound is then broken down by a sequence of enzyme-catalyzed reactions into two molecules of the three-carbon compound pyruvate. After absorption by the mitochondrion, pyruvate is broken down to carbon dioxide and water by a sequence of reactions that requires molecular oxygen.

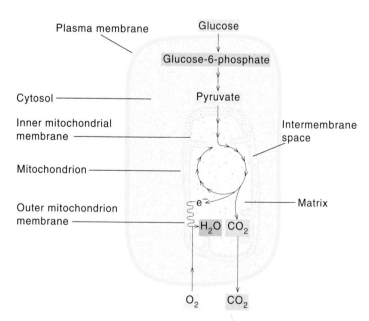

Synthetic reactions, such as the biosynthesis of amino acids in the bacterium *Escherichia coli,* are similarly organized into pathways (fig. 1.20). Frequently pathways have branchpoints. For example, the synthesis of the amino acids threo-

Figure 1.20

Synthesis of various amino acids from oxaloacetate. Each arrow represents a discrete biochemical step requiring a unique enzyme. Thus aspartic acid is produced in one step from oxaloacetate, whereas isoleucine is produced in five steps from threonine.

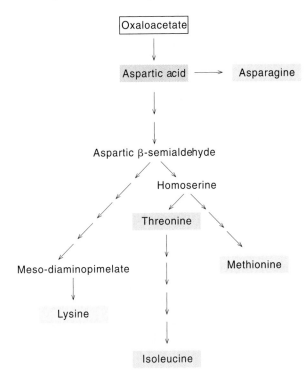

nine and lysine starts with oxaloacetate. After three steps, a branchpoint is reached with the formation of the organic compound aspartic-β-semialdehyde. One branch of this pathway leads to the synthesis of the amino acid lysine, and another branch leads to the synthesis of the amino acids methionine, threonine, and isoleucine.

To understand the role of each biochemical reaction we must identify its position in a pathway and also consider how that pathway interacts with others.

Biochemical Reactions Are Regulated

Hundreds of biochemical reactions take place even in the cells of relatively simple microorganisms. Living systems have evolved a sophisticated hierarchy of controls that permits them to maintain a stable intracellular environment. These controls ensure that substances required for maintenance and growth are produced in amounts that are adequate without being excessive. Biochemical controls have developed in such a way that the cell can make adjustments in response to a changing external environment. Adjustments are needed because the temperature, ionic strength, acid concentration, and concentration of nutrients present in the

external environment vary over much wider limits than could be tolerated inside the cell.

The rate of intracellular reactions is a function of the availability of substrates and enzymes. Enzyme activity is controlled at two different levels. First, the rate of a catalyzed reaction is regulated by the amount of the catalyzing enzyme present in the cell. Control of enzyme amounts is usually accomplished by regulating the rate of enzyme synthesis; in some cases the rate of enzyme degradation is also regulated. We can think of controls that regulate the total amount of enzyme present as coarse controls. They define the limits of possible enzyme activity as being anywhere from 0 to 100% of the full activity of the enzyme. Fine controls that act directly on enzymes are also present. Only certain special enzymes, called regulatory enzymes, are susceptible to this second type of regulation. Regulatory enzymes usually occupy key points in biochemical pathways, and their state of activity frequently is decisive in determining the utilization of the pathway.

The underlying principle in regulation is maintaining a favorable intracellular environment in the most economical manner. The cell makes products in the amounts that are needed. Each pathway is regulated in a somewhat different way, ensuring that biochemical energy and substrates are efficiently utilized.

Organisms Are Biochemically Dependent on One Another

Between 3 and 4 billion years ago the first self-replicating molecules appeared on earth. These entities had to have the capacity for extracting nutrients from the chemical compounds that existed in prebiotic times. We have some general notions about what types of substances were present at that time. One of the most important substances that was not present at that time in significant amounts was molecular oxygen, O_2. Currently this form of oxygen is required by all forms of life visible to the naked eye.

The O_2 used by most organisms is ultimately converted by them into CO_2. Oxygen is utilized at a rapid rate and it would soon disappear if it were not for special classes of photosynthetic organisms that are constantly producing more O_2 by the oxidation of water.

The oxygen story is an example of the dependence of one class of organisms on another for certain chemicals. A similar situation exists with the elements carbon and nitrogen, which must be converted from gaseous forms, CO_2 and N_2, to organic forms usable by most organisms. Reduced carbon compounds are constantly being lost by oxidation to gaseous CO_2. The supply of organic carbon compounds required by all forms of life is replenished by photosynthetic organisms; these include most plants and certain microorganisms. Similarly, nitrogen in organic molecules is constantly being lost to the atmosphere in the form of gaseous nitrogen. The reactions required for the conversion of nitrogen to a reduced form more usable to the majority of organisms occurs in only a limited number of microorganisms; yet without these nitrogen-fixing organisms life as we know it would soon vanish.

As we ascend the evolutionary tree, we find increasingly complex multicellular forms. Such organisms generally require more complex nutrients, which must ultimately be supplied to them by simpler living forms. Bacteria like *E. coli* can make all of their own amino acids from a reduced form of nitrogen, such as NH_3, and a reduced form of carbon, such as glucose. Humans, on the other hand, must receive most of their amino acids as nutrients. Humans and other complex organisms have gained new biochemical capacities, which permit them to synthesize the components associated with highly specialized differentiated tissues. At the same time, they have lost many of the biochemical systems required to survive on simpler nutrients.

Many biochemical reactions of great importance take place in only a limited number of organisms. This fact increases the complexity of the study of biochemistry. We must learn many reactions; we must also be aware of the biochemical potentials of different organisms. This is the only way we can understand the biochemical interdependency of organisms.

Information for the Synthesis of Proteins Is Carried by the DNA

DNA contains the genetic information transmitted to each daughter cell when cells divide. The DNA usually exists in the form of nucleoprotein (DNA-protein) complexes called chromosomes. A prokaryotic cell contains a single chromosome. Prior to cell division this chromosome duplicates and segregates so that an identical complement of DNA goes to each of two newly formed daughter cells.

Eukaryotic cells are more complex than prokaryotic cells and usually contain more DNA, which is partitioned between several chromosomes. In both prokaryotes and eukaryotes, almost all cells of the same organism contain the same number of chromosomes. In eukaryotes most of the chromosomes are localized in the nucleus. Thus the DNA is isolated from the main body of the cytoplasm—a unique feature of eukaryotes and the primary distinction between prokaryotes and eukaryotes. Some organelles, notably the mitochondria and the chloroplasts, contain a single circular chromosome.

Eukaryotic chromosomes are detectable by light microscopy at the stage just prior to cell duplication. At this stage, called mitosis, chromosomes appear as elongated refractile structures that can be seen to segregate in equal numbers and types to each of the daughter cells before cell division (fig. 1.21). Each chromosome carries hereditary (genetic) information necessary for the synthesis of specific compounds essential for cell maintenance, growth, and replication. Each chromosome contains a single very long DNA molecule composed of 10^6 or more nucleotides in a specific sequence. The sequence of nucleotides in the chromosomal DNA determines the sequence of amino acids in the protein polypeptide chains of the organism. The relationship between base sequences and resultant amino acids is known as the genetic code. Each grouping of three bases, called a triplet, represents a specific amino acid and is called a codon. The genetic code ensures that the organism's characteristics are reflected by the sequence of nucleotides in its DNA. When chromosomes replicate, the DNA replicates precisely, so that the same nucleotide sequence is passed along to each of the daughter cells resulting from mitosis and cell division.

The DNA does not transfer its genetic information directly to protein. Rather, this information passes through an intermediary, the messenger RNA (mRNA). The mRNA is made on a DNA template in the nucleus of a eukaryotic cell and then passes into the cytoplasm, where it serves in turn as a template for the synthesis of the polypeptide chain. The overall process of information transfer from DNA to mRNA (transcription) and from mRNA to protein (translation) is depicted in figure 1.22.

Biochemical Systems Have Been Evolving for Almost Four Billion Years

Biochemical systems are conservative and opportunistic. They tend to evolve one step at a time, using a readily accessible route that leads to an advantage. To appreciate biochemical systems today it is useful to have some understanding of how they came to be.

The earth was formed by a process of accretion about 4.6 billion years ago. Initially it was a molten mass lacking the gravitational pull to retain its gases at the prevalent elevated temperatures. And yet, within a mere 700 million years of the planet's birth, as calculated from the isotopic record of sediments, cellular life almost certainly existed. What raw materials were available to bring about this amazing turn of events? What were the sources of energy used to drive the necessary reactions? Where did the important reactions take place? Was it in the atmosphere, in the oceans, on dry land, or all three?

Table 1.4 shows a distribution of the major elements found in the earth's crust, the ocean water, and the human body. The composition of the human body, which is reason-

Figure 1.21

Mitosis and cell division in eukaryotes. After DNA duplication has occurred, mitosis is the process by which quantitatively and qualitatively identical DNA is delivered to daughter cells formed by cell division. Mitosis is traditionally divided into a series of stages characterized by the appearance and movement of the DNA-bearing structures, the chromosomes. (*a*) Premitosis. (*b*) through (*h*) Successive stages of mitosis. (*i*) Postmitosis.

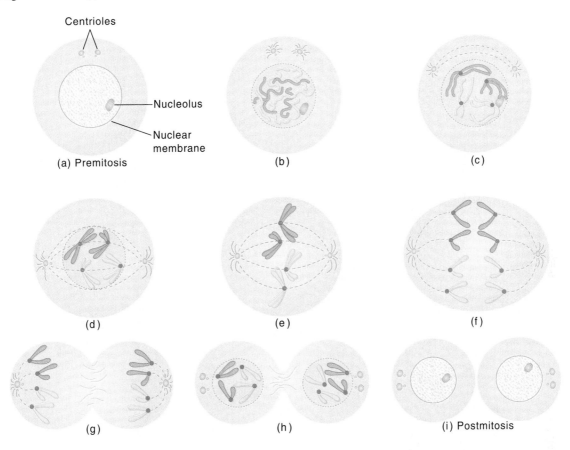

ably representative of living organisms, differs appreciably from that of the earth's crust. The four most abundant elements in the human body are hydrogen, carbon, nitrogen, and oxygen. Of these, only oxygen belongs to the class of elements in highest abundance that make up the earth's crust. Nevertheless, the elements needed to make living things were present in sufficient quantities in the earth's crust and its primitive oceans and the atmosphere.

The original water and air associated with the newly formed planet were lost because of the high temperatures. As the earth cooled, water and various gases on the surface and in the atmosphere were produced by an outgassing process. Today the earth is only about 0.5% water by weight, but because of water's low density, most of it is present on the earth's surface, where it has a major impact on the environment. Water cycles through its gaseous form in the atmosphere and its liquid form in the oceans and bodies of fresh water.

Geological evidence indicates that appreciable amounts of the total water mass have always been present as liquid water. Thus the temperature of the planet has for the most part been between 0° and 100°C, a range that is conducive to the formation of biomolecules and the origin and propagation of life. The earth has also been kind to living things in other ways. The buffering action of various clays and minerals is believed to have maintained the pH level of the oceans between 8.0 and 8.5, which is close to the pH inside living cells.

Unlike the temperature and pH of the surface water, the composition of the atmosphere has changed drastically since the origin of life. In fact, the processes taking place in living things are primarily responsible for these changes. Today's atmosphere, which is about one millionth of the mass of the earth itself, is mainly composed of nitrogen (78%) and oxygen (21%). Most of the remaining atmosphere is argon (0.9%), water (variable up to 4%), and car-

Figure 1.22

Transfer of information from DNA to protein. The nucleotide sequence in DNA specifies the sequence of amino acids in a polypeptide. DNA usually exists as a two-chain helical structure. The information contained in the nucleotide sequence of only one of the DNA chains is used to specify the nucleotide sequence of the messenger RNA molecule (mRNA). This sequence information is used in polypeptide synthesis. A three-nucleotide sequence in the mRNA molecule codes for a specific amino acid in the polypeptide chain. (Illustration copyright by Irving Geis. Reprinted by permission.)

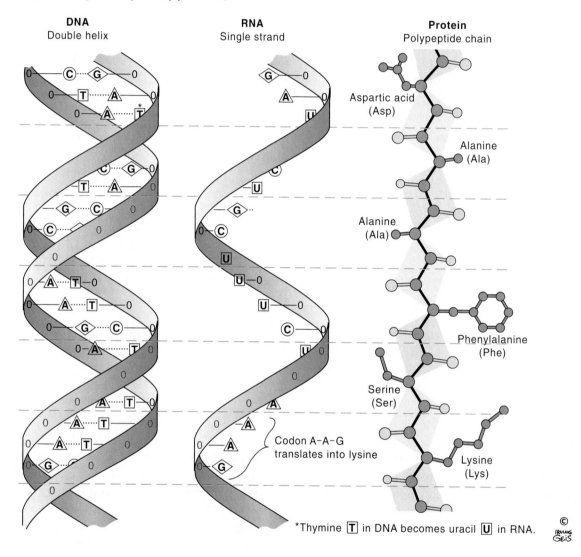

*Thymine T in DNA becomes uracil U in RNA.

bon dioxide (0.034%). The gases that made up the primitive atmosphere were quite different; especially conspicuous was the absence of gaseous oxygen. Most of the oxygen in the present atmosphere is due to the oxidation of water by photosynthetic organisms. The main forms of carbon and nitrogen in the primitive atmosphere were probably CO_2 and N_2 as they are today. In addition, and probably of great significance to the origin of life, small amounts of the more reduced forms of carbon and H_2 gas were present. Thus the primitive earth probably had a weakly reducing atmosphere as contrasted with today's highly oxidizing atmosphere. This situation was most favorable to the origin of life, because organic compounds that enter the biomass tend to be in a reduced state and they are readily oxidized in the presence of gaseous oxygen. It is generally believed that the first organics formed in this primitive atmosphere and then rained down to form larger bioorganic molecules in the liquid phase.

Table 1.4

Distribution of the 24 Elements Used
in Biological Systems[a]

Element	Atomic No.	Earth's Crust	Ocean	Human Body
Hydrogen (H)	1	2,882	66,200	60,562
Carbon (C)	6	56	1.4	10,680
Nitrogen (N)	7	7	<1	2,440
Oxygen (O)	8	60,425	33,100	25,670
Fluorine (F)	9	77	<1	<1
Sodium (Na)	11	2,554	290	75
Magnesium (Mg)	12	1,784	34	11
Silicon (Si)	14	20,475	<1	<1
Phosphorus (P)	15	79	<1	130
Sulfur (S)	16	33	17	130
Chlorine (Cl)	17	11	340	33
Potassium (K)	19	1,374	6	37
Calcium (Ca)	20	1,878	6	230
Vanadium (V)	23	4	<1	<1
Chromium (Cr)	24	8	<1	<1
Manganese (Mn)	25	37	<1	<1
Iron (Fe)	26	1,858	<1	<1
Cobalt (Co)	27	1	<1	<1
Nickel (Ni)	28	3	<1	<1
Copper (Cu)	29	1	<1	<1
Zinc (Zu)	30	2	<1	<1
Selenium (Se)	34	<1	<1	<1
Molybdenum (Mo)	42	<1	<1	<1
Iodine (I)	53	<1	<1	<1

[a] Amounts are given in atoms per 100,000.

The origin of life probably occurred in three phases (fig. 1.23): (1) The earliest phase was a period of chemical evolution during which the compounds needed for the nucleation of life must have been formed. These compounds include the most important class of biological macromolecules, the nucleic acids. In this phase of evolution, the synthesis of nucleic acids was "noninstructed." (2) As soon as some nucleic acids were present, physical forces between them must have led to an "instructed" synthesis, in which the already formed molecules served as templates for the synthesis of new polymers. It seems likely that "feedback loops" selected out certain nucleic acids for preferential synthesis. At some point during this period of instructed synthesis more nucleic acids and possibly protein macromolecules were formed. The products of this phase of mo-

Figure 1.23

The origin of life probably occurred in three overlapping phases: Phase I, chemical evolution, involved the noninstructed synthesis of biological macromolecules. In phase II, biological macromolecules self-organized into systems that could reproduce. In phase III, organisms evolved from simple genetic systems to complex multicellular organisms. The arrow pointing from left to right emphasizes the unidirectional nature of the overall process.

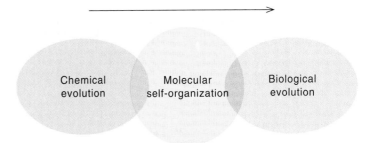

lecular self-organization must sooner or later have begun to resemble the complex organized units that we observe in the self-reproducing biosynthetic cycles of living cells. (3) In the final phase of the origin of life we find the beginnings of the divergent process of biological evolution, the development of the simplest single-celled organisms, and their differentiation into complex multicellular beings.

All Living Systems Are Related through a Common Evolution

The classical view of evolution, based on morphological differences among organisms, can be diagrammed as a branching tree in which all existing organisms are shown at the tips of the branches (fig. 1.24). An evolutionary tree starts from the simple ancestral prokaryotic cell, which branches off in three main directions into the archaebacteria, the eubacteria, and the eukaryotes. Each of these kingdoms has continued to branch in elaborate ways; we have shown only some of the main branchpoints in figure 1.24. Prokaryotes for the most part have remained as relatively undifferentiated single-celled organisms containing a single chromosome. By contrast, eukaryotes have changed dramatically. Although the organisms on many branches of the eukaryotic part of the evolutionary tree have remained as relatively undifferentiated single-celled forms, significant numbers of eukaryotic organisms have evolved into multicellular forms in which the individual cells of the total organism have differentiated to serve different functions. This process has given rise to plants, animals, and fungi.

A somewhat different view of biological evolution has arisen from a comparison of nucleic acid sequences in different organisms. The best-known sequences for such stud-

Figure 1.24

Classical evolutionary tree. All living forms have a common origin, believed to be the ancestral prokaryote. Through a process of evolution some of these prokaryotes changed into other organisms with different characteristics. The evolutionary tree indicates the main pathways of evolution.

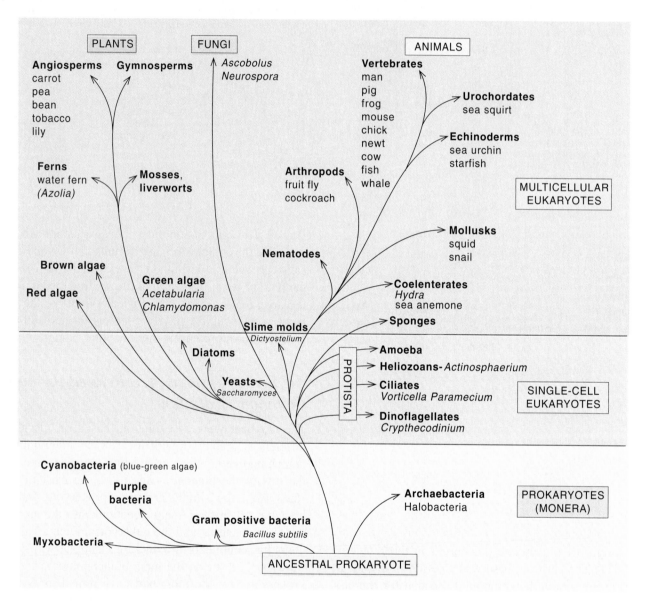

ies have come from the 16S ribosomal RNAs in different organisms. This comparative study has provided us with the evolutionary pattern shown in figure 1.25, in which distances are proportional to sequence differences. A most striking characteristic of this unrooted tree is the distinctness of the three primary kingdoms as evidenced by the large sequence distances that separate each of the kingdoms. Although we do not know the position of the root of the tree, it seems likely that the archaebacteria are closer to the common ancestor of all three kingdoms than are the eubacteria or the prokaryotes.

The organisms most familiar to us, the multicellular plants and animals, occupy a shallow domain within the eukaryotic line of descent. True, the developmental programs of the multicellular forms have generated an incredible diversity in form and function. Nevertheless, both bacteria and unicellular eukaryotes span far greater evolutionary histories. This fact is reflected in the greater biochemical diversity that we find in the unicellular microorganisms.

Figure 1.25

An evolutionary tree can be constructed by comparing the complete sequences of 21 different 16S and 16S-like ribosomal RNAs (rRNAs). The scale bar represents the number of accumulated nucleotide differences (mutations) per sequence position in the rRNAs of the various organisms. (Source: From N. R. Pace, G. J. Olsen and C. R. Woese, *Cell* 5:325, 1986.)

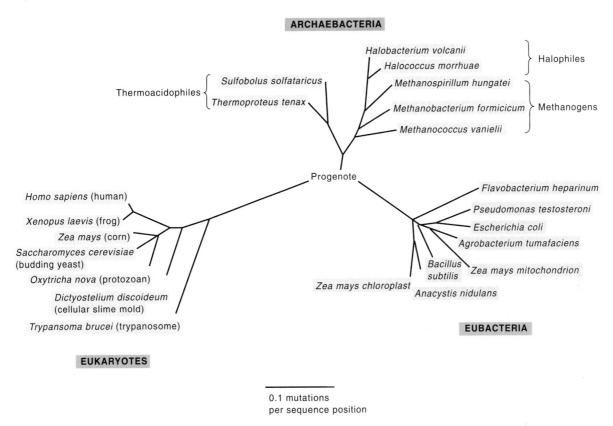

Summary

In this chapter we discussed the ways in which biochemistry parallels ordinary chemistry and those in which it is quite different. The chief points to remember are the following.

1. The basic unit of life is the cell, which is a membrane-enclosed, microscopically visible object.
2. Cells are composed of small molecules, macromolecules, and organelles. The most prominent small molecule is water, which constitutes 70% of the cell by weight. Other small molecules are present only in quite small amounts; they are precursors or breakdown products of macromolecules or coenzymes. There are four types of macromolecules: lipids, carbohydrates, proteins, and nucleic acids.

3. Noncovalent intermolecular forces largely determine the folded structure adopted by a macromolecule, particularly the relative affinity of different groupings on the macromolecule for water. In general, hydrophobic groupings are buried within the folded macromolecular structure, whereas hydrophilic groupings are located on the surface, where they can interact with water.
4. Biochemical reactions utilize a limited number of elements, most prominently carbon, hydrogen, oxygen, nitrogen, sulfur, and phosphorus. Many biochemical reactions are simple organic reactions.

5. Biochemical reactions are carried out under very mild conditions in aqueous solvent. The reactions can proceed under these conditions because of the highly efficient nature of protein enzyme catalysts.

6. Biochemical reactions frequently require energy. The most common source of chemical energy used is adenosine triphosphate (ATP). The splitting of a phosphate from the ATP molecule can provide the energy needed to make an otherwise unfavorable reaction proceed in the desired direction.

7. Biochemical reactions of different types are localized to different parts in the cell.

8. Biochemical reactions are frequently organized into multistep pathways.

9. Biochemical reactions are regulated according to need by controlling the amount and activity of enzymes in the system.

10. Most organisms depend on other organisms for their survival. Frequently this is because a given organism cannot make all of the compounds needed for its growth and survival.

11. The specific properties of any protein are due to the specific sequence of amino acids in its polypeptide chains. This sequence is determined by the genetic information carried by the sequence of DNA nucleotides. DNA transfers the information to messenger RNA, which serves as the template for protein synthesis.

12. Shortly after the earth was formed and had cooled to a reasonable temperature, chemical processes produced compounds that would be used in the development of living cells. Nucleic acids are the most important compounds for living cells, and it is believed that they played the central role in the origin of life.

13. Evolutionary trees based on morphology or biochemical differences indicate that all living systems are related through a common evolution.

Selected Readings

Becker, W. M., *The World of the Cell.* Menlo Park, Calif.: Benjamin/Cummings, 1986. A very readable cell biology book that could be referred to while taking biochemistry.

de Duve, C., *Blueprint for a Cell.* Burlington, N.C.: California Biological Supply Co., 1991. A short book on the origin of life that contains an excellent reference list.

Dickerson, R. E., Chemical evolution and the origin of life. *Sci. Am.* 239(3):70–86, 1978.

Doolittle, R. F., The genealogy of some recently evolved vertebrate proteins. *Trends Biochem. Sci.* 10:233–237, 1985.

Kimura, M., The neutral theory of molecular evolution. *Sci. Am.* 241(5):98–126, 1979.

Schopf, J. W., The evolution of the earliest cells. *Sci. Am.* 229(3):10–138, 1978. An authoritative account from a foremost geologist.

Stillinger, F. H., Water revisited. *Science* 209:451–457, 1980. A reminder of the central importance of water to the origin of life.

Wilson, A. C., The molecular basis of evolution. *Sci. Am.* 253(4):164–173, 1985.

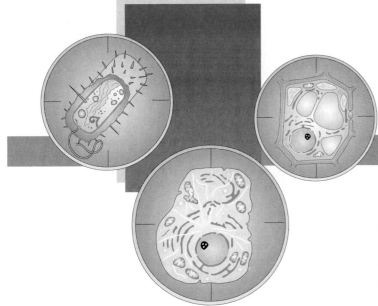

Thermodynamics in Biochemistry

Chapter Outline

Thermodynamic Quantities
> *The First Law of Thermodynamics: Any Change in the Energy of a System Requires an Equal and Opposite Change in the Surroundings*
> *The Second Law of Thermodynamics: In Any Spontaneous Process the Total Entropy of the System and the Surroundings Increases*
> *Free Energy Provides the Most Useful Criterion for Spontaneity*

Applications of the Free Energy Function
> *Values of Free Energy Are Known for Many Compounds*
> *The Standard Free Energy Change in a Reaction Is Related Logarithmically to the Equilibrium Constant*
> *Free Energy Is the Maximum Energy Available for Useful Work*
> *Biological Systems Perform Various Kinds of Work*
> *Favorable Reactions Can Drive Unfavorable Reactions*

ATP as the Main Carrier of Free Energy in Biochemical Systems
> *The Hydrolysis of ATP Yields a Large Amount of Free Energy*

All living systems obey the laws of thermodynamics: A pathway that requires energy usually consumes ATP while a pathway that releases energy usually produces ATP.

The primary usefulness of thermodynamics to biochemists lies in predicting whether particular chemical reactions could occur spontaneously. A simple illustration is to predict what compounds could possibly serve as energy sources for an organism. You are aware from everyday experience that oxidation of organic molecules by molecular oxygen releases energy. For example, wood or coal burns with a large output of heat. Similarly, organisms can obtain energy by oxidizing carbohydrates, fats, or proteins. Some organisms oxidize hydrocarbons, some oxidize reduced forms of sulfur, and others oxidize iron. But no organisms live by oxidizing molecular nitrogen, and the explanation lies in thermodynamics. The reaction cannot occur spontaneously. This example illustrates the importance of thermodynamics in controlling all life. Because organisms live by extracting chemical energy from their surroundings, thermodynamics is not an esoteric subject. It is a matter of life or death.

We say that thermodynamics determines whether a process "could" occur, because thermodynamics tells us only whether the process is possible, not whether it actually does occur in a finite period of time. The rate at which a thermodynamically possible reaction occurs depends on the detailed mechanism of the process. For a biochemical process to occur rapidly, appropriate enzymes must be available. The distinction between spontaneity and speed is more critical for biochemists than it is for chemists, because if a chemical reaction does not proceed rapidly a chemist can change the pressure or temperature or increase the concentration of the reactants. A living organism is under more rigid constraints: It must function at a fixed temperature and pressure and within a limited range of concentrations of reactants.

In this chapter we discuss thermodynamic quantities. We then expand the discussion to show how the concept of free energy is used in predicting biochemical pathways, and we explore the central role of ATP in providing energy for biochemical reactions.

Thermodynamic Quantities

The properties of a substance can be classified as either intensive or extensive. Intensive properties, which include density, pressure, temperature, and concentration, do not depend on the amount of the material. Extensive properties, such as volume and weight, do depend on the amount. Most thermodynamic properties are extensive including energy (E), enthalpy (H), entropy (S), and free energy (G).

Energy, enthalpy, entropy, and free energy are all properties of the state of a substance. This means that they do not depend on how the substance was made or how it reached a particular state. In a chemical reaction, it is the difference between the initial and final states that is important; the pathway taken to get the initial state to the final state has no bearing on whether the overall reaction releases or consumes energy (fig. 2.1).

The First Law of Thermodynamics: Any Change in the Energy of a System Requires an Equal and Opposite Change in the Surroundings

Of the thermodynamic properties we just listed, energy is probably the most familiar. Energy can be equated with the capacity to do work. The energy of a molecule includes the internal nuclear energies and the molecular electronic, translational, rotational, and vibrational energies. Electronic energies, which reflect the interactions among the electrons and nuclei, usually are much larger than the translational, rotational, and vibrational energies.

In biochemistry, we are concerned not so much with the absolute energies of molecules as with changes in energy that occur in the course of reactions. It is easier to evaluate a change in energy (ΔE) than to calculate the absolute energies of the reactants or products, because many of

Figure 2.1

The change in energy of a system depends only on the initial and final states, not on the path by which the system gets from one state to the other. This diagram illustrates the conversion of a phosphate ester of glucose (glucose-6-phosphate) to free glucose and inorganic phosphate ion (P_i) by two different pathways. Although the two routes proceed through intermediate compounds that differ in energy (A, B, C, and D), the overall energy change (ΔE) is the same. However, the work done and the amount of heat absorbed or released generally is not the same for the two paths.

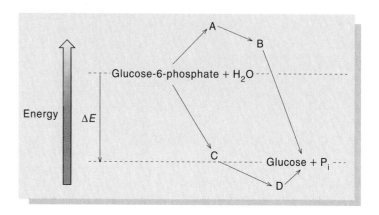

the terms that contribute to the total energy do not change much during a chemical reaction. Electronic energies usually dominate ΔE, as they do the total energies. Good estimates of the energy change resulting from a chemical reaction can usually be obtained by calculating the difference between the bond energies of the reactants and those of the products (see table 4.2).

The first law of thermodynamics says that the total amount of energy in the universe is constant. Energy can undergo transformations from one form to another; for example, the chemical energy of a molecule can be transformed into thermal, electric, or mechanical energy. But any change in the total energy of one part of the universe is matched by an equal and opposite change in another part, so that the overall energy of the universe remains constant.

To apply the first law to a chemical reaction, we must take into account all the energy changes that occur. This means including any changes in the surroundings, as well as in the system of interest (fig. 2.2). The system might be a reaction occurring in a test tube or a living cell; the surroundings are all the rest of the universe. In practical terms, however, we usually need to consider only the immediate surroundings, because only these are likely to be influenced by what happens in the system. According to the first law, the overall energy remains constant even though energy in some form may flow from the system to its surroundings or from the surroundings to the system.

Figure 2.2

A system and its surroundings. Heat flow into the system is designated as a positive quantity (q), and work that the system does on the surroundings is designated as a positive quantity (w). The first law of thermodynamics relates q and w to changes in the energy of the system. Any change in the energy of the system (ΔE_{sys}) is balanced by an opposite change in the energy of the surroundings (ΔE_{sur}), so that the overall energy change (ΔE_{tot}) is zero.

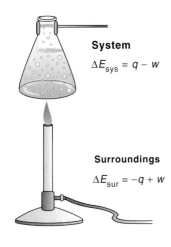

System

$\Delta E_{sys} = q - w$

Surroundings

$\Delta E_{sur} = -q + w$

$$\Delta E_{tot} = \Delta E_{sys} + \Delta E_{sur} = 0$$

If the amount of matter in a system is constant, there are only two means by which the system can gain or lose energy: transfer of heat or performance of work. The energy of a system increases if the system absorbs heat from its surroundings or if the surroundings do work on the system; it decreases if the system gives off heat to the surroundings or does work. A common way of stating the first law of thermodynamics is to equate the change in energy of the system (ΔE) to the difference between the heat absorbed by the system (q) and the work done by the system (w):

$$\Delta E = q - w \qquad (1)$$

Both q and w depend on the path of the reaction, but their difference ΔE is independent of the path and therefore defines a state function.

The relative energy of a compound can be measured in a bomb calorimeter, a device in which the compound is thoroughly combusted and the resulting heat is measured. Energies (and enthalpies, discussed next) are usually given in units of kilocalories per mole or kilojoules per mole. One kilojoule (kJ) is the amount of energy needed to apply 1 newton (N) of force over 1 km; 1 kilocalorie (kcal) is the heat needed to raise the temperature of 1 kg of water from 14.5° to 15.5°C. One kcal is equivalent to 4.184 kJ. Physicists generally prefer to use kilojoules; chemists, including biochemists, use both units.

It is customary to record energies of molecules with reference to the energies of their constituent elements. The standard energy of formation reported for a molecule, ΔE°_{f}, is equal to the energy change associated with the formation of the molecule from the elements in their standard states. The energy change in any other reaction can be obtained by subtracting the energies of formation of the reactants from the energies of formation of the products. The complex organic molecules found in cells have relatively weak chemical bonds, and thus higher energies, than H_2O, CO_2, and the other small molecules from which they are formed.

Enthalpy is a function of state that is closely related to energy but is usually more pertinent for describing the thermodynamics of chemical or biochemical reactions. The change in enthalpy (ΔH) is related to the change in energy (ΔE) by the expression

$$\Delta H = \Delta E + \Delta(PV) \qquad (2)$$

where $\Delta(PV)$ is the change in the product of the pressure (P) and volume (V) of the system. ΔH is the amount of heat absorbed from the surroundings if a reaction occurs at constant pressure and no work is done other than the work of expansion or contraction of the system. (The work done when a system expands by ΔV against a constant pressure P is $P\,\Delta V$. This type of work is generally not very useful in biochemical systems.) In most biochemical reactions, little change occurs in either pressure or volume, so the difference between ΔH and ΔE is relatively small.

In most chemical reactions that occur spontaneously, the enthalpy of the system decreases. If no work is done, the system gives off heat to the surroundings. But changes in enthalpy or energy do not provide a reliable way of determining whether a reaction can proceed spontaneously. For example, although LiCl and $(NH_4)_2SO_4$ both dissolve readily in water, the former process releases heat, whereas the latter absorbs heat. A mixture of solid $(NH_4)_2SO_4$ and water proceeds spontaneously to a state of higher enthalpy. This reaction is driven by an increase in the entropy of the system.

To assess the potential for a reaction to occur we must consider the entropy as well as the enthalpy.

The Second Law of Thermodynamics: In Any Spontaneous Process the Total Entropy of the System and the Surroundings Increases

The second law of thermodynamics is that the universe inevitably proceeds from states that are more ordered to states that are more disordered. This phenomenon is measured by a thermodynamic function called entropy, which is denoted by the symbol S. A reaction in which entropy increases (ΔS is positive) proceeds in preference to one in which entropy decreases.

Entropy is an index of the number of different ways that a system could be arranged without changing its energy. If a system could be arranged in Ω different ways, all with the same energy, the absolute entropy per molecule would be

$$S = k \ln \Omega \qquad (3)$$

where k is Boltzmann's constant ($k = 3.4 \times 10^{-24}$ cal/degree Kelvin). For a mole of substance,

$$S = Nk \ln \Omega = R \ln \Omega \qquad (4)$$

Here N is the number of molecules in a mole (6×10^{23}) and R is the gas constant ($R \approx 2$ cal/(degree K · mole). Quantitative values for entropies are usually given in entropy units (1 eu = 1 cal/degree K).

The underlying idea here is that the more ways a particular state could be obtained, the greater is the probability of finding a system in that state. A system that is highly disordered could be obtained in many different arrangements that are all energetically equivalent. Thus a state in which molecules are free to move about and rotate into many different orientations or conformations is favored over a state in which motion is more restricted. The second law makes the remarkably general assertion that the total entropy change in any reaction that occurs spontaneously must be greater than zero. Note, however, that this statement specifies the total entropy, which means that we must consider the entropy change in the surroundings as well as that in the system. The entropy of the system can decrease if the entropy of the surroundings increases by a greater amount.

The absolute entropies of small molecules can be calculated by statistical mechanical methods. Table 2.1 shows the results of such calculations for liquid propane. The largest contributions to the entropy come from the translational and rotational freedom of the molecule, and much smaller contributions from vibrations; electronic terms are insignificant. Although exact calculations of this type become intractable for large biological molecules, the relative sizes of the contributions from different types of motions are similar to those in small molecules. Thus entropy is associated primarily with translation and rotation. This relationship is very different from enthalpy, in which electronic terms are dominant and translational and rotational energies are comparatively small.

Statistical mechanical calculations show that the translational entropy of a molecule depends on $\frac{3}{2}R \ln M_r$ (plus some smaller terms), where R is the gas constant and M_r is the molecular weight. Suppose that a molecule Y undergoes

Table 2.1

Contributions to the Entropy of Liquid Propane at 231 K

	kcal/(degree K · mole)
Translational entropy	36.04
Rotational entropy	23.38
Vibrational entropy	1.05
Electronic entropy	0.00
Total	60.47

a dimerization reaction so that its molecular weight doubles:

$$2\,Y \longrightarrow Y_2$$

Intuitively, we expect dimerization to decrease the entropy because the two monomeric units can no longer move independently. We can calculate the effect quantitatively as a function of the molecular weight as follows. If M_r is the molecular weight of the monomer, then the change in translational entropy in going from the monomeric state to the dimer is approximately

$$\Delta S \approx \frac{3}{2}R \ln 2M_r - 2\left(\frac{3}{2}R \ln M_r\right)$$

$$= \frac{3}{2}R \ln 2 - \frac{3}{2}R \ln M_r$$

$$= -\frac{3}{2}R \ln \left(\frac{M_r}{2}\right) \qquad (5)$$

From this equation we see that the decrease of the translational entropy resulting from dimerization is a logarithmic function of the molecular weight.

Structural features that make molecules more rigid reduce rotational and vibrational contributions to entropy. Thus the formation of a double bond or ring decreases the entropy even when the molecular weight is unchanged. The formation of comparatively rigid macromolecular structures from flexible polypeptide or polynucleotide chains also requires an entropy decrease, although this can be offset by increases in the entropy of the surrounding water molecules (see chapter 4).

The entropy of a compound depends strongly on the physical state of the material. A gas has more translational and rotational freedom than a liquid, and a liquid has more freedom than a solid. As a result, entropy increases when a solid melts or a liquid vaporizes.

It can be shown that the increase in the entropy of a system that undergoes an isothermal, reversible process is

$$\Delta S = \frac{\Delta H}{T} \qquad (6)$$

where T is the absolute temperature in degrees kelvin. An isothermal process is one that occurs at constant temperature. A reversible process is one that proceeds infinitely slowly through a series of intermediate states in which the system is always at equilibrium. For any real process occurring at a finite rate, the system is not strictly at equilibrium. As a consequence the ΔS is usually somewhat larger than the value given by equation (6).

From equation (6), the entropy increase on vaporization or melting can be determined simply from the heat of vaporization divided by the boiling point, or the heat of fusion divided by the melting point. The entropy increase on vaporization of water is 26 eu/mole and that on melting of ice is 5.3 eu/mole. These values are consistent with our intuition that the increase in translational and rotational freedom is much greater in going from a liquid to a gas than in going from a solid to a liquid.

The entropy of a solution is increased by the mixing of solvents, and it is decreased by interactions among the solvent molecules or interactions of solutes with the solvent. The mixing of two miscible liquids is a thermodynamically favorable process because it increases the number of positions available to the molecules. The entropy change on going from the unmixed liquids to the mixed state can be calculated from the expression

$$\Delta S = n_a R \ln \frac{1}{X_a} + n_b R \ln \frac{1}{X_b}$$
$$= -n_a R \ln X_a - n_b R \ln X_b \qquad (7)$$

where n_a and n_b are the number of moles of A and B that are mixed, and X_a and X_b are the corresponding mole fractions in the final solution. Because X_a and X_b are always less than 1, $\ln X_a$ and $\ln X_b$ are negative. This means that the dilution of each component resulting from the mixing makes a positive contribution to the entropy. Equation (7) applies to the mixing of ideal solutions, in which no interactions occur among the molecules. Any intermolecular interactions decrease the entropy by restricting the system's translational and rotational freedom.

Solvation, the interaction of a solute with the solvent, makes an important negative contribution to the entropy of a solution. Solvation can take the form of hydrogen bonding to donor or acceptor groups on the solute, or of a looser clustering of solvent molecules oriented around the solute (fig. 2.3). In general, the entropy of solvation by water be-

Figure 2.3

The entropy decrease resulting from solvation. When a salt is dissolved in water, the entropy of the dissociated cations and anions increases because of the increased possibilities for translation and rotation. But at the same time the movement of water molecules becomes restricted in the vicinity of the ions. The net effect is frequently a decrease in the entropy of the solution. Such a decrease in entropy can occur if the solution releases heat to the surroundings, because this increases the entropy of the surroundings.

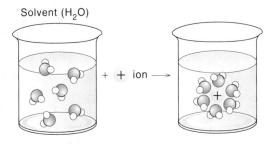

Solvent (H_2O)

comes more negative with an increase in the charge or polarity of the solute. Small ions are solvated more strongly than large ions with the same charge, and anions are solvated more strongly than cations.

It is noteworthy that enthalpy depends mainly on electronic interactions, while entropy depends mainly on translation and rotation; solvation affects both enthalpy and entropy. Enthalpies and entropies of solvation usually tend to oppose each other. For charged species, the more negative (favorable) the enthalpy of solvation, the more negative (unfavorable) the entropy of solvation.

From our earlier discussion, you might expect that the dissociation of a proton from a carboxylic acid, which increases the number of independent particles, would lead to an increase in entropy. However, this effect is more than counterbalanced by solvation effects. The charged anion and proton both ''freeze out'' many of the surrounding molecules of water (fig. 2.4). Thus the ionization of a weak acid decreases the number of mobile molecules and so leads to a decrease in entropy. The entropy of ionization of a typical carboxylic acid in water is about -22 eu/mole. The entropy of dissociation of a proton from a quaternary ammonium group ($R{-}NH_3^+ \longrightarrow R{-}NH_2 + H^+$) is usually smaller, because in this case the dissociation does not alter the number of charged species in the solution.

A different type of solvation effect occurs when an apolar molecule is added to water. The result is a decrease in entropy but not because of favorable interactions between the molecule and the solvent. The water orients on the surface of the apolar molecule to form a relatively rigid cage held together by hydrogen bonds (see fig. 1.10). This effect plays important roles in governing the folding of proteins and determining the structure of biological membranes.

Figure 2.4

An ionization reaction often decreases the entropy of a solution, instead of increasing it as one might at first expect, because clustering of water molecules around the ions can result in a net decrease in the number of free water molecules.

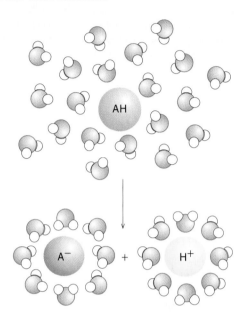

Figure 2.5

The entropy of binding of gas molecules to a solid catalyst is negative because of the restricted movements of the adsorbed molecules. By contrast, the entropy change on binding of a substrate to an enzyme is frequently positive. This effect arises because the restricted movement of the substrate is more than compensated for by the release of bound water from the enzyme and the substrate.

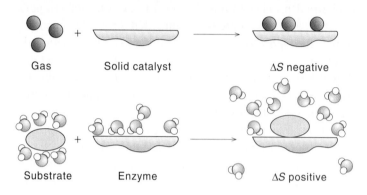

Substantial changes in entropy can occur when a small molecule binds to a protein or other macromolecule. Of particular interest is binding of a substrate or inhibitor to an enzyme. It is instructive to compare the entropy changes here with those that accompany the binding of gas molecules to the surface of a solid catalyst. When gas molecules adsorb on a solid surface, there is a large decrease in entropy (fig. 2.5). The translational entropy of the gas disappears, and the thermodynamics of the adsorbed molecules becomes more like that of a solid. This negative change in entropy is an obstacle to the industrial use of solid catalysts for reactions of gases. To make the reaction proceed, the unfavorable entropy change must be overcome by increasing the pressure or tailoring the catalyst so that the enthalpy of interaction with the gas is strongly negative. In contrast, the binding of a substrate to an enzyme frequently has a positive ΔS because water molecules are displaced when the substrate binds (fig. 2.5). Binding of an ester substrate to pepsin, for example, produces an entropy increase of 20.6 eu/mole, and binding of urea to urease produces an entropy increase of 13.3 eu/mole. Because of the favorable entropy change, the binding of the substrate may occur spontaneously even if ΔH is unfavorable.

Another important entropy effect that occurs when an enzyme and substrate combine can be called the chelation effect. This effect is best understood by discussing a relatively simple case of metal chelation. Cadmium ion tends to be quadrivalent, so if there is one amino group in a ligand molecule, as in methylamine, the cadmium can bind to four molecules. If there are two amino groups, as in ethylenediamine, the cadmium combines with two ligand molecules, as shown in table 2.2. Notice that the entropy change is much more unfavorable for the combination with four methylamines than for that with two ethylenediamines. Water molecules are released from the cadmium ion when the ligands bind, but the entropy increase from this release is about the same whether methylamine or ethylenediamine is added. The less favorable entropy change resulting from association with methylamine is due to the larger number of molecules that must attach to the cadmium in this case.

In general, the chelation effect means that a molecule with n points of attachment to another molecule binds more strongly than n molecules with one point of attachment, even though the enthalpy change upon binding at each point is the same. The chelation effect is important in substrate binding to enzymes, where there typically are multiple points of attachment. Several weak interactions can produce an overall tight binding because of the additive contributions of the small, favorable enthalpy changes and the lack of a proportional decrease in entropy. This effect is even greater in the binding of two proteins or the binding of a protein to a nucleic acid, because many points of interaction usually exist between these large molecules.

Standard Enthalpy and Entropy Changes on Forming Complexes between Cadmium Ion and Methylamine
or Ethylenediamine (en)

Reaction[a]	$\Delta H°$ (kcal/mole)	$\Delta S°$ (eu/mole)	$T \Delta S°$ (kcal/mole)	$\Delta G°$ (kcal/mole)
$Cd^{2+} + 4\,CH_3NH_2 \longrightarrow Cd(CH_3NH_2)_4^{2+}$	−13.7	−16.0	−4.77	−8.94
$Cd^{2+} + 2\,en \longrightarrow Cd(en)_2^{2+}$	−13.5	−3.3	−0.98	−12.50

[a] en = ethylenediamine; temperature = 25°C.

Source: Data from Spike and Parry, Thermodynamics of chelation I. The statistical factor in chelate ring formation, *J. Amer. Chem. Soc.* 75:2726, 1953.

The final column in table 2.2 indicates the changes in free energy accompanying the reactions of cadmium ion with the two amine compounds. Free energy is a function of both enthalpy and entropy; it provides the most useful indication of whether a reaction can proceed spontaneously, as explained in the next section.

Free Energy Provides the Most Useful Criterion for Spontaneity

We have seen that a system tends toward the lowest enthalpy and the highest entropy. The tendency for the enthalpy of a system to decrease can be explained simply by the second law of thermodynamics. If no work is done, an enthalpy decrease means a transfer of heat from the system to the surroundings, and if the surroundings are at a lower temperature, such a flow of heat will be driven by an increase in the entropy of the surroundings. But enthalpy changes do not afford a reliable rule for determining whether a reaction can proceed spontaneously, because the enthalpy of a system can increase if the entropy of the system also increases. The second law does provide a reliable rule, but its application is often difficult because it requires that we consider the entropy changes in the surroundings as well as in the system.

A more convenient function for predicting the direction of a reaction was discovered by Josiah Gibbs. He was the first to appreciate that in reactions occurring at equilibrium and constant temperature, the change in entropy of the system is numerically equal to the change in enthalpy divided by the absolute temperature. This relationship is the one already presented in equation (6). The equation can be transposed to

$$\Delta H - T \Delta S = 0 \qquad (8)$$

In search of a criterion for spontaneity, Gibbs proposed a new function called the free energy, defined by the equation

$$\Delta G = \Delta H - T \Delta S \qquad (9)$$

Here ΔH and ΔS are the changes in enthalpy and entropy in the system alone, not including the surroundings. For a reaction occurring at equilibrium, such as the melting of ice at 0°C, the change in free energy is zero. For the same reaction occurring at a higher temperature, say 10°C, the term $T \Delta S$ is larger, making ΔG negative. Ice at 10°C melts spontaneously. In the reverse reaction, conversion of water to ice at 10°C, there is a positive change in free energy. This process does not occur spontaneously. Gibbs proposed that a reaction can occur spontaneously if, and only if, ΔG is negative. If ΔG is zero, the system is in equilibrium and no net reaction occurs in either direction.

Free energy, like energy, enthalpy, and entropy, is a state function and an extensive property of a system. If the free energy change is favorable (negative), and a good pathway exists, a reaction does occur. If no pathway exists for the conversion, a catalyst may be added that provides an acceptable pathway. However, if the free energy change is unfavorable (positive), no catalyst can ever make the reaction proceed.

Applications of the Free Energy Function

The free energy function dominates most discussions of thermodynamics in biochemistry. Not only does the sign of ΔG determine the direction in which a reaction proceeds, but the magnitude of ΔG indicates just how far the reaction must proceed before the system comes to equilibrium. This is because the standard free energy change $\Delta G°$ has a simple relationship to the equilibrium constant. We elaborate on these points in the following sections. Despite its usefulness, however, many people find the free energy function difficult to grasp intuitively. The reason is that ΔG is a composite of enthalpic and entropic terms, which often make opposite contributions.

Values of Free Energy Are Known for Many Compounds

The standard free energy of formation of a compound, ΔG°_f, is the difference between the free energy of the compound in its standard state and the total free energies of the elements of which the compound is composed, again when the elements are in their standard states. The standard states usually are chosen to be the states in which the elements or molecules are stable at 25°C and 1 atmosphere pressure. For oxygen and nitrogen, these are the gases O_2 and N_2; for solid elements such as carbon, they are the pure solids. For most solutes, the standard states are taken to be 1 M solutions. However, in biochemistry the standard state for hydrogen ion in solution is usually defined as a 10^{-7} M solution because this is close to the concentration in most systems of interest to biochemists.

Standard free energies of formation are known for thousands of compounds. They usually are given in units of kcal/mole or kJ/mole. The values for a few compounds of biological interest are collected in table 2.3. By subtracting the sum of the free energies of formation of the reactants from the sum of the free energies of formation of the products, it is possible to calculate the standard free energy change in any reaction for which all the free energies of formation are known.

From the values listed in table 2.3, we can calculate the standard free energy change for the reaction

$$\text{Oxaloacetate}^{2-} + \text{H}^+ \,(10^{-7}\,\text{M}) \longrightarrow$$
$$\text{CO}_2(g) + \text{pyruvate}^-$$

as

$$\Delta G^\circ = -113.44 - 94.45 - (9.87 - 190.62)$$
$$= -7.4 \text{ kcal/mole}$$

The free energy change, when all the reactants and products are in their standard states (1 M oxaloacetate dianion and pyruvate anion, 10^{-7} M hydrogen ion, and 1 atm CO_2), is -7.4 kcal/mole. The negative value of ΔG° means that the reaction proceeds spontaneously under these conditions. However, some of the concentrations are not very realistic. At pH 7, carbon dioxide is present partly in the form of the bicarbonate anion, rather than as gaseous CO_2. To take this into account, we can add the standard free energy change

Table 2.3

Standard Free Energies of Formation of Some Compounds of Biological Interest

Substance	ΔG°_f (kcal/mole)	ΔG°_f (kJ/mole)
Lactate ions (1 M)	-123.76	-516
Pyruvate ions (1 M)	-113.44	-474
Succinate dianions (1 M)	-164.97	-690
Glycerol (1 M)	-116.76	-488
Water	-56.69	-280
Acetate anions (1 M)	-88.99	-369
Oxaloacetate dianions (1 M)	-190.62	-797
Hydrogen ions (10^{-7} M)	-9.87[a]	-41[a]
Carbon dioxide (gas)	-94.45	-394
Bicarbonate ions (1 M)	-140.49	-587

[a] This is the value for hydrogen ions at a concentration of 10^{-7} M. The free energy of formation at unit activity (1 M) is 0.

for the reaction of CO_2 with water to give the bicarbonate anion plus a proton:

$$\text{CO}_2(g) + \text{H}_2\text{O} \longrightarrow \text{HCO}_3^- + \text{H}^+$$

This calculation yields a correction of $-140.49 - 9.87 - (-56.69 - 94.45) = 0.8$ kcal/mole. The free energy change for the reaction of oxaloacetate to form pyruvate and 1 M bicarbonate ions instead of CO_2 is $-7.4 + 0.8 = -6.6$ kcal/mole.

The preceding calculation illustrates the point that the standard free energy change for a reaction can be found by adding or subtracting the free energies of any other reactions that combine to give the desired reaction. Another example is the calculation of the standard free energy of hydrolysis of ATP to adenosine diphosphate (ADP) and P_i at pH 7. This calculation can be done by combining the free energy change for the hydrolysis of glucose-6-phosphate with the free energy change for forming glucose-6-phosphate from glucose and ATP, as shown in table 2.4. We return to these reactions in a later section.

The Standard Free Energy Change in a Reaction Is Related Logarithmically to the Equilibrium Constant

In biochemistry we are most concerned with reactions occurring in aqueous solution. Suppose we have a chemical reaction with the stoichiometry

Table 2.4

Calculating the Standard Free Energy of ATP Hydrolysis ($\Delta G^{\circ\prime}$) by Adding the Free Energies of Two Other Reactions

Reaction[a]	$\Delta G^{\circ\prime}$ (kcal/mole)
Glucose + $ATP^{4-} \longrightarrow$ glucose-6-phosphate^{2-} + ADP^{3-} + H^+	−5.4
Glucose-6-phosphate^{2-} + $H_2O \longrightarrow$ glucose + HPO_4^{2-}	−3.0
ATP^{4-} + $H_2O \longrightarrow$ ADP^{3-} + HPO_4^{2-} + H^+	−8.4

[a] The values of $\Delta G^{\circ\prime}$ are for reactions at pH 7 in the absence of Mg^{2+}. In the presence of 10 mM Mg^{2+}, $\Delta G^{\circ\prime}$ for ATP hydrolysis is about −7.5 kcal/mole.

$$aA + bB \longrightarrow cC + dD$$

where a, b, c, and d refer to the moles of A, B, C, and D, respectively. The free energy change in the reaction is

$$\Delta G = G_{\text{final state}} - G_{\text{initial state}} \quad (10)$$

If the reaction occurs at constant temperature and pressure, equation (10) can be expressed as the difference ΔG° between the standard free energies of the products and reactants plus a correction for the concentrations:

$$\Delta G = \Delta G^\circ + RT \ln \frac{[C]^c[D]^d}{[A]^a[B]^b} \quad (11)$$

The last term in equation (11) is the correction for concentration and as such is an entropic contribution to ΔG. It is derived by using equation (7) to find the entropy changes associated with diluting the reactants and products from their standard states (1 M) to the actual concentrations in the solution. (In a rigorous treatment, we should use activities instead of concentrations in this formula, but for simplicity we ignore the difference, keeping in mind that it can be substantial in some cases.) If the concentrations of the reactants exceed those of the products, so that the ratio $[C]^c[D]^d/[A]^a[B]^b$ is less than 1, the logarithm is negative, making ΔG more negative than ΔG° and favoring the reaction in the forward direction. A concentration ratio greater than 1 favors the reverse reaction.

When the reaction comes to equilibrium,

$$\frac{[C]^c[D]^d}{[A]^a[B]^b} = K_{\text{eq}} \quad (12)$$

where K_{eq} is the equilibrium constant for the reaction. We also know that at equilibrium $\Delta G = 0$. Therefore, from equation (11),

$$\Delta G^\circ = -RT \ln K_{\text{eq}} \quad (13)$$

Table 2.5

Relationship between ΔG° and K_{eq} (at 25°C)

ΔG° (kcal/mole)[a]	K_{eq}
−6.82	10^5
−5.46	10^4
−4.09	10^3
−2.73	10^2
−1.36	10
0	1
1.36	10^{-1}
2.73	10^{-2}
4.09	10^{-3}
5.46	10^{-4}
6.82	10^{-5}

[a] ΔG° values at 25°C are calculated from the equation

$$\Delta G^\circ = -RT \ln K_{\text{eq}}$$
$$= -1.98 \times 298 \times 2.3 \log K_{\text{eq}}$$
$$= -1364 \log K_{\text{eq}}$$

Thus the standard free energy change for a reaction can be used to obtain the equilibrium constant. Conversely, if we know K_{eq}, we can find ΔG°. Because of the logarithmic relationship K_{eq} has a very steep dependence on ΔG° (table 2.5). A reaction that proceeds to 99% completion is, for most practical purposes, a quantitative reaction. It requires an equilibrium constant of 100 but a standard free energy change of only −2.7 kcal/mole, which is little more than half the standard free energy change for the formation of a hydrogen bond.

Equations (11) and (13) are two of the most important thermodynamic relationships for biochemists to remember. If the concentrations of reactants and products are at their equilibrium values, there is no change in free energy for the reactions going in either direction. Living cells, however, maintain some compounds at concentrations far from the equilibrium values, so that their reactions are associated with large changes in free energy. We expand on this point in chapter 11.

We mentioned that biochemists usually define the standard state of protons as 10^{-7} M and report values of free energy and equilibrium constants for solutions at pH 7. These values are designated by a prime and written as $\Delta G^{\circ\prime}$, $\Delta G'$ and K'_{eq}. Unprimed symbols are used to designate values based on a standard state of 1 M for protons (pH 0). For a reaction that releases one proton, the relationship between K'_{eq} and K_{eq} is $K'_{eq} = 10^7 K_{eq}$. In evaluating the standard free energies $\Delta G^{\circ\prime}$ and ΔG°, it is critical to use the equilibrium constants K'_{eq} and K_{eq}, respectively, because these can be very different quantities.

Free Energy Is the Maximum Energy Available for Useful Work

The free energy change gives a quantitative measure of the maximum amount of useful work that could be obtained from a reaction that occurs at constant temperature and pressure. By "useful" work, we mean work other than the unavoidable work of expansion or contraction against the fixed pressure of the surroundings. If ΔG is zero, the system is at equilibrium, which means that we could not obtain any useful work from the process. If ΔG is less than zero, the process could yield useful work as the system proceeds spontaneously toward equilibrium. If ΔG is greater than zero, the process is headed away from equilibrium, and we have to perform work on the system to drive it in this direction. The farther the reactants are from equilibrium, the larger the value of $-\Delta G$, and the larger the amount of work we might obtain from the reaction. However, $-\Delta G$ gives only the maximum amount of useful work. Remember that work and heat, unlike ΔG, ΔH, and ΔS, are not functions of state. The amount of work actually obtained depends on the path the process takes, and it can be zero even if $-\Delta G$ is large.

Biological Systems Perform Various Kinds of Work

To sustain and propagate life requires that cells do various types of work. This work takes three major forms, related to three broad categories of cellular activities:

1. Mechanical work: changes in location or orientation. Mechanical work is done whenever an organism, cell, or subcellular structure moves against the force of gravity or friction. As examples, consider the contracting muscles that propel a runner up a hill, the swimming of a flagellated protozoan in a pond, the migration of chromosomes toward the opposite poles of the mitotic spindle, and the movement of a ribosome along a strand of messenger RNA.

2. Concentration and electrical work: movements of molecules and ions across membranes. Concentration work, the movement of a molecule or ion across a membrane against a prevailing concentration gradient, establishes the localized concentrations of specific materials on which most essential life processes depend. Concentration work is sometimes referred to as osmotic work. Examples include the uptake of amino acids from the blood by muscle cells, pumping of sodium ions out of a marine microorganism, and movement of nitrate from the soil into the cells of a plant root. Electrical work is required to move a charged species across a membrane against an electric potential gradient. Although the most dramatic example of this is the generation of large potential differences in the electric organ of the electric eel, electrical work is done by almost all types of cells. It underlies the mechanisms of excitation of nerve and muscle cells and the conduction of impulses along axons.

3. Synthetic work: changes in chemical bonds. Synthetic work is necessary for the formation of the complex organic molecules of which cells are composed. As we have seen, these are in general molecules of higher enthalpy and lower entropy than the simple molecules available to organisms from their environment, so free energy must be expended in their synthesis. Synthetic work is most obvious during periods of growth of an organism, but it also occurs in nongrowing, mature organisms, which must continuously repair and replace existing structures. The continuous expenditure of energy to elaborate and maintain ordered structures that were created out of less-ordered raw materials is one of the most characteristic properties of living cells.

Favorable Reactions Can Drive Unfavorable Reactions

In this section we consider the value of the free energy function in understanding how energetically unfavorable reactions are coupled with energetically favorable ones to do synthetic work. In table 2.4, we made use of the principle that the free energies of all the components of a solution are additive. In general, if the free energy changes associated with two reactions A $\longrightarrow$ B and C $\longrightarrow$ D are ΔG_{AB} and ΔG_{CD}, the free energy change accompanying the combined process A + C $\longrightarrow$ B + D is simply $\Delta G_{AB} + \Delta G_{CD}$. It is this principle that allows living organisms to synthesize complex molecules with high enthalpies and low entropies. Thermodynamically unfavorable reactions can be driven by coupling them to favorable processes.

From equation (13) you can see that whereas free energy changes combine additively, equilibrium constants combine multiplicatively. If the equilibrium constants for the reactions A $\longrightarrow$ B and C $\longrightarrow$ D are K_{AB} and K_{CD}, the equilibrium constant for A + C $\longrightarrow$ B + D is $K_{AB}K_{CD}$.

There are numerous ways to achieve a coupling of favorable and unfavorable reactions. As an example, let's consider the formation of glucose-6-phosphate and water from glucose and inorganic phosphate ion (P_i):

$$\text{Glucose} + P_i \longrightarrow \text{glucose-6-phosphate} + H_2O \quad \textbf{(14)}$$

This reaction has an unfavorable $\Delta G^{\circ\prime}$ of +3.0 kcal/mole at 298 K (K_{eq} is 0.0062); the reaction does not occur spontaneously. On the other hand, the reaction

$$\text{ATP} + H_2O \longrightarrow \text{ADP} + P_i + H^+ \quad \textbf{(15)}$$

has a highly favorable $\Delta G^{\circ\prime}$ of about −8.4 kcal/mole ($K_{eq} \approx 1.35 \times 10^6$). If reactions (14) and (15) are combined to give the reaction

$$\text{Glucose} + \text{ATP} \longrightarrow \text{G-6-P} + \text{ADP} + H^+ \quad \textbf{(16)}$$

the overall $\Delta G^{\circ\prime}$ is 3.0 − 8.4, or −5.4 kcal/mole, and the overall equilibrium constant is 8.7×10^3. The combined reaction is thermodynamically favorable (fig. 2.6). A cell can use this reaction to synthesize glucose-6-phosphate, provided that it has a source of ATP.

To take advantage of such a thermodynamic combination of favorable and unfavorable processes, a cell must have a catalytic mechanism for actually linking the two reactions. The breakdown of ATP and ADP and P_i, reaction (15), would be fruitless if it occurred independently of reaction (14). In many cells, the combined reaction (16) is catalyzed by an enzyme that facilitates the transfer of phosphate from ATP directly to glucose (hexokinase). This is a common motif in biosynthetic processes. But the coupling mechanism does not have to be so direct.

Another common mechanism for coupling an unfavorable reaction to a favorable one is simply to arrange for one of the reactions to precede or follow the other.

Figure 2.6

The formation of glucose-6-phosphate (G-6-P) has a positive $\Delta G^{\circ\prime}$; the hydrolysis of ATP and ADP has a negative $\Delta G^{\circ\prime}$. If the two reactions are combined, the overall $\Delta G^{\circ\prime}$ is negative.

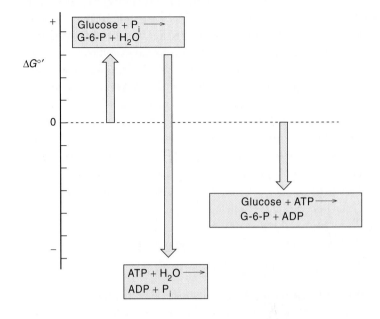

As an example, consider the following sequence of reactions:

$$\text{Acetyl-CoA} + \text{oxaloacetate} \xrightarrow{\ 1\ } \text{citryl-CoA} \xrightarrow{\ 2\ }$$
$$\text{citrate} + \text{coenzyme A} \quad \textbf{(17)}$$

The first step has a $\Delta G^{\circ\prime}$ of −0.05 kcal/mole, which is close to zero; it does not occur to any great extent unless the concentrations of acetyl-coenzyme A (acetyl-CoA) and oxaloacetate are greater than the concentration of citryl-CoA. The second step, however, has a highly favorable $\Delta G^{\circ\prime}$ of −8.4 kcal/mole. When the two steps are combined, $\Delta G^{\circ\prime}$ for the overall reaction is about −8.3 kcal/mole, and the equilibrium constant lies far in the forward direction. These two reactions are catalyzed by the enzyme citrate synthase, by a mechanism that ensures that they always occur together.

In the case of reactions that occur sequentially, with one step pulling or pushing the other, it is not necessary for the two steps to be catalyzed by the same enzyme, although this can help to speed up the overall sequence. For example, in many organisms the reactions shown in (17) are preceded by a reaction in which malate is converted to oxaloacetate:

$$\text{Malate}^{2-} + \text{NAD}^+ \longrightarrow$$
$$\text{oxaloacetate}^{2-} + \text{NADH} + H^+ \quad \textbf{(18)}$$

This step is catalyzed by a separate enzyme, malate dehydrogenase. It has a very unfavorable $\Delta G^{\circ\prime}$ of about $+7$ kcal/mole. When this reaction is followed by reaction (17), the combined $\Delta G^{\circ\prime}$ is about -1.3 kcal/mole, so the equilibrium constant for the overall sequence is favorable. You can view this simply as an illustration of the principle of mass action: (17) pulls (18) along by removing one of the products.

This last point deserves additional emphasis. It is important to keep in mind that what determines whether a process occurs spontaneously is ΔG, not ΔG°. As we saw in equation (11), the actual free energy change depends on the concentrations of the reactants and products. The hydrolysis of ATP (reaction 15), for example, has a $\Delta G^{\circ\prime}$ of about -8.4 kcal/mole, but the actual ΔG in the cytosol of living cells is typically more negative than this by between 5 and 6 kcal/mole because the concentration ratio $[ADP][P_i]/[ATP]$ is much less than 1.

ATP as the Main Carrier of Free Energy in Biochemical Systems

Virtually all living organisms use ATP for transferring free energy between energy-producing and energy-consuming systems. Processes that proceed with large negative changes in free energy, such as the oxidative degradation of carbohydrates or fatty acids, are used to drive the formation of ATP, and the hydrolysis of ATP is used to drive biosynthetic reactions and other processes that require increases in free energy. In the human body, about 2.3 kg of ATP is formed and consumed every day in the course of these reactions.

The Hydrolysis of ATP Yields a Large Amount of Free Energy

ATP can be hydrolyzed at two different sites, as shown in figure 2.7. Hydrolysis of the linkage between the β and γ phosphate groups yields ADP and P_i. Hydrolysis between the α and β phosphates gives adenosine monophosphate (AMP) and pyrophosphate ion ($HP_2O_7^{3-}$). The standard free energy change ($\Delta G^{\circ\prime}$) is about -8.4 kcal/mole in either case. The formation of AMP and pyrophosphate, however, can be pulled forward by hydrolysis of the pyrophosphate to give two equivalents of P_i. This secondary reaction has a $\Delta G^{\circ\prime}$ of about -7.9 kcal/mole and is catalyzed by pyrophosphatase enzymes present in most types of cells. The overall $\Delta G^{\circ\prime}$ of about -16.3 kcal/mole for breakdown of ATP to AMP and 2 P_i makes this process effectively irreversible ($K_{eq} \approx 1 \times 10^{12}$). Cells commonly use this series of reactions in biosynthetic processes in which a reversal is intolerable, such as in the synthesis of nucleic acids. The simpler hydrolysis to produce ADP and P_i allows some re-

versibility but has the advantage that less free energy is needed to resynthesize ATP from ADP than from AMP.

The standard free energies for the hydrolysis of ATP, ADP, or pyrophosphate depend on the pH, on the ionic strength, and also on the concentration of Mg^{2+}, which binds to both the reactants and the products and is required as a cosubstrate by most enzymes that use ATP. At pH 7, the principal ionic forms of ATP, ADP, and P_i have net charges of -4, -3, and -2, respectively (see fig. 2.7). Because the hydrolysis of ATP^{4-} to give $ADP^{3-} + HPO_3^{2-}$ is accompanied by release of a proton, raising the pH makes the hydrolysis more favorable. Increasing the Mg^{2+} concentration makes the hydrolysis less favorable. $\Delta G^{\circ\prime}$ decreases in magnitude from -8.4 kcal/mole in the absence of Mg^{2+} to about -7.7 kcal/mole in the presence of 1 mM Mg^{2+}, and to about -7.5 kcal/mole at 10 mM. The value -7.5 kcal/mole is probably close to the $\Delta G^{\circ\prime}$ under typical physiological conditions, and is used elsewhere in this text. But remember that the low ratio of $[ADP][P_i]$ to $[ATP]$ can make ΔG for hydrolysis of ATP in living cells substantially more negative than the $\Delta G^{\circ\prime}$. In resting cells, the enzymatic reactions that synthesize ATP usually are more than adequate to keep up with the processes that consume it.

Why is $\Delta G^{\circ\prime}$ for the hydrolysis of ATP so negative? The answer involves several different factors. At pH 7 in the presence of 10 mM Mg^{2+}, $\Delta H^{\circ\prime}$ and $-T\Delta S^{\circ\prime}$ for ATP hydrolysis are both negative, and the two terms make similar contributions to the overall value of $\Delta G^{\circ\prime}$. The favorable entropy change results partly from the fact that the proton released in the reaction is diluted into a solution with a very low proton concentration. In addition, solvent water molecules probably are less highly ordered around ADP and P_i than they are around ATP. The negative $\Delta H^{\circ\prime}$ results partly from the fact that the negatively charged oxygen atoms in ATP tend to repel each other. The phosphoric anhydride bond in ATP also is weakened by competition between the phosphorus atoms, which both tend to pull electrons away from the bridging oxygen. Another consideration is that the major products of ATP hydrolysis at pH 7 (ADP^{3-} and $HOPO_3^{2-}$) have a larger total number of resonance forms than the reactant ATP^{4-} does (fig. 2.8). The increased resonance lowers the energy of ADP^{3-} and $HOPO_3^{2-}$ relative to ATP^{4-}. The magnitudes of all of these effects vary with the pH, because protonation of the oxygen atoms in ATP, ADP, or P_i relieves some of the repulsive electrostatic interactions, decreases the contributions that some of the resonance forms make to the structure, and decreases the ordering of nearby water molecules. At very low pH, when ATP, ADP, and P_i are all fully protonated, ΔG° probably becomes slightly positive, favoring ATP formation rather than hydrolysis.

Figure 2.7

Alternative sites of ATP hydrolysis. The charged species shown are the main ones present at physiological pH and ionic strength. The phosphate groups of ATP are referred to as α, β, and γ as indicated. Under physiological conditions, ATP and ADP also bind Mg^{2+} (not shown).

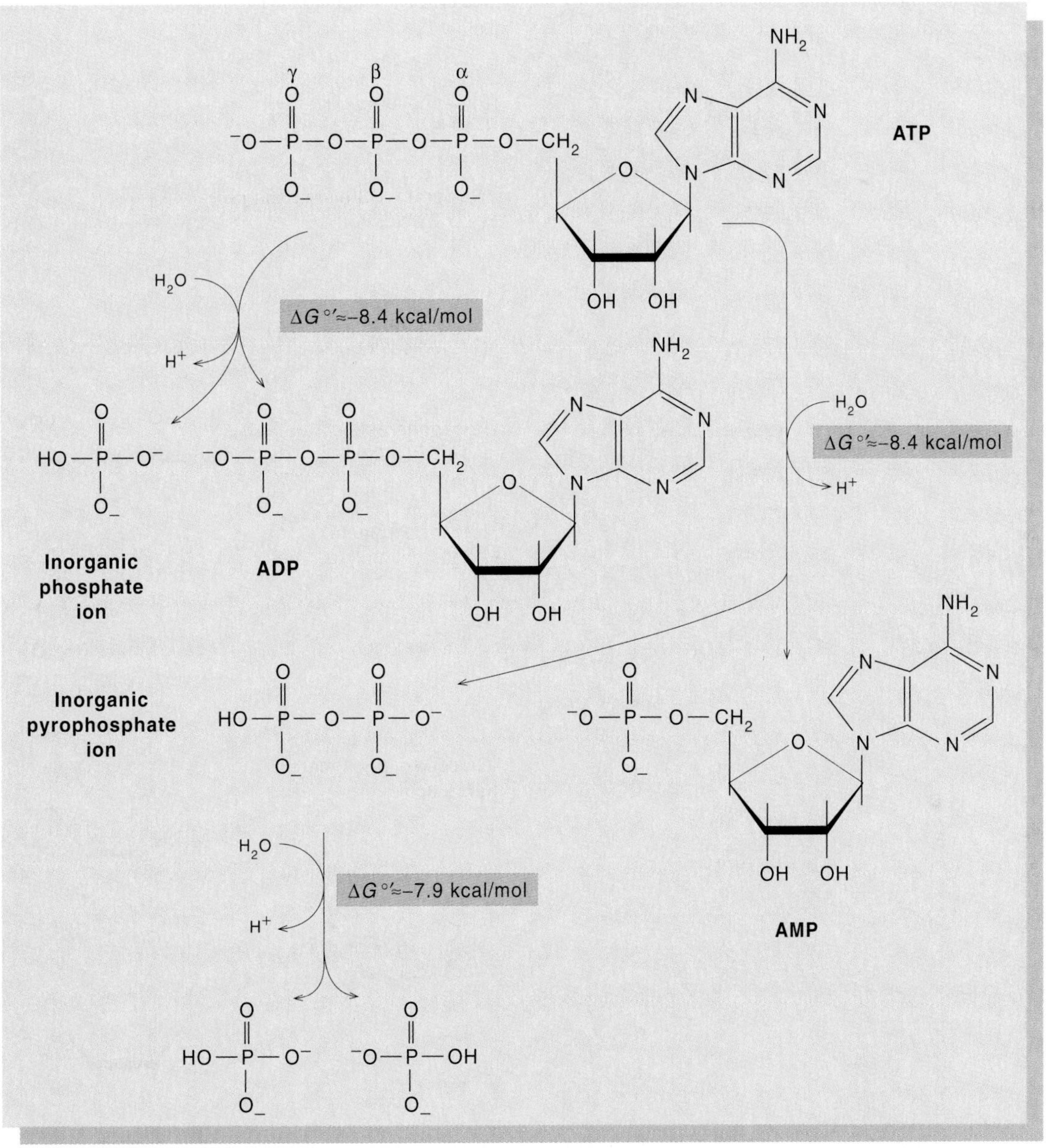

In addition to ATP and other nucleoside triphosphates, cells use a variety of other organic phosphate compounds in energy metabolism. These include acetyl phosphate, glycerate-1,3-bisphosphate, phosphoenolpyruvate, phosphocreatine, and phosphoarginine. Figure 2.9 shows the structures of these compounds and some simpler phosphate esters. The compounds are ranked in the figure in order of their standard free energies of hydrolysis, with the materials having the most negative values of $\Delta G^{\circ\prime}$ at the top. Given equal concentrations of reactants and products, any compound in the figure could be synthesized, in principle, at the expense of any of the compounds above it. Thus ADP could be phos-

Figure 2.8

The phosphate groups of ATP, ADP, and P_i can be written in a variety of resonance forms. This figure shows the major resonance forms of the β phosphate group in ATP and of the same group in ADP. The hydrolysis of ATP to ADP results in an increase in the number of resonance forms available to this group. This increase contributes to the negative $\Delta H°'$ of hydrolysis of ATP. At physiological pH there are also favorable contributions to $\Delta G°'$ from the release of a proton and from the disordering of water around the polyphosphate chain.

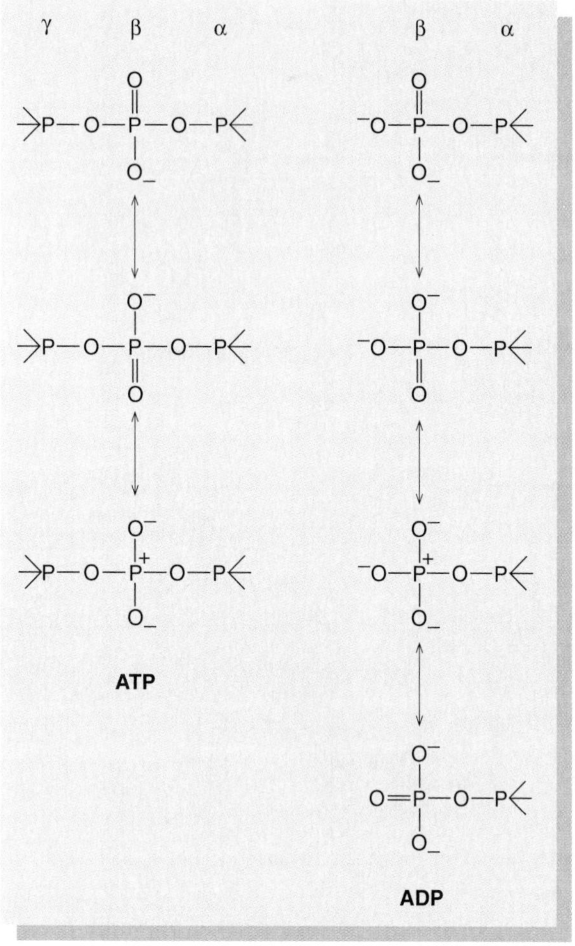

Figure 2.9

Standard free energies of hydrolysis for some common phosphorylated compounds. The $\Delta G°'$ value refers to hydrolysis of the phosphate group indicated by the symbol (P). Note that ATP occupies an intermediate position among these compounds. Given equal concentrations of the reactants and products, ADP can accept a phosphate group from any of the compounds above it, and ATP can donate a phosphate to the unphosphorylated forms of the compounds below it.

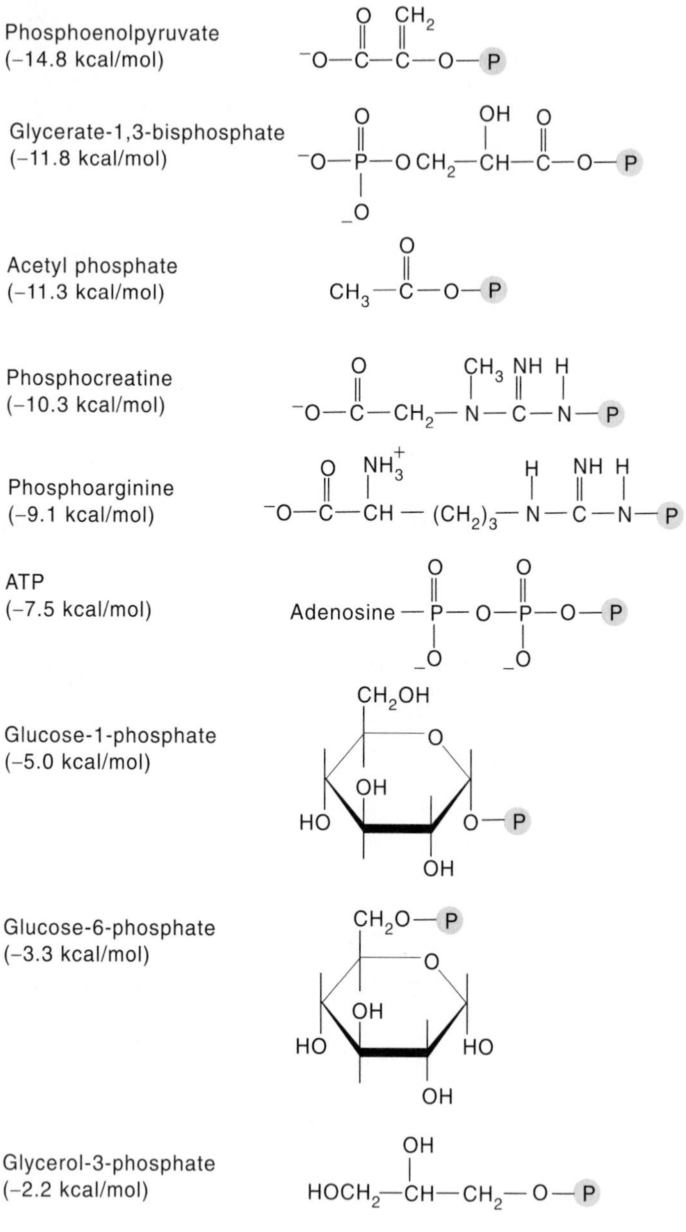

phorylated to ATP by phosphocreatine, glycerate-1,3-bisphosphate, or phosphoenolpyruvate, and ATP can be used to convert glucose to glucose-6-phosphate, or glycerol to glycerol phosphate. The reverse processes occur only if the ratio of reactants to products is sufficiently high, but this is not necessarily an unusual circumstance. In cells of some tissues, the reaction

$$\text{Creatine} + \text{ATP} \longrightarrow \text{creatine phosphate} + \text{ADP} + \text{H}^+ \qquad (19)$$

occurs readily in either direction in response to changes in the concentrations of the reactants and products. The same is true of the reaction

$$\text{Glycerate-3-phosphate} + \text{ATP} \longrightarrow \text{glycerate-1,3-bisphosphate} + \text{ADP} + \text{H}^+ \qquad (20)$$

Most of the phosphorylated materials shown in figure 2.9 participate in only a few biochemical reactions, and many of them are formed only in certain types of cells or organisms. ATP, in contrast, is formed by all living things, and it participates in literally hundreds of different enzymatic reactions. What is it about ATP that has made this molecule such a universal currency of free energy in biology? Several considerations are relevant here. First, $\Delta G^{\circ\prime}$ for hydrolysis of ATP to ADP is large enough so that this reaction releases a substantial amount of free energy, enough to drive many of the reactions that are important for biosynthetic pathways. At the same time, the $\Delta G^{\circ\prime}$ is small enough so that ATP itself can be synthesized readily at the expense of available nutrients. We touched on this point when discussing the relative merits of hydrolysis at the α-β or β-γ positions in ATP.

The second consideration returns us to the distinction between spontaneity and speed. Although the hydrolysis of ATP has a large negative $\Delta G^{\circ\prime}$, it also is critical that ATP is a relatively stable compound in aqueous solution. It does not hydrolyze rapidly under physiological conditions of pH and temperature. The hydrolysis, though far downhill thermodynamically, is slowed by a substantial activation barrier. But the barrier must be of such a nature that it can easily be overcome enzymatically. This feature allows the free energy of hydrolysis to be channeled quickly and selectively into reactions in which it is needed but to be conserved when energy is not in demand.

Third, the products of the hydrolysis of ATP provide opportunities for coupling to a wide variety of chemical reactions. We saw how the phosphate group is incorporated into glucose-6-phosphate, and later chapters provide many illustrations of similar enzymatic processes. We also will see reactions in which the pyrophosphate group or the adenylyl group (AMP) is incorporated into the products. Such a broad array of reactions could not be driven by a molecule that decomposed to release an inert material such as N_2.

Finally, the adenine and ribosyl groups of ATP, ADP, and AMP provide additional structural features that allow these molecules to bind to enzymes and thus to participate in regulating enzymatic activities. This may be part of the reason that no known organisms base their energy-transfer reactions entirely on inorganic pyrophosphate or other polyphosphate compounds without a nucleoside moiety.

Summary

In this chapter we discussed some principles of thermodynamics as they relate to biochemical reactions. The following points are of greatest importance.

1. Thermodynamics is useful in biochemistry for predicting whether a given reaction can occur and, if so, how much work a cell can obtain from the process.

2. The thermodynamic quantities energy, enthalpy, entropy, and free energy are properties of the state of a system. Changes in these quantities depend only on the difference between the initial and final states, not on the mechanism whereby the system goes from one state to the other.

3. Energy is the capacity to do work.

4. The first law of thermodynamics says that energy cannot be created or destroyed in a chemical reaction. If the energy of a system increases, the surroundings must lose an equivalent amount of energy, either by the transfer of heat or by the performance of work.

5. The energy of a molecule includes translational, rotational, and vibrational energy, as well as electronic and nuclear energy. Electronic terms usually account for most of the change in energy ΔE in a chemical reaction.

6. The change in enthalpy ΔH is given by the expression $\Delta H = \Delta E + \Delta(PV)$. For most biochemical reactions, ΔE and ΔH are nearly equal. The organic molecules found in cells generally have much higher enthalpies than the simpler molecules from which they are built.

7. In most reactions that proceed spontaneously, the enthalpy of the system decreases. If no work is done, the system gives off heat to the surroundings. But in some spontaneous reactions, heat is absorbed in the absence of work and the enthalpy of the system increases. Such reactions invariably show an increase in the entropy of the system.

8. Entropy is a measure of the order in a system: Systems that are highly ordered have low entropies. The entropy of a molecule depends mainly on translational and rotational freedom. Biological macromolecules generally have much lower entropies than their building blocks.

9. The second law of thermodynamics states that an overall increase must take place in the entropy of the system and its surroundings in any process that occurs spontaneously. An isolated system proceeds spontaneously to states of increasingly greater entropy (greater disorder).

10. The change in the free energy of a system is defined as $\Delta G = \Delta H - T \Delta S$, where T is the absolute temperature. A reaction at constant pressure and temperature can occur spontaneously if, and only if, ΔG is negative. The maximal amount of useful work that can be obtained from a reaction is equal to $-\Delta G$.

11. For a reaction in solution, ΔG depends on the standard free energy change ($\Delta G°$) and on the concentrations of the reactants and products. The standard free energy change is related to the equilibrium constant by the expression $\Delta G° = -RT \ln K_{eq}$. Increasing the concentration of the reactants relative to the concentration of the products makes ΔG more negative.

12. Reactions that are thermodynamically unfavorable can be coupled to favorable reactions. The coupling of the reactions may be direct or sequential, as in a biochemical pathway.

13. ATP is the main coupling agent for free energy in living cells. The free energy provided by the hydrolysis of ATP is used to drive many reactions that would not occur spontaneously by themselves.

14. Several features make ATP particularly well suited for its role. First, hydrolysis of ATP to ADP and P_i or to AMP and PP_i releases a considerable amount of free energy. Second, ATP does not hydrolyze rapidly by itself, but it can be hydrolyzed readily in enzymatically catalyzed reactions. This difference allows the free energy of hydrolysis to be channeled into reactions in which it is needed but to be conserved when energy is not in demand. Third, the products of the hydrolysis of ATP provide opportunities for coupling to a wide variety of chemical reactions. Finally, the adenine and ribosyl groups of ATP, ADP, and AMP provide additional structural features that allow these molecules to bind to a large number of enzymes and thus to participate in regulating enzymatic activities.

Selected Readings

Alberty, R. A., and F. Daniels, *Physical Chemistry,* 5th ed. New York: Wiley, 1975.

Cantor, C. R., and P. R. Schimmel, *Biophysical Chemistry*. San Francisco: Freeman, 1980.

Ingraham, L. L., and A. B. Pardee, Free Energy and Entropy in Metabolism. In D. M. Greenberg (ed.), *Metabolic Pathways,* vol. 1. New York: Academic Press, 1967.

Tinoco, I., Jr., K. Sauer, and J. C. Wang, *Physical Chemistry, Principles and Applications in Biological Sciences,* 2d ed. Englewood Cliffs, N.J.: Prentice-Hall, 1985.

Van Holde, K. E., *Physical Biochemistry,* 2d ed. Englewood Cliffs, N.J.: Prentice-Hall, 1985.

Problems

1. In some respects, thermodynamic considerations are more important to a biochemist than to a chemist. Why is this so?

2. Name three extensive and three intensive properties that relate to thermodynamic quantities. What is the basic difference between the two types of properties?

3. What is meant by a state function? Why is enthalpy a state function?

4. Why can we equate internal energy and enthalpy for most biochemical reactions?

5. In thermodynamics it is important to distinguish between the total system and the system being studied. Why?

6. Why do enthalpies and entropies of solvation tend to negate one another?

7. Why does the entropy of a weak acid decrease on ionization?

8. Transfer of a hydrophobic molecule (e.g., a hydrophobic amino acid and side chain) from an aqueous to a nonaqueous environment is entropically favorable. Explain.

9. As we will see in chapter 13, oxaloacetate is formed by the oxidation of malate.

$$\text{L-Malate} + \text{NAD}^+ \longrightarrow$$
$$\text{Oxaloacetate} + \text{NADH} + \text{H}^+$$

The reaction has a $\Delta G^{\circ\prime}$ of $+7.0$ kcal/mole. Suggest reasons that the reaction proceeds in the direction of oxaloacetate production in the cell.

10. Cite three factors that make ATP ideally suited to transfer energy in biochemical systems.

11. You wish to measure the $\Delta G^{\circ\prime}$ for hydrolysis of ATP,

$$\text{ATP} + \text{H}_2\text{O} \longrightarrow \text{ADP} + \text{P}_i + \text{H}^+$$

but the equilibrium for the hydrolysis lies so far toward products that analysis of the ATP concentration at equilibrium is neither practical nor accurate. However, you have the following data that allow calculation of the value indirectly.

$$\text{Creatine phosphate} + \text{ADP} + \text{H}^+ \longrightarrow$$
$$\text{ATP} + \text{Creatine } K'_{eq} = 59.5 \quad \textbf{(P1)}$$

$$\text{Creatine} + \text{P}_i \longrightarrow$$
$$\text{Creatine phosphate} + \text{H}_2\text{O}$$
$$\Delta G^{\circ\prime} = +10.5 \text{ kcal/mole} \quad \textbf{(P2)}$$

Assume that $2.3RT = 1.36$ kcal/mole
(a) Calculate the value of $\Delta G^{\circ\prime}$ for reaction (P1).
(b) Calculate the $\Delta G^{\circ\prime}$ for the hydrolysis of ATP.

12. The hydrolysis of lactose (D-galactosyl-β-(1,4) D-glucose) to D-galactose and D-glucose occurs with $\Delta G^{\circ\prime}$ of -4.0 kcal/mole.

(a) Calculate K'_{eq} for the hydrolytic reaction.
(b) What are the $\Delta G^{\circ\prime}$ and K'_{eq} for the synthesis of lactose from D-galactose and D-glucose?
(c) Lactose is synthesized in the cell from UDP-galactose plus D-glucose and is catalyzed by lactose synthase. Given that $\Delta G^{\circ\prime}$ of hydrolysis of UDP-galactose is -7.3 kcal/mole, calculate $\Delta G^{\circ\prime}$ and K'_{eq} for the reaction

$$\text{UDP-galactose} + \text{D-glucose} \longrightarrow \text{Lactose} + \text{UDP}$$

13. For each of the following reactions, calculate $\Delta G^{\circ\prime}$ and indicate whether the reaction is thermodynamically favorable as written.
(a) Glycerate-1,3-bisphosphate + Creatine $\longrightarrow$
    Phosphocreatine + 3-Phosphoglycerate
(b) Glucose-6-phosphate $\longrightarrow$ Glucose-1-phosphate
(c) Phosphoenolpyruvate + ADP $\longrightarrow$ Pyruvate + ATP

14. Assume that an individual needs 2,500 Calories per day (1 Calorie = 1,000 calories or 1 kcal) to meet energy requirements. For simplicity, consider that the energy needs of this individual are met with glucose (not that unrealistic, considering that most of the world meets these needs mainly with starches). Glucose, and carbohydrates in general contain about 4 Calories per gram. The ATP yield during catabolism is 30 moles of ATP per mole of glucose. What mass (in pounds) of K_4ATP are synthesized per day by this individual?

15. Proteins that serve as gene repressors frequently have two identical binding sites that interact with complementary sites on the DNA. If a repressor protein is cut in half without damaging either of its DNA binding sites, how is the binding to DNA affected? How is the binding affected if one of the two binding sites on the DNA is eliminated by changing the nucleotide sequence? Discuss the enthalpy, entropy, and free energy effects.

Protein Structure and Function

Chapter 3   The Building Blocks of Proteins: Amino Acids, Peptides, and Polypeptides   49

Chapter 4   The Three-Dimensional Structures of Proteins   72

Chapter 5   Functional Diversity of Proteins   101

Chapter 6   Methods for Characterization and Purification of Proteins   118

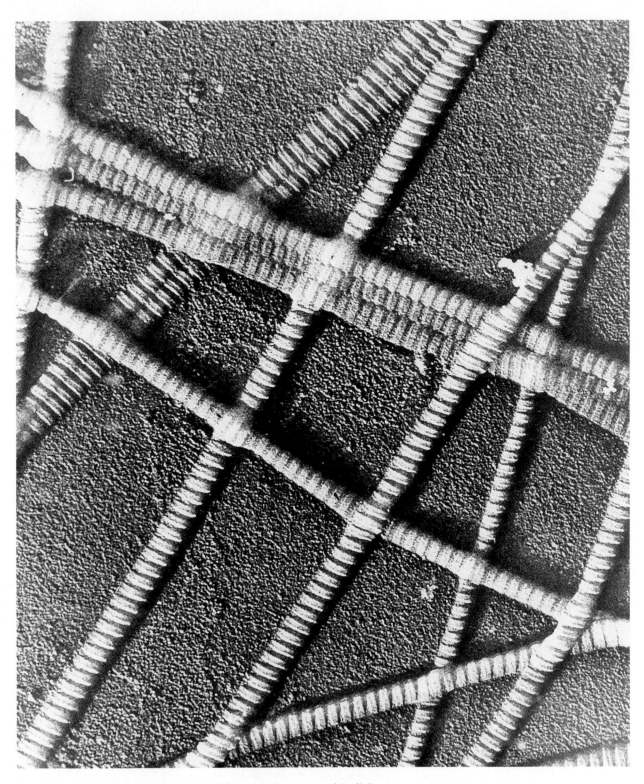

An electron micrograph of collagen fibrils from skin. The banded appearance of the fibrils arises from a staggered arrangement of collagen molecules. Each collagen molecule in turn is composed of three polypeptide chains that interact by hydrogen bonds. (Courtesy of Jerome Gross, Massachusetts General Hospital.)

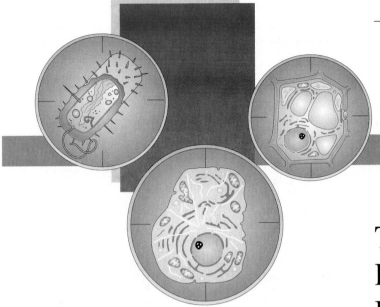

The Building Blocks of Proteins: Amino Acids, Peptides, and Polypeptides

Chapter Outline

Amino Acids
    Amino Acids Have Both Acid and Base Properties
    Aromatic Amino Acids Absorb Light in the Near-
      Ultraviolet
    All Amino Acids except Glycine Show Asymmetry
Peptides and Polypeptides
Determination of Amino Acid Composition of Proteins
Determination of Amino Acid Sequence of Proteins
Chemical Synthesis of Peptides and Polypeptides

Amino acids have common features that permit them to be linked together into polypeptide chains and uncommon features that give each polypeptide chain its unique character.

In the middle of the nineteenth century, the Dutch chemist Gerardus Mulder extracted a substance common to animal tissues and the juices of plants, which he believed to be ''without doubt the most important of all substances of the organic kingdom, and without it life on our planet would probably not exist.'' At the suggestion of the famous Swedish chemist Jons Jakob Berzelius, Mulder named this substance protein (from the Greek *proteios,* meaning ''of first importance'') and assigned to it a specific chemical formula $(C_{40}H_{62}N_{10}O_{12})$. Although he was wrong about the chemistry of proteins, he was right about their being indispensable to living organisms. The term ''protein'' endures.

Proteins are the most abundant of cellular components. They include enzymes, antibodies, hormones, transport molecules, and even components for the cytoskeleton of the cell itself. Proteins are also informational macromolecules, the ultimate heirs of the genetic information encoded in the sequence of nucleotide bases within the chromosomes. Structurally and functionally, they are the most diverse and dynamic of molecules and play key roles in nearly every biological process. Proteins are complex macromolecules with exquisite specificity; each is a specialized player in the orchestrated activity of the cell. Together they tear down and build up molecules, extract energy, repel invaders, act as delivery systems, and even synthesize the genetic apparatus itself.

In the first three chapters of part 2 we discuss the basic structural and chemical properties of proteins. In this chapter we concentrate on the structural and chemical properties of amino acids, peptides, and polypeptides—the building blocks of proteins. From our presentation you will learn the following:

1. Certain acidic and basic properties are common to all amino acids found in proteins except for the amino acid proline.

2. Side chains give amino acids their individuality. These side chains serve a variety of structural and functional roles.

3. The α-carboxyl group of one amino acid can react with the α-amino group of another amino acid to form a dipeptide.

4. Many amino acids, reacting in a similar way, can become linked to form a linear polypeptide chain.

5. The amino acid sequence in a polypeptide can be determined by a process of partial breakdown into manageable fragments, followed by stepwise analysis proceeding from one end of the chain to the other.

6. Polypeptide chains with a prespecified sequence can be synthesized by well-established chemical methods.

Amino Acids

Every protein molecule can be viewed as a polymer of amino acids. There are 20 common amino acids. Figure 3.1a shows the structure of a single amino acid. At the center is a tetrahedral carbon atom called the alpha (α) carbon (C_α). It is covalently bonded on one side to an amino group (NH$_2$) and on the other side to a carboxyl group (COOH). A third bond is always to a hydrogen, and the fourth bond is to a variable side chain (R). In neutral solution (pH 7), the carboxyl group loses a proton and the amino group gains one. Thus an amino acid in solution, while neutral overall, is a doubly charged species called a zwitterion (fig. 3.1b).

The structures of the 20 amino acids commonly found in proteins are listed in table 3.1. All of these amino acids except proline have a protonated α-amino group (—NH$_3^+$) attached to the α carbon. In proline one of the N—H linkages is replaced by an N—C linkage forming part of a cyclic

Figure 3.1

Amino acid anatomy. (*a*) Uncharged amino acid. (*b*) Doubly charged zwitterion. (Illustration copyright by Irving Geis. Reprinted by permission.)

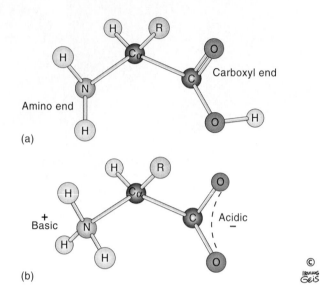

structure. Various ways of classifying amino acids according to their R groups have been proposed. In table 3.1 we divide the amino acids into three categories. The first category contains eight amino acids with relatively apolar R groups; the second category contains seven amino acids with uncharged polar R groups; and the third category contains five amino acids with R groups that normally exist in the charged state.

Amino acids are often abbreviated by three-letter symbols; when this proves to be too cumbersome (as in certain kinds of charts and figures), one-letter symbols are used. Both designations are given in table 3.1, together with the molecular weight (M_r) of each amino acid.

In addition to the 20 commonly occurring α-amino acids, a variety of other amino acids are found in minor amounts in proteins and in nonprotein compounds. The unusual amino acids found in proteins result from modification of the common amino acids. In a few cases these amino acids are incorporated directly into the polypeptide chains during synthesis. Most frequently the amino acid is modified after incorporation. The unusual amino acids found in nonprotein compounds are extremely varied in type and are formed by a number of different metabolic pathways (see chapter 21).

Table 3.1

Structure of the Twenty Amino Acids Found in Proteins

Group I. Amino Acids with Apolar R Groups			Group II. Amino Acids with Uncharged Polar R Groups			Group III. Amino Acids with Charged R Groups		
	R Groups	*Common Group*		*R Groups*	*Common Group*		*R Groups*	*Common Group*
Alanine Ala A M_r 89[a]			Glycine Gly G M_r 75			Aspartic acid Asp D M_r 133		
Valine Val V M_r 117			Serine Ser S M_r 105			Glutamic acid Glu E M_r 147		
Leucine Leu L M_r 131			Threonine Thr T M_r 119			Lysine Lys K M_r 146		
Isoleucine Ile I M_r 131			Cysteine Cys C M_r 121			Arginine Arg R M_r 174		
Proline Pro P M_r 115			Tyrosine Tyr Y M_r 181			Histidine (at pH 6.0) His H M_r 155		
Phenylalanine Phe F M_r 165			Asparagine Asn N M_r 132					
Tryptophan Trp W M_r 204			Glutamine Gln Q M_r 146					
Methionine Met M M_r 149								

[a] Molecular weights (M_r) in this text are expressed in units of grams per mole.

Amino Acids Have Both Acid and Base Properties

The charge properties of amino acids are very important in determining the reactivity of certain amino acid side chains and in the properties they confer on proteins. The charge properties of amino acids in aqueous solution may best be considered under the general treatment of acid–base ionization theory. We find this treatment useful at other points in the text as well.

Recall that water can be considered a weak acid (or a weak base) because it dissociates into a proton and a hydroxide ion, according to the equilibrium

$$H_2O \rightleftharpoons H^+ + OH^- \qquad (1)$$

The equilibrium expression for this reaction is

$$K_{eq} = \frac{[H^+][OH^-]}{[H_2O]} \qquad (2)$$

Because water dissociates to such a small extent, the concentration of undissociated water is high and does not vary significantly for chemical reactions in aqueous solution. Therefore, the denominator in this equation is effectively constant, with a value of 55.5. The constant K_w for the dissociation of water is redefined by the expression

$$K_w = [H^+][OH^-] = 10^{-14} \text{ (mole/l)}^2 \qquad (3)$$

at 25°C.

In pure water we expect equal amounts of H^+ ("hydrogen ion") and OH^- ("hydroxide ion"). From equation (3) we can calculate the concentration of H^+ or OH^- in pure water to be 10^{-7} M. Therefore, a solution with an H^+ concentration of 10^{-7} M is defined as neutral. An H^+ concentration greater than 10^{-7} M indicates an acidic solution; an H^+ concentration less than 10^{-7} M indicates a basic solution. Rather than deal with exponentials, it is convenient to express the H^+ concentration on a pH scale, the term "pH" being defined by the equation

$$pH = \log(1/[H^+]) = -\log[H^+] \qquad (4)$$

According to this definition a neutral solution has a pH of 7. Other values of pH and corresponding H^+ and OH^- concentrations are given in table 3.2.

The most common equilibria that biochemists encounter are those of acids and bases. The dissociation of an acid may be written as

$$HA \rightleftharpoons H^+ + A^- \qquad (5)$$

The equilibrium constant for this reaction is called the acid dissociation constant K_a written as

$$K_a = \frac{[H^+][A^-]}{[HA]} \qquad (6)$$

Table 3.2

The pH Scale

pH	$[H^+]$	$[OH^-]$
0	10^0	10^{-14}
1	10^{-1}	10^{-13}
2	10^{-2}	10^{-12}
3	10^{-3}	10^{-11}
4	10^{-4}	10^{-10}
5	10^{-5}	10^{-9}
6	10^{-6}	10^{-8}
7	10^{-7}	10^{-7}
8	10^{-8}	10^{-6}
9	10^{-9}	10^{-5}
10	10^{-10}	10^{-4}
11	10^{-11}	10^{-3}
12	10^{-12}	10^{-2}
13	10^{-13}	10^{-1}
14	10^{-14}	10^0

Strong acids in aqueous solution dissociate completely into anions and protons. The concentration of hydrogen ion $[H^+]$ is therefore equal to the total concentration C_{HA} of the acid HA that is added to the solution. Thus the pH of the solution of a strong acid is simply $-\log C_{HA}$.

The pH of the solution of a weak acid is a function of both the C_{HA} and the acid dissociation constant. The dissociation constant of a weak acid may be written in terms of the species present in the equation for the acid dissociation constant.

First solving equation (6) for $[H^+]$ gives

$$[H^+] = \frac{K_a[HA]}{[A^-]} \qquad (7)$$

Taking the logarithm of both sides and changing signs gives us

$$-\log[H^+] = -\log K_a + \log\frac{[A^-]}{[HA]} \qquad (8)$$

Substituting pH for $-\log[H^+]$ and pK_a for $-\log K_a$ in equation (8), we obtain the Henderson-Hasselbach equation:

$$pH = pK_a + \log\left[\frac{A^-}{HA}\right] = pK_a + \log\left[\frac{base}{acid}\right] \qquad (9)$$

Figure 3.2

The dependence of pH on the equivalents of base added to a typical weak acid. Note that at the pK_a, $[A^-] = [HA]$.

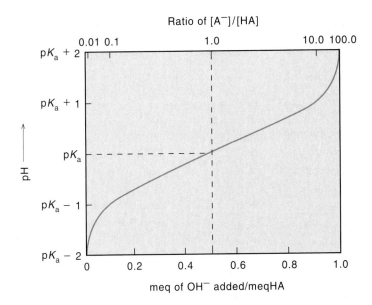

Figure 3.3

Titration curve of alanine. The predominant ionic species at each cardinal point in the titration is indicated.

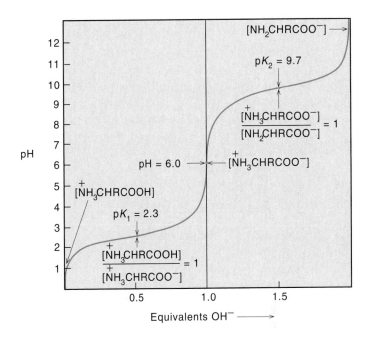

The Henderson-Hasselbach equation is useful for calculating the molar ratio of base (proton acceptor) to acid (proton donor) for a given pH and pK or for calculating the pK, given the ratio of base (proton acceptor) to acid (proton donor). It can be seen that when the concentration of anion or base is equal to the concentration of undissociated acid (i.e., when the acid is half neutralized), the pH of the solution is equal to the pK of the acid.

The values of pK for a particular molecule are determined by titration. A typical pH dependence curve for the titration of a weak acid by a strong base is shown in figure 3.2. The concentration of the anion equals the concentration of the acid when the acid is exactly half neutralized. Note that at this point on the curve, the pH is least sensitive to the quantity of added base (or acid). Under these conditions, the solution is said to be buffered. Biochemical reactions are typically highly dependent on the pH of the solution. Therefore, it is frequently advantageous to study reactions in buffered solutions. The ideal buffer is one that has a pK numerically equivalent to the working pH.

A simple amino acid with a nonionizable R group gives a complex titration curve with two inflection points. For an example, see the titration of alanine, shown in figure 3.3. At very low pH, alanine carries a single positive charge on the α-amino group. The first inflection point occurs at a pH of 2.3. This is the pK for titration of the carboxyl group, pK_1 ($-COOH \longrightarrow -COO^-$). At a pH of 6.0, alanine has an equal amount of positive and negative charge. This value is referred to as the isoelectric point (pI), or the isoe-

lectric pH. As the titration continues, a second inflection point is reached at a pH of 9.7. The pK at this point, pK_2, the $[-NH_3^+]$ and $[-NH_2]$ are equal.

Amino acids with an ionizable R group show even more complex titration curves, indicative of three ionizable groups (fig. 3.4). The pK for the ionizable side chain, pK_R, is usually readily distinguishable from the pK values for the ionizable α-carboxyl and α-amino groups, pK_1 and pK_2, respectively, because the α-amino groups have numerical values close to the comparable pK values of alanine (see fig. 3.3 and table 3.3). Note that the only ionizable R group with a pK_R in the vicinity of 7, where most biological systems function, is that for histidine. This means that although other ionizable groups are usually fully charged under biological conditions, the side chain of histidine can be fully charged, uncharged, or partially charged, depending on the precise situation. This variability has major implications for the way the histidine side chain functions in enzyme catalysis. The side chain can serve as either a proton donor or a proton acceptor (see discussion in chapter 8).

An additional point should be noted from table 3.3. Whereas the amino acid side chains (R groups) that are normally charged at physiological pH are restricted to five amino acids (aspartic acid, glutamic acid, lysine, arginine, and sometimes histidine), a number of potentially ionizable R groups are part of other amino acids. These include cysteine, serine, threonine, and tyrosine. The ionization reac-

Figure 3.4

Titration curves of glutamic acid, lysine, and histidine. In each case, the pK of the R group is designated pK_R.

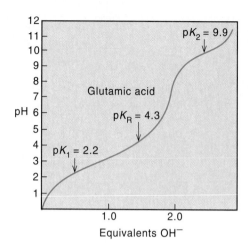

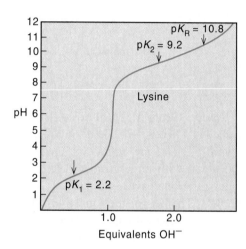

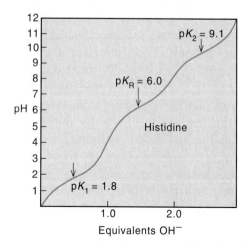

Table 3.3

Values of pK for the Ionizable Groups of the Twenty Amino Acids Commonly Found in Proteins

Amino Acid	pK_1 $(\alpha\text{—COOH})$	pK_2 $(\alpha\text{—NH}_3^+)$	pK_R (R Group)
Alanine	2.35	9.87	—
Arginine	1.82	8.99	12.48
Asparagine	2.1	8.84	—
Aspartic acid	1.99	9.90	3.90
Cysteine	1.92	10.78	8.33
Glutamic acid	2.10	9.47	4.07
Glutamine	2.17	9.13	—
Glycine	2.35	9.78	—
Histidine	1.80	9.33	6.04
Isoleucine	2.32	9.76	—
Leucine	2.33	9.74	—
Lysine	2.16	9.18	10.79
Methionine	2.13	9.28	—
Phenylalanine	2.16	9.18	—
Proline	1.95	10.65	—
Serine	2.19	9.21	≈ 13
Threonine	2.09	9.10	≈ 13
Tryptophan	2.43	9.44	—
Tyrosine	2.20	9.11	10.13
Valine	2.29	9.74	—

tions for all of the potentially ionizable side chains are indicated in figure 3.5.

The acidic and basic groups within a protein can be titrated just like free amino acids to determine their number and their pK_a values. A titration curve for β-lactoglobulin is shown in figure 3.6. This protein contains 94 potentially ionizable groups. The protein is positively charged at low pH and negatively charged at high pH. At intermediate pH values a point is found where the sum of the positive side-chain charges exactly equals the sum of the negative charges, so that the net charge on the protein is zero. This value, as we have noted, is the isoelectric point (pI) of the protein; for β-lactoglobulin the pI is about 5.2. The isoelectric point is not an invariant quantity. The binding of charged species present in the solution could raise or lower the pI, depending on their charge.

Figure 3.5

Equilibrium between charged and uncharged forms of amino acid side chains.

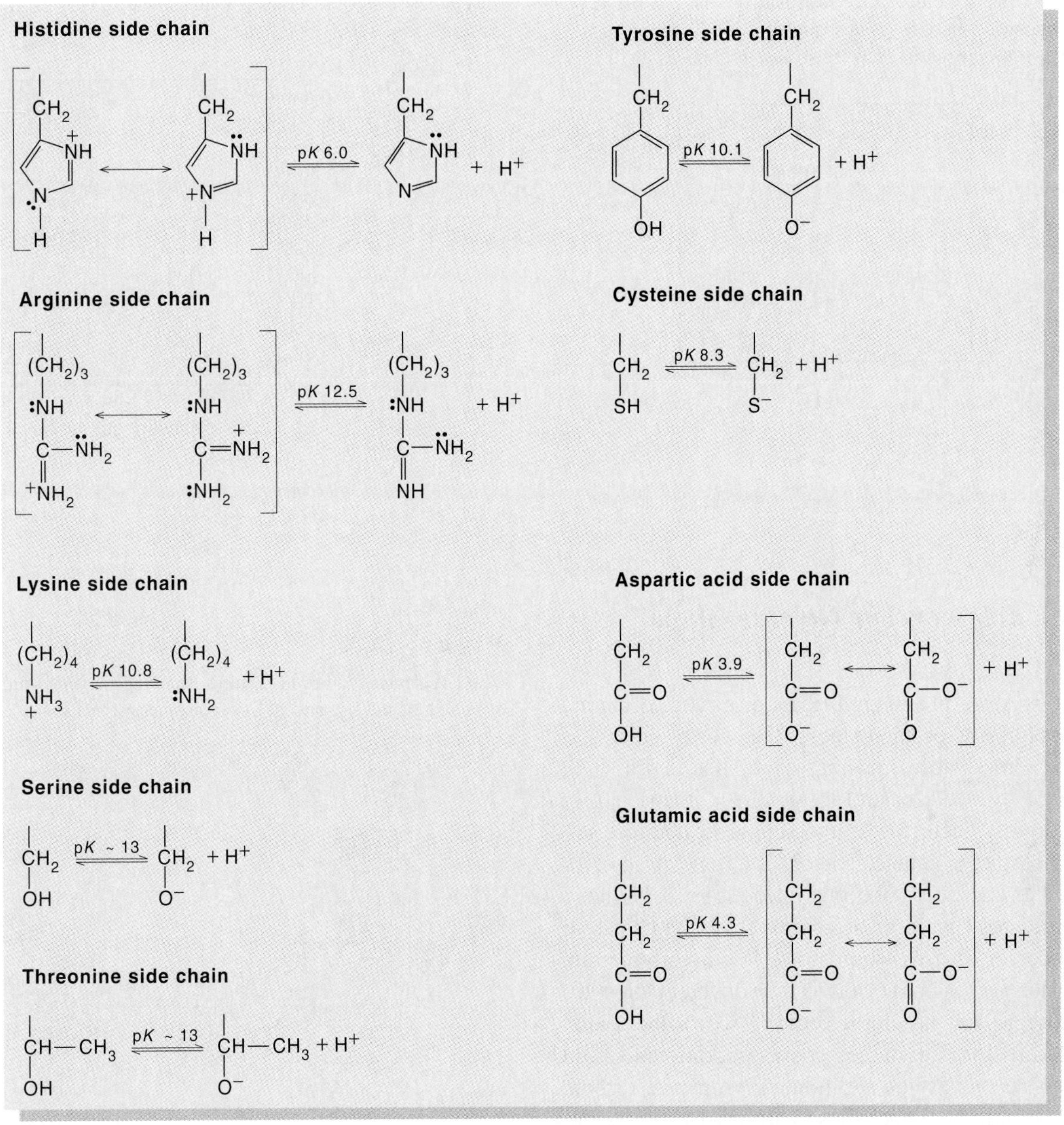

Aromatic Amino Acids Absorb Light in the Near-Ultraviolet

The aromatic amino acids phenylalanine, tyrosine, and tryptophan all possess absorption maxima in the near-ultraviolet (fig. 3.7). These absorption bands arise from the interaction of radiation with electrons in the aromatic rings. The near-ultraviolet absorption properties of proteins are determined solely by their content of these three aromatic amino acids. In solution, UV absorption can be quantified with the help of a conventional spectrophotometer and used as a measure of the concentration of proteins (see Methods of Biochemical Analysis 3A).

Figure 3.6

Titration curve of β-lactoglobulin. At very low values of pH (<2) all ionizable groups are protonated. At a pH of about 7.2 (indicated by horizontal bar) 51 groups (mostly the glutamic and aspartic amino acids and some of the histidines) have lost their protons. At pH 12 most of the remaining ionizable groups (mostly lysine and arginine amino acids and some histidines) have lost their protons as well.

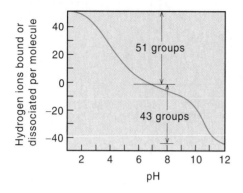

Figure 3.7

Ultraviolet absorption spectra of tryptophan (Trp), tyrosine (Tyr), and phenylalanine (Phe) at pH 6. The molar absorptivity is reflected in the extinction coefficient, with the concentration of the absorbing species expressed in moles per liter. (Source: From D. B. Wetlaufer, *Adv. Protein Chem.* 17:303–390, 1962.)

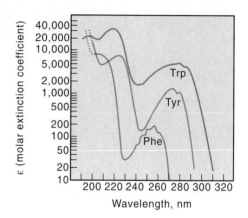

All Amino Acids except Glycine Show Asymmetry

One of the most striking and significant properties of amino acids is their chirality, or handedness. The word ''chiral'' is related to the Greek word meaning hand. Just as the right hand is related to the left hand by a mirror image, so, in general, a naturally occurring amino acid is related to a stereoisomer by its mirror image. The observation is true of 19 out of the 20 amino acids; the one exception is glycine.

The chirality of amino acids stems from the chiral, or asymmetric, center, the α-carbon atom. The α-carbon atom is a chiral center if it is connected to four different substituents. Thus glycine has no chiral center. Two of the amino acids, isoleucine and threonine, possess additional chiral centers because each has one additional asymmetric carbon. You should be able to locate these carbons by simple inspection.

Two structures that constitute a stereoisomeric pair are referred to as enantiomers. The two enantiomers for alanine are illustrated in figure 3.8. These two isomers are called L-alanine and D-alanine, according to the way in which the substituents are arranged about the asymmetric carbon atom. The naming by D and L (for ''dextrorotatory'' and ''levorotatory''; see chapter 6) refers to a convention established by Emil Fischer many years ago. According to this convention all amino acids found in proteins are of the L form. Some D-amino acids are found in bacterial cell walls and certain antibiotics.

Figure 3.8

The covalent structure of alanine, showing the three-dimensional structure of the Ⓛ and Ⓓ stereoisomeric forms.

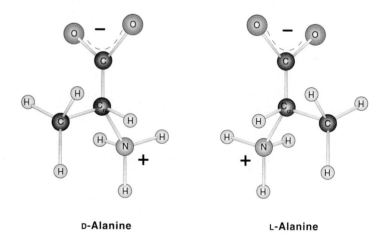

D-Alanine   L-Alanine

Peptides and Polypeptides

Amino acids can link together by a covalent peptide bond between the α-carboxyl end of one amino acid and the α-amino end of another. Formally, this bond is formed by the loss of a water molecule, as shown in figure 3.9. The peptide bond has partial double-bond character owing to resonance effects; as a result, the C—N peptide linkage and all of the atoms directly connected to C and N lie in a planar configuration called the amide plane. In the following chap-

Figure 3.9

Formation of a dipepetide from two amino acids. (*a*) Two amino acids. (*b*) A peptide bond (CO—NH) links amino acids by joining the α-carboxyl group of one with the α-amino group of another. A water molecule is lost in the reaction. It is conventional to draw dipeptides and polypeptides so that their free amino terminus is to the left and their free carboxyl terminus is to the right. The amide plane refers to six atoms that lie in the same plane. (Illustration copyright by Irving Geis. Reprinted by permission.)

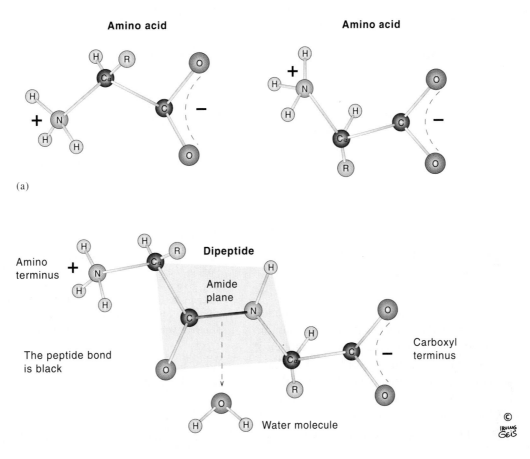

ter we see that this amide plane, by limiting the number of orientations available to the polypeptide chain, plays a major role in determining the three-dimensional structures of proteins.

Any number of amino acids can be joined by successive peptide linkages, forming a polypeptide chain. The polypeptide chain, like the dipeptide, has a directional sense. One end, called the N-terminal, or amino-terminal, end, has a free α-amino group, whereas the other end, the C-terminal, or carboxyl-terminal, end, has a free α-carboxyl group. The sequence of main-chain atoms from the N-terminal end to the C-terminal end is N—C_α—C—N and so on, and in the opposite direction it is C—C_α—N—C and so on. Short polypeptide chains, up to a length of about 20 amino acids, are called peptides or oligopeptides if they are fragments of whole polypeptide chains. A small protein mole-

cule may contain a polypeptide chain of only 50 amino acids; a large protein may contain chains of 3,000 amino acids or more. One of the larger single polypeptide chains is that of the muscle protein myosin, which consists of approximately 1,750 amino acid residues. Figure 3.10 shows a section of a polypeptide chain as a linear array with α carbons and planar amides alternating as repeating units of the main chain. Different side chains are attached to each α carbon.

In addition to the covalent peptide bonds formed between adjacent amino acids within a polypeptide chain, covalent disulfide bonds can be formed within the same polypeptide chain or between different polypeptide chains (fig. 3.11). Such disulfide linkages have an important stabilizing influence on the structures formed by many proteins (see chapter 4).

Figure 3.10

A polypeptide chain, with the backbone shown in color and the amino acid side chains in outline. The polypeptide chain is oriented so that the C-terminal end (not shown) is to the left. (Illustration copyright by Irving Geis. Reprinted by permission.)

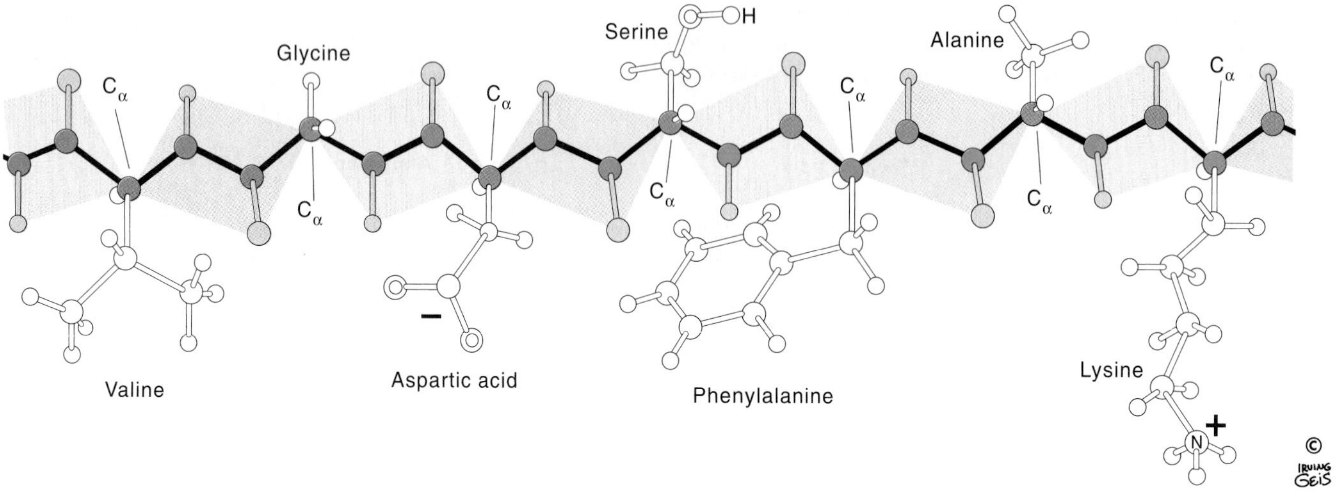

Determination of Amino Acid Composition of Proteins

Each protein is uniquely characterized by its amino acid composition and sequence. A protein's amino acid composition is defined simply as the number of each type of amino acid composing the polypeptide chain. To discover a protein's amino acid composition it is necessary to (1) break down the polypeptide chain into its constituent amino acids, (2) separate the resulting free amino acids according to type, and (3) measure the quantities of each amino acid.

Cleavage of the peptide bonds is usually achieved by boiling the protein in 6-N HCl; this treatment causes hydrolysis of the peptide bonds and the consequent release of free amino acids (fig. 3.12). Although acid hydrolysis is the most frequently used means of breaking a protein into its constituent amino acids, it results in the partial destruction of the indole ring of tryptophan. Consequently, the amount of tryptophan in the protein must be estimated by an alternative method (e.g., spectroscopic absorption) when using acid hydrolysis. In addition, acid hydrolysis results in the loss of ammonia from the side-chain amide groups of glutamine and asparagine, with the consequent production of

Figure 3.11

Disulfide bonds can form between two cysteines. The cysteines can exist in the cytosol as free amino acids (as shown), in which case they give rise to cystine, or they can be on polypeptide chains. In the latter instance, they can be on the same polypeptide chains or different polypeptide chains. In either case the formation of covalent disulfide bonds stabilizes structural relationships.

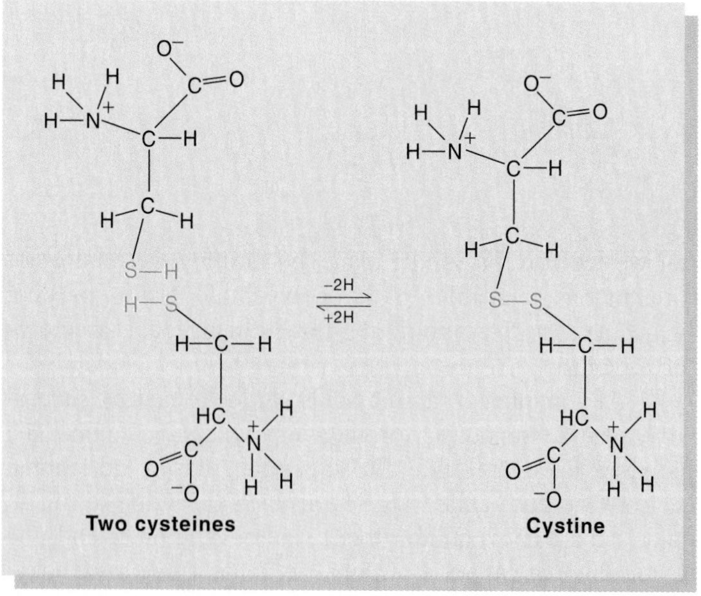

Figure 3.12

Acid hydrolysis of a protein or polypeptide to yield amino acids.

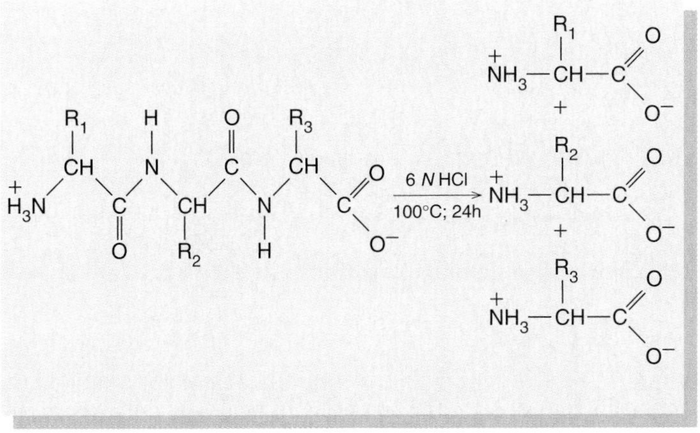

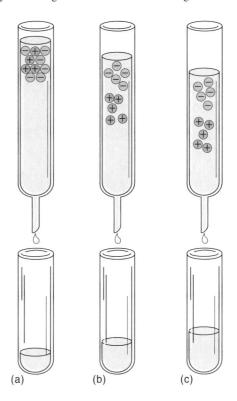

Figure 3.13

Migration of aspartic acid ⊖ and lysine ⊕ through a column with a higher affinity for aspartic acid. Views show the column at successively increasing time intervals after starting the elution.

(a)   (b)   (c)

glutamic and aspartic acids. Therefore, estimates of amino acid composition based on acid hydrolysis show glutamine and glutamic acid combined and measured as glutamic acid. Similarly, asparagine and aspartic acid are combined and measured as aspartic acid.

Separation of amino acids for quantitative analytical purposes is usually achieved by ion-exchange chromatography. The general efficacy of chromatographic techniques is based on a difference in affinity between each compound to be separated and an immobile phase or resin. The resin consists of some relatively chemically inert polymer, which has weakly basic side-chain constituents that are positively charged at pH 7. If we were to add some of this resin to a solution containing free aspartic acid and lysine at pH 7, the negatively charged aspartic acid would have a higher affinity for the resin than would the positively charged lysine. If we then pump a solution of these two amino acids through a column containing such a positively charged resin, the progress of the aspartic acid through the column would be retarded relative to the lysine, owing to the greater affinity of the aspartic acid for the resin (fig. 3.13).

Ion-exchange resins exist that have differential binding affinities for all the naturally occurring amino acids. Such resins are effective in separating a solution of amino acids into its components. We must emphasize that the details of the forces responsible for the differential binding of amino acids to an ion-exchange resin are quite complicated and depend additionally on side-chain polarity, on subtle differences in the pK values of α-amino and α-carboxyl groups, on solvation effects, and on other factors. To enhance the separation properties of the column, such separation techniques frequently exploit changes in the pH of the solution buffer (eluting buffer) used to remove the compounds of interest. For example, a column might initially be run with the eluting buffer at a pH that results in some amino acids being so strongly bound to the resin that they are essentially immobile. However, after the separation and elution of the less strongly bound amino acids, the pH of the eluting buffer can be appropriately shifted to lessen the charge difference between the resin and the strongly bound amino acids. These amino acids can then be eluted and separated according to the newly established pattern of resin-binding affinities.

Quantitative determination of the separated amino acids is achieved by their reaction with ninhydrin to produce

Figure 3.14

Reaction of ninhydrin with an amino acid yields a colored complex. The ninhydrin reaction permits qualitative location of amino acids in chromatography and quantitative assay of separated amino acids.

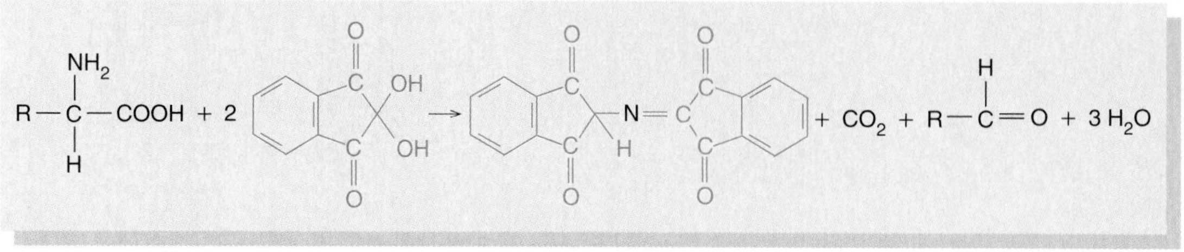

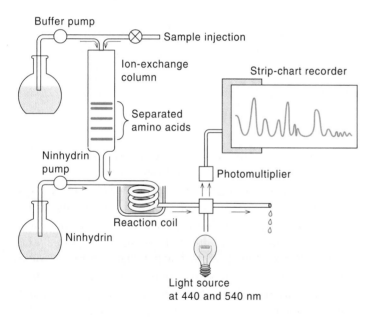

Figure 3.15

Schematic diagram of an amino acid analyzer. The amino acids are passed through an ion-exchange column and thereby separated. Eluted fractions are mixed and reacted with ninhydrin. The intensity of the resulting colored product is measured in a spectrophotometer, and the results are displayed on a recording chart.

a colored reaction product. This product is measured spectrophotometrically. As shown in figure 3.14, the ninhydrin reaction abstracts an amino group from each amino acid, so that the amount of colored product formed is proportional to the amount of amino acid initially present.

Currently, measurements of amino acid composition are usually carried out on an amino acid analyzer, a device that automates the previously described operations. As illus-

trated in figure 3.15, the amino acid analyzer consists of an ion-exchange column through which the appropriate eluting buffer is pumped after the amino acids are introduced at the top of the column. As the separated amino acids emerge, they are mixed with ninhydrin solution and passed through a heated coil of tubing to allow the formation of the colored ninhydrin reaction product. The separated ninhydrin reaction products then pass through a cell that measures their optical absorbance at 540 and 440 nm and plots the results on a strip-chart recorder. The absorbance is measured at two wavelengths because proline, which is substituted at its amino group, forms a different ninhydrin reaction product, with an absorption maximum that is correspondingly different from that of the remaining amino acids.

Usually the amino acid analyzer is first standardized by running through it a sample containing known quantities of amino acids to account for any differences in their ninhydrin reaction properties. In this way it is possible to relate directly the amount of amino acid present to the amount of colored product formed, as measured by the area under the "peak" produced on the strip-chart recorder (see fig. 3.15). Similarly, the amino acid hydrolysate of a protein of unknown composition can be run through the analyzer, and the relative peak areas can be used to estimate the ratios of the different amino acids present.

Conversion of the relative ratios of amino acids into an estimate of actual composition requires some additional information concerning the protein's molecular weight; for example, an analysis giving relative ratios of Ala (1.0), Gly (0.5), and Lys (2.0) could correspond to composition Ala_2-Gly-Lys_4 or any multiple thereof. The required information is usually available, and in any case, an estimation of composition based on a minimum molecular weight of the protein is always possible.

Determination of Amino Acid Sequence of Proteins

The most important properties of a protein are determined by the sequence of amino acids in the polypeptide chain. This sequence is called the primary structure of the protein. We know the sequences for thousands of peptides and proteins, largely through the use of methods developed in Fred Sanger's laboratory and first used to determine the sequence of the peptide hormone insulin in 1953. Knowledge of the amino acid sequence is extremely useful in a number of ways: (1) it permits comparisons between normal and mutant proteins (see chapter 5); (2) it permits comparisons between comparable proteins in different species and thereby has been instrumental in positioning different organisms on the evolutionary tree (see fig. 1.24); (3) finally and most important, it is a vital piece of information for determining the three-dimensional structure of the protein.

Determining the order of amino acids involves the sequential removal and identification of successive amino acid residues from one or the other free terminal of the polypeptide chain. However, in practice it is extremely difficult to get the required specific cleavage reaction of the desired products to proceed with 100% yield. This obstacle becomes significant when sequencing long polypeptides, because the fraction of the total material of minimum polypeptide chain length becomes constantly smaller as the successive removal of terminal residues continues. Conversely, the amino acid released from the polypeptide chain becomes increasingly contaminated with amino acids released from previously unreacted chains.

Because of this fundamental chemical limitation, the polypeptide chain must be broken down into sequences short enough for the chemistry to produce reliable results. The short sequences are then reassembled to obtain the overall sequence. The steps actually involved in protein sequencing (fig. 3.16) are

1. purification of the protein
2. cleavage of all disulfide bonds
3. determination of the terminal amino acid residues
4. specific cleavage of the polypeptide chain into small fragments in at least two different ways
5. independent separation and sequence determination of peptides produced by the different cleavage methods
6. reassembly of the individual peptides with appropriate overlaps to determine the overall sequence

The first step, protein purification, is discussed in chapter 5. Once the protein is pure, sequence analysis can begin, with cleavage of the disulfide bonds. Cleavage is achieved by oxidizing the disulfide linkages with performic acid (fig. 3.17). Sometimes this step results in the production of two or more polypeptide chains, in which case the individual chains must be separated.

The third step is to determine the polypeptide chain end groups. If the polypeptide chains are pure, then only one N-terminal and one C-terminal group should be detected. The amino-terminal amino acid can be identified by reaction with fluorodinitrobenzene (FDNB) (fig. 3.18). Subsequent acid hydrolysis releases a colored dinitrophenol (DNP)-labeled amino-terminal amino acid, which can be identified by its characteristic migration rate on thin-layer chromatography or paper electrophoresis. A more sensitive method of end-group determination involves the use of dansyl chloride (see Methods of Biochemical Analysis 3B).

Chemical methods for carboxyl end-group determination are considerably less satisfactory. Treatment of the peptide with anhydrous hydrazine at 100°C results in conversion of all the amino acid residues to amino acid hydrazides except for the carboxyl-terminal residue, which remains as the free amino acid and can be isolated and identified chromatographically. Alternatively, the polypeptide can be subjected to limited breakdown (proteolysis) with the enzyme carboxypeptidase. This results in release of the carboxyl-terminal amino acid as the major free amino acid reaction product. The amino acid type can then be identified chromatographically.

Step 4 involves breaking down the polypeptide chain into shorter, well-defined fragments for subsequent sequence analysis. Fragmentation can be achieved by the use of endopeptidases, which are enzymes that catalyze polypeptide chain cleavage at specific sites in the protein. Figure 3.19 shows the specificity of four endopeptidases commonly used for this purpose. Another specific chemical method for polypeptide chain cleavage involves reaction with cyanogen bromide. This reaction cleaves specifically at the methionine residues, with the accompanying conversion of free carboxyl-terminal methionine to homoserine lactone (fig. 3.20). Although this methionine reaction product differs from the 20 naturally occurring amino acids, it is nevertheless readily identified by subsequent conversion to homoserine.

Peptides resulting from cleavage of the intact protein are generally separated by column chromatography. The isolated peptides may then be analyzed (step 5) to determine both their amino acid composition and their sequence. Se-

F i g u r e 3 . 1 6

Steps involved in the sequence determination of the B chain of
insulin. Amino acids are represented here by their single-letter codes
(see table 3.1).

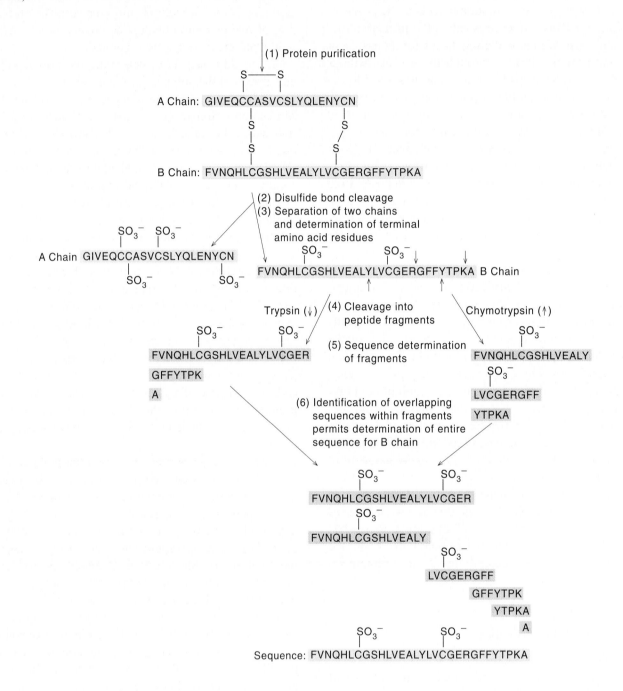

Figure 3.17

Disulfide cleavage reactions. Prior to sequence analysis inter- and intrachain disulfide linkages are irreversibly cleaved by one of the two procedures shown.

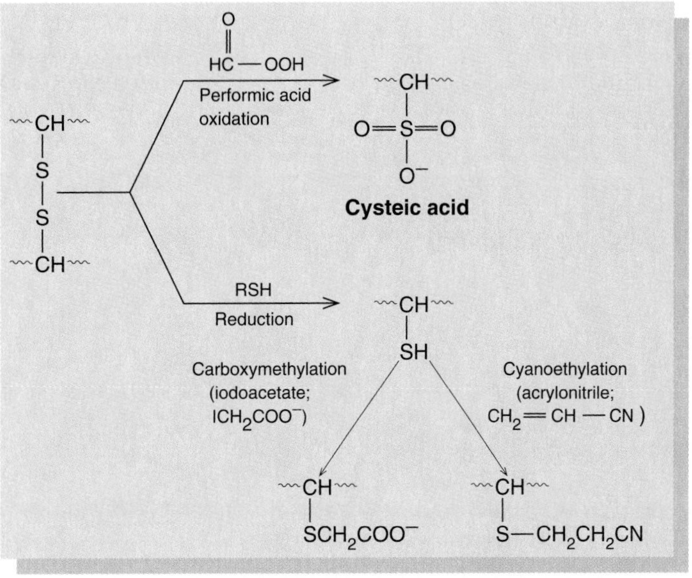
HC—OOH
Performic acid
oxidation
Cysteic acid
RSH
Reduction
Carboxymethylation
(iodoacetate;
ICH2COO−)
Cyanoethylation
(acrylonitrile;
CH2=CH—CN)
SCH2COO−
S—CH2CH2CN

Figure 3.18

Polypeptide chain end-group analysis. (*a*) Amino-terminal group identification. A more sensitive method, the dansyl chloride method, is described in Methods of Biochemical Analysis 3B. (*b*) Carboxyl-terminal group identification. Identification of this amino acid is considerably more difficult.

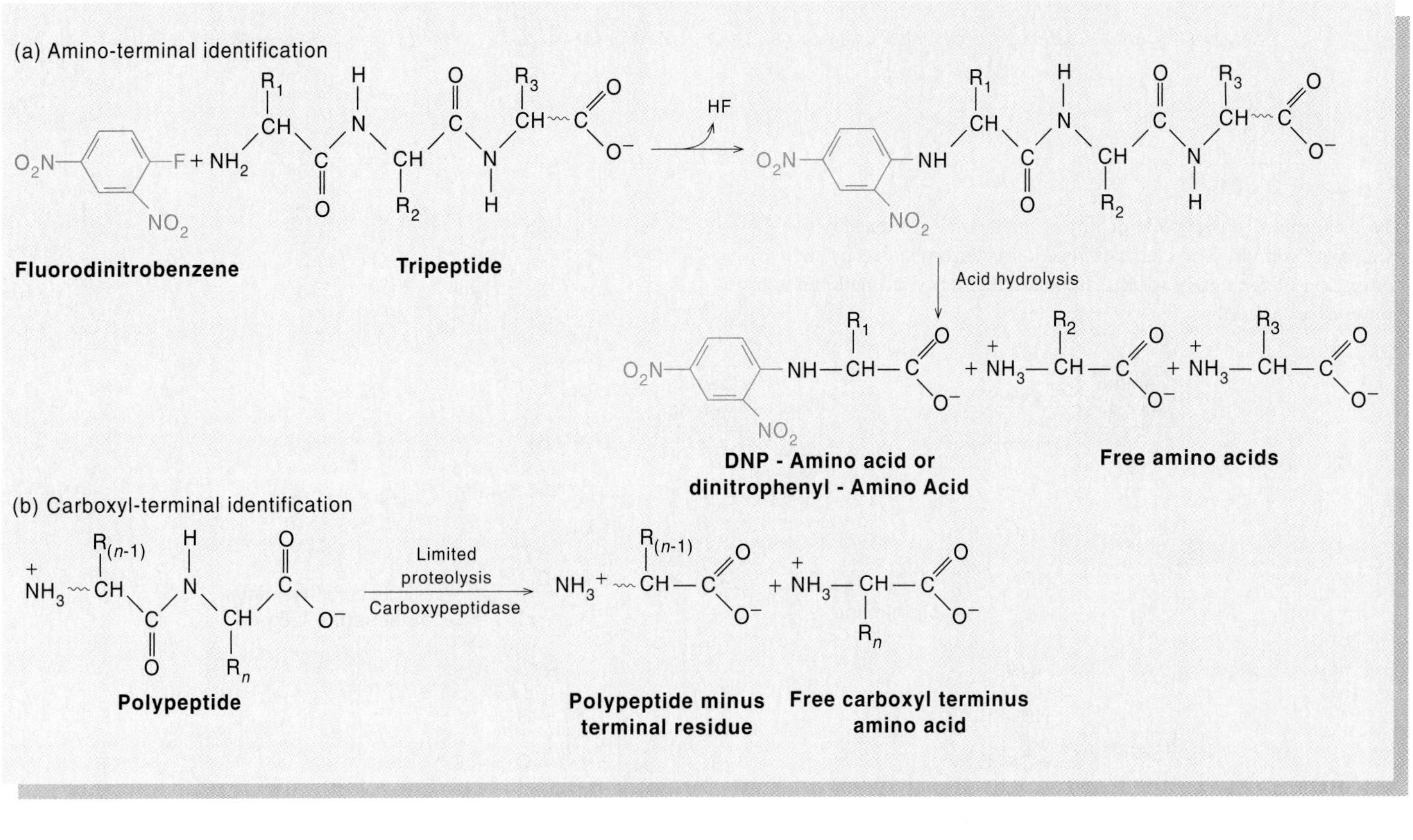
(a) Amino-terminal identification
Fluorodinitrobenzene
Tripeptide
HF
Acid hydrolysis
DNP - Amino acid or
dinitrophenyl - Amino Acid
Free amino acids
(b) Carboxyl-terminal identification
Polypeptide
Limited
proteolysis
Carboxypeptidase
Polypeptide minus
terminal residue
Free carboxyl terminus
amino acid

Figure 3.19

Site of action of some endopeptidases used for polypeptide chain cleavage prior to sequence analysis. Of the four different enzymes used, trypsin is used most frequently because of its high specificity.

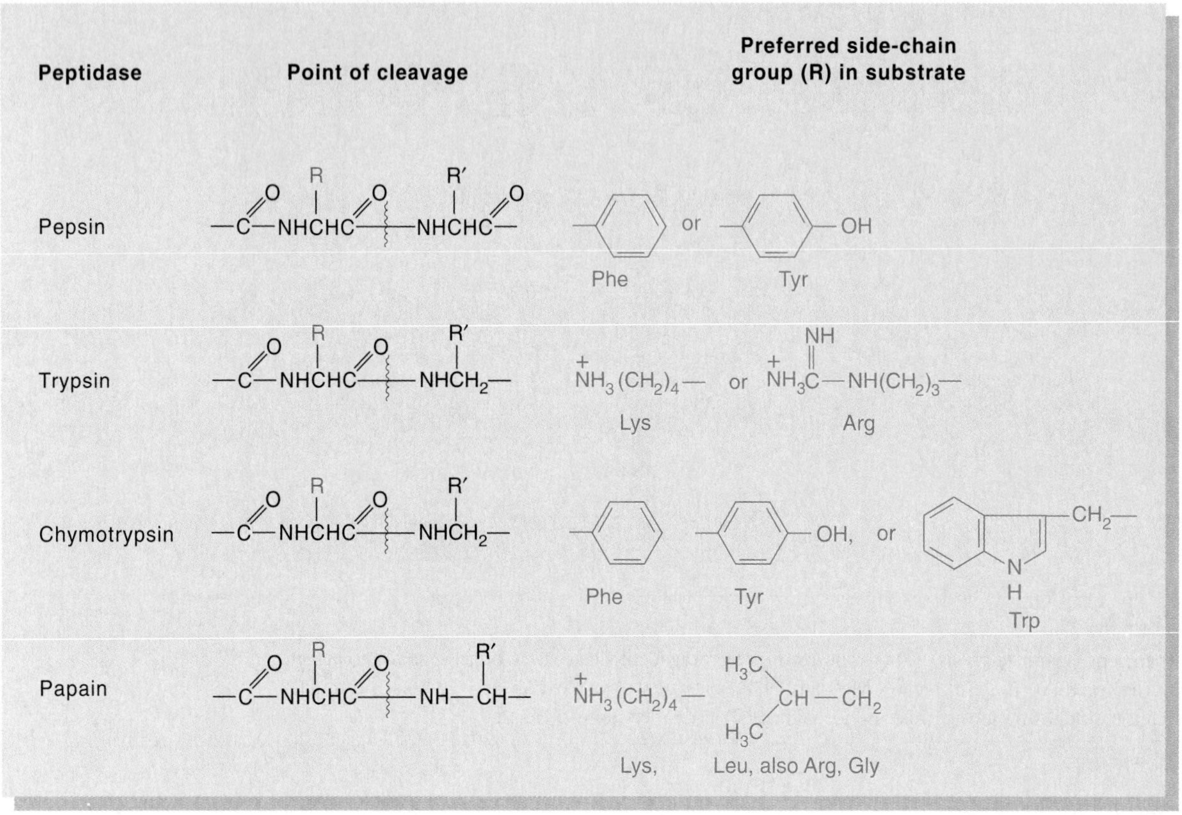

Figure 3.20

The cleavage of polypeptide chains at methionine residues by cyanogen bromide. The cleavage reaction is accompanied by the conversion of the newly formed free carboxyl-terminal methionine to homoserine lactone.

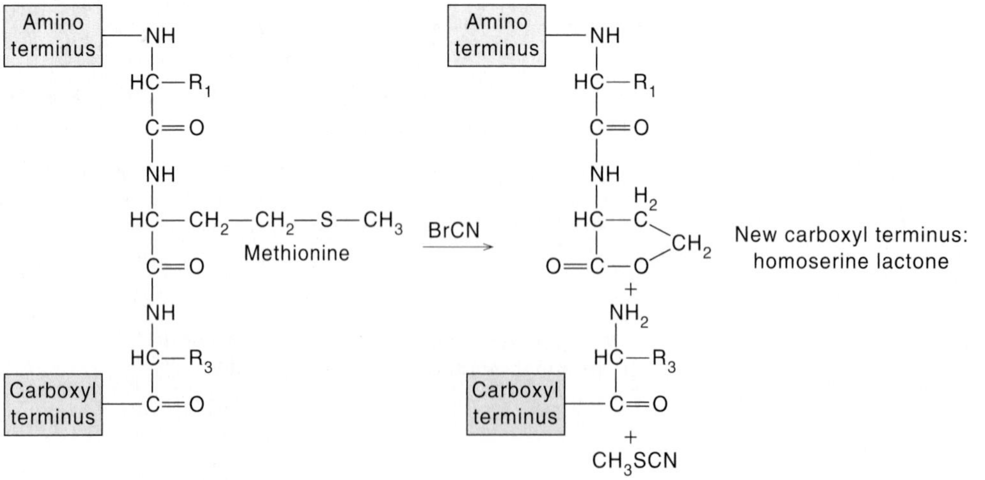

Figure 3.21

The Edman degradation method for polypeptide sequence determination. The sequence is determined one amino acid at a time, starting from the amino-terminal end of the polypeptide. First the polypeptide is reacted with phenylisothiocyanate to form a polypeptidyl phenylthiocarbamyl derivative. Gentle hydrolysis releases the amino-terminal amino acid as a phenylthiohydantoin (PTH), which can be separated and detected spectrophotometrically. The remaining intact polypeptide, shortened by one amino acid, is then ready for further cycles of this procedure. A more sensitive reagent, dimethylaminoazobenzene isothiocyanate, can be used in place of phenylisothiocyanate. The chemistry is the same.

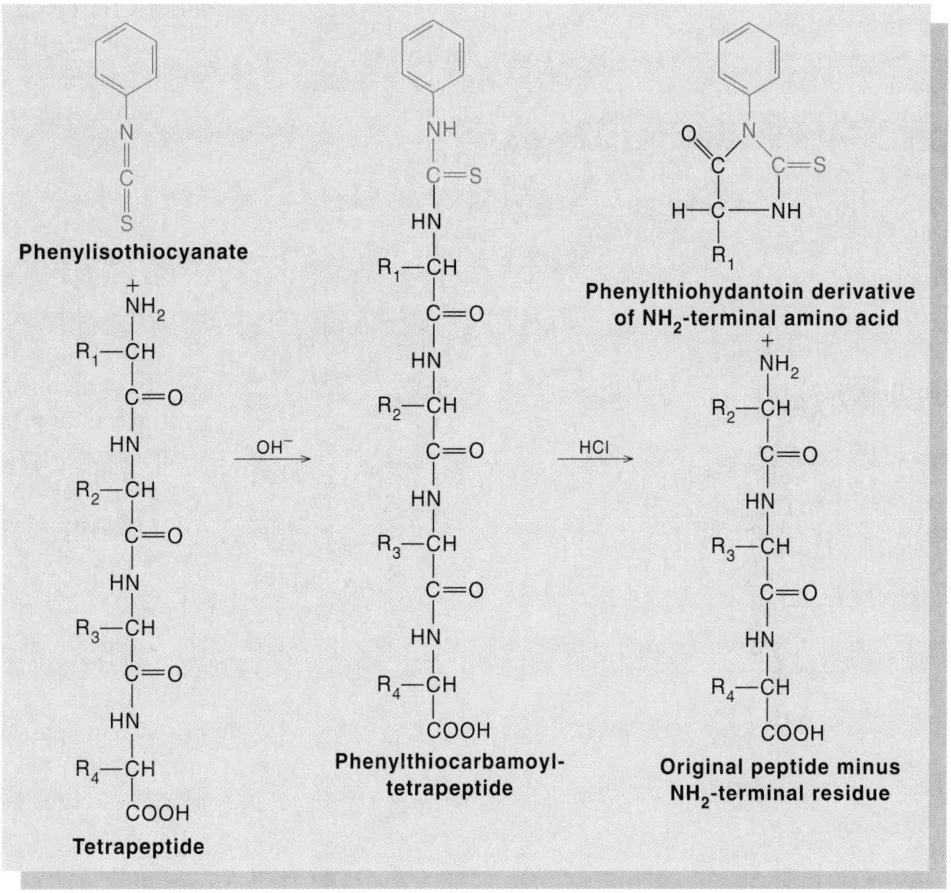

quence determination involves the stepwise removal and identification of successive amino acids from the polypeptide amino terminal by means of the Edman degradation (fig. 3.21). This process is carried out by reacting the free amino-terminal group with phenylisothiocyanate to form a peptidyl phenylthiocarbamyl derivative. Gentle hydrolysis with hydrochloric acid releases the amino-terminal amino acid as a phenylthiohydantoin (PTH) derivative. The remaining intact peptide, shortened by one amino acid, is then ready for further cycles of this procedure. The PTH-amino acid can be identified by its properties on thin-layer chromatography (fig. 3.22). High pressure liquid chromatography (HPLC) is the method of choice where quantitative results on small amounts of material are required. The general use of HPLC is described in chapter 6.

Devices called sequenators are available that automate the Edman degradation procedure. The success of these devices depends in large part on the technical innovation of covalently linking the peptide to be sequenced to glass beads. Attachment of the peptide through its carboxyl-terminal group to this immobile phase facilitates the complete removal of potentially contaminating reaction products during successive stages of the degradation.

Finally, having established the sequences of the individual peptides, it is necessary only to establish how they are connected together in the intact protein (step 6). It is at this stage that we see why the preceding sequence analysis was performed on peptides obtained by two different specific cleavage methods. This approach makes it possible to piece together the overall sequence, because the two sets of results produce overlapping sequences. That is, the free amino and carboxyl residues of peptides originally interconnected in the intact protein and liberated by one specific cleavage method recur in the internal sequences of the peptides liberated by a second specific method.

Once the protein's primary sequence has been determined, the location of disulfide bonds in the intact protein can be established by repeating a specific enzymatic cleavage on another sample of the same protein in which the disulfide bonds have not previously been cleaved. Separation of the resulting peptides shows the appearance of one new peptide and the disappearance of two other peptides, when compared with the enzymatic digestion product of the material whose disulfide bonds have first been chemically cleaved. In fact, these difference techniques are generally useful in the detection of sites of mutations in protein mole-

Figure 3.22

Thin-layer chromatography of amino acid–phenylthiohydantoin derivatives on silica gel plates. (*a*) Separation is done in a 98:2 mixture of chloroform and ethanol. (*b*) This is followed by further separation using an 88:2:10 mixture of chloroform, ethanol, and methanol. More sophisticated procedures, using column chromatography, give superior resolution and improved sensitivity. Automated sequencers always use such procedures. A general description of the use of columns is given in chapter 6.

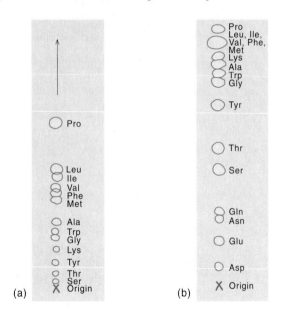

cules of previously known sequence, because a single substitution generally affects the chromatographic properties of only a single peptide released during proteolytic digestion.

Great progress has been made in recent years in devising procedures for sequencing the DNA that encodes for proteins (see chapter 27). Knowing the sequence of coding triplets in DNA allows us to read off the amino acid sequence of the corresponding protein. Nevertheless, such studies have produced the remarkable observation that some eukaryotic DNA sequences coding for proteins are not continuous but instead contain untranslated intervening DNA sequences. Although these results have profound implications for protein evolution, they obviously confound the general applicability of DNA-sequencing methods for the purposes of protein primary structure determination. In cases like this, the usual solution has been to isolate the mRNA for the protein and use this to make a DNA carrying the same sequence. This procedure circumvents the intervening sequence problem because the mRNA carries only the coding sequences (see chapter 28).

Chemical Synthesis of Peptides and Polypeptides

Knowledge about the structure–function interrelationships in proteins and peptides has encouraged biochemists to develop techniques for synthesizing peptides and proteins with predetermined sequences. To synthesize a peptide in the laboratory, we must overcome several problems related to preventing undesired groups from reacting. The amino and carboxyl groups that are to remain unlinked must be blocked; so must all reactive side chains.

After blocking those groups to be protected, the carboxyl group is activated. It is of interest that carboxyl-group activation is also employed in natural biosynthesis in the cell (see chapter 29). After peptide synthesis, the protecting groups must be removed by a mild method. The overall process—comprising protection, activation, coupling, and unblocking—is shown in figure 3.23.

An important variation of the usual methods of peptide synthesis involves attaching a protected (*t*-butoxycarbonyl group) amino acid to a solid polystyrene resin, removal of the amino protecting group, condensation with a second protected amino acid, and so on. In the last step, the finished peptide is cleaved from the resin. This method (outlined in figure 3.24) has the advantage that cumbersome purification between steps, often resulting in serious losses, is replaced by mere washing of the insoluble resin. Because each reaction is essentially quantitative, very long peptides, and even proteins, can be synthesized by this method. Indeed, Li synthesized a 39-amino-acid protein hormone, adrenocorticotropic hormone, by this method, and Robert Merrifield synthesized bovine pancreatic ribonuclease, which contains 129 amino acids in a single polypeptide chain. A number of variants of ribonuclease that contain one or more changes in amino acid sequence also have been made by this method. The importance of the Merrifield process was underscored by the awarding of a Nobel Prize to Merrifield in 1984.

Figure 3.23

Schematic diagram illustrating the chemical method for peptide synthesis. First the amino acids to be linked are selected. The carboxyl group and the amino group that are to be excluded from peptide synthesis are protected (steps 1 and 1′). Next the amino acid containing the unprotected carboxyl group is carboxyl-activated (step 2). This amino acid is mixed and reacted with the other amino acid (step 3). Protecting groups are then removed from the product (step 4).

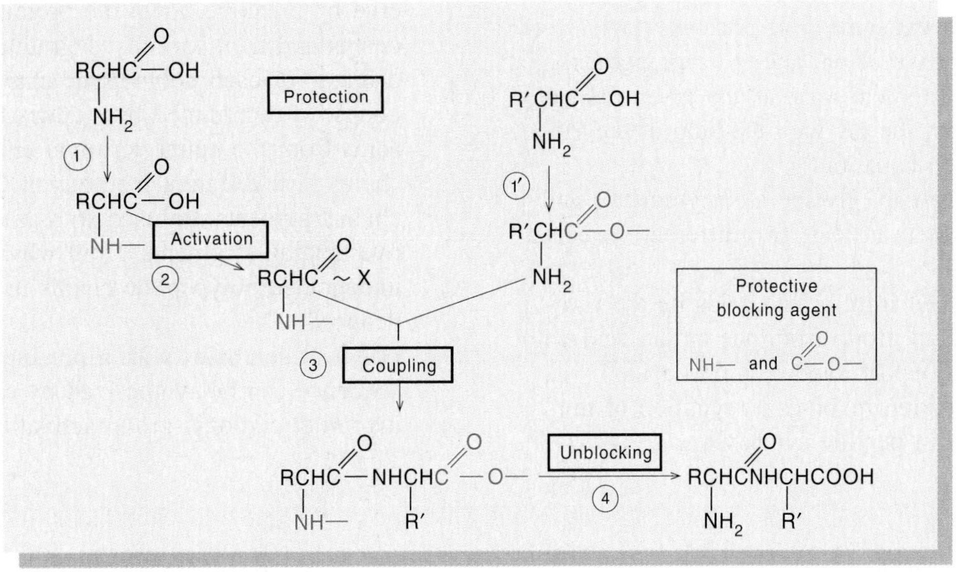

Figure 3.24

Merrifield procedure for solid-state dipeptide synthesis. (1) Polymer is activated. (2) Amino acid containing a *t*-butoxycarbonyl (BOC)-protecting group is carboxyl-linked to the polymer. This amino acid will be the carboxyl-terminal amino acid in the final peptide. (3) The BOC protecting group is removed from the polymer-linked amino acid. (4) A second amino acid, containing a BOC on its α-amino group and a dicyclohexylcarbodiimide (DCC)-activated group, is reacted with the column-bound amino acid to form a dipeptide. (5) The dipeptide is released from the polymer and the BOC-protecting group by adding hydrogen bromide (HBr) in trifluoroacetic acid.

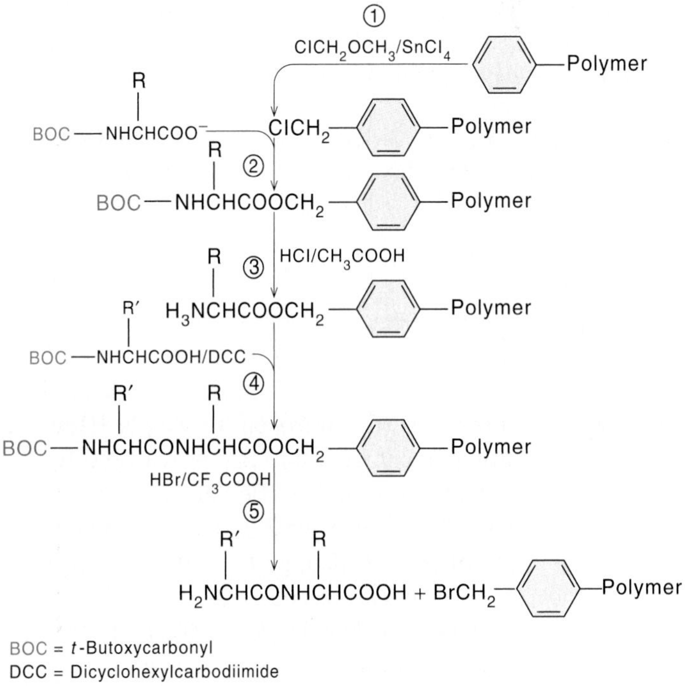

BOC = *t*-Butoxycarbonyl
DCC = Dicyclohexylcarbodiimide

Summary

In this chapter we dealt with some of the fundamental properties of amino acids and polypeptide chains. The following points are especially important.

1. Nineteen of the 20 amino acids commonly found in proteins have a carboxyl group and an amino group attached to an α-carbon atom; they differ in the side chain attached to the same α carbon.
2. All amino acids have acidic and basic properties. The ratio of base to acid form at any given pH can be calculated from the pK with the help of the Henderson-Hasselbach equation.
3. All amino acids except glycine are asymmetric and therefore can exist in at least two different stereoisomeric forms.
4. Peptides are formed from amino acids by the reaction of the α-amino group from one amino acid with the α-carboxyl group of another amino acid.
5. Polypeptide formation involves a repetition of the process involved in peptide synthesis.
6. The amino acid composition of proteins can be discovered by first breaking down the protein into its component amino acids and then separating the amino acids in the mixture for quantitative estimation.
7. The amino acid sequences of proteins can be discovered by breaking down the protein into polypeptide chains and then partially degrading the polypeptide chains. For each polypeptide chain fragment, the sequence is determined by stepwise removal of amino acids from the amino-terminal end of the polypeptide chain. Two different methods of forming polypeptide chain fragments are used so as to produce a map of overlapping fragments, from which the sequence of undegraded polypeptide chains in the proteins can be deduced.
8. Polypeptide chains with a predetermined amino acid sequence can be synthesized by chemical methods involving carboxyl-group activation.

Selected Readings

Barrett, G. C. (ed.), *Chemistry and Biochemistry of Amino Acids.* New York: Chapman and Hall, 1985. A recent and authoritative volume on this classical subject.

Gray, W. R., End group analysis using dansyl chloride. *Methods. Enzymol.* 25:121–138, 1972. This volume of *Methods in Enzymology* contains several chapters on end-group analysis.

Hunkapiller, M. W., J. E. Strickler, and K. J. Wilson, Contemporary methodology for protein structure determination. *Science* 226:304–311, 1984.

Kent, S. B. H., Chemical synthesis of peptides and proteins. *Ann. Rev. Biochem.* 57:957–989, 1988. Comprehensive and up-to-date.

Merrifield, B., Solid phase synthesis. *Science* 232:341–347, 1986.

Sanger, R., Sequences, sequences and sequences. *Ann. Rev. Biochem.* 57:1–28, 1988.

Problems

1. A typical protein is 16% nitrogen by weight. How well does this percentage value match with the formula for proteins proposed by Berzelius?
2. (a) A 10mM solution of a weak monocarboxylic acid has a pH of 3.00. Calculate the values for K_a and pK_a for this carboxylic acid.
   (b) You add 0.06 g of NaOH ($M_r = 40$) to 1,000 ml of the acid solution in part (a). Calculate the final pH assuming no volume change.
3. A buffer was prepared by dissolving 3.71 g of citric acid and 2.91 g of KOH in water and diluting to a final volume of 250 ml. What is the pH of this buffer? What is the [H^+]? Use 3.14, 4.77, and 6.39 for the pK_a's of citric acid.
4. Given the pK_a values in the text, predict how the titration curves for glutamic acid and glutamine differ.
5. Calculate the isoelectric point for histidine, aspartic acid, and arginine. Calculate the fractional charge for

each ionizable group on aspartate at pH equal to the pI. Do these calculations verify the isoelectric point of aspartic acid?

6. Which of the naturally occurring amino acid side chains are charged at pH 2? pH 7? pH 12? (Consider only those amino acids whose side chains have >10% charge at the pH examined.)

7. Amino acids are sometimes used as buffers. Indicate the appropriate pH value(s) of buffers containing aspartic acid, histidine, and serine.

8. Polyhistidine is insoluble in water at pH 7.8 but is soluble at pH 5.5. Explain this observation. Would you expect the polymer to be soluble at pH 10?

9. As indicated in the text, 19 of the 20 amino acids have a chiral α carbon with only glycine (R = H) lacking chirality. Two of the protein amino acids have a second chiral carbon. Can you identify them?

10. A mixture of alanine, glutamic acid, and arginine was chromatographed on a weakly basic ion-exchange column (positively charged) at pH 6.1. Predict the order of elution of the amino acids from the ion-exchange column. Are the amino acids separated from each other? Explain.

    Suppose you have a weakly acidic ion-exchange column (negatively charged), also at pH 6.1. Predict the order of elution of the amino acids from this column. Propose a strategy for separating the amino acids using one or both columns. Explain your rationale. (Assume only ionic interactions between the amino acids and the ion-exchange resin.)

11. You have a peptide that is a potent inhibitor of nerve conduction and you wish to obtain its primary sequence. Amino acid analysis reveals the composition to be Ala(5); Lys; Phe. Reaction of the intact peptide with FDNB releases free DNP-alanine on acid hydrolysis. ϵ-DNP-lysine (but not α-DNP-lysine) is also found. Tryptic digestion gives a tripeptide (composition Lys, Ala(2)) and a tetrapeptide (composition Ala(3), Phe). Chymotryptic digestion of the intact peptide releases a hexapeptide and free alanine. Derive the peptide sequence.

12. From a rare fungus you have isolated an octapeptide that prevents baldness, and you wish to determine the peptide sequence. The amino acid composition is Lys(2), Asp, Tyr, Phe, Gly, Ser, Ala. Reaction of the intact peptide with FDNB yields DNP-alanine plus 2 moles of ϵ-DNP-lysine on acid hydrolysis. Cleavage with trypsin yields peptides the compositions of which are (Lys, Ala, Ser) and (Gly, Phe, Lys), plus a dipeptide. Reaction with chymotrypsin releases free aspartic acid, a tetrapeptide with the composition (Lys, Ser, Phe, Ala), and a tripeptide the composition of which, following acid hydrolysis, is (Gly, Lys, Tyr). What is the sequence?

Measurement of Ultraviolet Absorption in Solution

The general quantitative relationship that governs all absorption processes is called the Beer-Lambert law:

$$I = I_0 10^{-\epsilon cd}$$

where I_0 is the intensity of the incident radiation, I is the intensity of the radiation transmitted through a cell of thickness d (in centimeters) that contains a solution of concentration c (expressed either in moles per liter or in grams per 100 ml), and ϵ is the extinction coefficient, a characteristic of the substance being investigated (see fig. 3.7).

Light absorption is measured by a spectrophotometer as shown in the illustration. The spectrophotometer usually is capable of directly recording the absorbance A, which is related to I and I_0 by the equation

$$A = \log_{10}\left(\frac{I_0}{I}\right)$$

Hence $A = \epsilon cd$, and A is a direct measure of concentration. We can see from figure 3.7 that the ϵ values are largest for tryptophan and smallest for phenylalanine.

Because protein absorption maxima in the near-ultraviolet (240–300 nm) are determined by the content of the

Figure 1

Schematic diagram of a spectrophotometer for measuring light absorption. Laboratory instruments for making measurements are much more complex than this, but they all contain the same basic components: a light source, a monochromator, a sample, and a detector. λ is the wavelength of the light, I_0 and I are the incident light intensity and the transmitted light intensity, respectively, and d is the thickness of the absorbing solution.

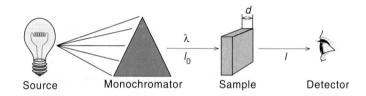

Source   Monochromator   Sample   Detector

aromatic amino acids and their respective values, most proteins have absorption maxima in the 280-nm region. By contrast, absorption in the far-ultraviolet (around 190 nm) is shown by all polypeptides regardless of their aromatic amino acid content. The reason is that absorption in this region is due primarily to the peptide linkage.

The Dansyl Chloride Method for N-Terminal Amino Acid Determination

The dansyl chloride method provides an alternative to the Sanger method for N-terminal amino acid determination. Because it is considerably more sensitive than the Sanger method, it has become the method of choice. The reaction is diagrammed in the illustration. A polypeptide is treated with dansyl chloride to give an *N*-dansyl peptide derivative. This derivative is hydrolyzed to yield a highly fluorescent *N*-dansyl-amino acid, which is detected chromatographically.

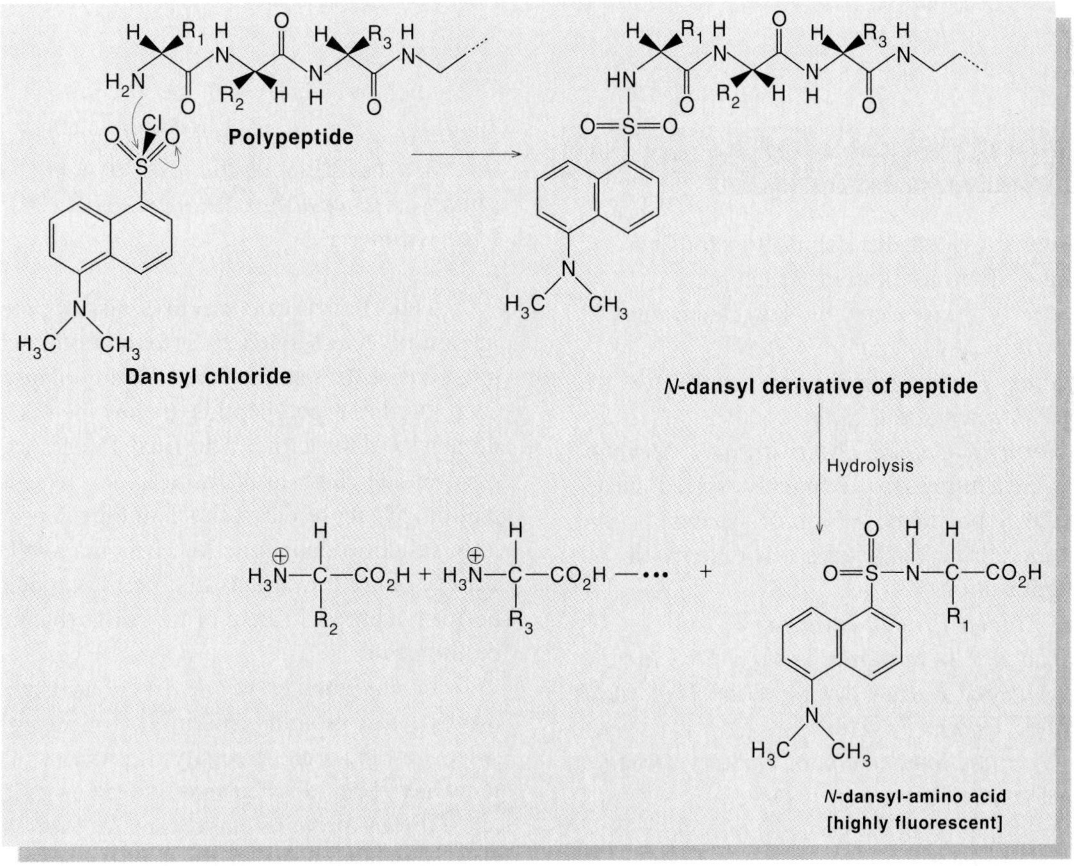

Polypeptide

Dansyl chloride

N-dansyl derivative of peptide

Hydrolysis

**N-dansyl-amino acid
[highly fluorescent]**

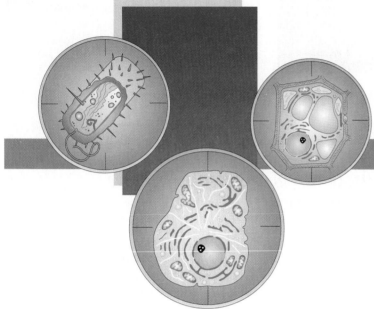

The Three-Dimensional Structures of Proteins

Chapter Outline

Pauling and Corey Provided the Foundation for Our
Understanding of Fibrous Protein Structures
 *The Structure of the α-Keratins Was Determined
 with the Help of Molecular Models*
 *The β-Keratins Form Sheetlike Structures with
 Extended Polypeptide Chains*
 Collagen Forms a Unique Triple-Stranded Structure
Globular Protein Structures Are Extremely Varied and
Require a More Sophisticated Form of Analysis
Folding of Globular Proteins Reveals a Hierarchy of
Structural Organization
 Visualizing Folded Protein Structures
 Primary Structure Determines Tertiary Structure
 *Secondary Valence Forces Are the Glue That Holds
 Polypeptide Chains Together*
 Domains Are Functional Units of Tertiary Structure
 Predicting Protein Tertiary Structure
 *Quaternary Structure Involves the Interaction of
 Two or More Proteins*

Proteins adopt the most stable folded structures; this is a function of the way in which the individual amino acid residues interact with each other and with water.

The enormous structural diversity of proteins begins with the amino acid sequences of polypeptide chains. Each protein consists of one or more unique polypeptide chains, and each of these polypeptide chains is folded into a three-dimensional structure. The final folded arrangement of the polypeptide chain in the protein is referred to as its conformation. Most proteins exist in unique conformations exquisitely suited to their function. It is the availability of a wide variety of conformations that permits proteins as a group to perform a broader range of functions than any other class of biomolecules.

In this chapter we deal primarily with the structural properties of proteins, and in the following chapter we consider the functional diversity of proteins. Traditionally proteins have been divided into two groups: fibrous and globular. Fibrous proteins aggregate to form highly elongated structures having the shape of fibers or sheets. Each protein unit that makes up these aggregated structures is built from a repeating structural motif, giving the molecules a simple structure that is relatively easy to analyze. By contrast, globular proteins are more complex, containing one or more polypeptide chains folded back on themselves many times to give an approximately spherical shape. We consider the relatively simple fibrous proteins first.

Figure 4.1

Resonance and the planar structure of the peptide bond. (*a*) Two major hybrids contribute to the structure of the peptide bond. In structure 1 the C—N bond is a single bond with no overlap between the nitrogen lone electron pair and the carbonyl carbon. The carboxyl carbon is sp^2-hybridized and is therefore planar, whereas the nitrogen is sp^3-hybridized and pyramidal. By contrast, in structure 2 there is a double bond between the amide nitrogen and the carbonyl carbon; also, the nitrogen atom bears a charge of +1 and the carbonyl oxygen bears a charge of −1. Both the carboxyl carbon and the amide nitrogen are sp^2-hybridized, both are planar, and all six atoms lie in the same plane. (*b*) The structure of the peptide bond is a compromise between the two resonating hybrids, structures 1 and 2. (*c*) Dimensions of the peptide bond and surrounding linkages. The C—N bond length of 1.325 Å is significantly less than the length of a single C—N bond, 1.47 Å. (Illustration for part *c* copyright by Irving Geis. Reprinted by permission.)

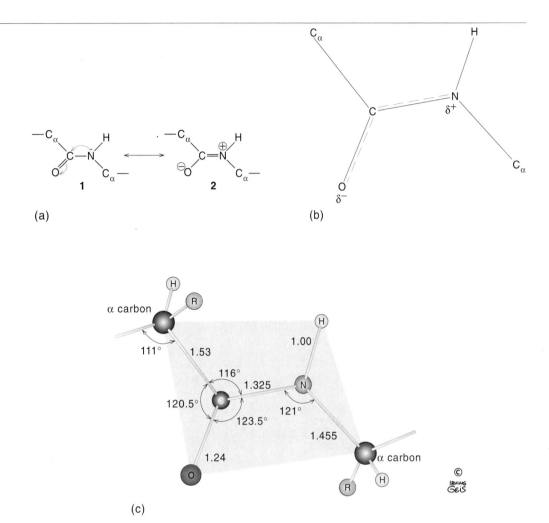

(a)

(b)

(c)

Pauling and Corey Provided the Foundation for Our Understanding of Fibrous Protein Structures

Linus Pauling and Robert Corey examined the structures of crystals formed by amino acids and short peptides before they ventured into the world of proteins. From their crystallographic investigations of amino acids and peptides, they formulated two rules that describe the ways in which amino acids and peptides interact with one another to form noncovalently bonded crystalline structures. These rules laid the foundations for our understanding of how amino acids in protein polypeptide chains interact with one another.

Rule number one was that the peptidyl C—N linkage and the four atoms to which the C and the N atoms are directly linked always form a planar structure as though the C—N linkage is a double bond rather than a single bond as normally written. Pauling reasoned that the C—N linkage is a resonating structure with partial double-bond character

(fig. 4.1), so that it locks the peptide grouping into a planar conformation. This property is extremely important because it greatly reduces the flexibility in the polypeptide chain. With this grouping in a rigid conformation, the only flexibility remaining in the polypeptide backbone results from rotation about the carbon that joins adjacent peptide planar groups (fig. 4.2).

The second rule that Pauling and Corey formulated was that peptide carbonyl and amino groups always form the maximum number of hydrogen bonds. Recall (see chapter 1) that in water hydrogen bonds are formed between a partially unshielded proton from one molecule and an oxygen atom that originates from another molecule (see fig. 1.7). Nitrogen atoms can also serve as H bond acceptors. The attraction between the H bond donor and the H bond acceptor is strongest along the lone pair orbital axis of the acceptor atom. As a rule the angle between an N or O acceptor and an N—H or O—H donor is close to 180° (fig. 4.3). Thus a hydrogen bond brings two interacting groups close

Figure 4.2

Basic dimensions of a dipeptide. The conformational degrees of freedom of a polypeptide chain are restricted to rotations about the single-bond connections between the adjacent planar transpeptide groups to C_α, that is, the C_α—C_2 and C_α—N_1 single bonds. The corresponding rotations are represented by ψ and ϕ, respectively, which have values of 180° for the fully extended configuration shown. (Illustration copyright by Irving Geis. Reprinted by permission.)

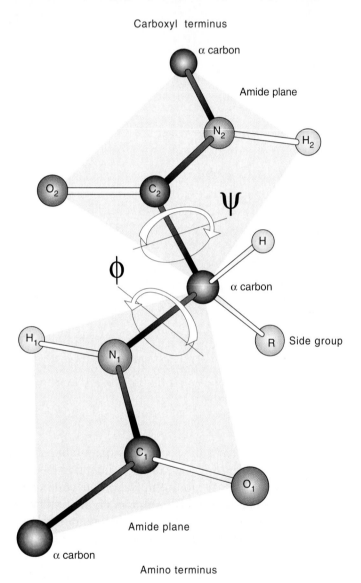

Carboxyl terminus

α carbon

Amide plane

Side group

Amide plane

α carbon

Amino terminus

Figure 4.3

Major hydrogen-bond donor and acceptor groups found in proteins. Note that the angle between the O acceptor and the N—H donor is 180° in the hydrogen-bond complex.

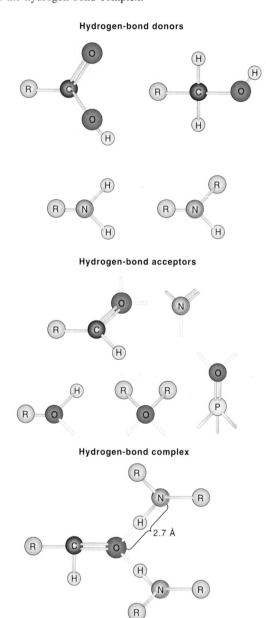

Hydrogen-bond donors

Hydrogen-bond acceptors

Hydrogen-bond complex

2.7 Å

together and orients them in a certain way. This second rule further limits the number of conformations available to polypeptide chains.

Following their investigations on amino acid and peptide crystals Pauling and Corey turned their attention to the x-ray diffraction patterns of a number of fibrous proteins. A vast number of fibrous proteins exist in nature, but the majority of them give diffraction patterns that fall into one of three types: the α pattern (see Methods of Biochemical Analysis 4A), the β pattern, and the collagen pattern. Fiber diffraction data only give information about the repeating units of a protein structure because of the lack of three-dimensional order in the fibers; but because fibrous protein polypeptide chains are arranged in simple repetitious units the overall structure could be deduced. The first type of

Table 4.1

Radii for Covalently Bonded and Nonbonded Atoms

Covalent Bond Radii (in Å)				Van der Waals Radii (in Å)			
Element	*Single Bond*	*Double Bond*	*Triple Bond*	*Element*			
Hydrogen	0.30			Hydrogen	1.2		
Carbon	0.77	0.67	0.60	Carbon	2.0		
Nitrogen	0.70			Nitrogen	1.5		
Oxygen	0.66			Oxygen	1.4		
Phosphorus	1.10			Phosphorus	1.9		
Sulfur	1.04			Sulfur	1.8		

diffraction pattern, which was observed for a subgroup of the keratins, was consistent with a helical arrangement of the polypeptide chains; this pattern indicated that there are two regularly repeating units along the helix axis: one with a repeat of 5.4 Å and one with a repeat of 1.5 Å.

The Structure of the α-Keratins Was Determined with the Help of Molecular Models

Space-filling molecular models were used in an attempt to build structures compatible with the x-ray data. These models were designed so that covalently linked atoms were accurately spaced according to known dimensions (table 4.1). Moreover, the individual atoms made as hard spheres were of a size so that nonbonded atoms could get no closer to one another than their van der Waals radii would normally allow (see table 4.1). It should be recalled from chemistry that the average separation between two nonbonded atoms is known as their van der Waals separation. If nonbonded atoms get closer than this they repel each other greatly, and favorable interaction between them falls less rapidly if the distance is larger than this. Pauling and Corey tried to arrange the polypeptide chains so as to maximize the number of peptide hydrogen bonds in a way that was consistent with the x-ray diffraction data. By trial and error they came to the conclusion that the most acceptable structure was a right-handed helical coil (fig. 4.4). This structure, known as the alpha (α) helix, has a rigid, regularly repeating backbone structure with an advance of 1.5 Å per amino acid residue along the helix axis explaining the corresponding spot in the diffraction (see Methods of Biochemical Analysis 4A). The diffraction spots resulting from the helix could be explained by

a spacing of 5.4 Å between adjacent turns of the polypeptide backbone along the helix axis. In fig. 4.4 we see three ways of representing this structure. In (*a*) we see a ball-and-stick model in which the planar groups are highlighted and the hydrogen bonds are represented as dashed lines. In (*b*) we see a space-filling model similar to the type that Pauling and Corey used. And in (*c*) we see a wire model. The latter two models are shown in both side and top views.

The space-filling model (*b*) shows that the α helix is a tightly packed structure with no unfilled cavities whether one looks at a profile or down the helix. The wire model (*c*) illustrates the helical structure best. But for purposes of discussion, the ball-and-stick model (*a*) is the most suitable. Careful inspection shows that the polypeptide backbone follows the path of a right-handed helical spring in which each residue's carbonyl group forms a hydrogen bond with the amide NH group of the residue four amino acids further along the polypeptide chain. All residues in the α helix have nearly identical conformations so they lead to a regular structure in which each 360° of helical turn incorporates approximately 3.6 amino acid residues and rises 5.4 Å along the helix axis direction. This rise accounts for the observed advance per amino acid residue along the helix axis of 5.4/3.6 = 1.5 Å.

Although alternative helical arrangements having different hydrogen-bonding patterns and different geometries are possible, the α helix is by far the most commonly observed. The unique stability of the α helix is related to the formation of regularly arranged hydrogen bonds between all the carbonyl and amino groups and to the tight packing achieved in folding the chain to form the structure.

Fibrous proteins in which the α helix is a major structural component are found in hair, scales, horns, hooves, wool, beaks, nails, and claws—proteins referred to as α-

Figure 4.4

Three ways of projecting the α-helix. (*a*) This simple ball-and-stick model highlights the planar peptides. The interpeptide hydrogen bonds are shown by dashed lines, and the amino acid side chains are indicated by R groups. Approximately two turnings of the helix are shown. There are about 3.6 residues per turn. In (*b*) and (*c*) we see identical projections of a side view using space-filling and wire models, respectively. The space-filling models use van der Waals radii for the atoms. (Illustration for part *a* copyright by Irving Geis. Reprinted by permission.)

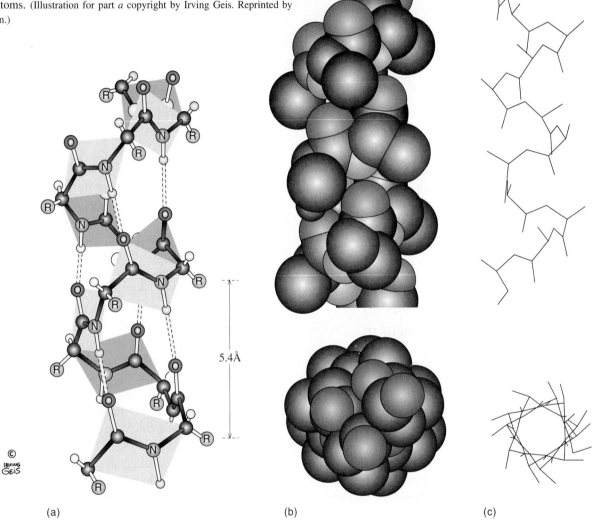

5.4Å

(a)   (b)   (c)

keratins. When individual α helices aggregate in this side-by-side fashion, they usually form long cables in which the individual helices are spirally twisted so that the resulting cable has an overall left-handed twist (fig. 4.5). The formation of a cable with twisted α helices results from optimization of packing among the amino acid side-chain residues between helices. In figure 4.6 we see how the side-chain residues of an α helix are arranged in a spiral fashion so that residues falling on the same side of a helix do not lie along a line parallel to the helix axis. As a consequence the packing of helices is optimized when helices interact at an angle of about 18°. Obviously, if α helices involved in such a packing interaction were straight, they would soon separate. However, their packing interaction can be preserved if the helices are slightly twisted around each other, with the resultant formation of the left-twisted cable structure characteristic of the keratins. The coiled-coil character of such fibers consequently represents a trade-off between some local deformations that coil the α helix and the optimization of extended side-chain packing interactions in the cable as a whole.

Figure 4.5

The assembly of hair α-keratin from one α helix to a protofibril, to a microfibril, and finally, to a single hair. (Illustration copyright by Irving Geis. Reprinted by permission.)

α helix

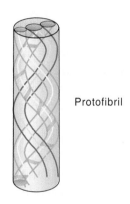

Protofibril

Microfibril

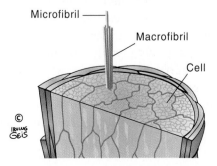

Microfibril

Macrofibril

Cell

Figure 4.6

Coiling of α helices in α-keratins. Residues on the same side of an α helix form rows that are tilted relative to the helix axis. Packing helices together in fibers is optimized when the individual helices wrap around each other so that rows of residues pack together along the fiber axis. Helices in coiled coil (c) are oriented in parallel.

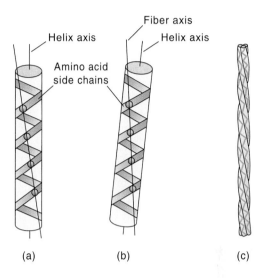

The springiness of hair and wool fibers results from the tendency of the α-helical cables to untwist when stretched and spring back when the external force is removed. In many forms of keratin, the individual α helices, or fibers, are covalently linked by disulfide bonds formed between cysteine residues of adjacent polypeptide chains. In addition to giving added strength to the fibers, the pattern of these covalent interactions serves to influence and fix the extent of curliness in the hair fiber as a whole. Chemical reactions and mechanical processes involving reductive cleavage, reorganization, and reoxidation of these interhelix disulfide bonds form the basis of the "permanent wave."

The β-Keratins Form Sheetlike Structures with Extended Polypeptide Chains

Pauling and Corey noticed that certain fibrous proteins give radically different diffraction patterns and behave macroscopically as sheets rather than fibers. Diffraction patterns of fibrous proteins from different sources reveal a repeat along the direction of the extended polypeptide chain of either 13.0 or 14.0 Å, suggesting two closely related structures. Pauling and Corey interpreted these patterns as resulting from extended polypeptide chains lying side by side either in a parallel or an antiparallel fashion (fig. 4.7). Although both of these structures exist in nature, the antiparallel sheet is more common in fibrous proteins. This structure

Figure 4.7

Two forms of the β-sheet structure: (a) the antiparallel and (b) the parallel β sheet. The advance per two amino acid residues is indicated for each structure.

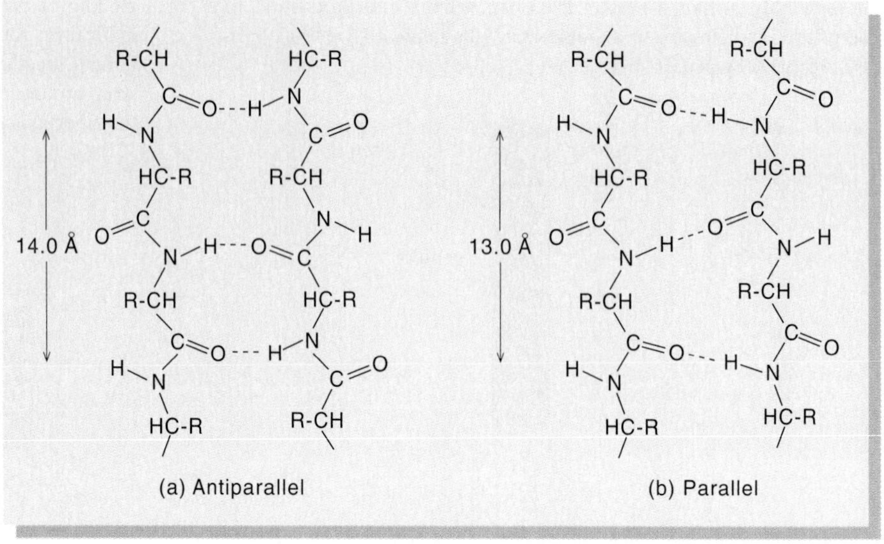

(a) Antiparallel   (b) Parallel

Figure 4.8

The antiparallel β sheet. This structure is composed of two or more polypeptide chains in the fully extended form, with hydrogen bonds formed between the chains. Hydrogen bonds are shown as dashed lines. (Illustration copyright by Irving Geis. Reprinted by permission.)

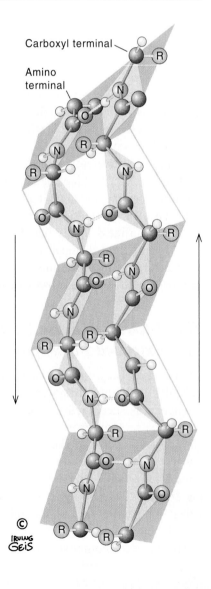

Carboxyl terminal

Amino terminal

is known as the antiparallel β-pleated sheet (fig. 4.8). Regular hydrogen bonds form between the peptide backbone amide NH and carbonyl oxygen groups of adjacent chains. The β sheet can be extended into a multistranded structure simply by adding successive chains in the appropriate directions to the sheet. Parallel and antiparallel β-pleated sheets are both composed of polypeptide chains that have conformations pointing alternate R groups to opposite sides of the sheet but that have their peptide planes nearly in the sheet plane to allow for good interchain hydrogen bonding. Nevertheless, the chain conformation that produces the best interchain hydrogen bonding in parallel sheets is slightly less extended than that for the antiparallel arrangement. As a result, the parallel sheet has both a shorter repeat period 6.5 Å (versus 7.0 for the antiparallel structure) and a more pronounced pleat.

The best known β-keratin structure in nature is that found in certain silks. Silks are composed of stacked antiparallel β sheets (fig. 4.9). Sequence analysis of silk proteins shows them to be composed largely of glycine, serine, and alanine, in which every alternate residue is glycine. Since the side-chain groups of a flat antiparallel sheet point alternately upward and downward from the plane of the sheet, all the glycine residues are arranged on one surface of each sheet and all the substituted amino acids are on the other. Two or more such sheets can consequently be packed intimately together to form an arrangement of stacked sheets in which two adjacent glycine-substituted or alanine-substituted sheet surfaces interlock with each other (fig. 4.9b). Owing to the extended conformations of the polypeptide chains in the β sheets and the interlocking of the side chains between sheets, silk is a mechanically rigid material that resists stretching.

Figure 4.9

The three-dimensional architecture of silk (*a*). The side chains of one sheet nestle quite efficiently between those of neighboring sheets (*b*). (Illustration copyright by Irving Geis. Reprinted by permission.)

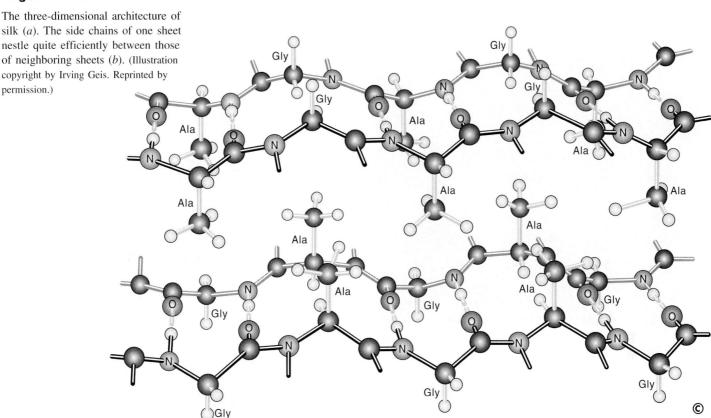

(a)

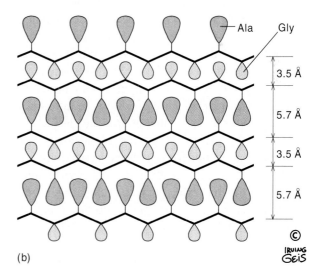

(b)

Collagen Forms a Unique Triple-Stranded Structure

Collagen is a particularly rigid and inextensible protein that serves as a major constituent of tendons and connective tissues. In the electron microscope it can be seen that collagen fibrils have a distinctive banded pattern with a periodicity of 680 Å (fig. 4.10). These fibrils are of varying thicknesses depending on their source and the mode of preparation. Individual fibrils are composed of collagen molecules 3,000 Å long that aggregate in a staggered side-by-side fashion (fig. 4.11).

Collagen has a most unusual amino acid composition in which glycine, proline, and hydroxyproline are the dominant amino acids. Further characterization of the polypeptide chains shows that these amino acids are arranged in a repetitious tripeptide sequence, Gly-X-Y, in which X is frequently a proline and Y is frequently a hydroxyproline. This unusual amino acid sequence and the unique diffraction pattern of collagen were strong indications for a totally different type of fibrous protein. Pauling attempted to determine the structure of collagen by his molecular model approach but failed. At a meeting in Cambridge, England, in 1958 where the correct structure was presented, Pauling jested

Figure 4.10

An electron micrograph of collagen fibrils from skin. (Courtesy of Jerome Gross, Massachusetts General Hospital.)

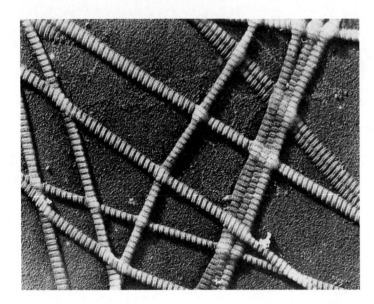

that the structure he had proposed might be more stable than the one found in nature. The natural structure was found by Ramachandran. The repeating proline residue excluded the possibility that the polypeptide chains in collagen could adopt either an α-helical or a β-sheet conformation. Instead, individual collagen polypeptide chains assume a left-handed helical conformation and aggregate into three-stranded cables with a right-handed twist (fig. 4.12). When viewed down the polypeptide chain axis (fig. 4.13*b*), the successive side-chain groups point toward the corners of an equilateral triangle. The glycine at every third residue is required because there is no room for any other amino acid inside the triple helix where the glycine R groups are located. The three collagen chains do not form hydrogen bonds among residues of the same chain. Instead, the collagen chains within each three-stranded cable form interchain hydrogen bonds. This produces a highly interlocked fibrous structure that is admirably suited to its biological role, which is to provide rigid connections between muscles and bones as well as structural reinforcement for skin and connective tissues.

Although living organisms contain additional types of fibrous proteins, as well as polysaccharide-based structural motifs, we focused here on the three arrangements that are the most widely distributed. Two of these, the α-keratins and the β-keratins incorporate polypeptide secondary structures that also commonly occur in globular proteins. Colla-

Figure 4.11

The banded appearance of collagen fibrils in the electron microscope arises from the schematically represented staggered arrangement of collagen molecules (*above*) that results in a periodically indented surface. *D*, the distance between cross striations is $\approx$680 Å so that the length of a 3,000-Å-long collagen molecule is 4.4*D*. (Line art courtesy of Karl A. Piez. Photomicrograph © Michael C. Webb/Visuals Unlimited.)

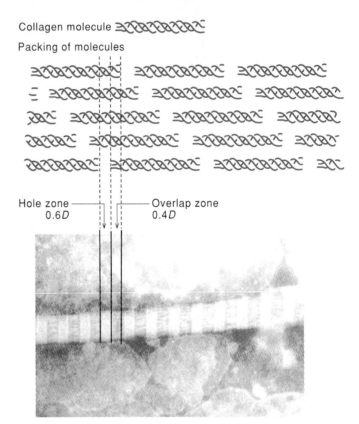

gen, in contrast, is a protein that evolution has developed to play more specialized roles.

Globular Protein Structures Are Extremely Varied and Require a More Sophisticated Form of Analysis

It is not surprising that repeating structures with long-range order were the first protein structures to be understood. The demands on the available technology were minimal. Much more sophisticated technology was required to interpret the diffraction patterns of proteins that have shorter range repetition in their structures (see Methods of Biochemical Analysis 4B). Among the crystallographers who struggled over

Figure 4.12

The triple helix of collagen. (Illustration copyright by Irving Geis. Reprinted by permission.)

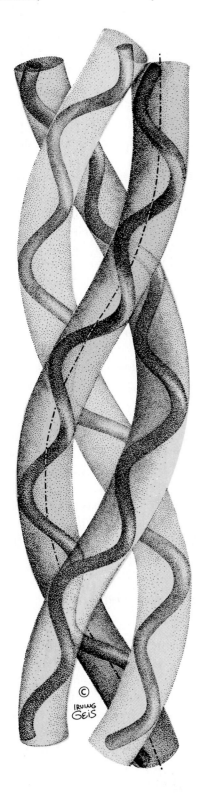

Figure 4.13

The basic coiled-coil structure of collagen. Three left-handed single-chain helices wrap around one another with a right-handed twist. (a) Ball-and-stick single-collagen chain. (b) View from the top of the helix axis. Note that glycines are all on the inside. In this structure the C=O and N—H groups of glycine protrude approximately perpendicularly to the helix axis so as to form interchain hydrogen bonds. (Illustration for part a copyright by Irving Geis. Reprinted by permission.)

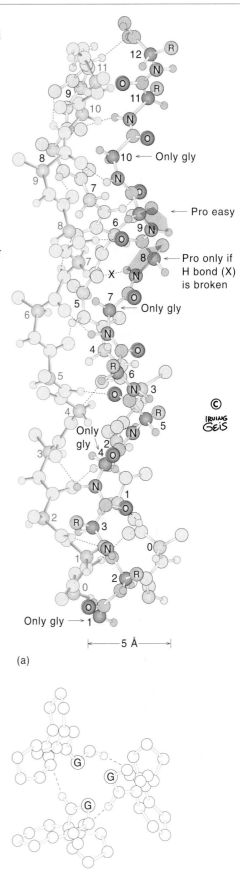

(a)

(b)

these structures were two who emerged as the major pioneers: John Kendrew and Max Perutz. In the 1950s they led their research teams to determine the first of such structures—myoglobin and hemoglobin. Since that time enormous advances in protein chemistry and computer technology have systematized the necessary research and greatly reduced the amount of work and time required to determine a protein structure. Accurate structural determinations have now been made for more than 500 different proteins. From this wealth of data, patterns of structure are becoming apparent that suggest, among other things, that the overall folding arrangements of proteins may some day be predictable from the amino acid sequences of the polypeptide chains.

Folding of Globular Proteins Reveals a Hierarchy of Structural Organization

In the analysis of fibrous protein structures little mention was made of the importance of their interaction with water in determining the final folded structures of the proteins. This is because most of the side chains in fibrous proteins are exposed to water except when they interact with each other to form multimolecular aggregates. Then the relative affinity between other similar side chains and water becomes a major issue. In the case of globular proteins the interaction of amino acid side chains with water is a major issue from the start because globular proteins have many of their amino acid side chains buried in the interior of their folded structures. Hence, in our analysis of the structure of globular proteins we must be aware of the structural considerations that are important in the determination of fibrous proteins but also of additional considerations, first raised in chapter 1, that relate to the interaction of the amino acid side chains with water.

We may think of the structure of globular proteins at four levels (fig. 4.14). Ultimately, the entire three-dimensional structure is determined by a protein's amino acid sequence, or primary structure. At a higher level of organization, the regularly repeating geometry of the hydrogen-bonding groups of the polypeptide backbone leads to the formation of regular hydrogen-bonded secondary structures. Secondary structural elements include regions of α helix, β sheets, and bends. Association between elements of secondary structure in turn results in the formation of structural domains, the properties of which are determined both by chiral characteristics of the polypeptide chain and by the nature of the amino acid side chains. The side chains of the residues in each domain pack together in a way that tends to optimize favorable interactions between side chains and between side chains and the surrounding water.

Further association of domains results in the formation of the protein's tertiary structure—the overall folding of the polypeptide chain in three dimensions. Finally fully folded protein subunits can pack together to form quaternary structures.

Visualizing Folded Protein Structures

Because globular proteins have nonrepeating structures, it is essential to have a means for displaying the entire three-dimensional structure in sufficient detail, and yet not too much detail, so that the overall structural design can be appreciated.

The primary data of protein crystallography yield a three-dimensional electron-density map, which must be interpreted in terms of a three-dimensional model of all atom positions in the protein. Such modeling is usually done by computer graphics.

In figure 4.15 we see four different presentations that show selected parts of the protein ribonuclease and that highlight features of special interest. Each method of presentation has its advantages. The space-filling model (see fig. 4.15a) is excellent for displaying the volume occupied by molecular constituents and the shape of the outer surface. Figure 4.15b shows a stereo pair of a space-filling model. When the pair is viewed with stereo glasses, the three-dimensional illusion is striking. Figure 4.15c shows the polypeptide chain, with N and O atoms labeled and with dotted lines representing hydrogen bonds. In addition, all α-carbon positions are numbered. Figure 4.15d shows an abstraction of the polypeptide chain in which the β strands are characterized as flat arrows and the α helices as spiral ribbons. This simplified style has proved useful in classifying and comparing proteins according to the folding patterns of their secondary and tertiary structures.

Primary Structure Determines Tertiary Structure

Throughout our discussion of protein structures we assumed that structures form because they represent the most stable way of arranging the polypeptide chains. The first direct support for this notion for a globular protein came from the studies of Christian Anfinsen. Anfinsen unfolded pancreatic ribonuclease, an enzyme containing 124 amino acid residues with four disulfide bridges, and then found conditions under which it could be refolded into its original native structure. First, the enzyme was denatured in a solution containing a hydrogen-bond-breaking reagent (urea) and 2-mercaptoethanol (a thiol reagent that reduces disulfides to sulfhydryls), thus cleaving the cova-

Figure 4.14

Hierarchies of protein structures. (Illustration for part *c* copyright by
Irving Geis. Reprinted by permission.)

(a) Primary structure (amino acid sequence in the protein chain)

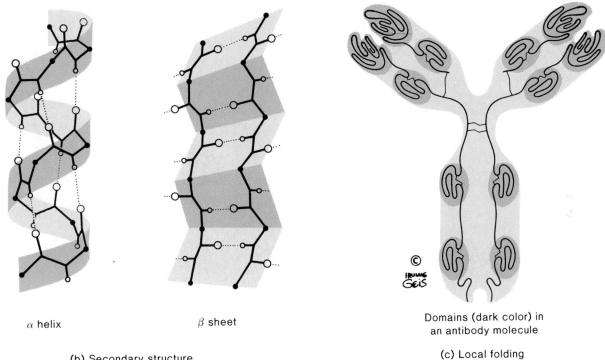

α helix   β sheet   Domains (dark color) in
                                              an antibody molecule

(b) Secondary structure   (c) Local folding

 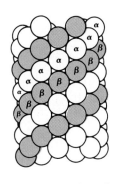

One complete protein chain   The four separate chains   α (white) and β (color)
(β chain of hemoglobin)   of hemoglobin assembled   tubulin molecules in a
                                into an oligomeric protein   microtubule

(d) Tertiary structure   (e) Quaternary structure   (f) Quaternary structure

Figure 4.15

Four ways of representing the three-dimensional structure of the protein pancreatic ribonuclease. (*a*) Space-filling model. (*b*) The stereo pair of space-filling models illusion can be seen without stereo glasses by the following method. With your eyes about 10 inches from the page, stare at the drawings below as if you were looking straight ahead at a far-away object. A double image forms and the central pair drifts together and fuses. Then the illusion becomes apparent. Adjust the page, if necessary, so that a horizontal line is perfectly parallel with the eyes. As an aid to seeing one image with each eye, use two cardboard tubes from toilet-paper rolls. Close the right eye and focus the left eye on the left image. Do the same for the other eye. Now, with both eyes together, the three-dimensional illusion should appear. (*c*) Ball and stick model. Dotted lines represent hydrogen bonds, and numbers indicate α-carbon positions. (*d*) Ribbon model. The β-strands are represented by the flat red arrows. The α helices are shown as spiral ribbons. (Illustrations for parts *a–d* copyright by Irving Geis. Reprinted by permission.)

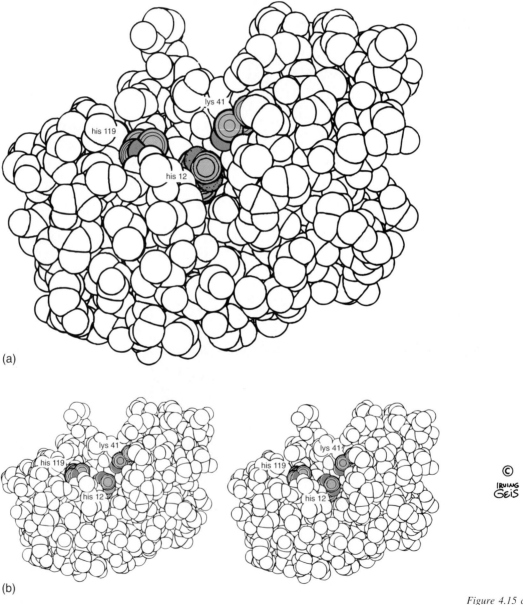

(a)

(b)

Figure 4.15 continued on facing page.

Continued

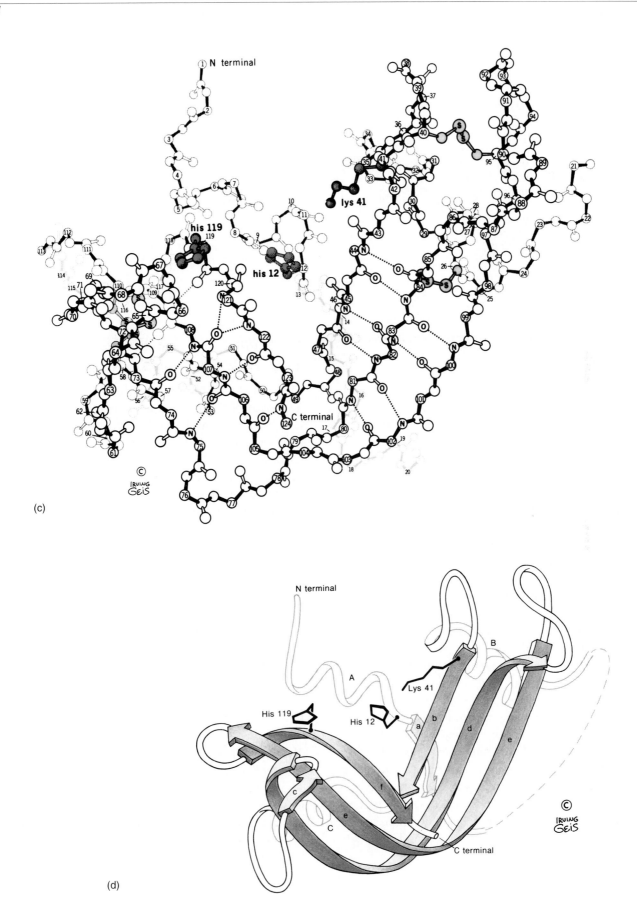

(c)

(d)

Figure 4.16

Schematic representation of an experiment to demonstrate that the information for folding into a biologically active conformation is contained in the protein's amino acid sequence.

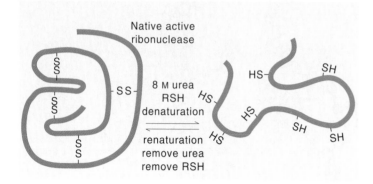

Native active
ribonuclease

8 M urea
RSH
denaturation

renaturation
remove urea
remove RSH

HS   SH
HS
HS
HS
SH

Table 4.2

Bond Energies between Some Atoms
of Biological Interest

Energy Values for Single Bonds (kcal/mole)					
C—C	82	C—H	99	S—H	81
O—O	34	N—H	94	C—N	70
S—S	51	O—H	110	C—O	84
Energy Values for Multiple Bonds (kcal/mole)					
C=C	147	C=N	147	C=S	108
O=O	96	C=O	164	N≡N	226

lent cross-links (fig. 4.16). These conditions have been used since as a general means of denaturing proteins in a manner that completely disrupts the conformation without causing precipitation. The reduced, denatured ribonuclease was enzymatically inactive. Renaturation was carried out by removing the urea, which resulted in refolding of the protein. Finally the mercaptoethanol was removed to permit the air oxidation of the reduced disulfides back to disulfide crosslinks. The result of this series of manipulations was an almost complete recovery of the enzymatic activity of the original ribonuclease molecule.

Thus it appears that the information for folding a protein to the native conformation is embodied in the amino acid sequence, because of the many possible disulfide-paired ribonuclease isomers, only one was formed in major yield. Further studies have given similar results with many other proteins.

Despite the elegance and simplicity of Anfinsen's experiment, the current indications are that special proteins accelerate the folding process and may return unfolded or incorrectly folded proteins to their native states. For example, two types of polypeptide-chain-binding (PCB) proteins have been discovered: proteins related to the "heat shock" protein with a subunit molecular weight of 70,000 (designated hsp70) and the GroEL family of proteins. The hsp70 proteins appear to be ubiquitous, occurring in bacteria, mitochondria, and the eukaryotic cytosol. They are especially abundant in the endoplasmic reticulum. The GroEL proteins are limited to bacteria, mitochondria, and chloroplasts. We have more to say about the function of some of these proteins in chapter 29.

Secondary Valence Forces Are the Glue That Holds Polypeptide Chains Together

The studies of Anfinsen not only showed that folding can be a spontaneous process predetermined by the primary amino acid sequence, but they suggested that folded structures are thermodynamically more stable. Thus the folding of globular proteins should be understandable in terms of the types of forces that exist between polypeptide chains and water.

When forces involved in determining protein conformation are being considered, we can usually ignore covalent bond energies, despite their large magnitude (table 4.2). This is because covalent bonds (except for disulfide bonds) are not made or broken when polypeptide chains fold into their native three-dimensional conformations. The bonds that are affected (made or broken) on folding are, by and large, of the noncovalent type involving secondary valence forces. Typically the intermolecular bond energies between noncovalently linked atoms range in value from 0.1 to 6 kcal/mol. The intermolecular forces between noncovalently linked atoms may be grouped into four categories: hydrophobic effects, electrostatic forces, van der Waals forces, and hydrogen bonds (H bonds).

Hydrogen Bonds. Many characteristics of H bonds have already been discussed because of the major role they play in stabilizing fibrous proteins. Additional aspects of H bonds to be considered relate more closely to the structures of globular proteins. Evidence for the existence of an H bond comes from the observation of a decreased distance between donor and acceptor groups forming the H bond.

Thus from the van der Waals radii given in table 4.1, we can calculate the distances between nonbonded H and O atoms (2.6 Å) and between nonbonded H and N atoms (2.7 Å). When an H bond is present, this distance is usually reduced by about 0.8 Å in both cases. Some important H-bond donors and acceptors are shown in figure 4.3.

Polypeptides carry a number of H-bond donor and acceptor groups, both in their backbone structure and in their side chains. Water also contains a hydroxyl H-donor group and an oxygen H-acceptor group for making H bonds (see fig. 1.7). Formation of the maximum number of H bonds between a polypeptide chain and water requires the complete unfolding of the polypeptide chain. However, it is not obvious that such an unfolding would result in a net energy gain. The reason is that water is a highly H-bonded structure, and for every H bond formed between water and protein, an H bond within the water structure itself must be broken. The compromise followed by most proteins is to maximize the number of intramolecular H bonds between the backbone peptide groups but to keep most of the potential H-bond-forming side chains of the protein near the protein–water interface where they can interact directly with water.

Van der Waals Forces. Van der Waals interactions are of two types: one attractive and one repulsive. Attractive van der Waals forces involve interactions among induced dipoles that arise from fluctuations in the electron charge densities of neighboring nonbonded atoms. Such interactions amount to 0.1–0.2 kcal/mol; despite their small size, the large number of such interactions that occur when molecules come close together makes such interactions quite significant. Van der Waals forces favor close packing in folded protein structures.

Repulsive van der Waals interactions occur when noncovalently bonded atoms or molecules come very close together. An electron–electron repulsion arises when the charge clouds between two molecules begin to overlap. If two molecules are held together exclusively by van der Waals forces, their average separation is governed by a balance between the van der Waals attractive and repulsive forces. This distance is known as the van der Waals separation. Some van der Waals radii for biologically important atoms are given in table 4.1. The van der Waals separation between two nonbonded atoms is given by the sum of their respective van der Waals radii.

Hydrophobic Effects. Hydrogen bonds and van der Waals forces are of major importance in determining the secondary structures formed by fibrous proteins. To understand the complex folded structures found in globular proteins additional types of interactions between amino acid side chains

and water must also be considered. The so-called hydrophobic effects, that lead to the interaction of hydrophobic groups in proteins, are the hardest type of noncovalent interactions to appreciate. Whereas H bonds and van der Waals forces relate primarily to enthalpic factors, hydrophobic effects relate primarily to entropic factors. Furthermore, the entropic factors mainly concern the solvent not the solute.

As a first step to understanding the nature of these effects it is useful to consider the interaction between a small hydrophobic molecule, like hexane and water. It is tempting to ascribe the low water solubility of hexane to van der Waals attractive forces between these small hydrophobic molecules. However, thermodynamic measurements indicate that this is not the case. The hydrophobic molecule hexane has a small favorable enthalpy for solution in water. In spite of this, hexane is poorly soluble in water due to an unfavorable entropic factor. What is this mysterious factor? Owing to the weak enthalpic interactions between hexane and water, the water withdraws slightly in the region of the apolar hydrophobic molecule and forms a relatively rigid hydrogen-bonded network with itself (for example, see the clathrate structure illustrated in fig. 1.10). The network effectively restricts the number of possible orientations of water molecules directly surrounding the dissolved hexane molecules. This ordering of water constitutes an energetically unfavorable entropic effect.

The same type of entropic effect plays a major role in directing the folding of globular proteins. About half of the amino acid side chains in proteins are hydrophobic (e.g., alanine, valine, isoleucine, leucine, and phenylalanine). Entropic effects strongly favor internal locations for these side chains where they are free from contacts with water (e.g., see fig. 1.12, which shows the location of polar and apolar side chains in cytochrome *c*).

The native folded state of a globular protein reflects a delicate balance between opposing energetic contributions of large magnitude. Whereas entropic factors favor the packing of hydrophobic side chains into the interior regions of globular proteins, enthalpic factors favor placing hydrophilic side chains on the surface of the protein where they can interact with water. In the limited number of cases where data are available, the overall entropy of folding appears to be slightly negative (unfavorable) and the overall enthalpy is also slightly negative (favorable). On balance, folding is opposed by the entropy change but favored by the enthalpy change and occurs in most cases because the latter factor outweighs the former.

Electrostatic Forces. Electrostatic forces are of three main types: charge–charge interactions, charge–dipole interactions, and dipole–dipole interactions. The energy of interaction between two charges Q_1 and Q_2 is proportional to the

Table 4.3

Energy Dependence of Interaction on the Distance of Separation of the Interacting Species

Range of Interaction	Type of Interaction
$1/R$	Charge–charge
$1/R^2$	Charge–dipole
$1/R^3$	Dipole–dipole
$1/R^6$	Van der Waals (dipole-induced dipole) attractive forces
$1/R^{12}$	Van der Waals repulsive forces

product of the charges and inversely proportional to the distance R between them (table 4.3):

$$\text{energy of interaction} \propto \frac{Q_1 Q_2}{R}$$

In solution this interaction is reduced by the dielectric constant of the surrounding medium:

$$\text{energy of interaction} \propto \frac{Q_1 Q_2}{\epsilon R}$$

If the two charges in question are buried within a protein, their interaction energy can be substantially increased because the dielectric constant in the regions inaccessible to water is much lower than the dielectric constant of water. The long-range nature of charge–charge interactions has led to the speculation that such forces can be important in accelerating interaction between proteins, between proteins and nucleic acids, and between proteins and small molecules, such as coenzymes and substrates (see chapter 10).

Favorable charge–charge interactions between oppositely charged amino acids are less significant in determining protein folding than are the ion–dipole interactions between the charged groups of amino acid side chains and water. With very few exceptions side chains containing charged groups as well as polar side chains are located on the protein surface at the protein–water interface.

Domains Are Functional Units of Tertiary Structure

Within a single folded chain or subunit, contiguous portions of the polypeptide chain often fold into compact local units called domains, each of which might consist, for example of a four-helix cluster or a ''barrel'' or an antiparallel β sheet (fig. 4.17). Sometimes the domains within a protein are very

Figure 4.17

Examples of three types of structural domains found in many proteins: (*a*) the four-helix cluster; (*b*) the β barrel; and (*c*) the antiparallel β sheet. (Reprinted by permission of Jane S. Richardson.)

FOUR HELIX CLUSTER

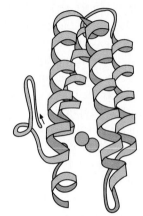

(a)   **Myohemerythrin**

β-BARREL SHAPE

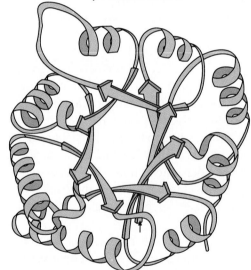

(b)   **Triose phosphate isomerase**

ANTIPARALLEL β SHEET

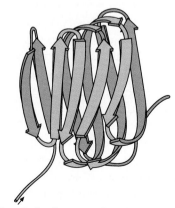

(c)   **Tomato bushy stunt virus domain 3**

Figure 4.18

Papain, a protein in which the domains are very different from each other. (Reprinted by permission of Jane S. Richardson.)

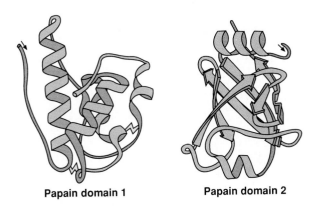

Papain domain 1   **Papain domain 2**

Figure 4.20

Schematic backbone drawing of the elastase molecule, showing the similar β-barrel structures of the two domains. (Reprinted by permission of Jane S. Richardson.)

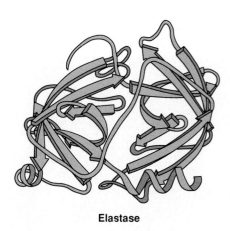

Elastase

Figure 4.19

Rhodanese domains 1 and 2 as an example of a protein with two domains that resemble each other extremely closely. Rhodanese is a liver enzyme that detoxifies cyanide by catalyzing the formation of thiocyanate from thiosulfate and cyanide. (Reprinted by permission of Jane S. Richardson.)

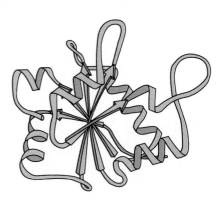

Rhodanese domain 1

Rhodanese domain 2

different from each other, as within the protease papain (fig. 4.18), but often they resemble each other very closely, as in rhodanese (fig. 4.19).

The separateness of two domains within a subunit varies all the way from independent globular domains joined only by a flexible length of polypeptide chain, to domains with tight and extensive contact and a smooth globular surface for the outside of the entire subunit, as in the proteolytic enzyme elastase (fig. 4.20). An intermediate level of domain separateness, characterized by a definite neck or cleft between the domains, is found in phosphoglycerate kinase (fig. 4.21).

Domains as well as subunits can serve as modular bricks to aid in efficient assembly of the native conformation. Undoubtedly, the existence of separate domains is important in simplifying the protein-folding process into separable, smaller steps, especially for very large proteins. There is no strict upper limit on folding size. Indeed, known domains vary in size all the way from about 40 residues to more than 400. Furthermore, it has been estimated that there may be more than a thousand basically different types of domains.

Another important function of domains is to allow for movement. Completely flexible hinges would be impossible between subunits because they would simply fall apart. However, flexible hinges can exist between covalently linked domains. Limited flexibility between domains is often crucial to substrate binding, allosteric control (discussed in chapter 9), or assembly of large structures. In hexokinase, the two domains within the individual subunits hinge toward each other on binding of the substrate glucose, enclosing it almost completely (fig. 4.22). In this manner glucose can be bound in an environment that excludes water

Figure 4.21

The dumbbell domain organization of phosphoglycerate kinase, with a relatively narrow neck between two well-separated domains. (Copyright 1994 by the Scripps Research Institute/Molecular Graphics Images by Michael Pique using software by Yng Chen, Michael Connolly, Michael Carson, Alex Shah, and AVS, Inc. Visualization advice by Holly Miller, Wake Forest University Medical Center.)

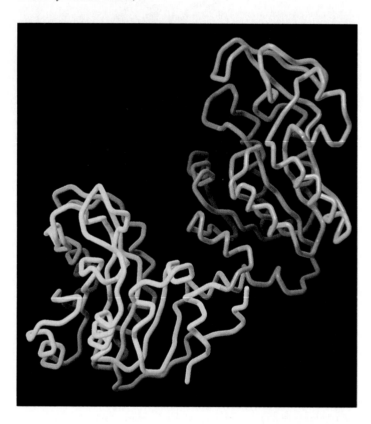

Figure 4.22

Schematic representation of the change in conformation of the hexokinase enzyme on binding substrate. E and E' are the inactive and active conformations of the enzyme, respectively. G is the sugar substrate. Regions of protein or substrate surface excluded from contact with solvent are indicated by a crinkled line. Figure 8.3 presents a more detailed view of the hexokinase molecule. (Source: From W. S. Bennett and T. A. Steitz, Glucose-induced conformational changes in yeast hexokinase, *Proc. Natl. Acad. Sci. USA* 75:4848, 1978.)

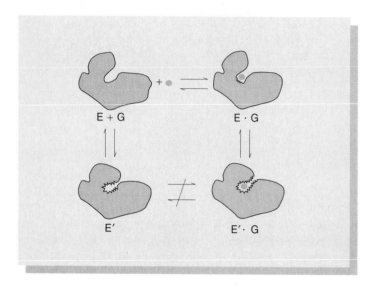

Figure 4.23

Relative probabilities that any given amino acid occurs in the α-helical, β-sheet, or β-hairpin-bend secondary structural conformations.

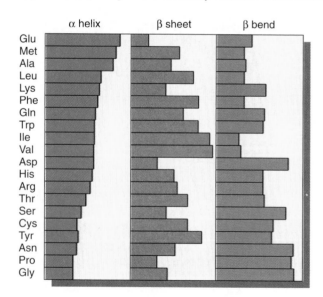

as a competing substrate (see chapter 12 for further details on the hexokinase reaction).

Predicting Protein Tertiary Structure

We began the discussion of globular protein tertiary structure by pointing out that the secondary and tertiary structure is determined by the primary structure and that this is probably a reflection of the fact that the native folded conformation is the most stable structure that can be formed. If this is so, then it should be possible to predict a protein's structure from its primary sequence. At this juncture, such predictions remain an elusive goal. However, most proteins are made of a limited number of domains, which tend to reappear in many different proteins. Since this is the case, it may be possible to predict the structures of many proteins in the future by using the information accumulated from x-ray diffraction studies of related proteins.

It is clear that certain amino acids tend to form particular secondary structures. As shown in figure 4.23 glutamic acid, methionine, and alanine appear to be the strongest α-helix formers, whereas valine, isoleucine, and tyrosine are the most probable β-sheet formers. Proline, glycine, asparagine, aspartic acid, and serine occur most frequently in so-called β-bend conformations—an unfolded segment that permits a sharp change in direction.

This type of information is of value in the prediction of secondary structural regions of proteins from their amino acid sequences. The observed frequencies of occurrence of an amino acid in a given conformation provide estimates of the probabilities that the same amino acid behaves similarly in a sequence the actual secondary structure of which is unknown. To predict the secondary structure from the sequence, it is consequently necessary only to plot the probabilities for the individual amino acids sequentially, or better, to plot a local average over a few adjacent residues. Plotting such an average accounts for the cooperative nature of secondary-structure formation. The sequences Gly-Pro-Ser and Ala-His-Ala-Glu-Ala, for example, give high joint probabilities for being, respectively, in β-bend and α-helical conformations. However, comparisons of predicted versus directly observed polypeptide conformations give mixed results. This situation is a consequence of two facts: that several amino acids are somewhat ambiguous in their secondary-structure-forming tendencies, and that strong β-bend formers occasionally turn up in the middle of α helices.

Quaternary Structure Involves the Interaction of Two or More Proteins

Although many globular proteins function as monomers, biological systems abound with examples of more complex protein assemblies (table 4.4). The higher order organization of globular subunits to form a functional aggregate is referred to as quaternary structure. Protein quaternary structures can be classified into two fundamentally different types. The first involves the assembly of proteins (sometimes referred to as subunits because they constitute a part of the final structure) that have very different structures. Examples range from the hormone insulin, which has two different subunits, to complex assemblies such as ribosomes, which contain 20 or more nonidentical protein subunits in addition to one or more RNA components. The organization of quaternary structures depends on the specific nature of the interactions between each molecular subunit and its neighbors. Each intermolecular interaction generally occurs only once within a given aggregate arrangement, so that the overall complex structure has a highly irregular geometry.

A second, commonly observed pattern of quaternary structure is for a molecular aggregate to have multiple cop-

Table 4.4

Molecular Weight and Subunit Composition of Selected Proteins

Protein	Molecular Weight	Number of Subunits	Function
Glucagon	3,300	1	Hormone
Insulin	11,466	2	Hormone
Cytochrome c	13,000	1	Electron transport
Ribonuclease A (pancreas)	13,700	1	Enzyme
Lysozyme (egg white)	13,900	1	Enzyme
Myoglobin	16,900	1	Oxygen storage
Chymotrypsin	21,600	1	Enzyme
Carbonic anhydrase	30,000	1	Enzyme
Rhodanese	33,000	1	Enzyme
Peroxidase (horseradish)	40,000	1	Enzyme
Hemoglobin	64,500	4	Oxygen transport
Concanavalin A	102,000	4	Unknown
Hexokinase (yeast)	102,000	2	Enzyme
Lactate dehydrogenase	140,000	4	Enzyme
Bacteriochlorophyll protein	150,000	3	Enzyme
Ceruloplasmin	151,000	8	Copper transport
Glycogen phosphorylase	194,000	2	Enzyme
Pyruvate dehydrogenase (E. coli)	260,000	4	Enzyme
Aspartate carbamoyltransferase	310,000	12	Enzyme
Phosphofructokinase (muscle)	340,000	4	Enzyme
Ferritin	440,000	24	Iron storage
Glutamine synthase (E. coli)	600,000	12	Enzyme
Satellite tobacco necrosis virus	1,300,000	60	Virus coat
Tobacco mosaic virus	40,000,000	2,130	Virus coat

Figure 4.24

Tobacco mosaic virus structure (TMV). (*a*) Diagram of TMV structure, an example of a helical virus. The nucleocapsid (protein shell) is composed of a helical assembly of 2,130 identical protein subunits (protomers) with the RNA of the virus spiraling on the inside. (*b*) An electron micrograph of the negatively stained helical capsid (400,000×). In negative staining the virus is immersed in a pool of a heavy-metal salt that is much more electron-dense than the virus. The result is that the darker portions of the figure that surround the virus appear denser than those parts of the figure where the less dense nucleoprotein is located. (© Dennis Kunkel/Phototake.)

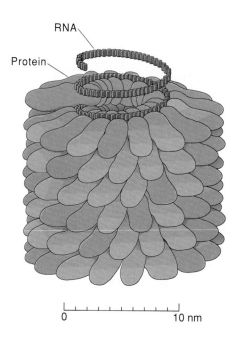

(a)

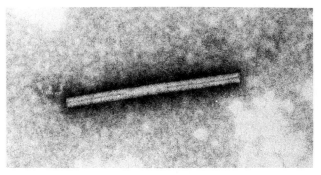

(b)

Figure 4.25

The structure of the capsid (protein shell) for an icosahedral virus such as tomato bushy stunt virus. Pentons (P) are located at the 12 vertices of the icosahedron. Hexons (H), of which there are 20, form the edges and faces of the icosahedron. Each penton is composed of five protein subunits and each hexon is composed of six protein subunits. In all, the structure contains 180 protein subunits.

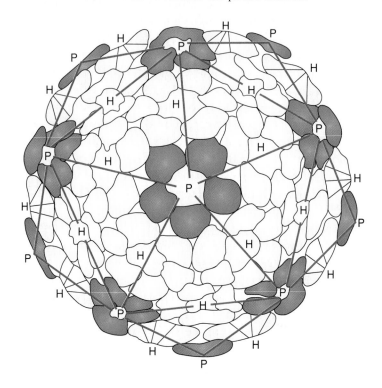

In a typical rodlike virus, the helical aggregation of the protein-coat subunits form a cylindrical container for the virus's nucleic acid (fig. 4.24). The assembly of symmetrical aggregates in a polyhedral virus reflects the structural stabilization that occurs when all the subunits interact in geometrically similar ways, that is, essentially like the atoms in a salt crystal. However, one of the surprising results from x-ray crystallography is that such assemblies often contain distinctly different types of quaternary interactions, even between chemically identical subunits. In the 180-subunit polyhedral protein coat of tomato bushy stunt virus, for example, groups of 5 subunits are in contact around a fivefold symmetry axis, whereas other groups of 6 subunits have a distinct but similar contact around a three-fold axis (fig. 4.25). The versatility that permits such non-equivalent associations allows assembly of larger and more complex structures and may be even more common in biological structures too large to have been examined crystallographically.

ies of the same kind of subunits. Owing to the recurrence of specific structural interactions between the subunits, such aggregates typically form regular geometric arrangements. Some outstanding examples of such aggregates are the rod-like viruses and the polyhedral viruses.

Summary

In this chapter we introduced some of the basic principles that govern protein structure. The discussion of protein structures begun in this chapter is continued in many other chapters in this text in which we consider structures designed for specific purposes. In chapter 5 we examine the protein structures for two systems: the protein that transports oxygen in the blood and the proteins that constitute muscle tissue. In chapters 8 and 9 we discuss structures of specific enzymes. In chapters 17 and 24 we consider proteins that interact with membranes. In chapters 30 and 31 we study regulatory proteins that interact with specific sites on the DNA. And finally, in supplement 3 we examine the structures of immunoglobin molecules.

In this chapter we also introduced the subject of the three-dimensional structures of proteins. This is an important subject to keep in mind throughout the text. Our discussion focused on the following points.

1. Most proteins may be divided into two groups: fibrous and globular. Fibrous proteins usually serve structural roles. Globular proteins function as enzymes and in many other capacities.

2. The three most prominent groups of fibrous proteins are the α-keratins, the β-keratins, and collagen.

3. The α-keratins are composed of right-handed helical polypeptide chains in which all the peptide NH and carbonyl groups form intramolecular hydrogen bonds. When these helical coils interact, they form left-handed coiled coils.

4. The β-keratins consist of extended polypeptide chains in which adjacent polypeptides are oriented in either a parallel or an antiparallel fashion. Sheets formed from such extended polypeptide chains may be stacked on top of one another.

5. Collagen fibrils are composed of extended polypeptide chains that are coiled in a left-handed manner. Three of these chains interact by hydrogen bonding and coil together into a right-handed cable. Collagen fibrils are composed of a staggered array of many such cables interacting in a side-by-side manner.

6. The structures of fibrous proteins are determined by the amino acid sequence, by the principle of forming the maximum number of hydrogen bonds, and by the steric limitations of the polypeptide chain, in which the peptide grouping is in a planar conformation.

7. X-ray diffraction provides data from which we can deduce the dimensions of the polypeptide chains in proteins. The use of x-ray techniques is, however, limited to molecules that can be oriented to achieve two- or three-dimensional order.

8. Fibrous proteins may achieve two-dimensional order, but they usually do not achieve three-dimensional order. Therefore, the diffraction pattern of fibrous proteins gives information about the regularly repeating elements along the long axis of the fibers but tells us very little about the orientation of amino acid side chains.

9. Many globular proteins can be crystallized to achieve three-dimensional order. Study of the crystals of a globular protein can lead to a complete determination of its three-dimensional structure.

10. The forces that hold globular proteins together are the same as those that hold fibrous proteins together, but there is less emphasis on regularity and more emphasis on burying the hydrophobic regions in the interior of the protein.

11. The secondary structures found in the keratins recur in smaller patches in globular proteins. Such regions of secondary structure are folded into a seemingly endless array of tertiary structures.

12. Tertiary structures can be understood in terms of a limited number of domains.

13. Quaternary structures are formed between nonidentical subunits to give irregular macromolecular complexes or between identical subunits to give geometrically regular structures.

Selected Readings

Anfinsen, B. C., Principles that govern the folding of protein chains. *Science* 181:223–230, 1973. Nobel Prize recounting by the man who showed that proteins fold spontaneously into their native structures.

Baron, M., D. G. Norman, and I. D. Campbell, Protein modules. *Trends Biochem. Sci.* 16:13–17, 1991.

Branden, Carl, and John Tooze, *Introduction to Protein Structure.* New York and London: Garland Publishing, 1991.

Cantor, C. R., and P. R. Schimmel, *Biophysical Chemistry,* vols. 1, 2, and 3. New York: Freeman, 1980. Includes several chapters (2, 5, 13, 17, 20, and 21) on the principles of protein folding and conformation.

Chothia, C., Principles that determine the structures of proteins. *Ann. Rev. Biochem.* 53:537–572, 1984.

Chothia, C., and A. V. Finkelstein, The classification and origins of protein folding patterns. *Ann. Rev. Biochem.* 59:1007–1039, 1990.

Chothia, C., and A. Leak, Helix movements in proteins. *Trends Biochem. Sci.* 10:116–118, 1985.

Cohen, C., and D. A. D. Parry, α-Helical coiled coils—a widespread motif in protein. *Trends Biochem. Sci.* 11:245–248, 1986.

Creighton, T. E., *Proteins, Structures and Molecular Principles.* New York: Freeman, 1984. Very readable and reasonably comprehensive.

Dorit, R. L., L. Schoenbach, and W. Gilbert, How big is the universe of exons? *Science* 250:1377–1381, 1990. Predicts that there are between 1,000 and 7,000 different kinds of domains in all proteins found in nature.

Farber, G. K., and G. A. Petsko, The evolution of α/β barrel enzymes. *Trends Biochem. Sci.* 15:228–234, 1990.

Fasman, G. D., Protein conformation prediction. *Trends Biochem. Sci.* 14:295–299, 1989.

Fersht, A. R., The hydrogen bond in molecular recognition. *Trends Biochem. Sci.* 12:301–304, 1987.

Hogle, J. M., M. Chow, and D. J. Filman, The structure of polio virus. *Sci. Am.* 256(3):42–49, 1987.

Karplus, M., and J. A. McCannon, The dynamics of proteins. *Sci. Am.* 254(4):42–51, 1986. A reminder that proteins are not rigid inflexible structures.

Pauling, L., *The Nature of the Chemical Bond,* 3d ed. Ithaca: Cornell University Press, 1960. A classic on molecular structure.

Pauling, L., and R. B. Corey, Configurations of polypeptide chains with favored orientations around single bonds: two new pleated sheets. *Proc. Natl. Acad. Sci. USA* 37:729–740, 1953. Classic paper.

Pauling, L., R. B. Corey, and H. R. Branson, The structure of proteins: two hydrogen-bonded helical configurations of the polypeptide chain. *Proc. Natl. Acad. Sci. USA* 27:205–211, 1951. Another classic paper.

Richardson, J. S., and D. C. Richardson, The *de novo* design of protein structures. *Trends Biochem. Sci.* 14:304–309, 1989.

Rose, C. D., A. R. Geselowizt, G. J. Lesser, R. H. Lee, and M. H. Zehfus, Hydrophobicity of amino acid residues in globular proteins. *Science* 229:834–838, 1985.

Rossman, M. G., and P. Argos, Protein folding. *Ann. Rev. Biochem.* 50:497–532, 1981.

Rossman, M. G., and J. E. Johnson, Icosahedral RNA virus structure. *Ann. Rev. Biochem.* 58:533–573, 1989.

Sali, A., J. P. Overington, M. S. Johnson, and T. L. Bundell, From comparisons of protein sequences and structures to protein modelling and design. *Trends Biochem. Sci.* 15:235–240, 1990.

Tonegawa, S., The molecules of the immune system. *Sci. Am.* 253(4):122–130, 1985.

Valegard, K., L. Liljas, K. Fridborg, and T. Unge, The three-dimensional structure of the bacterial virus MS2. *Nature* 345:36–41, 1990.

Wuthrich, K., Protein structure determination in solution by nuclear magnetic resonance spectroscopy. *Science* 243:45–50, 1989. The most effective technique for determining protein fine structure in cases where x-ray diffraction cannot be used.

Wright, P. E., What can two-dimensional NMR tell us about proteins? *Trends Biochem. Sci.* 14:255–259, 1989.

Yang, J. T., Protein secondary structure and circular dichroism: a practical guide. *Chemtracts, Biochem. Mol. Biol.* 1:484–490, 1990.

Problems

1. The principal force driving the folding of some proteins is the movement of hydrophobic amino acid side chains out of an aqueous environment. Explain.

2. Outline the hierarchy of structural organization in proteins.

3. What is the role of loops or short segments of "random" structure in a protein whose structure is primarily α-helical?

4. What are some consequences of changing a hydrophilic residue to a hydrophobic residue on the surface of a globular protein? What are the consequences of changing an interior hydrophobic residue to a hydrophilic residue in the protein?

5. Some proteins are anchored to membranes by insertion of a segment of the N terminus into the hydrophobic interior of the membrane. Predict (guess) the probable structure of the sequence (Met-Ala-(Leu-Phe-Ala)$_3$-(Leu-Met-Phe)$_3$-Pro-Asn-Gly-Met-Leu-Phe). Why would this sequence be likely to insert into a membrane?

6. Suppose that every other Leu residue in the peptide shown in problem 5 was changed to Asp. Would this necessarily alter the secondary structure? Explain whether insertion into the membrane would be altered.

7. If you had several helical springs, how could you determine whether each spring was right- or left-handed?

8. "Left- and right-handed α helices of polyglycine are equally stable." Defend or refute this statement.

9. Urea

$$\underset{H_2NCNH_2}{\overset{\displaystyle O}{\overset{\displaystyle \|}{}}}$$

and guanadinium chloride ($[(H_2N)_2C{=}NH_2]^+Cl^-$) are commonly used denaturants (cause loss of protein conformation). Provide explanations for how these substances might disrupt protein structures.

10. Many proteins (e.g., important metabolic enzymes) are insoluble in water and are found "attached" to membranes within cells. What amino acid residues do you expect to find on the "side" of the protein that "attaches" to the membrane?

11. Often the enzymes mentioned in problem 10 are purified with the aid of detergents such as sodium dodecylsulfate

$$CH_3(CH_2)_{11}{-}O{-}\underset{\displaystyle O}{\overset{\displaystyle O}{\overset{\displaystyle \|}{\underset{\displaystyle \|}{S}}}}{-}O^-{\ }^+Na$$

What is the function of the detergent?

12. It might be argued that in protein structure, as in everyday life, it is a "right-handed world." Use examples of protein structure discussed in this chapter to support this contention.

X-Ray Diffraction: Applications to Fibrous Proteins

X-ray diffraction played a major role in the discovery of the structure of fibrous proteins. In most cases the fibers under study are oriented in two dimensions by stretching. In this analysis we illustrate how the technique is used to study the α form of the synthetic polypeptide poly-L-alanine.

A stretched fiber containing many poly-L-alanine molecules is suspended vertically and exposed to a collimated monochromatic beam of CuK_α x-rays, as shown in figure 1(a). Only a small percentage of the x-ray beam is diffracted; most of the beam travels through the specimen with no change in direction. A photographic film is held in back of the specimen. A hole in the center of the film allows the incident undiffracted beam to pass through.

Coherent diffraction occurs only in certain directions specified by Bragg's law: $2d \sin \theta = n\lambda$, where d is the distance between identical repeating structural elements, θ is the angle between the incident beam and the regularly spaced diffracting planes, λ is the wavelength of x-rays used, and n is the order of diffraction, which may equal any integer but is usually strongest for $n = 1$. For small θ, $\sin \theta \approx \theta$ and $d \approx 1/\theta$, so that a spot far out on the photographic film is indicative of a repeating element of small dimension.

Figure 1(b) shows the diffraction pattern obtained when the fiber axis is normal to the beam. Note the strong off-vertical reflection at 5.4 Å (arrow). A different diffraction pattern (c) is obtained when the fiber axis is inclined to the beam at 31°. Note the strong reflection at 1.5 Å in the upper part of the diagram (arrow).

Figure 1

(a) Experimental arrangement for obtaining x-ray diffraction pattern shown in (b). (b) Diffraction pattern of the α form of a cluster of poly-L-alanine molecules oriented vertically. (c) Diffraction pattern of the same fiber bundle with the fiber axis inclined to the beam at 31°. (b and c Brown & Trotter, 1956. Source: C. H. Bamford et al., *Synthetic Polypeptides*. Copyright © 1956, Academic Press, Orlando, Fla.)

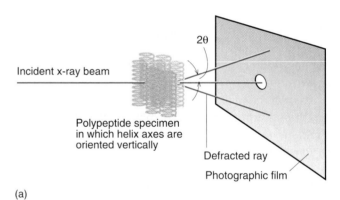

Incident x-ray beam

2θ

Polypeptide specimen in which helix axes are oriented vertically

Defracted ray

Photographic film

(a)

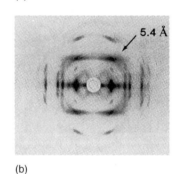

5.4 Å

(b)

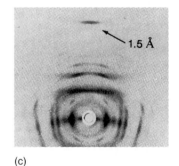

1.5 Å

(c)

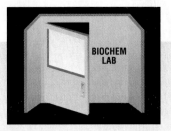

Three-Dimensional Protein Structure Can Be Analyzed by X-Ray Diffraction of Protein Crystals

The quality of the information potentially available through x-ray diffraction depends on the degree or extent of order of the protein molecules in a given sample. A sample under investigation usually consists of a hydrated purified protein. If the individual protein molecules are packed in a random order relative to one another, the distances between atoms or groups of atoms can be derived, but we cannot tell how these spacings are ordered in three dimensions.

If the protein is fibrous, it is often possible to learn its two-dimensional order. For example, when a fibrous sample consisting of long α helices is stretched, the helix axes become oriented in the direction of stretching. The resulting fibrous bundle gives a characteristic x-ray diffraction pattern indicating an ordered molecular arrangement along the helix axis. As we have seen, the pitch of the α helix (5.4 Å) and the advance per residue along the helix axis (1.5 Å) were detected by this technique. But the orientation of specific atoms in the helix cannot be determined from diffraction patterns of stretched fibers because of the lack of three-dimensional order. To deduce the correct three-dimensional structure for the α helix (and the β sheet), it was necessary to work with molecular models. (Currently, computer programs are available for such purposes.) By trial and error, model structures were built that were consistent with the steric limitations of the polypeptide chain that had the helix pitch and the advance per residue indicated by the diffraction pattern of stretched fibers.

The most information about a protein's structure is obtained from ordered three-dimensional protein crystals; this is the main interest of x-ray crystallographers. The goal in x-ray crystallography is to obtain a three-dimensional image of a protein molecule in its native state at a sufficient level of detail to locate its individual constituent atoms. The way this is done can most easily be appreciated by considering the more familiar problem of how we obtain a magnified image of an object in a conventional light microscope. In a light microscope, light from a point source is projected on the object we wish to examine. When the light waves hit the object, they are scattered so that each small part of the ob-

Figure 1

Schematic diagram of the procedures followed for image reconstruction in light microscopy (top) and x-ray crystallography (bottom).

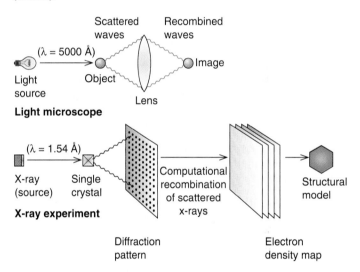

ject essentially serves as a new source of light waves. The important point is that the light waves scattered from the object contain information about its structure. The scattered waves are collected and recombined by a lens to produce a magnified image of the object (fig. 1).

Given this picture, we might ask what prevents us from simply putting a protein molecule in place of our object and viewing its magnified image. The basic problem here is one of resolution. The resolution, or extent of detail, that can be recovered from any imaging system depends on the wavelength of light incident on the object. Specifically, the best resolution obtainable equals $\lambda/2$, or one-half the wavelength of the incident light. Because λ lies in the range of 4,000–7,000 Å for visible light, a visible-light microscope clearly does not have the resolving power to distinguish the atomic structural detail of molecules. What we need is a form of incident radiation with a wavelength comparable to interatomic distances. X-rays emitted from excited metal atoms, with wavelengths in the range of one to a few angstroms, would be most suitable.

However, simply replacing a visible-light source with an x-ray source does not solve all the problems. For example to get a three-dimensional view of a protein, some provision must be made for looking at it from all possible angles, an obvious impossibility when dealing with a single molecule. Furthermore, when x-rays interact with proteins, very few of the rays are scattered. Most x-rays pass through the protein, but a relatively large number of them interact destructively with the protein, so that a single molecule would be destroyed before scattering enough x-rays to form a useful image. Both these problems are overcome by replacing a single protein molecule with an ordered three-dimensional array of many molecules that scatters x-rays essentially as if it were one molecule. The ordered array of protein molecules forms a single crystal, so the general technique is called protein x-ray crystallography.

The problems do not end here, because although the protein crystal readily scatters incident x-rays, no lens materials are available that can recombine the scattered x-rays to produce an image. Instead, the best that can be done is to directly collect the scattered x-rays in the form of a diffraction pattern. Although recording the diffraction pattern results in loss of some important information, experimental techniques have been developed for recovering the lost information. Eventually the scattered waves can be mathematically recombined in a computational analog of a lens. By collecting the diffraction pattern of the crystal in many orientations, it is possible to construct a three-dimensional image of the protein molecule.

Crystals suitable for protein x-ray studies may be grown by a variety of techniques, which generally depend on solvent perturbation methods for rendering proteins insoluble in a structurally intact state. The trick is to induce the molecules to associate with each other in a specific fashion to produce a three-dimensionally ordered array. A typical protein crystal useful for diffraction work is about 0.5 mm on a side and contains about 10^{12} protein molecules (an array 10^4 molecules long along each crystal edge). Note especially that, because protein crystals are from 20 to 70% solvent by volume, crystalline protein is in an environment that is not substantially different from free solution.

The x-ray radiation usually employed for protein crystallographic studies is derived from the bombardment of a copper target with high-voltage (50 kV) electrons, producing characteristic copper x-rays with $\lambda = 1.54$ Å. Figure 2 shows, in schematic fashion, the x-ray diffraction pattern from a protein crystal. Several features about this pattern bear explanation. First, as you can see, the diffraction pattern consists of a regular lattice of spots of different intensities. The spots are due to destructive interference of waves

Figure 2

Schematic view of an x-ray diffraction pattern. The spacing of the spots is reciprocally related to the dimensions of the repeating unit cell of the crystal. The symmetry of the spots (e.g., the mirror planes in the sample shown) and the pattern of missing spots (alternating spots along the mirror axes) give information on how molecules are arranged in the unit cell. Information concerning the structure of the molecule is contained in the intensities of the spots. Spots closest to the center of the film arise from large-scale or low-resolution structural features of the molecule, whereas those farther out correspond to progressively more detailed features. Circles show 5-Å and 3-Å regions of resolution. Mirror axes are labeled m. Spacing of vertically oriented spots b^* and horizontally oriented spots a^* are reciprocally related to b and a, the dimensions of the unit cell.

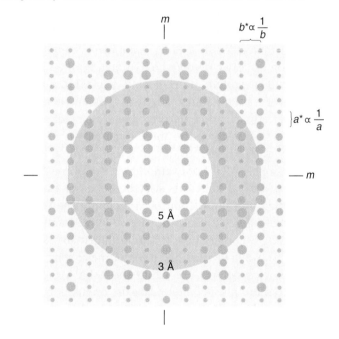

scattered from the repeating unit of the crystal. For the crystal with the diffraction pattern shown, the repeating unit (or crystal unit cell) contains four symmetrically arranged protein molecules. Corresponding symmetrical features appear in the spot intensity pattern. Further, the lattice spacing of the diffraction spots is inversely proportional to the actual dimensions of the crystal's repeating unit or unit cell. Consequently, both the crystal's unit-cell dimensions and general molecule packing arrangement can be derived from inspection of the crystal's diffraction pattern.

Information concerning the detailed structural features of the protein is contained in the intensities of the diffraction spots. All the atoms in the protein structure make individual contributions to the intensity of each diffraction spot. Therefore, to deduce the three-dimensional structure, all the spots must be measured, either by scanning the x-ray films with a densitometer or by measuring the diffraction spots individually with a scintillation counter.

Initial studies of a protein's tertiary structure are generally carried out at low resolution, that is, using intensity data near the origin (center) of the diffraction pattern. Diffraction data near the origin reflect large-scale structural features of the molecule, whereas those nearer the edge correspond to progressively more detailed features. Figure 3 provides examples of electron-density maps calculated at different resolutions to show how various levels of structural detail appear at different degrees of resolution.

A powerful aspect of protein crystallography is that once the native structure is known, various cofactors or enzyme substrate analogs can be bound to the molecule in the crystal. By simply measuring the diffraction intensities, we can compute a new map that allows direct and explicit examination of the structural interactions between the native protein and its substrate or cofactor molecules. Detailed analysis of these interactions has provided much of the foundation for our current understanding of many protein catalytic and functional properties.

Figure 3

Views of crystallographic electron-density maps, showing how the structural detail revealed depends on the resolution of the data used to compute the maps. The actual molecular structure is inserted in its true position in the electron-density maps.

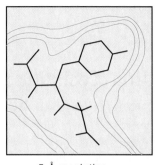

5-Å resolution

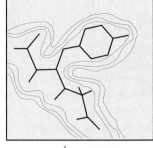

3-Å resolution

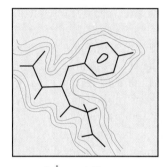

2-Å resolution

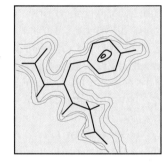

1.5-Å resolution

Radiation Techniques for Examining Protein Structure

The Anfinsen experiment raises the question: Must we always use x-ray diffraction to determine a protein's structure? In the case of ribonuclease the structure is known from x-ray diffraction, but at the time of Anfinsen's investigation it was not. He relied heavily on the fact that the reconstituted enzyme had the same activity as the native enzyme. This remains an acceptable approach as long as one is working with an enzyme that has a demonstrable catalytic activity, but what of the many proteins that do not?

Because of such limitations, x-ray diffraction is still the most powerful tool for determining protein structure. However, apart from the enormous amount of work it requires to use, x-ray diffraction suffers from two disadvantages: (1) it requires that a protein be available in the crystalline form, which is not always the case; and (2) there is no absolute assurance that a protein's structure in the crystalline form is the same as its structure in solution, which may be more like the environment in the cell.

Fortunately, there are at least a dozen other techniques, less fussy in their demands, that may be used to investigate protein structure either in the solid state or the solution state (table 4C). The information obtained by these procedures is very extensive. As you can see from the table, it covers virtually every aspect of protein structure, with considerable overlap between what is yielded by different techniques. Excellent references explaining the use of other radiation techniques are given at the end of this chapter.

Table 4C

Radiation Techniques for Examining Protein Structure

Technique	Information Obtained
(a) X-ray diffraction	Detailed atomic structure
(b) Nuclear magnetic resonance spectrometry (NMR)	Structure of specific sites; ionization state of individual residues
(c) Electron paramagnetic resonance (EPR)	Structure of specific sites; this includes structures of small molecules whether they be carbohydrates, lipids, small proteins or complexes between segments of DNA and proteins
(d) Spectrometry	
(e) Optical absorption spectroscopy	Measurement of concentrations or rate of reactions
(f) Infrared absorption spectroscopy	Type and extent of secondary structure
(g) Light scattering	Molecular weight and size
(h) Ultraviolet absorption spectroscopy	Concentration and conformation
(i) Fluorescence	Proximity between specific sites
(j) Raman scattering	Structure of specific sites
(k) Optical rotatory dispersion (ORD)	Type and extent of secondary structure
(l) Circular dichroism (CD)	Type and extent of secondary structure
(m) Fluorescence polarization	Molecular weight, shape, flexibility, and orientation of secondary-structure units

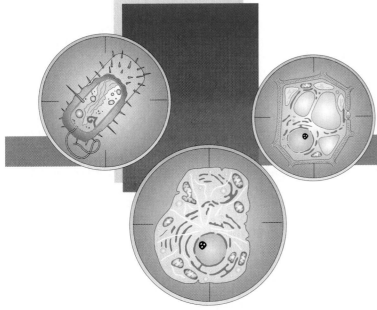

Functional Diversity of Proteins

Chapter Outline

Hemoglobin Is an Allosteric Oxygen-Binding Protein
The Binding of Certain Factors to Hemoglobin Influences Oxygen Binding in a Negative Way
X-Ray Diffraction Studies Reveal Two Conformations for Hemoglobin
Changes in Conformation Are Initiated by Oxygen Binding
Alternative Theories on How Hemoglobins and Other Allosteric Proteins Work
Muscle Is an Aggregate of Proteins Involved in Contraction

Each protein is exquisitely suited to carry out a specific function.

Now that some of the major types of protein structures have been described it is appropriate to turn to the question of how protein structure relates to protein function. To explore this question, two protein systems, hemoglobin and the actin-myosin complex, are examined in detail.

Hemoglobin Is an Allosteric Oxygen-Binding Protein

Hemoglobin is the principal soluble protein in the red blood cells (erythrocytes). It is an oxygen-binding protein, the main task of which is to pick up oxygen as the blood passes through the lungs and to release it in the capillaries of other tissues. A closely related protein, myoglobin, serves to store oxygen temporarily in the muscles and other peripheral tissues, where the oxygen is consumed by cellular metabolism. Myoglobin consists of a single polypeptide subunit with a molecular weight of 16,900 and a bound cofactor, heme. Hemoglobin has an $\alpha_2\beta_2$ subunit structure, with a heme bound to each of the four subunits. The α and β subunits are structurally very similar to each other and to the single subunit of myoglobin. (The α subunits have 141 amino acids; the β subunits, 146.) The heme

Figure 5.1

The heme pocket. The helices of hemoglobin (and myoglobin) form a hydrophobic pocket for the heme and provide an environment where the iron atom reversibly binds oxygen. The chemical structure of heme is shown in figure 5.10 and is described in atomic detail in chapters 10 and 14. (Illustration copyright by Irving Geis. Reprinted by permission.)

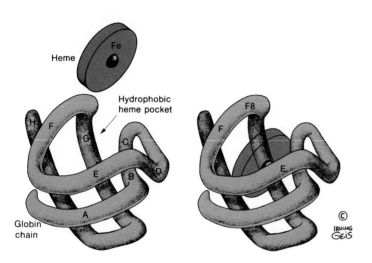

cofactor in each subunit consists of a porphyrin (protoporphyrin IX) with a central iron atom. (The structure of heme is shown in fig. 5.1 and is discussed in more detail in chapters 10 and 14.)

Figure 5.1 shows the overall folding plan of the polypeptide chains of hemoglobin and myoglobin. Each chain has eight stretches of α helix, which are labeled A–H. These helical segments fold together to form a pocket that surrounds the heme. The amino acid side chains lining the heme-binding pocket are predominantly apolar, and thus are well suited to binding the hydrophobic porphyrin ring. This apolar, water-free environment allows the Fe^{2+} iron atom to bind O_2 reversibly while minimizing the opportunity for oxidation of the iron to the +3 state.

Although the components of myoglobin and hemoglobin are remarkably similar, their physiological responses are very different. On the basis of weight, each molecule binds about the same amount of oxygen at high oxygen tensions (pressures). At low oxygen tensions, however, hemoglobin releases bound oxygen much more readily. These differences are reflected in the oxygen-binding curves of the purified proteins in aqueous solution (fig. 5.2). The oxygen-binding curve for myoglobin (Mb) is hyperbolic in shape, as is expected for a simple one-to-one association of myoglobin and oxygen:

$$Mb + O_2 \rightleftharpoons MbO_2$$

$$K_f = \frac{[MbO_2]}{[Mb][O_2]} = \text{equilibrium formation constant}$$

Figure 5.2

Equilibrium curves measure the affinity for oxygen of hemoglobin and of the simpler myoglobin molecule. Myoglobin, a protein of muscle, has just one polypeptide chain and resembles a single subunit of hemoglobin. The vertical axis gives the amount of oxygen bound to one of these proteins, expressed as a percentage of the total amount that can be bound. The horizontal axis measures the partial pressure of oxygen in a mixture of gases with which the solution is allowed to reach equilibrium. For myoglobin, the equilibrium curve is hyperbolic. Myoglobin absorbs oxygen readily but becomes saturated at a low pressure. The hemoglobin curve is sigmoidal. Initially, hemoglobin is reluctant to take up oxygen, but its affinity increases with oxygen uptake. At arterial oxygen pressure, both molecules are nearly saturated, but at venous pressure, myoglobin gives up only about 10% of its oxygen, whereas hemoglobin releases roughly half. At any partial pressure, myoglobin has a higher affinity than hemoglobin, which allows oxygen to be transferred from blood to muscle.

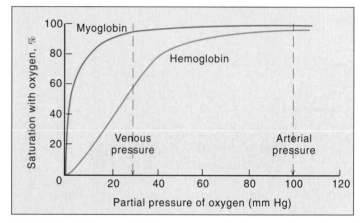

If y is the fraction of myoglobin molecules saturated, and if the oxygen concentration is expressed in terms of the partial pressure of oxygen $[O_2]$,[a] then

$$K_f = \frac{y}{[1-y][O_2]} \qquad \text{and} \qquad y = \frac{K_f[O_2]}{1 + K_f[O_2]} \quad \textbf{(1)}$$

This is the equation of a hyperbola, as shown in figure 5.2.

Hemoglobin (Hb) behaves differently. Its sigmoidal binding curve can be fitted by an association-constant expression with a greater-than-first-power dependence on the oxygen concentration:

$$K_f = \frac{[HbO_2]}{[Hb][O_2]^n} \qquad \text{and} \qquad y = \frac{K_f O_2{}^n}{1 + K_f O_2{}^n} \quad \textbf{(2)}$$

Under physiological conditions the value of n is around 2.8. This indicates that the binding of oxygen molecules to the

---

[a] Partial pressure is usually indicated by a lower-case p to the left. For simplicity the p has been deleted.

four hemes in hemoglobin does not occur independently: Binding to any one heme is affected by the state of the other three. The first oxygen attaches itself with the lowest affinity, and successive oxygens are bound with a somewhat higher affinity, and so on. The exact value of n in equation 2 is a function of other factors discussed later on. In general, a value of $n > 1$ indicates cooperative binding of O_2 (or positive cooperativity), a value of $n < 1$ indicates anticooperative binding (or negative cooperativity), and a value of $n = 1$ indicates no cooperativity.

The cooperative binding of oxygen by hemoglobin is ideally suited to the conditions involved in oxygen transport. In the lungs, where the oxygen tension is relatively high, hemoglobin can become nearly saturated with oxygen. In the tissues, however, where the oxygen tension is relatively low, hemoglobin can release about half its oxygen (see fig. 5.2). If myoglobin were used as the oxygen transporter, less than 10% of the oxygen would be released under similar conditions. The positive cooperativity associated with oxygen binding to hemoglobin is a special case of allostery in which the binding of "substrate" to one site stimulates or inhibits the binding of "substrate" to another site on the same multisubunit protein. Allostery is also commonly observed in regulatory proteins (chapter 9).

The Binding of Certain Factors to Hemoglobin Influences Oxygen Binding in a Negative Way

The combination of Hb with O_2 is not only a function of the oxygen tension but also depends on the pH, CO_2 concentration, and glycerate-2,3-bisphosphate (GBP or BPG). GBP (fig. 5.3) binds preferentially to the deoxygenated form of hemoglobin with a dissociation constant of about $10^{-5} M^{-1}$. Its dissociation constant with HbO_2 is about $10^{-3} M^{-1}$. Since the concentration of GBP and hemoglobin are both about 5 mM in the erythrocyte, we expect most of the deoxy form to be complexed with GBP and much of the oxyhemoglobin to be free of GBP. The net effect of the GBP is to shift the oxygen-binding curve to higher oxygen tensions (fig. 5.4). This shift is not sufficient to lower the binding of oxygen at the high oxygen tensions in the capillary tissues of the lungs, but it is sufficient to cause a substantially greater release of oxygen at the lower oxygen tensions that exist in the peripheral tissues.

Carbon dioxide (CO_2) and hydrogen ions (H^+) both have negative effects on oxygen binding also. These effects are closely related and together are known as the Bohr effect. A substantial amount of CO_2 generated by decarboxylation reactions of intermediary metabolism diffuses from cells through interstitial fluid into the blood plasma, with

Figure 5.3

The structure of glycerate-2,3-bisphosphate, an allosteric effector for hemoglobin oxygen release.

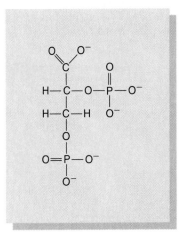

Figure 5.4

Oxygen-binding curve for hemoglobin as a function of the partial pressure of oxygen. Two curves are shown: one in the absence and one in the presence of glycerate-2,3-bisphosphate (GBP). GBP decreases the affinity between oxygen and hemoglobin, as shown by the displacement of the binding curve to high oxygen concentrations in its presence.

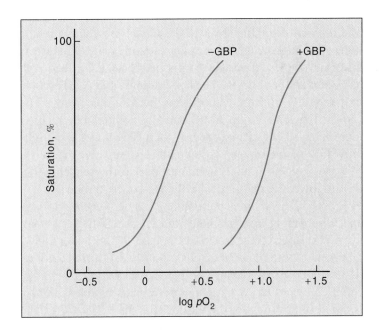

only a small fraction becoming hydrated as carbonic acid. This is because the nonenzymatic hydration of CO_2 to form H_2CO_3 is a slow reaction, with a half-time of about 10 s. On

Figure 5.5

Transport of oxygen (O_2) and carbon dioxide (CO_2) in the circulatory system. In most tissues O_2 is released and CO_2 is withdrawn by the red blood cells; in the lungs these processes are reversed.

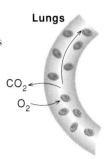

Lungs

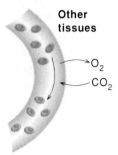

Other tissues

O_2

CO_2

CO_2

O_2

Lungs

$$HCO_3^- + H^+ \xrightarrow{\hspace{1cm}} H_2CO_3 \underset{\text{anhydrase}}{\overset{\text{Carbonic}}{\rightleftharpoons}} CO_2 + H_2O$$

$$O_2 + HHb^+ \rightleftharpoons H^+ + HbO_2$$

$$HHb^+ + O_2 + HCO_3^- \rightleftharpoons HbO_2 + CO_2 + H_2O$$

Other tissues

$$H^+ + HbO_2 \rightleftharpoons HHb^+ + O_2$$

$$CO_2 + H_2O \underset{\text{anhydrase}}{\overset{\text{Carbonic}}{\rightleftharpoons}} H_2CO_3 \rightleftharpoons H^+ + HCO_3^-$$

$$HbO_2 + CO_2 + H_2O \rightarrow HHb^+ + O_2 + HCO_3^-$$

entry into erythrocytes, hydration of CO_2 occurs in a reaction that is catalyzed rapidly by carbonic anhydrase, an enzyme that is localized in the erythrocytes. At the pH of blood (about 7.4) the carbonic acid largely dissociates into H^+ and HCO_3^-:

$$CO_2 + H_2O \underset{\text{anhydrase}}{\overset{\text{Carbonic}}{\rightleftharpoons}} H_2CO_3 \rightleftharpoons H^+ + HCO_3^-$$

The release of protons through these two consecutive reactions explains how increasing concentrations of CO_2 cause a lowering of pH. Lowering the pH decreases the affinity of hemoglobin for oxygen, because protons, like GBP, bind preferentially to deoxyhemoglobin. For example, at pH 7.6 and an oxygen tension of 40 mm Hg, hemoglobin retains more than 80% of its oxygen; if the pH is lowered to 6.8, only 45% of the oxygen is retained. The negative effect of CO_2 on oxygen binding is mainly due to the tendency of CO_2 to lower the pH (i.e., raise the H^+ concentration); however, HCO_3^- also binds preferentially to deoxyhemoglobin, and this binding also promotes the release of O_2.

The Bohr effect is closely related to the major roles that hemoglobin plays in disposing of the CO_2 produced in tissues, and in controlling the blood pH. While oxygen is being delivered to the tissues in the venous blood, the CO_2 is being removed from the tissues (fig. 5.5). The CO_2 that diffuses into the erythrocytes is rapidly converted into carbonic acid, which in turn dissociates into H^+ and HCO_3^-. The protons produced by this dissociation would lower the pH and reverse the dissociation if it were not for the buffer-

ing action of the hemoglobin. Loss of oxygen decreases the acid dissociation constant of the hemoglobin so that it picks up the excess protons. This serves two purposes: it helps to keep the pH constant and it enables the blood to absorb more CO_2. This CO_2 is ultimately released when the blood reaches the lungs and the hemoglobin again becomes oxygenated. Another important point is that the majority of the HCO_3^- produced in the erythrocyte diffuses into the venous blood. Indeed, the pH of the blood system is controlled within narrow limits by the buffering action of the bicarbonate and the hemoglobin with minor assistance from other proteins in the blood.

One must marvel at the way various factors work in concert so that hemoglobin can be useful in multiple roles: oxygen deliverer, carbon dioxide remover, and pH stabilizer. From the explanation we have given it should be clear why the hemoglobin is confined to cells in the blood rather than being present as a free plasma protein. The intracellular carbonic anhydrase and GBP of the erythrocytes are essential for efficient hemoglobin function.

X-Ray Diffraction Studies Reveal Two Conformations for Hemoglobin

X-ray diffraction studies on fully oxygenated hemoglobin and deoxygenated hemoglobin have shown that the molecule is capable of existing in at least two states, with significant differences in tertiary and quaternary structures (fig. 5.6). Further studies on partially oxygenated hemoglobin may indicate additional intermediate structures between

Figure 5.6

Three-dimensional structure of oxy- and deoxyhemoglobin as determined by x-ray crystallography. This is a view down the twofold symmetry axis, with the β chains on top. In the oxy–deoxy transformation (quaternary motion) $\alpha_1\beta_1$ and $\alpha_2\beta_2$ dimers move as units relative to each other. This allows glycerate-2,3-bisphosphate to bind to the larger central cavity in the deoxy conformation. A close-up of the binding site is shown in figure 5.9. (Illustration copyright by Irving Geis. Reprinted by permission.)

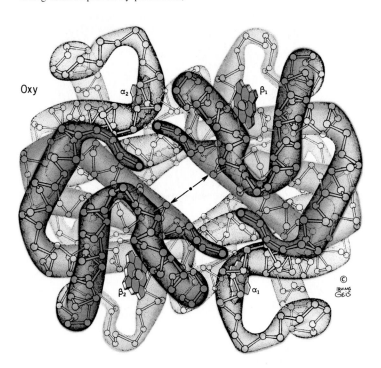

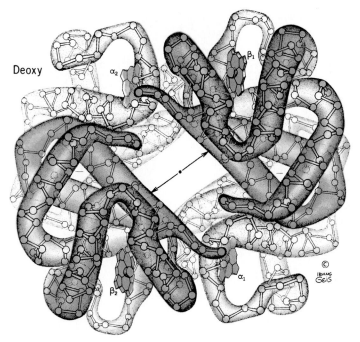

these two extremes. Presently the two-state model serves as a useful conceptual framework for explaining the cooperative nature of oxygen binding.

The hemoglobin tetramer is composed of two identical halves (dimers), with the $\alpha_1\beta_1$ subunits in one dimer and the $\alpha_2\beta_2$ subunits in the other. The subunits within the dimers are tightly held together while the dimers themselves are capable of motion with respect to each other (fig. 5.7). The interface between the movable dimers contains a network of salt bridges and hydrogen bonds when hemoglobin is in the deoxy conformation (fig. 5.8). The quaternary transformation that takes place on binding of oxygen causes the breakage of these bonds.

The effects of H^+, HCO_3^-, and glycerate-2,3-bisphosphate on oxygen binding can be understood in terms of their stabilizing effect on the deoxy conformation. The decreased oxygen binding as the pH is lowered from 7.6 to 6.8 suggests the involvement of histidine side chains, because these

are the only side-chain groups in proteins that commonly have a pK_a in this range. Certain histidines in the charged form make salt linkages that contribute to the stability of the deoxy form (see fig. 5.8). As the pH is lowered, these histidines tend to become charged, which increases the stability of the deoxy form. Such a change should inhibit a structural transition to the oxy form and thereby lower the affinity of the protein for oxygen. Similarly, glycerate-2,3-bisphosphate binds most strongly to the deoxy form (fig. 5.9) and thereby discourages the transition to the oxy form, which lowers the affinity for oxygen. Bicarbonate ions bind to the amino terminal α-amino groups in hemoglobin; this binding also favors the deoxy conformation. Binding of HCO_3^- is favored by the high HCO_3^- concentration created by the carbonic anhydrase reaction in the red cells. Hemoglobin thus serves as a carrier of HCO_3^- from tissues to the lungs, where CO_2 is discharged.

Figure 5.7

The deoxy–oxy shift on binding oxygen in one hemoglobin molecule. The projection shown in this figure is approximately perpendicular to the one shown in figure 5.6. The $\alpha_1\beta_1$ dimer moves as a unit relative to the $\alpha_2\beta_2$ dimer. The interface between the two dimers is crucial to the cooperativity effect in hemoglobin. The interface is not visible in this figure (see fig. 5.8). (Illustration copyright by Irving Geis. Reprinted by permission.)

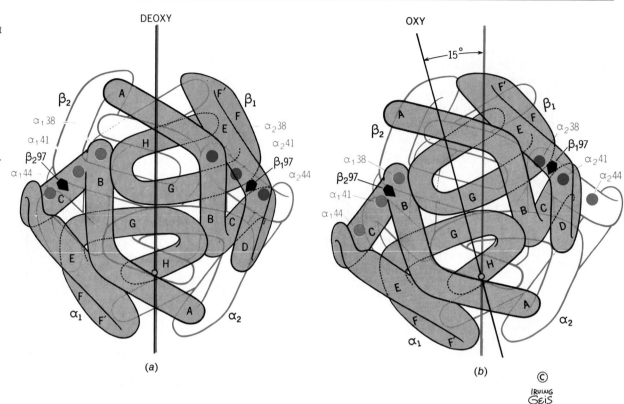

(a)

(b)

Figure 5.8

(a) The $\alpha_1\beta_2$ (and $\alpha_2\beta_1$) interface is shown schematically at the lower left and in detail (b). This is the regulatory zone of the hemoglobin molecule, which contains crucial hydrogen bonds and salt bridges. (b) All the hydrogen bonds and salt bridges shown here (dotted lines) exist only in the deoxy state, with the exception of $\alpha_1 41$–$\beta_2 40$ and $\alpha_1 94$–$\beta_2 102$, which exist only in the oxy state. Only the α carbons are shown in the backbone structure of the hemoglobin, except in the region of the interface. (Illustration for part b copyright by Irving Geis. Reprinted by permission.)

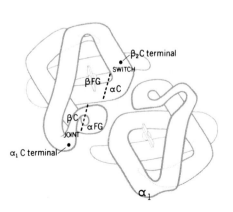

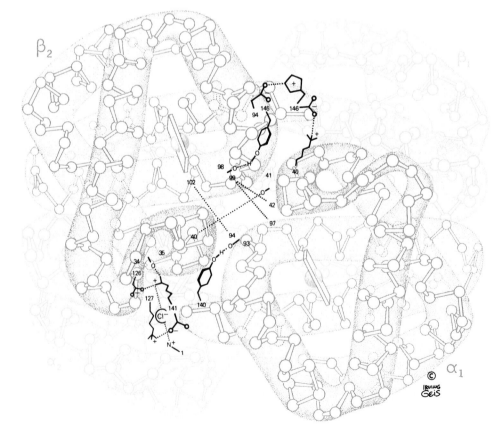

Figure 5.9

The binding of glycerate-2,3-bisphosphate in the central cavity of deoxyhemoglobin between β chains. The surrounding positively charged residues are the amino terminal, His 2, Lys 82, and His 143. (Illustration copyright by Irving Geis. Reprinted by permission.)

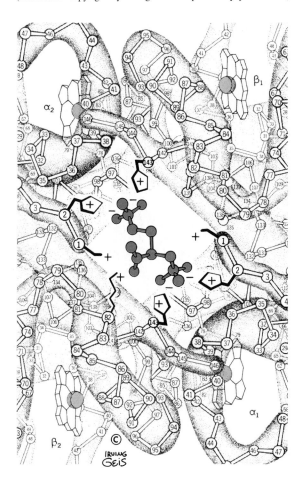

Figure 5.10

A close-up view of the iron–porphyrin complex with the F helix in deoxyhemoglobin. Note that the iron atom is displaced slightly above the plane of the porphyrin. (Illustration copyright by Irving Geis. Reprinted by permission.)

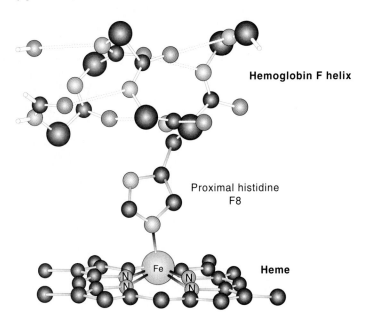

Figure 5.11

Downward movement of the iron atom and the complexed polypeptide chain on binding oxygen. The structure is shown before (black) and after (blue) binding oxygen. Movement of His F8 is transmitted to valine FG5, straining and breaking the hydrogen bond to the penultimate tyrosine. Only the α chain is shown here.

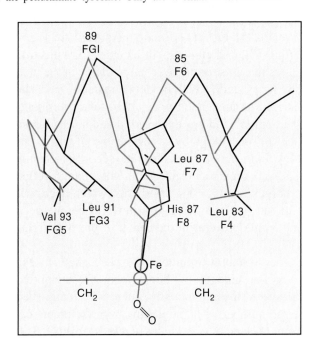

Changes in Conformation Are Initiated by Oxygen Binding

The oxygen binding at the heme group itself initiates the tertiary- and quaternary-structure changes that are responsible for the cooperative effects seen in hemoglobin. The heme group contains an Fe^{2+} ion located near the center of a porphyrin. This Fe^{2+} makes four single bonds to the nitrogens in the porphyrin ring, and a fifth bond to a histidine side chain of the F helix, histidine F8 (fig. 5.10). When oxygen is present it binds at the sixth coordination position of the iron, on the other side of the heme. Movement of the iron on oxygen binding (or release) pulls the F8 histidine and the F helix to which it is covalently attached (fig. 5.11). The tertiary-structure change in the F helix of one subunit causes strains elsewhere in this subunit and in the other

Figure 5.12

Invariant residues in the α and β chains of mammalian hemoglobin. The blue dots, indicating the positions of invariant residues, line the heme pockets as well as the crucial $\alpha_1\beta_2$ interface. The invariant residues have been found in about 60 species. There are 43 invariant positions in the hemoglobin molecule. (Illustration copyright by Irving Geis. Reprinted by permission.)

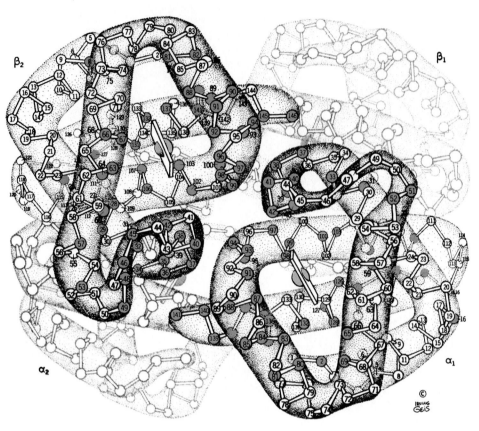

three subunits as well. The resulting structural reorganization favors the binding of additional oxygen at other unoccupied sites in the tetramer.

The Fe^{2+} ion is well suited to its job in hemoglobin, not only because it has a natural affinity for O_2, but also because it changes its electron structure in a significant way when it binds O_2. Fe^{2+} is normally paramagnetic as a result of the four unpaired electrons in its outer ($3d$) electron orbitals. As such, it is too large to sit precisely in the plane of a porphyrin, as studies with model compounds have shown. Fe^{2+} is also paramagnetic when it is pentacoordinated in deoxyhemoglobin, and as expected, it is displaced from the plane of the porphyrin by a few tenths of an angstrom. When O_2 binds, the Fe^{2+} becomes hexacoordinated and diamagnetic (no unpaired electrons). This change reflects a major reorganization of its $3d$ orbitals, which decreases the radius of the Fe^{2+} so that it can move to an energetically more favorable position in the center of the porphyrin (see fig. 5.11).

The structural arguments advanced here to explain oxygen binding by hemoglobin are supported by comparisons of the amino acid sequences of α and β chains for a large number of hemoglobins from different species. Data from 60 species of α chains and 66 species of β chains

reveal 43 invariant positions in the hemoglobin molecule. These invariants are plotted on a map of the hemoglobin structure in figure 5.12. Here the invariant positions are shown by blue dots. In a sense, the positions of these dots provide a diagram of the working machinery of hemoglobin, because they line the heme pockets, where oxygen is bound, and the crucial $\alpha_1\beta_2$ interface, which changes its orientation when oxygen binds. Electrostatic forces and hydrogen bonds stitch this inferface together in the deoxy conformation when the molecule gives up its oxygen to the tissues. If changes occur, resulting from mutation at any of these positions, then trouble is likely to develop. Figure 5.13 shows the positions of pathological mutations in hemoglobin that result from changes in the heme pockets or in the $\alpha_1\beta_2$ interface.

As a rule, the invariant amino acids are the only critical loci for hemoglobin function. One striking exception occurs at position 6 of the β chain. If a hydrophobic valine residue is substituted for glutamic acid (a charged side chain), disaster results. The specific consequence of this alteration is to cause hemoglobin tetramers to aggregate when they are in the deoxy state. The aggregates form long fibers that stiffen the normally flexible red blood cell. The resulting distortion of the red cells leads to capillary occlu-

Figure 5.13

Positions of mutations in hemoglobin that produce a pathological condition. Comparison with figure 5.12 shows that these mutations, in general, show the same pattern as the distribution of the invariant positions. Dark circles around a numbered position indicate positions of abnormal residues, solid black dot indicates the valine β_6 mutation in sickle cell anemia. Heavy black circles indicate a hemoglobin that (is easily) oxidized to the Fe^{3+} form (methemoglobin), and a jagged black perimeter indicates unstable hemoglobin. Dark blue indicates increased oxygen affinity; light blue indicates decreased oxygen affinity. (Illustration copyright by Irving Geis. Reprinted by permission.)

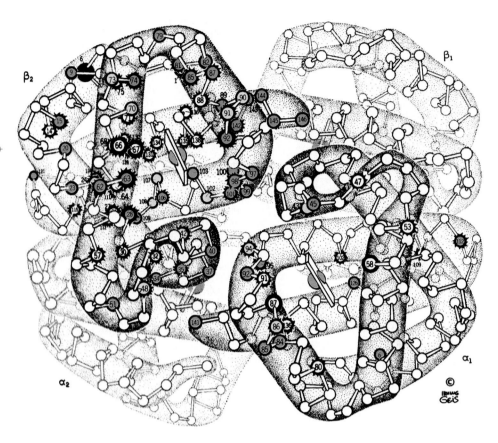

sion, which prevents proper delivery of oxygen to the tissues. This pathological condition is known as sickle cell anemia.

Alternative Theories on How Hemoglobins and Other Allosteric Proteins Work

We have spent a great deal of time describing the function of hemoglobin because it is the best understood regulatory protein and provides us with a model system for understanding in general terms how other allosteric proteins work. The first indication that hemoglobin was an allosteric protein came from the sigmoidal shape of its oxygen-binding curve (see fig. 5.2). Allosteric proteins are usually composed of two or more subunits. Different ligands may bind to quite different sites or to quite similar sites as in the case of hemoglobin.

Two quite different models were proposed about 25 years ago to explain the unusual nature of the hemoglobin oxygen-binding curve (fig. 5.14). These models can also be used as a starting point for discussing other allosteric proteins. The first model, introduced by Jacque Monod, Jeffery Wyman, and Pierre Changeux in 1965, is called the symmetry model. In this model hemoglobin can exist in only two conformations, one with all four of the subunits within a given tetramer in the low-affinity form and one with all four subunits in the high-affinity form (see fig. 5.14a). In this model, the hemoglobin molecule is always symmetrical, meaning that all the subunits are either in one state or the other and that all of the binding sites within a given tetramer have identical affinities. The binding of oxygen to one of the subunits favors the transition to the high-affinity form. The greater the number of oxygens binding to the tetramer, the more likely is the transition from the low-affinity form to the high-affinity form.

The second model, proposed by Koshland, Nemethy, and Filmer in 1966, is referred to as the sequential model (see fig. 5.14b). In this model the binding of an oxygen molecule to a given subunit causes that subunit to change its conformation to the high-affinity form. Because of its molecular contacts with its neighbors, the change increases the probability that another subunit in the same molecule will switch to the high-affinity form and bind a second oxygen more readily. The binding of the second oxygen has the same type of enhancing effect on the remaining unoccupied oxygen binding sites.

Figure 5.14

Alternative models for hemoglobin allostery. (*a*) In the symmetry model hemoglobin can exist in only two states. (*b*) In the sequential model hemoglobin can exist in a number of different states. Only the subunit binding oxygen must be in the high-affinity form.

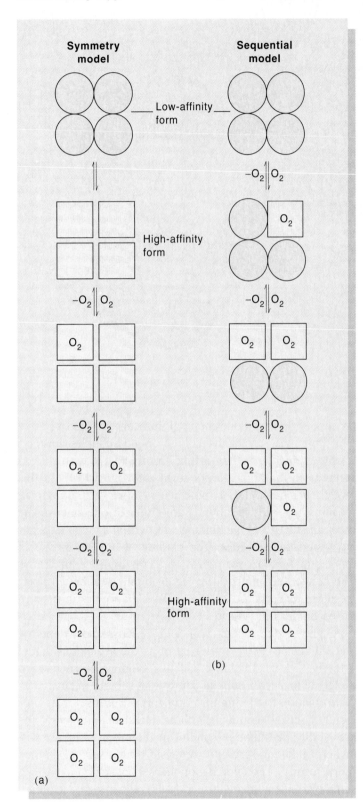

Either of these models (or something in between) could account for the sigmoidal oxygen-binding curve of hemoglobin, and either is consistent with the fact that deoxygenated and fully oxygenated hemoglobin have different conformations. The only way to discriminate rigorously between the two models is to obtain structural information on partially oxygenated hemoglobin. This information is still lacking, so no final judgment can yet be made. Even when the situation is fully resolved for hemoglobin, there is no assurance that other allosteric proteins work in the same way.

Muscle Is an Aggregate of Proteins Involved in Contraction

Vertebrate skeletal muscle represents a remarkable example of a supermolecular aggregate capable of undergoing a reversible reorganization. Voluntary muscle tissue is arranged into fibers that are surrounded by an electrically excitable membrane called the sarcolemma (fig. 5.15). Each fiber is composed of many myofibrils, which when viewed in the light microscope present a striated and banded appearance. As shown in figure 5.16*a* myofibril exhibits a longitudinally repeating structure called the sarcomere. This 23,000-Å long repeating unit is characterized by the appearance of several distinct bands: the less optically dense band being referred to as the I band, and the denser one as the A band. Furthermore, a dense line appears in the center of the I band, called the Z line; and a dense narrow band somewhat similar in appearance also occurs in the center of the A band, called the M line. Adjacent to the M line are regions of the A band that appear less dense than the remainder and are termed the H zone.

Transverse sections of the sarcomere reveal that these patterns result from the interdigitation of two sets of filaments (see fig. 5.16). For example, when a sarcomere is sectioned in the I band, a somewhat disordered arrangement of thin filaments (each about 70 Å in diameter) is seen. In contrast, when sectioned in the H zone, a hexagonal array of thick filaments (each about 150 Å in diameter) is apparent. The substantive observation is that a transverse section in the dense region of the A band shows a regularly packed array of interdigitating thick and thin filaments. It was this observation that led Hugh Huxley (1990) to propose that the process of muscle contraction involved sliding of the thick and thin filaments past each other (fig. 5.17).

Figure 5.15

The hierarchy of muscle organization. A voluntary muscle such as the bicep is a composite of many fibers connected to tendons at both ends. Each muscle fiber is composed of several myofibrils that are surrounded by an electrically excitable membrane (sarcolemma).

Myofibrils exhibit longitudinally repeating structures called sarcomeres. The fine structure of the sarcomere is described in figure 5.16.

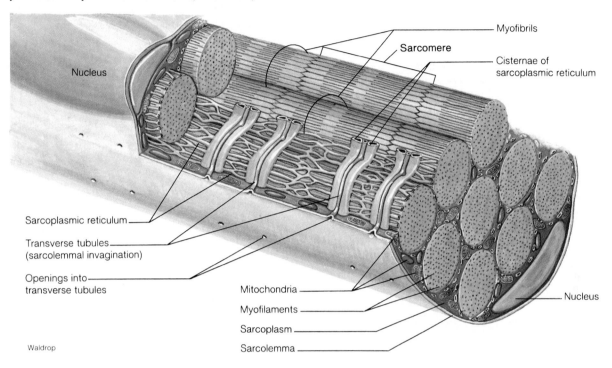

Figure 5.16

Electron micrograph of a striated muscle sarcomere showing the appearance of filamentous structures when cross-sectioned at the locations illustrated below. (Electron micrograph courtesy of Dr. Hugh Huxley, Brandeis University.)

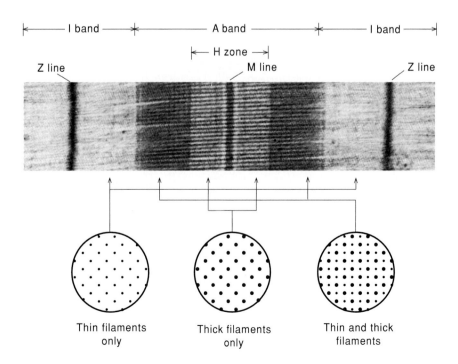

Figure 5.17

The sliding-filament model of muscle contraction. During contraction, the thick and thin filaments slide past each other so that the overall length of the sarcomere becomes shorter.

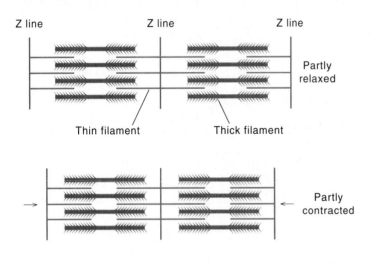

Table 5.1

Principal Proteins of Vertebrate Skeletal Muscle

Protein	M_r	Subunits	Function
Myosin	510,000	2 × 223,000 (heavy chains) 22,000–18,000 (light chains)	Major component of thick filaments
Actin	42,000	One type	Major component of thin filaments
Tropomyosin	64,000	2 × 32,000	Rodlike protein that binds along the length of actin filaments
Troponin	69,100	30,500 (TN-T) 20,800 (TN-I) 17,800 (TN-C)	Complex of three protein subunits involved in the regulation of muscle contraction

Figure 5.18

A molecular view of muscle structure. (*a*) Segment of actin-tropomyosin-troponin. (*b*) Segment of myosin. (*c*) Integration of thin filaments (actin) and thick filaments (myosin) in a muscle fiber.

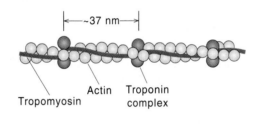

(a)

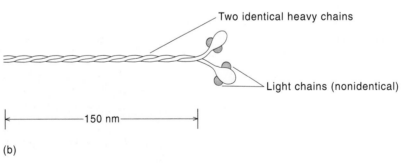

(b)

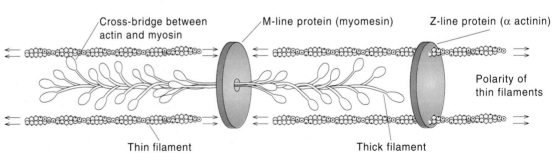

(c)

Subsequent analyses have shown that the thin filaments are composed of three proteins (fig. 5.18 and table 5.1). The main filamentous structure consists of an aggregate of globular actin molecules that takes on the form of a right-handed double helix. The individual actin molecules have a molecular weight of 42,000. Every turn of the actin helix incorporates 14 actin molecules and two molecules of a 360-Å long filamentous molecule of tropomyosin (TM) that fits into the two grooves of the actin double helix. The TM molecule is a dimer of two identical α-helical chains, which wind around each other in a coiled coil. Each TM dimer spans seven actin monomers, and a succession of TM dimers extends the full length of the thin filament. Two molecules of troponin (TN) bind to the actin filament at each helical repeat. Troponin is a complex of three nonidentical subunits: TN–C (a calcium-binding subunit) TN–T (a TM-binding subunit), and TN–A (an ''inhibitory'' subunit). The TN–TM proteins form a regulatory complex the properties of which are discussed below.

Thick filaments are composed of myosin, a large molecule containing two identical heavy chains (223 kD) and two different light chains (22 and 18 kD). The structural organization of myosin is illustrated in figure 5.18. The molecule has two identical globular head regions that incorporate the light chains and a significant fraction of the heavy chains. The tails of the heavy chains form very long α helices that wrap around each other to form left-handed coiled coils. The long α-helical structure is favored by the absence of proline over a region of more than a thousand residues and by the abundance of helix formers, such as leucine, alanine, and glutamate (see fig. 4.23). The coiled coil is favored by repeating seven-residue units in which every first and fourth residue has hydrophobic side chains that interact best in the coiled-coil conformation (see fig. 4.6). The individual myosin molecules can be cleaved into fragments by partial degradation with various proteases. Separation of such fragments has made it possible to demonstrate that globular head regions contain binding sites for actin and that the head pieces catalyze the hydrolysis of adenosine triphosphate (ATP). The α-helical coiled coils form the backbone of the thick filament. They also form an arm that can provide a flexible extension or hinge connecting the globular head to the body of the thick filament. The thick filament contains many myosin molecules oriented in a staggered fashion (see fig. 5.18).

Given this marvelous piece of molecular architecture, the question remains as to how it works. The answer lies in the observation that actin cyclically binds the globular myosin head group to form cross-bridges in a reaction that depends on the myosin-catalyzed hydrolysis of ATP. The cyclic binding and release of actin from myosin is driven by the energy-releasing hydrolysis of ATP in a manner that causes rearrangement of the actin–myosin cross-bridges. When muscle is completely relaxed, there is a minimum number of cross-bridges and the muscle is fully stretched. When the muscle is activated and under tension, it contracts and more cross-bridges are formed as the region of overlap between actin and myosin increases. At each stage of the contraction process it is essential to break the existing bridges with the help of ATP hydrolysis before new ones can be formed. It is important to realize that although ATP allows new bridges to be formed, the ATP is required to break the existing bridges, not to form the new ones. Cross-bridge formation is energetically favored in the absence of ATP.

A likely scenario for the contraction process is shown in figure 5.19. The ATP that binds to myosin causes the bridges to break or weaken. This bound ATP is rapidly hydrolyzed to ADP and P_i, but the hydrolysis products are not immediately released by the myosin. The P_i is released first, but its release requires effective contact with the actin—the regulated step in muscular contraction. Once the P_i has been released, a strong bridge forms between the myosin and the actin. This step is followed by a structural change in the myosin that leads to the translocation of the myosin relative to the actin filament and finally to ADP dissociation. The translocation step is referred to as the power stroke in muscular contraction because it is at this point that the energy ultimately donated by the ATP is expended in the form of a complex structural change. Further contraction requires fresh ATP followed by bridge dissociation and ATP hydrolysis. If conditions are right, the P_i dissociates and the bridges reform; this time they reform at points further along the actin. Cyclic repetition of this process results in a net increase in the number of actin–myosin bridges and further contraction of the sarcomere.

As indicated above, the process of contraction is regulated or triggered by the TN–TM system. Because voluntary muscles are under the conscious control of the animal, one expects a signal from the central nervous system to initiate the process of contraction. A nerve impulse communicated to the muscle causes a depolarization of the sarcolemma membrane that surrounds the muscle fibers, which in turn causes a release of Ca^{2+} from the endoplasmic reticulum in cytoplasm of the muscle cell (fig. 5.20). The Ca^{2+} ions form a complex with the TN–C component of the troponin molecule (see table 5.1). This process induces changes within the TM complex that overcome the inhibitory effect of the TN–I subunit. Then, through TN–T, a signal is sent to TM that triggers the contraction event. The precise nature of this signal from troponin to tropomyosin is unclear; it appears to involve a movement of the tropomyosin on the actin surface. This movement evidently leads to an allosteric transition of the actin that encourages

Figure 5.19

Steps in the contraction process. Because contraction is a cyclical process, the choice of a starting point is somewhat arbitrary. Five frames are shown; the first two and the last two frames are identical to make the cyclical nature of the process clear. In the first frame (*a*), the myosin head groups contain the hydrolysis products of a single ATP molecule, ADP and P_i. A structural transition in the actin leads to contact between the actin and the myosin and the release of P_i. The release of the P_i is the rate-limiting step in muscular contraction.

In the second frame (*b*), strong bridges form between actin and myosin. This is followed by a structural alteration in the myosin molecules and an effective translocation of the thick filament relative to the thin filament in (*c*). During this process the ADP is released. After the translocation step, the bridge structure is broken by the binding of ATP, which is rapidly hydrolyzed to ADP and P_i. Each thick filament has about 500 myosin heads, and each head cycles about five times per second in the course of a rapid contraction.

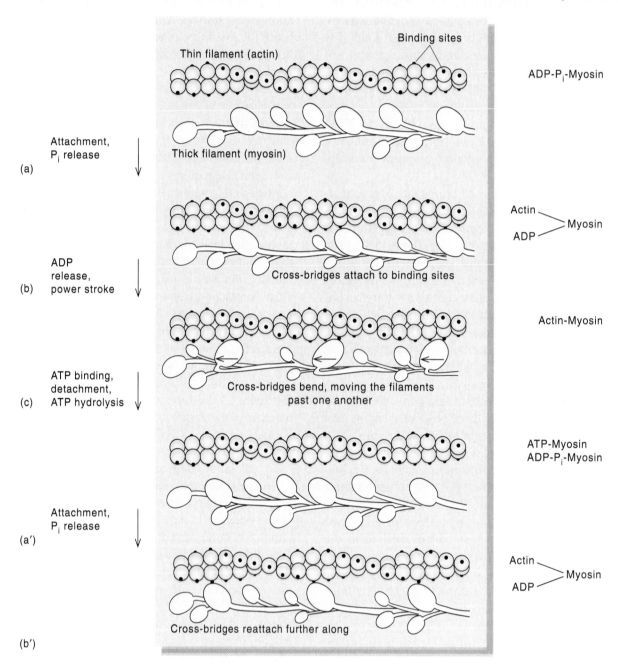

Figure 5.20

The effect of calcium on muscle contraction. Binding of calcium to the TN–TM–actin complex produces a shift in the location of TM, which produces an allosteric transition in actin. The allosteric transition in actin facilitates the release of P_i from myosin, which strengthens the interaction between actin and myosin.

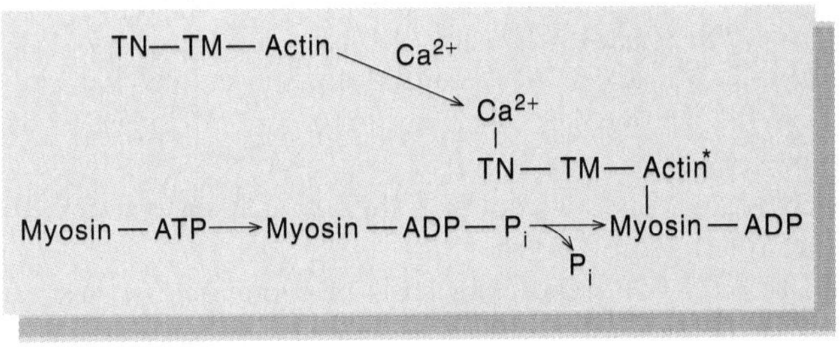

TN—TM— Actin Ca^{2+}

Ca^{2+}
|
TN — TM — Actin*
|
Myosin — ATP ⟶ Myosin — ADP — P_i ⟶ Myosin — ADP
P_i

more favorable contact between the complementary binding sites on actin and myosin. As a result, a strong bridge is formed and the P_i is released from the myosin.

It is noteworthy that for some time after death, muscles enter a state of rigor in which they are fully contracted and the maximum number of bridges are formed. Rigor is probably due to the depletion of ATP and a considerable discharge of calcium from the sarcoplasmic reticulum. In living tissue, the cytosolic Ca^{2+} concentration is restored to resting levels within 30 ms of receiving a signal, and the myofibrils relax.

Summary

In this chapter we considered the relationship between structural and functional properties for two protein systems.

1. Hemoglobin is a tetramer made of two almost identical subunits. The function of hemoglobin is threefold: to transport O_2 from the lungs to the tissues where it is consumed, to transport CO_2 from the tissues where it is produced to the lungs where it is expelled, and to maintain the blood pH over a narrow range. Cooperative interactions among the subunits allow hemoglobin to pick up the maximum amount of oxygen at high oxygen tensions in the lung tissue and to deliver the maximum amount of oxygen to the oxygen-consuming tissues.

2. Muscle is an aggregate of several different proteins. Its main protein components are organized as overlapping filaments of two types: thin filaments, composed mainly of actin molecules, and thick filaments, composed of myosin molecules. The process of muscular contraction entails a sliding of the two types of filaments past each other. In a fully contracted myofibril the actin and myosin filaments show a maximum overlap with each other. The contraction process involves the breakage and reformation of bridges between the actin and myosin molecules in a reaction that requires the expenditure of ATP.

Selected Readings

Akers, G. K., M. C. Doyle, D. Myers, and M. A. Daugherty, Molecular code for cooperativity in hemoglobin. *Science* 255:54–63, 1992. A new model for hemoglobin allostery.

Allen, R. D., The microtubule as an intracellular engine. *Sci. Am.* 238(2):42–49, 1987.

Baldwin, J., Structure and cooperativity of haemoglobin. *Trends Biochem. Sci.* 5:224–228, 1980.

Berg, H. C., How bacteria swim. *Sci. Am.* 233(2):36–44, 1975.

Cantor, C. R., and P. Schimmel, *Biophysical Chemistry*. New York: Freeman, 1980. Especially see volume 2, entitled *Techniques for Study of Biophysical Structure and Function*.

Caplan, A. I., Cartilage. *Sci. Am.* 251(4):84–94, 1984.

Dickerson, R. E., and I. Geis, *Hemoglobin*. Menlo Park, Calif.: Benjamin/Cummings, 1983. A magnificent presentation of every facet of hemoglobin biochemistry and genetics.

Doolittle, R., Proteins. *Sci. Am.* 253(4):88–96, 1985. Overview emphasizing evolutionary considerations.

Eisenberg, D., and D. Crothers, *Physical Chemistry and Its Applications to the Life Sciences*. Menlo Park, Calif.: Benjamin/Cummings, 1979.

Eyre, D. R., M. A. Pdaz, and P. M. Gallop, Cross-linking in collagen and elastin. *Ann. Rev. Biochem.* 53:717–748, 1984.

Gething, M. J., and J. Sambrook, Protein folding in the cell. *Nature* 355:33–45, 1992.

Huxley, H. E., Sliding filaments and molecular motile systems. *J. Biol. Chem.* 265:8347–8352, 1990.

Hynes, R. O., Fibronectins. *Sci. Am.* 254(6):42–51, 1986.

Ingram, V. M., Gene mutation in human haemoglobin: the chemical difference between normal and sickle-cell haemoglobin. *Nature* 180:326–328, 1957. A classic paper.

Karplus, M., and J. A. McCammon, The dynamics of proteins. *Sci. Am.* 254(4):42–51, 1986. A reminder that proteins are dynamic structures.

Lawn, R. M., and G. A. Vehar, The molecular genetics of hemoglobin. *Sci. Am.* 254(3):48–65, 1986.

Martin, G. R., R. Timpl, P. K. Muller, and K. Kuhn, The genetically distinct collagens. *Trends Biochem. Sci.* 10:285–287, 1985. A brief, authoritative account of the most abundant protein found in vertebrates.

Pauling, L., H. A. Itano, S. J. Singer, and I. C. Wells, Sickle-cell anemia: a molecular disease. *Science* 110:543–548, 1949. A classic paper.

Pollard, T. D., and J. A. Cooper, Actin and actin-binding proteins. *Ann. Rev. Biochem.* 55:987–1036, 1986. Discusses structure and function.

Rayment, I., H. M. Holden, M. Whittaker, C. B. Yohn, M. Lorenz, K. C. Holmes, and R. A. Milligan, Structure of the Actin-Myosin Complex and Its Implications for Muscle Contraction. *Science* 261:58–65, 1993.

Salemme, R., Structure and function of cytochromes *c*. *Ann. Rev. Biochem.* 46:299–329, 1977.

Steinert, P. M., and D. R. Roop, Molecular and cellular biology of intermediate filaments. *Ann. Rev. Biochem.* 57:593–626, 1988.

Tonegawa, S., The molecules of the immune system. *Sci. Am.* 253(4):122–130, 1985.

Wilson, A. C., The molecular basis of evolution. *Sci. Am.* 253(4):164–170, 1985.

Problems

1. In the latter half of the 1800s one of the first suggestions that proteins were large molecules came from ashing experiments in which hemoglobin was converted to Fe_2O_3. These procedures suggested a molecular mass of hemoglobin greater than 15,900, a number unheard of at that time. Why did the researchers of the past century, who were excellent analytical chemists, deviate so significantly from the molecular mass of 64,500 now known for hemoglobin? What quantity of Fe_2O_3 results from the ashing of 1.00 g of hemoglobin?

2. Carbonic acid in the blood readily dissociates into hydrogen and bicarbonate ions. If the serum pH of 7.4 equals that inside the erythrocytes, what percentage of the carbonic acid is ionized? Use a value of 6.4 for the first pK_a of carbonic acid.

3. In addition to oxygen, hemoglobin subunits can also carry carbon dioxide. This is performed by covalent addition of CO_2 to the N termini of the hemoglobin chains to produce a carbonate structure. Propose reactions for this process utilizing (a) CO_2 and (b) HCO_3^-.

4. Figure 5.13 indicates locations of mutations that have been shown to produce pathological conditions. The majority of the types of mutations that have been discovered in human hemoglobins have been mutations in which either amino acid residues bearing charges are replaced with ones with no charge or in which uncharged amino acids are replaced with charged amino acids. Do you think this represents a basic biological principle or is it an artifact of the detection process? Explain.

5. Sickle-cell anemia becomes most apparent during a sickle-cell crisis when the soft tissues are often acutely painful. Using information provided in the text on the molecular-cellular effects of the sickling of red blood cells, can you provide an explanation for the origin of the pain?

6. The hemoglobin present in a fetus is analogous to the $\alpha_2\beta_2$ tetramer of the adult, but the two β chains have been replaced with comparable γ chains. Considering the relevant biology, which hemoglobin type (adult or fetal) do you expect to have the greater affinity for oxygen?

7. Use the information presented in problem 6 above to propose a possible "future genetic engineering solution" to sickle-cell anemia. Are there any deleterious ramifications of your proposal?

8. Carbon 2 in glycerate-2,3-bisphosphate is in the D configuration. Would you expect the L configuration of glycerate-2,3-bisphosphate to have the same effect on the biochemistry of hemoglobin? Why or why not?

9. Figures 5.10 and 5.11 show a histidyl F8 residue interacting with the heme iron on deoxy- and oxyhemoglobin. Would you expect the imidazole nitrogen of the histidyl group to be protonated or unprotonated during this interaction? Why?

10. Would you expect the blood–hemoglobin system to transport more moles of O_2 or CO_2? Why?

11. The protein tropomyosin (TM) is composed of two identical chains of α helix (see table 5.1) that are in turn twisted around each other in a helical structure. Consider the average amino acid residue weight to be 105 daltons and use typical α-helix dimensions (fig. 4.4) to calculate the length of each chain in TM. Explain any discrepancy observed between the calculated length and the observed length of 360 Å.

12. Bryan Allen made aviation history by pedaling the *Gossamer Albatross* from near Folkestone, England, to Cap Gris Nez, France, from 4:51 AM to 7:40 AM on June 12, 1979. During this flight he continually produced about a third of a horsepower (*Nat. Geographic* 156:5,640, 1979). Considering that the energy available from the hydrolysis of ATP is 7.5 kcal/mole (see fig. 2.9), determine the number of moles of ATP required for this flight. For simplicity, assume that the muscles are 50% efficient in the conversion of chemical energy into mechanical energy and that one horsepower equals the energy expenditure of 178 cal/s.

13. What is the cause of rigor after death?

14. Explain how muscular contraction is regulated.

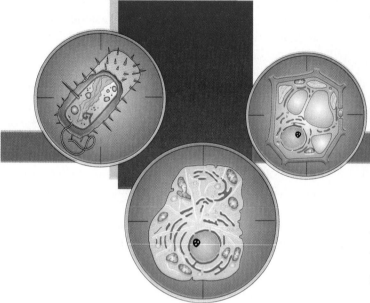

Methods for Characterization and Purification of Proteins

Chapter Outline

Methods of Protein Fractionation and Characterization
 Differential Centrifugation Divides a Sample into Two Fractions
 Differential Precipitation Is Based on Solubility Differences
 Column Procedures Are the Most Versatile and Productive Purification Methods
 Electrophoresis Is Used for Resolving Mixtures
 Sedimentation and Diffusion Are Used for Size and Shape Determination
Protein Purification Procedures
 Purification of an Enzyme with Two Catalytic Activities
 Purification of a Membrane-Bound Protein

Proteins can be isolated and characterized according to their size, shape, and charge.

In the previous three chapters we described the structures of amino acids and proteins, and in two cases we examined how these structures relate to their function. Some of the methods for structure determination were also discussed (e.g., sequence analysis in chapter 3 and x-ray diffraction in chapter 4). To analyze the structure of a protein we must isolate it from the complex mixture in which it exists in whole cells. The primary object of this chapter is to acquaint you with techniques used for protein purification. Because these procedures are often used for protein characterization as well, they will add to your repertoire of methods for protein characterization.

In the first part of this chapter methods for protein fractionation are discussed in isolation. Success in protein purification depends on picking a number of procedures and combining them in an effective order. Two examples are given in which trains of procedures are combined to purify specific proteins from crude whole-cell extracts.

Methods of Protein Fractionation and Characterization

Many more techniques for protein analysis exist than can be covered in a single chapter. The emphasis here is on some of the more popular procedures and the principles involved in their use.

Differential Centrifugation Divides a Sample into Two Fractions

A typical crude broken-cell preparation contains disrupted cell membranes, cellular organelles, and a large number of soluble proteins, all dispersed in an aqueous buffered solution. The membranes and the organelles can usually be separated from one another and from the soluble proteins by differential centrifugation. Differential centrifugation divides a sample into two fractions: the pelleted fraction, or sediment, and the supernatant fraction, that is, the fraction that is not sedimented. The two fractions may then be separated by decantation.

According to its purpose, differential centrifugation involves the use of different speeds and different times of centrifugation (table 6.1). For example, if the protein of interest is in the mitochondrial fraction, the crude cell lysate is centrifuged first at 4,000 × g for 10 min to remove cell membranes, nuclei, and (in the case of plant material) chloroplasts. The supernatant from this step contains, among other elements, the mitochondria, and is decanted and recentrifuged at 15,000 × g for 20 min to obtain a sediment primarily containing mitochondria. If ribosomes instead of mitochondria are the goal, then the crude lysate is centrifuged at 30,000 × g for 30 min, and the resulting supernatant is decanted and centrifuged at 100,000 × g for 180 min to obtain a ribosomal sediment. If obtaining the soluble protein fraction is the goal, then the entire lysate is centrifuged at 100,000 × g for 180 min, and the resulting supernatant, containing the soluble protein, is carefully decanted for further processing.

Differential Precipitation Is Based on Solubility Differences

Once a crude extract of protein has been made, it is common to separate this mixture into different fractions by a precipitation step. The solubility of a protein reflects a delicate balance between different energetic interactions—both internally within the protein and between the protein and the surrounding solvent. Consequently, the choice of solvent can affect both the solubility and the structure of a protein.

Table 6.1

Sedimentation Conditions for Different Cellular Fractions

Fraction Sedimented	Centrifugal Force (× g)	Time (min)
Cells (eukaryotic)	1,000	5
Chloroplasts; cell membranes; nuclei	4,000	10
Mitochondria; bacteria cells	15,000	20
Lysosomes; bacterial membranes	30,000	30
Ribosomes	100,000	180

A Minimum of Solubility Occurs at the Isoelectric Point Proteins typically have charged amino acid side chains on their surfaces that undergo energetically favorable polar interactions with the surrounding water. The total charge on the protein is the sum of the side-chain charges. However, the actual charge on the weakly acidic and basic side-chain groups also depends on the solution pH. In fact, the acidic and basic groups within the protein can be titrated just like free amino acids (see fig. 3.6) to determine their number and their pK values.

Proteins tend to show a minimum solubility at their isoelectric pH—a fact that is apparent for β-lactoglobulin (fig. 6.1). The decrease in solubility at the isoelectric pH reflects the fact that the individual protein molecules, which all have similar charges at pH values away from their isoelectric points, cease to repel each other. Instead, they coalesce into insoluble aggregates.

Salting In and Salting Out Proteins also show a variation in solubility that depends on the concentration of salts in the solution. These frequently complex effects may involve specific interactions between charged side chains and solution ions or, particularly at high salt concentrations, may reflect more comprehensive changes in the solvent properties. For the effect of salt concentration on the solubility of β-lactoglobulin, again see figure 6.1. Most globulins are sparingly soluble in pure water. The increase in solubility that occurs after adding salts such as sodium chloride is often referred to as salting in.

The effects of four different salts on the solubility of hemoglobin at pH 7 can be seen in figure 6.2. All four salts produce the salting-in effect with this protein; two of them, sodium sulfate and ammonium sulfate, also produce a greatly decreased solubility of the protein at high salt con-

Figure 6.1

Solubility of β-lactoglobulin as a function of pH and ionic strength. The isoelectric pH (pI) for this protein is about 5.2, which corresponds to the point of minimum solubility.

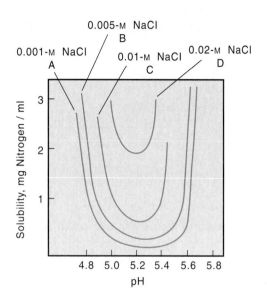

Figure 6.2

Solubility of horse carbon monoxide hemoglobin in different salt solutions. The addition of a moderate amount of salt (salting in) is required to solubilize this protein. At high concentrations, certain salts compete more favorably for solvent, decreasing the solubility of the protein and thus leading to its precipitation (salting out). (Source: E. J. Cohn and J. T. Edsall, *Proteins, Amino Acids, and Peptides as Ions and Dipolar Ions.* Copyright © 1942, Reinhold, New York, N.Y.)

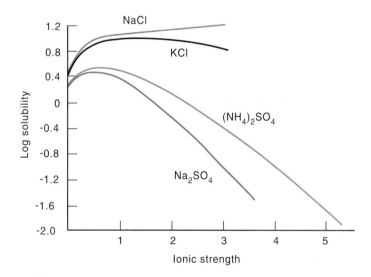

centrations. This result is called salting out and occurs with salts that effectively compete with the protein for available water molecules. At high salt concentrations, the protein molecules tend to associate with each other because protein–protein interactions become energetically more favorable than protein–solvent interactions.

Each protein has a characteristic salting-out point, a fact we can exploit to make protein separations in crude extracts. For this purpose $(NH_4)_2SO_4$ is the most commonly used salt because it is very soluble and is generally effective at lower concentrations than many other salts.

Typically, the desired protein precipitates over a range of salt concentrations. If it precipitates in the range of 20–30% by weight, we first add ammonium sulfate to a concentration slightly below 20% and then centrifuge to remove by sedimentation any proteins that precipitate in the 0–20% range. To the supernatant from this centrifugation we then add more ammonium sulfate, to 30%. Centrifugation at this point brings down the desired protein, as well as other proteins that precipitate in this range of salt concentrations. The supernatant is discarded and the sediment saved for further purification.

Column Procedures Are the Most Versatile and Productive Purification Methods

Column procedures are the most effective and most varied of purification methods. Common to all column procedures is the use of a glass cylinder with an opening at the top and bottom. The cylinder is filled with a column of hydrated material, and the protein sample is applied to the top of the column. Then further buffer is passed through the column. In most column procedures, proteins bind differentially to the column material, and a change in the elution buffer causes them to be eluted differentially according to their degree of affinity for the column material. The eluant exiting from the bottom of the column is collected in equal-sized fractions with the help of an automatic fraction collector (fig. 6.3). Each fraction is analyzed, and fractions containing an appreciable amount of the desired protein are pooled for further purification.

Many variations on this basic procedure are in common use:

1. In gel-exclusion chromatography a cross-linked dextran without any special attached functional groups is used for the column substrate (fig. 6.4). Large molecules flow more rapidly through this type of column than small ones. The dextrans have different degrees of cross-linking, making them effective over different size ranges.

2. Ion-exchange chromatography makes use of the fact that proteins differ enormously in their affinity for positively or negatively charged columns. However, the cross-linked resins used in amino acid fractionation (see chap-

Figure 6.3

Collecting fractions during column chromatography. Column material and elution procedure are chosen to effect optimal separation of the desired protein.

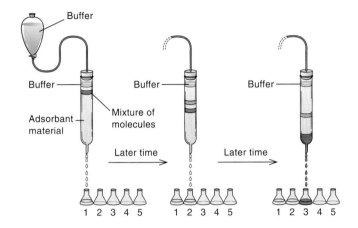

Figure 6.4

Polydextran column showing separation of small and large molecules. The column material is immersed in solvent, which penetrates the gel particles. A separation is initiated by layering a small sample containing different-sized proteins on the top of the column. This sample is pushed through the column by opening the stopcock at the bottom and adding further solvent at the top to keep up with the flow. As shown, the small protein molecules can penetrate the gel particles but the big ones cannot. Therefore the big proteins move through the column much more rapidly, and a separation of the two proteins results.

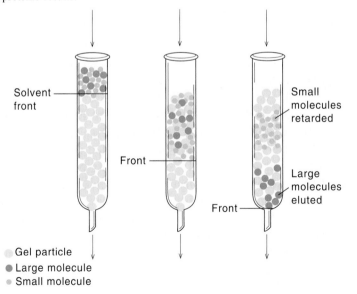

Table 6.2

Some Column Materials for Ion-Exchange Chromatography of Proteins

Matrix	Functional Groups on Column
Phosphocellulose (PC)	$-PO_3^-$
Carboxymethyl cellulose (CMC)	$-CH_2-COO^-$
Diethylaminoethyl cellulose (DEAE)	$-(CH_2)_2-\overset{+}{N}H\begin{smallmatrix}CH_2-CH_3 \\ CH_2-CH_3\end{smallmatrix}$

ter 3) are rarely used for protein separations because proteins are too large to penetrate the resin beads. Instead, finely divided celluloses containing either positively or negatively charged groups are most commonly used (table 6.2). The affinity of a protein for the column material is proportional to the salt concentration required to release the protein from the material. Typically, a column is loaded with protein solution at a low ionic strength so that most of the proteins bind to the column. After loading, elution is initiated by gradually increasing the salt concentration of the elution buffer. Proteins are eluted in the order of increasing affinity.

3. Finely divided celluloses may also be used in the column in conjunction with attached hydrophobic groups such as octyl alcohol. In this case, it is proteins with exposed hydrophobic centers that bind to the column with varying affinities. These proteins may be eluted in order of decreasing affinity for the column by increasing the level of free octyl alcohol in the eluting buffer.

4. Affinity chromatography makes use of chemical groups that have a special binding affinity for the proteins that are being sought. For example, many enzymes bind preferentially to cofactors, such as adenosine triphosphate or pyridine nucleotides. Often, the separation of such enzymes from other proteins can be achieved on a column that has one of these cofactors attached. As the mixture passes through the column, proteins that have specific binding affinities for the cofactor bind to the column, and other proteins pass through. The cofactor-binding protein can then be eluted with a solution containing the same soluble cofactor.

5. High-performance liquid chromatography (HPLC) is not so much a new type of chromatography as a new way of applying old chromatographic techniques, using extremely high pressures. The same principles are in-

Figure 6.5

Gel electrophoresis for analyzing and sizing proteins. (*a*) Apparatus for slab-gel electrophoresis. Samples are layered in the little slots cut in the top of the gel slab. Buffer is carefully layered over the samples, and a voltage is applied to the gel for a period of usually 1–4 h. (*b*) After this time the proteins have moved into the gel at a distance proportional to their electrophoretic mobility. The pattern shown indicates that different samples were layered in each slot. (*c*) Results obtained when a mixture of proteins was layered at the top of the gel in phosphate buffer, pH 7.2, containing 0.2% SDS. After electrophoresis the gel was removed from the apparatus and stained with Coomassie Blue. The protein and its molecular weight are indicated next to each of the stained bands. (*d*) The logarithm of the molecular weight against the mobility (distance traveled) shows an approximately linear relationship. (Source: Data of K. Weber and M. Osborn.)

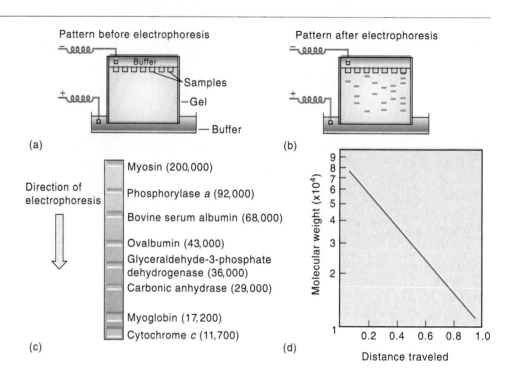

volved, but column materials usually consist of more finely divided particles made of physically stronger materials, which can withstand pressures of 5,000–10,000 psi without changing their structure. The column apparatus itself must also be designed to withstand high pressures. The main advantage of HPLC is that superior resolution of eluted substances can be achieved.

Electrophoresis Is Used for Resolving Mixtures

Gel electrophoresis is the best way to analyze mixtures and assess purity. Gel electrophoresis separates proteins according to their size and their charge. It is almost always performed in aqueous solution supported by a gel system. The gel is a loosely cross-linked network that functions to stabilize the protein boundaries between the protein and the solvent, both during and after electrophoresis, so that they may be stained or otherwise manipulated.

The most popular method of electrophoretic separation by gels employs sodium dodecyl sulfate (SDS). This method not only gives an index of protein purity but yields an estimate of the protein subunit molecular weights. The mixture of proteins to be characterized is first completely denatured by adding SDS (a detergent) and mercaptoethanol and by briefly heating the mixture. Denaturation is caused by the association of the apolar tails of the SDS molecules with protein hydrophobic groups. Any cystine disulfide

bonds are cleaved by a disulfide interchange reaction with mercaptoethanol.

The resulting unfolded polypeptide chains have relatively large numbers of SDS molecules bound to them. The success of the technique for molecular weight estimation depends on two facts: (1) Each bound SDS molecule contributes one negative charge to the denatured protein complex, so that the charge of the protein in its native state is effectively masked by the more numerous charged groups of the associated detergent molecules. (2) The total number of detergent molecules bound is proportional to the polypeptide chain length or, equivalently, the protein's molecular weight. As a result, the SDS-denatured protein molecules acquire net negative charges that are approximately proportional to their molecular weights (fig. 6.5).

Another electrophoretic method, called isoelectric focusing, is frequently used for characterizing proteins based on differences in their isoelectric points. The apparatus usually consists of a narrow tube containing a gel and a mixture of ampholytes, which are small molecules with positive and negative charges. The ampholytes have a wide range of isoelectric points, and are allowed to distribute in the column under the influence of an electric field. This step creates a pH gradient from one end of the gel to the other, as each particular ampholyte comes to rest at a position coincident with its isoelectric point. At this stage, a solution of proteins is introduced into the gel. The proteins migrate in the electric field until each reaches a point at which the pH resulting from the ampholyte gradient exactly equals its own isoelectric point. Isoelectric focusing provides a way of both accurately determining a protein's isoelectric point and effecting separations among proteins, the isoelectric points of which may differ by as little as a few hundredths of a pH unit.

An elegant extension of the electrophoretic method of separation involves combining isoelectric focusing with SDS gel electrophoresis to produce a two-dimensional electrophoretogram. This technique is most valuable for the analysis of very complex mixtures. First, the sample is run in a one-dimensional pH gradient gel (isoelectric focusing). The resulting narrow strip of gel, containing the partially separated mixture of proteins, is placed alongside a square slab of SDS gel. An electrical field is imposed so that the sample moves at right angles to its motion in the first gel. Figure 6.6 shows the separation of total *E. coli* protein into more than 1,000 different components.

Although electrophoresis is the method of choice for assessing protein purity, it is not frequently used as a purification step, at least not yet. This is because of the small amounts that can usually be analyzed and because of the denaturing conditions that are often used in electrophoretic analysis.

Sedimentation and Diffusion Are Used for Size and Shape Determination

We have seen that sedimentation by centrifugation is used for protein purification. The methods of centrifugation can be used quantitatively for the assessment of protein size and shape.

In a centrifugal force field, protein molecules slowly migrate toward the bottom of a centrifuge tube at a rate that is proportional to their molecular weight (fig. 6.7). The rate of sedimentation may be recorded by optical methods that do not interfere with the operation of the centrifuge. From this rate, we can obtain the sedimentation constant s. This constant equals the rate at which a molecule sediments, divided by the gravitational field (angular acceleration in a spinning rotor), as defined by the equation

$$s = \frac{dx/dt}{\omega^2 x} \qquad (1)$$

where dx/dt is the rate at which the particle travels at distance x from the center of rotation, ω is the angular velocity of the rotor in radians per second (hence $\omega^2 x$ is the angular acceleration), and t is the time of centrifugation in seconds. The sedimentation constant is usually given in Svedberg units (S); one $S = 10^{-13}\ s$.

From the sedimentation constant we can obtain the molecular weight, provided we have certain other information. This additional information includes the frictional co-

Figure 6.6

Two-dimensional SDS isoelectric-focusing gel electrophoresis. First the sample is run in a one-dimensional pH gradient, partially separating the sample along a strip of gel. Then the strip of gel containing the sample is placed alongside an SDS gel, and the proteins are permitted to further separate by moving in the second dimension, at right angles to the first separation. Sample shown is total *E. coli* protein; individual proteins are detected by autogradiography. (Source: Photograph provided by Patrick O'Farrell. See O'Farrell, in *J. Biol. Chem.* 250:4007, 1975.)

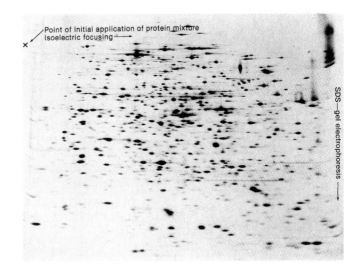

Figure 6.7

Apparatus for analytical ultracentrifugation. (*a*) The centrifuge rotor and method of making optical measurements. (*b*) The optical recordings as a function of centrifugation time. As the light-absorbing molecule sediments, the solution becomes transparent.

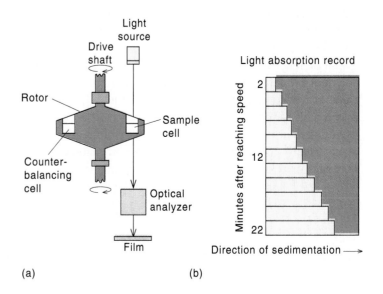

Table 6.3

Physical Constants of Some Proteins

Protein	Molecular Weight	Diffusion Constant ($D \times 10^7$)	Sedimentation Constant (S)	pI[a] (Isoelectric)
Cytochrome c (bovine heart)	13,370	11.4	1.17	10.6
Myoglobin (horse heart)	16,900	11.3	2.04	7.0
Chymotrypsinogen (bovine pancreas)	23,240	9.5	2.54	9.5
β-Lactoglobulin (goat milk)	37,100	7.5	2.9	5.2
Serum albumin (human)	68,500	6.1	4.6	4.9
Hemoglobin (human)	64,500	6.9	4.5	6.9
Catalase (horse liver)	247,500	4.1	11.3	5.6
Urease (jack bean)	482,700	3.46	18.6	5.1
Fibrinogen (human)	339,700	1.98	7.6	5.5
Myosin (cod)	524,800	1.10	6.4	—
Tobacco mosaic virus	40,590,000	0.46	198	—

[a] pI = −log I = the isoelectric point, that is, the pH at which the protein carries no net charge (see fig. 6.1).

efficient (f) of the protein and the density of the protein. The coefficient f is bigger for larger proteins and, for proteins of the same molecular weight, it is larger for elongated, rodlike molecules. The density of the protein is important because of the buoyancy factor, $1 - \bar{v}_p \rho_s$, which takes into account the density difference between solvent (ρ_s) and the volume of water displaced per gram of protein ($\bar{v}_p$). The equation that relates s, M, and f is

$$s = \frac{M(1 - \bar{v}_p \rho_s)}{Nf} \qquad (2)$$

where N is Avogadro's number. Thus, in order to estimate the molecular weight from the sedimentation constant we must have a means of determining $\bar{v}_p$ and f.

For most proteins $\bar{v}_p$ is about 0.75 ml/g, so its value does not present much of a problem. The frictional coefficient, however, is a sensitive function of the shape, varying over a wide range, and we must usually know its value if we need a serious estimate of the molecular weight. The value of f is usually found by working with the diffusion constant D, which is related to the frictional constant by

$$D = \frac{RT}{Nf} \quad \text{or} \quad f = \frac{RT}{ND} \qquad (3)$$

where R is the gas constant and T is the absolute temperature. Substituting this expression for f in equation (2) and transposing leads to

$$M_r = \frac{RTs}{D(1 - \bar{v}_p \rho_s)} \qquad (4)$$

The diffusion constant D is a function of both molecular weight and shape. It can be measured by observing the spread of an initially sharp boundary between the protein solution and a solvent as the protein diffuses into the solvent layer. Once we know the value of the diffusion constant, we can combine the information with the sedimentation data and calculate the molecular weight of the protein.

Representative sedimentation and diffusion data, together with the calculated molecular weights, are presented in table 6.3. The sedimentation constants are usually larger the greater the molecular weight. Diffusion constants are usually inversely proportional to the molecular weights. Exceptions arise because of proteins with unusual shapes. For example, the globular protein urease and the rodlike protein myosin have similar molecular weights. Yet their sedimentation and diffusion constants both differ by about a factor of 3. Rapid diffusion can be a highly desirable property for an enzyme that must pass rapidly from one point to another. Such a protein benefits from a globular shape, which is quite common for enzymes. By contrast, a rodlike protein can be advantageous for creating a cytoskeletal boundary within the cytoplasm. Myosin is used for just such purposes in many cell types.

Protein Purification Procedures

To characterize proteins, we must usually first isolate the protein under study from the complex mixture of proteins found in the organism. Once we decide to purify a particular protein, we must weigh several factors. For example, how much material is needed? What level of purity is required? The starting material should be readily available and should contain the desired protein in relative abundance. If the protein is part of a larger structure, such as the nucleus, the mitochondria, or the ribosome, then it is advisable to isolate the large structure first from a crude cell extract.

Purification must usually be performed in a series of steps, using different techniques at each step. Some purification techniques are more useful when handling large amounts of material, whereas others work best on small amounts. A purification procedure is arranged so that the techniques that are best for working with large amounts are used during early steps in the overall purification. The suitability of each purification step is evaluated in terms of the amount of purification achieved by that step and the percent recovery of the desired protein.

Combining techniques introduces new considerations and new problems. If two purification techniques each give 10-fold enrichment for the desired protein when executed independently on a crude extract, it does not mean that they give 100-fold enrichment when combined. In general, they give somewhat less. As a rule, purification techniques that combine most effectively usually are based on different properties of the protein. For example, a technique based on size fractionation is more effectively combined with a technique based on charge than with another technique based on size fractionation.

Throughout the purification process we must have a convenient means of assaying for the desired protein, so we can know the extent to which it is being enriched relative to the other proteins in the starting material. In addition, a major concern in protein purification is stability. Once the protein is removed from its normal habitat, it becomes susceptible to a variety of denaturation and degradation reactions. Specific inhibitors are sometimes added to minimize attack by proteases on the desired protein. During purification it is usual to carry out all operations at 5°C or below. This temperature control minimizes protease degradation problems and decreases the chances of denaturation.

In their natural habitat, proteins are usually surrounded by other proteins and organic factors. When these are removed or diluted as during purification, the protein becomes surrounded by water on all sides. Proteins react differently to a pure aqueous environment; many are destabilized and rapidly denatured. A common remedial measure is to add 5%–20% glycerol to the purification buffer. The organic surface of the glycerol is believed to simulate the environment of the protein in the intact cell. Two other ingredients that are most frequently added to purification buffers are mercaptoethanol and ethylenediaminetetraacetate (EDTA). The mercaptoethanol inhibits the oxidation of protein —SH groups, and the EDTA chelates

divalent cations. Divalent cations, even in trace amounts, can lead to aggregation problems or activate degradative enzymes.

The following two examples of purification show how various techniques can be effectively combined to produce purified proteins with minimum effort and loss of activity.

Purification of an Enzyme with Two Catalytic Activities

The last two steps in the biosynthesis of the mononucleotide uridine 5′-monophosphate (UMP) are catalyzed by (1) orotate phosphoribosyltransferase (OPRTase) and (2) orotate 5′-monophosphate decarboxylase (OMPDase).

Mary Ellen Jones and her colleagues set out to purify the enzyme or enzymes involved in these two reactions. Their main goal was to determine whether the two reactions are carried out by one protein or more than one. Their findings indicated that the two reactions were both catalyzed by the same enzyme, consisting of a single polypeptide chain. To demonstrate this fact, it was necessary to monitor both enzyme activities at each step in the purification and show that both activities copurified. For this purpose, Jones used specific enzyme assays for both enzyme activities. All fractions were assayed for both enzymatic activities at each stage of the purification.

The main data associated with the purification are summarized in table 6.4. This table indicates the total protein obtained in each step, the number of enzyme units[a] for each enzyme, and the ratio of enzyme units to total protein, called the specific activity. In the absence of enzyme inactivation, the specific activity should be directly proportional to the enrichment. The percent recovery refers to the amount of enzyme activity in the indicated fraction, as compared with the amount present in fraction 1. This number is usually less than 100%. The apparent losses may reflect actual losses of enzyme during purification, or they may reflect inactivation (usually due to unknown causes) of the enzyme during purification.

The nine steps involved in the purification of UMP synthase from starting tissue are outlined in figure 6.8. All steps were carried out at 0–5°C. About 200 g of Ehrlich ascites cells, a mammalian tumor rich in the desired enzymes, was suspended in buffer and processed in a tissue homogenizer, which mechanically breaks down the tissue

[a]Enzyme units are proportional to the amount of enzyme activity. The relationship between enzyme units and absolute amount of enzyme need not concern us here.

Table 6.4

Outline of Purification of UMP Synthase from Ehrlich's Ascites Carcinoma

Fraction	Volume (ml)	Protein (mg)	OMPDase[a]			OPRTase[a]			Ratio of OMPDase to OPRTase
			Units[b]	Sp. Act.[c]	Percent Recovery	Units[b]	Sp. Act.[c]	Percent Recovery	
1. Streptomycin fraction	1040	11,700	40.4	0.0034		20.5	0.0018		2.0
2. Dialyzed $(NH_4)_2SO_4$ fraction	144	311	24.3	0.0078	60	8.7	0.0028	42	2.8
3. Affinity column eluate (concentrated)	0.475	0.51	4.0	7.8[d]	10	0.35	0.69	3.3	11.4

[a] OMPDase = orotate 5'-monophosphate decarboxylase; OPRTase = orotate phosphoribosyltransferase.

[b] Units refer to total amount of enzyme activity.

[c] Specific activity refers to the units of enzyme activity divided by the total protein.

[d] This value represents a 2,300-fold enrichment from fraction 1.

and the cell membranes (step 1). Then EDTA and an —SH reagent were added to this total cell lysate. Solid streptomycin sulfate was also added with stirring (step 2). Streptomycin sulfate aggregates nucleic acids so that they may be more easily removed by centrifugation. The resulting slurry was subjected to high-speed centrifugation, and the resulting supernatant (see table 6.4, fraction 1) was carefully decanted for further processing (step 3).

Preliminary experiments had shown that the desired enzymes were in the 18.5%–28% $(NH_4)_2SO_4$ fraction. This knowledge served as the basis for the next three steps. First 239 g of solid $(NH_4)_2SO_4$ was added to 1040 ml of supernatant (step 4). The resulting precipitate was removed by centrifugation (step 5). Then an additional 120 g of $(NH_4)_2SO_4$ was added to the supernatant (step 6). The resulting slurry was centrifuged. This time the supernatant was discarded after centrifugation, leaving the sediment containing the enzyme activity for further processing (step 7). The sediment was resuspended in a dilute buffer for column chromatography (see table 6.4, fraction 2). Inspection of table 6.4 indicates only about a twofold increase in specific activity between fractions 1 and 2.

The main purification was achieved by two column steps, carried out in series. The first column was an affinity column containing an analog of UMP, 6-azauridine 5'-monophosphate (azaUMP), covalently attached to an agarose column support system. In dilute buffer, greater than 99% of the protein in fraction 2 is retained on this column. After thorough rinsing of the column with dilute buffer, 5×10^{-5} M azaUMP was added to the buffer. This addition resulted in the elution of the UMP synthase (step 8). Then the column eluant carrying the two enzyme activities associ-

ated with UMP synthase was resuspended in pure buffer and passed over a phosphocellulose column (step 9). Phosphocellulose was chosen because the negatively charged phosphate groups result in the retention of proteins by electrostatic attraction alone; they also resemble the phosphate groups in the naturally occurring enzyme substrate, OMP. Thus, the phosphocellulose column may be thought of as an ion-exchange column and an affinity column combined. In ordinary ion-exchange chromatography, the protein, after column loading, is eluted by increasing the ionic strength with a simple inorganic salt. In this example, Jones used a more specific method to elute the enzyme, which involved adding 10^{-5} M azaUMP and 2×10^{-5} M OMP to the original loading buffer. The addition does not substantially increase the ionic strength of the buffer. Therefore the only phosphocellulose-bound proteins that are likely to be eluted by this treatment are those with an especially high affinity for either of these nucleotides (azaUMP or OMP). This fact should greatly favor selective elution of those enzymes that carry specific sites for binding these nucleotides. The two column steps together resulted in an enzyme preparation (see table 6.4, fraction 3) that was approximately 2,300-fold purified over the starting material, as measured by the increase in specific activity.

The final product (see table 6.4, fraction 3) was examined by SDS gel electrophoresis and found to contain a single band with an estimated molecular weight of 51,000. The denaturing conditions of an SDS gel are expected to dissociate a multisubunit protein. Hence the SDS gel result indicated that the enzyme contains one type of polypeptide chain, but it does not tell us whether the enzyme contains one or more of these chains. Sedimentation analysis on a

Figure 6.8

Outline of purification scheme for UMP synthase from Ehrlich's ascites tumor cells of mice.

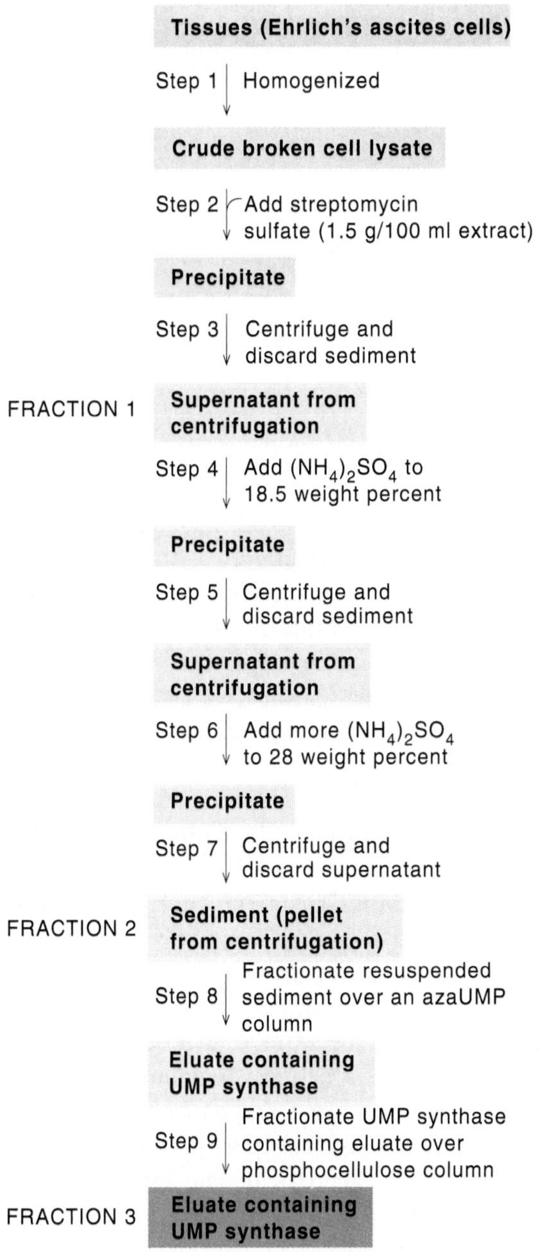

The conclusion drawn from these results was that both of the enzyme activities associated with UMP synthase—OMPDase and OPRTase—are contained in a single protein. The basis for the conclusion was that both enzyme activities are always present in the same fractions throughout the multistep purification. However, inspection of table 6.4 indicates a possible objection to this interpretation. In the columns showing percent recovery, it can be seen that substantial amounts of enzyme activity are lost for both enzymes during purification, but that considerably more activity is lost for the OPRTase. These losses could be due to actual loss of enzymes during purification or to some sort of inactivation of the enzyme sites. The preferential loss of OPRTase activity is emphasized by the last column in table 6.4, which gives the ratio of the two enzyme activities in the different fractions. Considered alone, these data could indicate that a separate catalytic unit of OPRTase is lost during purification. Jones thinks this is unlikely for two reasons: (1) the activity appears in no fractions other than with OMPDase during purification, and (2) the OPRTase activity is notably unstable. It appears, then, that both enzyme activities exist at distinct sites on a single protein. The greater loss in activity of one enzyme activity over the other is attributed to a greater sensitivity of one reaction site over the other on the enzyme surface.

Purification of a Membrane-Bound Protein

The second purification procedure we examine illustrates an unusual approach to the purification of a membrane-bound protein. The lactose carrier protein of *E. coli* is normally tightly bound to the plasma membrane. This protein is involved in the active transport of the disaccharide lactose across the cytoplasmic membrane. When lactose carrier protein is present, the intracellular concentration of lactose can achieve levels 1,000-fold higher than those found in the external medium. Ron Kaback devised a simple yet elegant procedure for the purification of this protein.

Purification of the membrane-bound lactose carrier protein is a very different problem from the purification of the soluble OMP synthase. Both the approach to purification and the assays for the protein during purification are quite novel. The assay involves reconstituting a transport system with membranes that are free of lactose carrier protein, then adding the partially purified carrier protein and radioactively labeled lactose. The activity in this assay system is proportional to the transport of radioactive lactose across the membrane in the cell-free reconstituted system.

The results of the purification steps are tabulated in table 6.5, and the purification procedure is outlined in

sucrose density gradient in a nondenaturing buffer indicated a single band with an estimated molecular weight of about 50,000. These two results taken together demonstrate that the enzyme in its native state contains a single polypeptide chain. The purity of the enzyme was also confirmed by isoelectric focusing and two-dimensional electrophoresis, with isoelectric focusing in the first direction followed by SDS gel electrophoresis in the second direction.

Table 6.5

Purification of the Lactose Carrier Protein

Fraction	Protein (mg)	Percent Recovery (Total Protein)	Percent Recovery (Carrier Protein)	Purification Factor
1. Membrane fraction	12.5	100	100	1.0
2. Urea-extracted membrane	5.6	45	76	1.7
3. Urea/cholate-extracted membrane	2.6	21	61	2.9
4. Octylglucoside extract	0.4	3.2	38	12
5. DEAE[a] column peak	0.056	0.4	14	35

[a] DEAE = diethylaminoethyl cellulose

figure 6.9. In this procedure advantage was taken of the fact that the carrier protein in its native state is firmly bound to the cytoplasmic membrane. Thus, the first step consisted of isolating these membranes from the rest of the cell constituents. Starting from the membrane fraction only 35-fold purification was required to achieve pure carrier protein. This rapid result was possible because a special strain of E. coli, containing about 100 times the normal carrier protein, was used as starting material for the purification. The high initial content was engineered by putting the carrier protein gene on a multicopy plasmid, which was then inserted into the cell—a procedure described in chapter 27.

Bacterial cells are much tougher than mammalian cells, requiring a more stringent procedure for cell disruption. In this case, the cells were placed in a so-called French pressure cell and bled through an orifice from very high pressures ($\approx$ 10,000 psi) to atmospheric pressure (step 1). Under these conditions the cells literally explode, fragmenting their membranes and releasing the cytoplasmic contents. The membranes were pelleted by a brief centrifugation, leaving a supernatant containing DNA, ribosomes, and cytoplasmic protein, which was removed by decantation (step 2). The pelleted membrane fraction was next resuspended and extracted with 5 M urea and then reextracted with 6% sodium cholate. The pellet obtained by high-speed centrifugation from these two extractions still contained most (61%) of the carrier protein in the more rapidly sedimenting membrane fraction (fraction 3), although 79% of the total membrane-bound protein was released by these treatments (steps 3 and 4).

At this point, the carrier protein was released from a suspension of the membranes by addition of the hydrophobic reagent octylglucoside in the presence of E. coli phospholipid (step 5). It is believed that the octyl part of octylglucoside competes effectively with the membrane for binding to hydrophobic centers on the carrier protein. The E. coli phospholipid facilitates dissociation of the carrier protein from the membrane fraction. The solubilized octylglucoside-containing extract was fourfold enriched in carrier protein after a high-speed centrifugation to remove residual membrane and membrane-bound proteins.

Finally, the octylglucoside-containing extract was passed over a positively charged diethylaminoethyl (DEAE) sepharose column (sepharose is a form of cross-linked polydextran) in a buffer containing 10 mM potassium phosphate, 20 mM lactose, and 0.25 mg of washed E. coli lipid per milliliter. The carrier protein passed through this column as a symmetrical peak of protein (step 6). Most of the remaining protein in the extract adsorbed to the positively charged column. The fractions containing the bulk of the carrier protein activity were judged to be pure by SDS gel electrophoretic analysis. The purified protein contained a single polypeptide chain with an estimated molecular weight of 33,000.

The two purifications described here are as different as the two proteins involved. No two purifications are exactly alike, but the principles of purification, as stated at the outset, are quite similar. Fortunately, an almost endless variety of purification techniques exists. This variety is both helpful and challenging because a great deal of knowledge and creativity are required to exploit it. In addition to professional expertise, the two most important things required to make a purification possible are an unambiguous assay for the protein in question and a means of stabilizing the protein during purification.

Figure 6.9

Outline of purification procedure for lactose
carrier protein from *E. coli.*

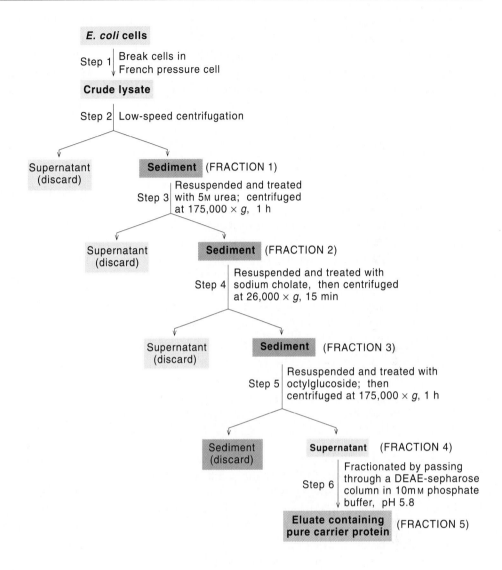

E. coli cells

Step 1 | Break cells in French pressure cell

Crude lysate

Step 2 | Low-speed centrifugation

Supernatant (discard)

Sediment (FRACTION 1)

Step 3 | Resuspended and treated with 5M urea; centrifuged at 175,000 × *g*, 1 h

Supernatant (discard)

Sediment (FRACTION 2)

Step 4 | Resuspended and treated with sodium cholate, then centrifuged at 26,000 × *g*, 15 min

Supernatant (discard)

Sediment (FRACTION 3)

Step 5 | Resuspended and treated with octylglucoside; then centrifuged at 175,000 × *g*, 1 h

Sediment (discard)

Supernatant (FRACTION 4)

Step 6 | Fractionated by passing through a DEAE-sepharose column in 10mM phosphate buffer, pH 5.8

Eluate containing pure carrier protein (FRACTION 5)

Summary

In this chapter we considered the different means of purification that enable us to study individual proteins in isolation. The following points are the most important.

1. Whereas proteins must be studied *in vivo* in their normal habitat, to characterize them in great detail they must also be isolated in pure form. Protein purification is a complex art, and a great variety of purification methods are usually applied in sequence for the purification of any protein.

2. Frequently the first step in protein purification is differential sedimentation of broken cell parts. In this way soluble proteins may be separated from organelle-sequestered proteins.

3. Following differential sedimentation, proteins may be separated into crude fractions by the addition of increasing amounts of $(NH_4)_2SO_4$. Specific proteins characteristically precipitate in a limited range of salt concentrations.

4. Column procedures are useful in fractionating proteins with different affinity properties and sizes. By the use of different column materials in conjunction with specific eluting solutions, highly purified protein preparations can be obtained.

5. Gel electrophoresis, which separates proteins according to their size and charge, can be used in purification or more frequently as a means of assaying the

purity during a purification. Isoelectric focusing, which separates proteins according to their isoelectric points, can be used for the same purpose.
6. The molecular weights of soluble proteins can be roughly estimated by SDS gel electrophoresis or can be rigorously determined by sedimentation techniques.

7. The purification of two proteins, UMP synthase from mammalian tumor cells, and lactose carrier protein from *E. coli* bacteria is described in detail to illustrate how different fractionation methods can be combined most effectively.

Selected Readings

Cantor, C. R., and P. Schimmel, *Biophysical Chemistry,* New York: W. H. Freeman, 1980. See especially volume 2, entitled *Study of Biophysical Structure and Function.*

Chaif, B. T., and S. B. H. Kent, Weighing naked proteins: Practical, high-accuracy mass measurement of peptides and proteins. *Science* 257:1885–1893, 1992. Proteins with molecular masses of as much as 100 kd or more can be analyzed at picomole sensitivities to give simple mass spectra corresponding to the intact molecule. This development has allowed unprecedented accuracy in the determination of protein molecular weights.

Eisenberg, D., and D. Crothers, *Physical Chemistry and Its Applications to the Life Sciences,* Menlo Park, Calif.: Benjamin/Cummings, 1979.

Methods in Enzymology, New York: Academic Press. A continuing series of over 250 volumes that discusses most methods at the professional level but is still understandable for students.

Scopes, R., *Protein Purification: Principles and Practice,* 2d ed. New York: Springer-Verlag, 1987. A recent general treatment of this subject.

Problems

1. Why are "salting out" procedures often used as an initial purification step following the production of a crude extract by centrifugation?
2. Rarely is DEAE-cellulose used above a pH of about 8.5. Can you provide a reason(s) why?
3. Given that the only structural difference between phosphorylase a and phosphorylase b is that phosphorylase a has a covalently bound phosphate on serine 14, do you expect phosphorylase a and phosphorylase b to elute as a single peak on DEAE-cellulose chromatography? What if gel filtration was utilized?

4. A method for the purification of 6-phosphogluconate dehydrogenase from *E. coli* is summarized in the table. For each step, calculate the specific activity, percentage yield, and degree of purification (*n*-fold). Indicate which step results in the greatest purification. Assume that the protein is pure after gel (Bio-Gel A) exclusion chromatography. What percentage of the initial crude cell extract protein was 6-phosphogluconate dehydrogenase?

Table for Problem 4

Step	Volume (ml)	Total Protein (mg)	Total Units	Specific Activity (U/mg)	Yield (%)	Purification (*n*-fold)
Cell extract	2,800	70,000	2,700			
((NH$_4$)$_2$SO$_4$) fractionation	3,000	25,400	2,300			
Heat treatment	3,000	16,500	1,980			
DEAE chromatography	80	390	1,680			
CM-cellulose	50	47	1,350			
Bio-Gel A	7	35	1,120			

5. Although used effectively in the 6-phosphogluconate dehydrogenase isolation procedure, heat treatment cannot be used in the isolation of all enzymes. Explain.

6. Assume that the isoelectric point (pI) of 6-phosphogluconate dehydrogenase is 6. Explain why the buffer used in the DEAE-cellulose chromatography must have a pH greater than 6 but less than 9 for the enzyme to bind to the DEAE resin.

7. Will the 6-phosphogluconate dehydrogenase bind to the CM-cellulose in the same buffer pH range used with the DEAE-cellulose? Explain. In what pH range would you expect the dehydrogenase to bind to CM-cellulose? Explain.

8. Examine the isolation procedure shown in problem 4 and explain why gel exclusion chromatography is used as the final step rather than as the step following the heat treatment.

9. A student isolated an enzyme from anaerobic bacteria and subjected a sample of the protein to SDS polyacrylamide gel electrophoresis. A single band was observed on staining the gel for protein. His adviser was excited about the result, but suggested that the protein be subjected to electrophoresis under nondenaturing (native) conditions. Electrophoresis under nondenaturing conditions revealed two bands after the gel was stained for protein. Assuming the sample had not been mishandled, offer an explanation for the observations.

10. A salt-precipitated fraction of ribonuclease contained two contaminating protein bands in addition to the ribonuclease. Further studies showed that one contaminant had a molecular weight of about 13,000 (similar to ribonuclease) but an isoelectric point 4 pH units more acidic than the pI of ribonuclease. The second contaminant had an isoelectric point similar to ribonuclease but had a molecular weight of 75,000. Suggest an efficient protocol for the separation of the ribonuclease from the contaminating proteins.

11. You have a mixture of proteins with the following properties:

Protein 1: M_r 12,000, pI = 10
Protein 2: M_r 62,000, pI = 4
Protein 3: M_r 28,000, pI = 8
Protein 4: M_r 9,000, pI = 5

Predict the order of emergence of these proteins when a mixture of the four is chromatographed in the following systems:
(a) DEAE-cellulose at pH 7, with a linear salt gradient elution.
(b) CM-cellulose at pH 7, with a linear salt gradient elution.
(c) A gel exclusion column with a fractionation range of 1,000–30,000 M_r, at pH 7.

12. You wish to purify an ATP-binding enzyme from a crude extract that contains several contaminating proteins. To purify the enzyme rapidly and to the highest purity, you must consider some sophisticated strategies, among them affinity chromatography. Explain how affinity chromatography can be applied to this separation, and explain the physical basis of the separation.

Catalysis

Chapter 7   Enzyme Kinetics   135
Chapter 8   How Enzymes Work   154
Chapter 9   Regulation of Enzyme Activities   175
Chapter 10   Vitamins and Coenzymes   198

Electron micrograph of a longitudinal section of a skeletal muscle fiber showing a number of myofibrils. Muscle protein is an unusual example of a structural protein with enzymatic activity. (Courtesy of Dr. Hugh Huxley)

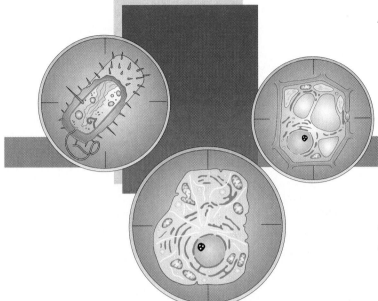

Enzyme Kinetics

Chapter Outline

The Discovery of Enzymes
Enzyme Terminology
Basic Aspects of Chemical Kinetics
    *A Critical Amount of Energy Is Needed for the
    Reactants to Reach the Transition State*
    *Catalysts Speed up Reactions by Lowering the Free
    Energy of Activation*
Kinetics of Enzyme-Catalyzed Reactions
    *Kinetic Parameters Are Determined by Measuring
    the Initial Reaction Velocity as a Function of the
    Substrate Concentration*
    *The Henri-Michaelis-Menten Treatment Assumes
    That the Enzyme–Substrate Complex Is in
    Equilibrium with Free Enzyme and Substrate*
    *Steady-State Kinetic Analysis Assumes That the
    Concentration of the Enzyme–Substrate Complex
    Remains Nearly Constant*
    *Kinetics of Enzymatic Reactions Involving Two
    Substrates*
    *Effects of Temperature and pH on Enzymatic
    Activity*
Enzyme Inhibition
    Competitive Inhibitors Bind at the Active Site
    *Noncompetitive and Uncompetitive Inhibitors Do
    Not Compete Directly with Substrate Binding*
    *Irreversible Inhibitors Permanently Alter the
    Enzyme Structure*

*An enzyme catalyzed reaction proceeds rapidly
under mild conditions because it lowers the
activation energy for a reaction; enzymes are
usually highly specific for the reactions they
catalyze.*

Most of this book is concerned with the reactions that
occur in living cells. These reactions are catalyzed by enzymes. In this and the following three chapters we focus on
the ways in which enzymes function. The present chapter
deals with the kinetics of enzyme-catalyzed reactions; the
next two explore the mechanisms of enzymatic catalysis.
Finally in chapter 10 we examine the small cofactors that
work together with many enzymes.

A catalyst is a substance that accelerates a chemical
reaction without itself undergoing any net change. The rate
enhancements achieved by many enzymes are extraordinarily high. For example, carbonic anhydrase, an enzyme
found in red blood cells, catalyzes the reaction

$$CO_2 + H_2O \longrightarrow H_2CO_3 \qquad (1)$$

In the presence of the enzyme, this reaction occurs about
10^7 times as rapidly as it does in the absence of the enzyme.
One molecule of carbonic anhydrase can hydrate about 10^6
molecules of CO_2 a second.

Kinetic analysis is one of the most basic topics of enzymology. Such studies reveal not only how fast an enzyme
can function, but also its preferences for various reactants
(or substrates as they usually are called), the effect of substrate concentration on the reaction rate, and the sensitivity

of the enzyme to specific inhibitors or activators. By studying the alterations in rate under different conditions we often gain clues to the mechanism of the reaction.

The Discovery of Enzymes

The existence of catalysts in biological materials was recognized as early as 1835 by Jöns Jakob Berzelius, the Swedish chemist who discovered several elements, introduced the way of writing chemical symbols, and coined the term "catalysis." Berzelius noted that potatoes contained something that catalyzed the breakdown of starch, and he suggested that all natural products are formed under the influence of such catalysts. But the chemical nature of biological catalysts was unknown, and it remained a mystery for many years. In the period between 1850 and 1860, Louis Pasteur demonstrated that fermentation, the anaerobic breakdown of sugar to CO_2 and ethanol, occurred in the presence of living cells and did not occur in a flask that was capped after any cells that it contained had been killed by heat. Then in 1897, Eduard Buchner discovered by accident that fermentation was catalyzed by a clear juice that he had prepared by grinding yeast with sand and filtering out the unbroken cells. Looking for a way to preserve the juice, Buchner had added sugar. It probably was a disappointment to him that the sugar was broken down rapidly and the mixture frothed with CO_2. But Buchner's discovery made it possible to explore metabolic processes such as fermentation in a greatly simplified system, without having to deal with the complexities of cell growth and multiplication, and without the barriers imposed by cell walls or membranes. Arthur Harden and William Young soon showed that yeast extracts contained two different types of molecules, both of which were necessary for fermentation to occur. Some were small, dialyzable, heat-stable molecules such as inorganic phosphate; others were much larger, nondialyzable molecules—the enzymes—that were destroyed easily by heat.

Although early investigators surmised that enzymes might be proteins, this remained in dispute until 1927, when James Sumner succeeded in purifying and crystalizing the enzyme urease from beans. In the 1930s, John Northrop isolated and characterized a series of digestive enzymes, generalizing Sumner's conclusion that enzymes are proteins. Since then thousands of different enzymes have been purified, and the structures of many of them have been solved to atomic resolution; almost all of these molecules have proved to be proteins. Surprisingly, however, recent work has shown that some RNA molecules also have enzymatic activity.

Figure 7.1

The six main classes of enzymes and the reactions they catalyze.

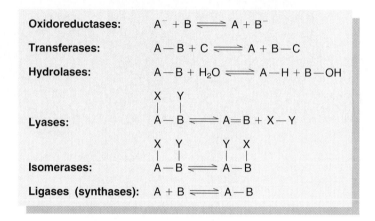

Enzyme Terminology

Enzymes often are known by common names obtained by adding the suffix "-ase" to the name of the substrate or to the reaction that they catalyze. Thus, glucose oxidase is an enzyme that catalyzes the oxidation of glucose; glucose-6-phosphatase catalyzes the hydrolysis of phosphate from glucose-6-phosphate; and urease catalyzes the hydrolysis of urea. Common names also are used for some groups of enzymes. For example, an enzyme that transfers a phosphate group from ATP to another molecule is usually called a "kinase," instead of the more formal "phosphotransferase."

A systematic scheme for classifying enzymes was adopted in 1972 by the International Union of Biochemistry. In this scheme, each enzyme is designated by four numbers that indicate the main class, subclass, subsubclass, and the serial number of the enzyme in its subsubclass. The six main classes (fig. 7.1) are (1) oxidoreductases, (2) transferases, (3) hydrolases, (4) lyases, (5) isomerases, and (6) ligases, or synthases. Oxidoreductases catalyze oxidation–reduction reactions. Transferases catalyze the transfer of a functional group from one molecule to another. Hydrolases catalyze bond cleavage by the introduction of water. Lyases catalyze the removal of a group to form a double bond or the addition of a group to a double bond. Isomerases catalyze intramolecular rearrangements, and ligases catalyze reactions that join two molecules.

Many enzymes require additional small molecules called cofactors for their activity. Cofactors can be simple inorganic ions such as Mg^{2+} or complex organic molecules known as coenzymes. The cofactor usually binds tightly to a special site on the enzyme. An enzyme lacking an essential cofactor is called an apoenzyme, and the intact enzyme with the bound cofactor is called the holoenzyme.

Basic Aspects of Chemical Kinetics

Before we delve into enzyme kinetics, we need to discuss some of the basic principles that apply to the kinetics of both enzymatic and nonenzymatic reactions. Let's first consider a nonenzymatic reaction that converts a single reactant (R) into a product (P):

$$R \longrightarrow P \qquad (2)$$

To measure the velocity of the reaction (v), we plot the concentration of R as a function of time (fig. 7.2). The rate at any particular time (t) is

$$v = -\frac{d[R]}{dt} \qquad (3)$$

where $d[R]/dt$ is the slope of the plot at that time. The minus sign is needed because v is defined by convention to be a positive number, whereas $d[R]/dt$ is always negative ([R] decreases with time). However, we might equally well choose to measure the increase in the concentration of the product as a function of time, in which case we express the rate as $v = d[P]/dt$.

For a simple reaction of this type, the rate at any given time usually is found to be proportional to the remaining concentration of the reactant:

$$v = k[R] \qquad (4)$$

The proportionality constant k is the rate constant. The rate constant is independent of the concentration of the reactant, but it can depend on other parameters, such as temperature or pH, and, as we shall see, it may be altered by a catalyst. It has dimensions of reciprocal seconds (s^{-1}). Combining equations (3) and (4), we have

$$\frac{d[R]}{dt} = -k[R] \qquad (5)$$

A reaction of this type is said to follow first-order kinetics because the rate is proportional to the concentration of a single species raised to the first power (fig. 7.2). An example is the decay of a radioactive isotope such as ^{14}C. The rate of decay at any time (the number of radioactive disintegrations per second) is simply proportional to the amount of ^{14}C present. The rate constant for this extremely slow nuclear reaction is $8 \times 10^{-12} \, s^{-1}$. Another example is the initial electron-transfer reaction that occurs when photosyn-

Figure 7.2

The kinetics of a first-order reaction in which a single reactant (R) is converted irreversibly to a product (P). The concentrations of R and P are plotted as functions of time. The rate (v) at any given time can be obtained from the slope of either curve:

$$v = -\frac{d[R]}{dt} = \frac{d[P]}{dt}$$

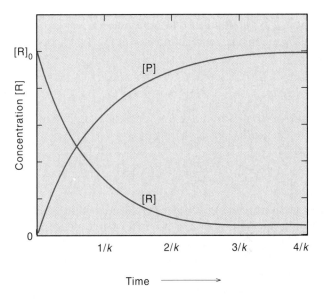

thetic organisms are excited with light (see chapter 15). In this case there actually are two reactants, the electron donor and the acceptor, but they are held close together on a protein so that they react as a unit; the excited complex simply decays spontaneously to a more stable state. This process occurs with a rate constant of $3 \times 10^{11} \, s^{-1}$, which makes it one of the fastest reactions known.

A slightly more complicated situation arises if the reaction is reversible:

$$R \underset{k_{-1}}{\overset{k_1}{\rightleftharpoons}} P \qquad (6)$$

Here k_1 is the rate constant for the forward reaction, and k_{-1} is that for the back reaction. In a reversible reaction, the rate equation becomes

$$\frac{d[R]}{dt} = -k_1[R] + k_{-1}[P] \qquad (7)$$

The term $-k_1[R]$ is the same as in equation (5); the term $k_{-1}[P]$ describes the formation of R from P. In this case, [R] and [P] proceed from their initial values to their final, equilibrium values ($[R]_{eq}$ and $[P]_{eq}$), which generally are not zero. At equilibrium, $d[R]/dt$ must go to zero, which requires that

$$k_1[R]_{eq} = k_{-1}[P]_{eq} \qquad (8)$$

Rearranging this expression gives

$$\frac{[P]_{eq}}{[R]_{eq}} = \frac{k_1}{k_{-1}} = K_{eq} \qquad (9)$$

where K_{eq} is the equilibrium constant.

Now consider a reaction involving two reactants, A and B:

$$A + B \xrightarrow{\ k\ } P \qquad (10)$$

The rate of such a bimolecular reaction usually depends on the concentrations of both reactants:

$$\frac{d[P]}{dt} = k[A][B] \qquad (11)$$

The reaction is said to follow second-order kinetics because its rate is proportional to a product of two concentrations. The kinetics are first order in either [A] or [B] alone, but second order overall. The rate constant for a second-order reaction has the dimensions of $M^{-1}s^{-1}$.

A Critical Amount of Energy Is Needed for the Reactants to Reach the Transition State

The equilibrium constant K_{eq} for a reaction is strictly a function of thermodynamic factors. The most pertinent thermodynamic quantity here is the change in free energy that occurs in the reaction ΔG. As we saw in chapter 2, ΔG and K_{eq} are related by the expression $\Delta G° = -RT \ln K_{eq}$. If the free energy of the products is less than the free energy of the reactants ($\Delta G < 0$), then the equilibrium constant favors the formation of products; if the free energy of the products exceeds that of the reactants ($\Delta G > 0$), the reverse reaction is favored. We noted above that the equilibrium constant for a one-step reaction is equal to the ratio of the rate constants in the forward and reverse directions [equation (9)]. However, the value of K_{eq} does not tell us how long it takes the reaction to reach equilibrium because it says nothing about the magnitudes of the individual rate constants. These may be very large or very small. The overall ΔG for a reaction therefore indicates only whether the reaction is possible, not how rapidly the reaction occurs. To understand the kinetics, we must look into the mechanism of the reaction.

The rate at which two molecules react depends partly on how frequently the molecules collide. Collisions occur as a result of random diffusion of the reactants in the solution. The number of collisions per second is proportional to the product of the two concentrations, and the second-order rate constant k for the reaction includes a proportionality factor for this relationship. The rate constant also includes a factor that gives the fraction of the collisions that are effective. If every collision were to result in a reaction, this second factor would be 1 and the rate constant for the reaction of two small molecules in aqueous solution would be about $10^{11} M^{-1}s^{-1}$. Because of their large masses, proteins diffuse relatively slowly, so the frequency at which a protein and a small molecule collide is lower than the collision frequency for two small molecules. As a consequence, a reaction between a protein and a small molecule has a maximum rate constant on the order of 10^8 to $10^9 M^{-1}s^{-1}$.

But not every collision results in a reaction. What determines the fraction of the collisions that are effective? A partial answer is that the colliding species must have a certain critical energy in order to surmount a barrier that separates the reactants from the products. This is illustrated schematically in figure 7.3. The surface in figure 7.3a represents the energy of a system in which a proton can be bound to either of two molecules. Suppose the proton is initially on molecule A, and we are interested in how rapidly it moves to molecule B. As the proton moves from one place to the other, its electrostatic interactions with molecule A become less favorable, and the interactions with B improve. The energy of the system goes through a maximum when the proton is at an intermediate position. At this point, the system is said to be in the transition state. The probability that a collision leads to a reaction depends, in part, on the probability that the molecules collide with enough energy to reach this state.

The amount of free energy required to reach the transition state is called the activation free energy, $\Delta G^{\ddagger}$. From equation (13) of chapter 2, we can equate $\Delta G^{\ddagger}$ to $-RT \ln K^{\ddagger}$, where $K^{\ddagger}$ is an equilibrium constant for the formation of the transition state from the reactants. The fraction of the reactants that are in the transition state at any given moment is given approximately by $K^{\ddagger}$, or $e^{-\Delta G^{\ddagger}/RT}$. We therefore can write the overall rate constant for a reaction as

$$k = k_0 e^{-\Delta G^{\ddagger}/RT} \qquad (12)$$

Here k_0 is an intrinsic rate constant for conversion of the transition state into the products and typically depends on the nuclear vibration frequency of a bond that is being formed or broken.

Figure 7.3

Schematic energy diagrams for a reaction in which a proton moves from one molecule to another. In drawing (a), coordinates in the plane at the bottom represent the location of the proton. The free energy of the system for a particular set of coordinates is represented by the distance of the cuplike surface above the plane. The two minima in the energy surface indicate the positions of the proton when it is bound optimally to one molecule or the other. The best route along the surface from one of these minima to the other goes through a pass, or saddle point. Drawing (b) shows a plot of the free energy as a function of distance along the optimal route over this pass. The activation free energy of the reaction ($\Delta G^{\ddagger}_a$) is the difference between the energies at the pass and at the starting point.

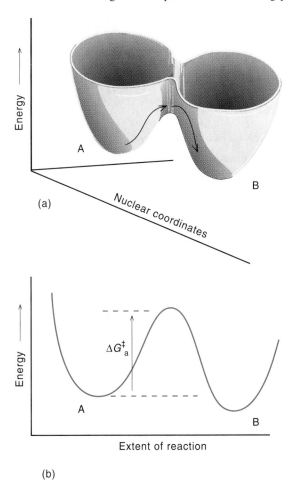

(a)

(b)

Catalysts Speed up Reactions by Lowering the Free Energy of Activation

Equation (12) indicates that, if $\Delta G^{\ddagger}$ is greater than zero as is most often the case, a reaction can be sped up either by decreasing $\Delta G^{\ddagger}$ or by raising the temperature. Little can usually be done to change k_0. For living organisms, it gener-

Figure 7.4

An enzyme speeds up a reaction by decreasing $\Delta G^{\ddagger}$. The enzyme does not change the free energy of the substrate (S) or product (P); it lowers the free energy of the transition state. The two vertical arrows indicate the activation free energies ($\Delta G^{\ddagger}$) of the catalyzed and uncatalyzed reactions.

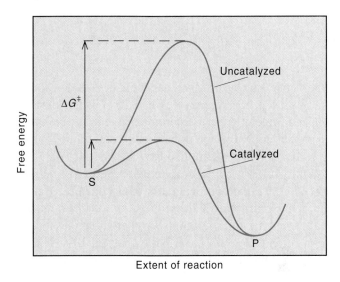

ally is impractical to raise the temperature because many organisms have little or no control over the ambient temperature and can survive only within a narrow range of temperatures. Also, changing the temperature is not a very selective way to control reaction rates because most reactions speed up when the temperature is raised. The remaining alternative is to decrease $\Delta G^{\ddagger}$.

Because rate constants depend exponentially on $-\Delta G^{\ddagger}/RT$ [equation (12)], small changes in $\Delta G^{\ddagger}$ will have a large effect on the reaction rate. At physiological temperatures, it takes a decrease of only 1.36 kcal/mole to speed up a reaction by a factor of 10, and a decrease by 8.16 kcal/mole increases the rate by a factor of 10^6. These are relatively modest free energy changes because forming a single H bond can release anywhere from 4 to 10 kcal/mole. In addition, because $\Delta G^{\ddagger}$ depends on the structures of the reactants and products and on the detailed nature of the reaction, it affords an opportunity to control reaction rates with a great deal of specificity. Enzymes, then, must work by decreasing the activation free energies for the specific reactions that they catalyze (fig. 7.4). They do this in a variety of ways that we discuss in detail in the next chapter. Our main concern here is to describe in more general terms how the kinetic properties of enzymatic reactions are measured and characterized.

It should be kept in mind that although an enzyme can increase the rate at which a reaction occurs, it (or any other catalyst) cannot alter the overall equilibrium constant. Since K_{eq} is equal to the ratio of the rate constants for the forward and reverse processes, the catalyst increases *both* of these rate constants without changing their ratio.

Kinetics of Enzyme-Catalyzed Reactions

Kinetic analysis was used to characterize enzyme-catalyzed reactions even before enzymes had been isolated in pure form. As a rule, kinetic measurements are made on purified enzymes *in vitro*. But the properties so determined must be referred back to the situation *in vivo* to ensure they are physiologically relevant. This is important because the rate of an enzymatic reaction can depend strongly on the concentrations of the substrates and products, and also on temperature, pH, and the concentrations of other molecules that activate or inhibit the enzyme. Kinetic analysis of such effects is indispensable to a comprehensive picture of an enzyme.

Kinetic Parameters Are Determined by Measuring the Initial Reaction Velocity as a Function of the Substrate Concentration

The usual procedure for measuring the rate of an enzymatic reaction is to mix enzyme with substrate and observe the formation of product or disappearance of substrate as soon as possible after mixing, when the substrate concentration is still close to its initial value and the product concentration is small. The measurements usually are repeated over a range of substrate concentrations to map out how the initial rate depends on concentration. Spectrophotometric techniques are used commonly in such experiments because in many cases they allow the concentration of a substrate or product in the mixture to be measured continuously as a function of time.

Measurements of reactions that occur in less than a few seconds require special techniques to speed up the mixing of the enzyme and substrate. One way to achieve this is to place solutions containing the enzyme and the substrate in two separate syringes. A pneumatic device then is used to inject the contents of both syringes rapidly into a common chamber that resides in a spectrophotometer for measuring the course of the reaction (fig. 7.5). Such an apparatus is referred to as a "stopped-flow" device because the flow stops abruptly when the movement of the pneumatic driver is arrested. In this type of apparatus it is possible to make kinetic measurements within about 1 ms after mixing of enzyme and substrate.

Figure 7.5

The stopped-flow apparatus for measuring enzyme-catalyzed reactions very soon after mixing enzyme and substrate.

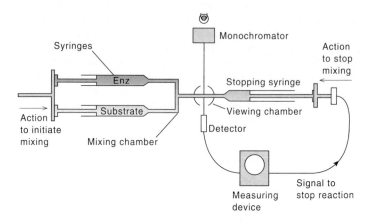

Equations (5) and (10) imply that the velocity of an uncatalyzed reaction increases indefinitely with an increase in the concentration of the reactants. With enzyme-catalyzed reactions, something very different is observed. The rate usually increases linearly with substrate concentration at low concentrations, but then levels off and becomes independent of the concentration at high concentrations (fig. 7.6). The explanation for this hyperbolic dependence on substrate concentration is straightforward. For an enzyme to affect $\Delta G^{\ddagger}$, the substrate must bind to a special site on the protein, the active site (fig. 7.7). At very low concentrations of substrate, the active sites of most of the enzyme molecules in the solution are unoccupied. Increasing the substrate concentration brings more enzyme molecules into play, and the reaction speeds up. At high concentrations, on the other hand, most of the enzyme molecules have their active sites occupied, and the observed rate depends only on the rate at which the bound reactants are converted into products. Further increases in the substrate concentration then have little effect.

The Henri-Michaelis-Menten Treatment Assumes That the Enzyme–Substrate Complex Is in Equilibrium with Free Enzyme and Substrate

The hyperbolic saturation curve that is commonly seen with enzymatic reactions led Leonor Michaelis and Maude Menten in 1913 to develop a general treatment for kinetic analysis of these reactions. Following earlier work by Victor Henri, Michaelis and Menten assumed that an enzyme–substrate complex (ES) is in equilibrium with free enzyme

Figure 7.6

The reaction velocity v as a function of the substrate concentration [S] for an enzyme-catalyzed reaction. At high substrate concentrations the reaction velocity reaches a limiting value, V_{max}. K_m is the substrate concentration at which the rate is half maximal.

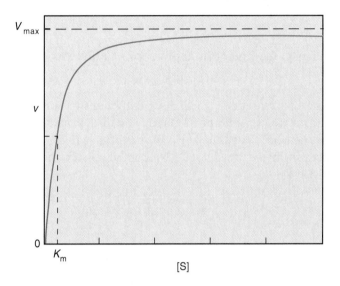

Figure 7.7

Thermolysin is an enzyme that hydrolyzes peptide bonds. Close-up view of the active site of thermolysin (beige) illustrating the enzyme-substrate interaction between a tight binding substrate analog, phosphoramidon (yellow), and the enzyme. (Based on the crystal structure described by D. E. Tronrud, A. F. Monzingo, and B. W. Matthews. Copyright 1994 by the Scripps Research Institute/Molecular Graphics Images by Michael Pique using software by Yng Chen, Michael Connolly, Michael Carson, Alex Shah, and AVS, Inc. Visualization advice by Holly Miller, Wake Forest University Medical Center.)

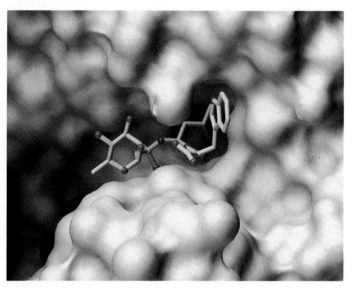

(E) and substrate (S), and that the formation of products (P) proceeds only through the ES complex:

$$E + S \underset{k_{-1}}{\overset{k_1}{\rightleftharpoons}} ES \xrightarrow{k_2} E + P \qquad (13)$$

Their objective was to relate the reaction rate to observable quantities and interpretable molecular parameters.

Because the slow step in the reaction described by equation (13) is assumed to be formation of E and P from ES, the velocity of the reaction should be

$$v = \frac{d[P]}{dt} = k_2[ES] \qquad (14)$$

A maximum velocity (V_{max}) is obtained when all of the enzyme is in the form of the enzyme–substrate complex. Since the total concentration of enzyme [E_t] is equal to [E] + [ES],

$$V_{max} = k_2[E_t] = k_2([E] + [ES]) \qquad (15)$$

Dividing equation (14) by equation (15) gives

$$\frac{v}{V_{max}} = \frac{[ES]}{[E] + [ES]} \qquad (16)$$

The fraction on the right-hand side can be evaluated by making use of the dissociation constant of the ES complex in equation (13), K_s:

$$K_s = \frac{[E][S]}{[ES]} \quad \text{and} \quad [ES] = \frac{[E][S]}{K_s} \qquad (17)$$

Substituting this value of [ES] into equation (16) and rearranging, gives

$$v = \frac{V_{max}[S]}{K_s + [S]} \qquad (18)$$

Equation (18) is the Henri-Michaelis-Menten equation, which relates the reaction velocity to the maximum velocity, the substrate concentration, and the dissociation constant for the enzyme–substrate complex. Usually substrate is present in much higher molar concentration than enzyme, and the initial period of the reaction is examined so that the free substrate concentration [S] is approximately equal to the total substrate added to the reaction mixture.

Steady-State Kinetic Analysis Assumes That the Concentration of the Enzyme–Substrate Complex Remains Nearly Constant

Rather than discussing the implications of equation (18) at this point, it is useful to develop a more general expression that avoids the assumption that the enzyme–substrate complex is in equilibrium with free enzyme and substrate. To develop this expression, we introduce the concept of the steady state, which was first pro-

Figure 7.8

Concentrations of free enzyme (E), substrate (S), enzyme–substrate complex (ES), and product (P) over the time course of a reaction. The shaded portion of the top graph is shown in expanded form in the bottom graph. After a brief initial period (usually less than a few seconds) the concentration of ES remains approximately constant for an extended period. The steady-state approximation is applicable during this second period. Most measurements of enzyme kinetics are made in the steady state.

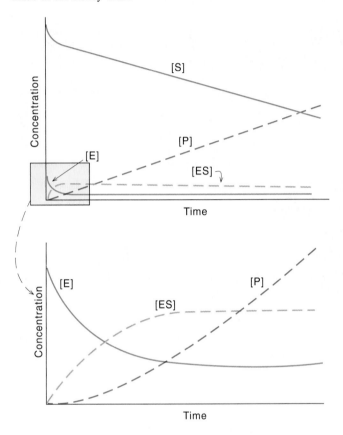

posed by G. E. Briggs and J. B. S. Haldane in 1925. The steady state constitutes the time interval when the rate of the reaction is approximately constant. The system usually reaches a steady state soon after enzyme and substrate are mixed, following a brief period when the concentration of the enzyme–substrate complex (ES) builds up (fig. 7.8). The ES concentration then remains almost constant for the duration of the steady state, while the concentrations of the substrate and product continue to change substantially.[a]

Let's now write out the reaction of equation (13) in more detail.

$$E + S \underset{k_{-1}}{\overset{k_1}{\rightleftharpoons}} ES \underset{k_{-2}}{\overset{k_2}{\rightleftharpoons}} E + P \qquad (19)$$

In this more complete scheme, the ES complex forms from E and S with rate constant k_1. ES can either dissociate again with rate constant k_{-1} or go on to P with k_2. If we confine

ourselves to measuring the initial rate of the reaction in the steady state, we can continue to neglect regeneration of ES from E and P (the step involving k_{-2}) because the concentration of P will be too small for this back-reaction to occur at a significant rate. For the rate of formation of ES, v_f, we then can write

$$v_f = k_1[E][S] \qquad (20)$$

Similarly, the rate of disappearance of ES, v_d, is

$$v_d = k_{-1}[ES] + k_2[ES] \qquad (21)$$

If the concentration of ES is virtually constant during the steady state, the rates of formation and disappearance of ES must be nearly equal, $v_f = v_d$. We therefore can describe the situation in the steady state by combining equations (20) and (21).

$$k_1[E][S] = (k_{-1} + k_2)[ES] \qquad (22)$$

or

$$\frac{[E][S]}{[ES]} = \frac{k_{-1} + k_2}{k_1} = K_m \qquad (23)$$

The constant K_m defined in equation (23) is called the Michaelis constant and is one of the key parameters in enzyme kinetics. It is a simple matter to proceed from this point to an expression comparable to the Henri-Michaelis-Menten equation (18), but with K_m in place of K_s. First, rearranging equation (23) gives

$$[ES] = \frac{[E][S]}{K_m} \qquad (24)$$

With this expression for [ES] we can follow the same procedure that led to equation (18), this time arriving at the Briggs-Haldane equation for the reaction velocity.

$$v = \frac{V_{max}[S]}{[S] + K_m} \qquad (25)$$

Equation (25) is identical to (18) except that the more complex constant K_m has replaced K_s. The term "Michaelis-Menten equation" is often used for either expression.

For the purpose of graphical representation of experimental data, it is convenient to rearrange equation (25). Taking the reciprocals of both sides of equation (25) gives

$$\frac{1}{v} = \frac{1}{V_{max}} + \frac{K_m}{V_{max}} \frac{1}{[S]} \qquad (26)$$

---

[a] The steady state is no stranger to the living cell. Most reactions in a living cell are in a steady state most of the time. Thus, the steady state used originally as a convenience by kineticists is also an appropriate way to analyze enzymatic reactions *in vivo*.

Figure 7.9

A plot of the reciprocal of the rate ($1/v$) as a function of the reciprocal of the substrate concentration ($1/[S]$) fits a straight line. Extrapolating the line to its intercept on the ordinate (infinite substrate concentration) gives $1/V_{max}$. Extrapolating to the intercept on the abscissa gives $-1/K_m$.

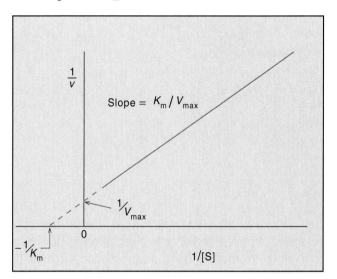

Table 7.1

The Michaelis Constants for Some Enzymes

Enzyme and Substrate	K_m (M)
Catalase	
$\quad$ H_2O_2	1.1
Hexokinase	
$\quad$ Glucose	1.5×10^{-4}
$\quad$ Fructose	1.5×10^{-3}
Chymotrypsin	
$\quad$ N-Benzoyltyrosinamide	2.5×10^{-3}
$\quad$ N-Formyltyrosinamide	1.2×10^{-2}
$\quad$ N-Acetyltyrosinamide	3.2×10^{-2}
$\quad$ Glycyltyrosinamide	1.2×10^{-1}
Aspartate aminotransferase	
$\quad$ Aspartate	9.0×10^{-4}
$\quad$ α-Ketoglutarate	1.0×10^{-4}
Fumarase	
$\quad$ Fumarate	5.0×10^{-6}
$\quad$ Malate	2.5×10^{-5}

This expression indicates that a plot of $1/v$ versus $1/[S]$ fits a straight line with a slope of K_m/V_{max} (fig. 7.9). Such a plot is known as a Lineweaver-Burk, or double-reciprocal, plot. The intercept of the line on the ordinate occurs at $1/v = 1/V_{max}$, and the intercept on the abscissa occurs at $1/[S] = -1/K_m$. V_{max} and K_m thus can be determined readily from the graph.

Significance of the Michaelis Constant, K_m. The Michaelis constant K_m has the dimensions of a concentration (molarity), because k_{-1} and k_2, the two rate constants in the numerator of equation (23), are first-order rate constants with units expressed per second (s^{-1}), whereas the denominator k_1 is a second-order rate constant with units of $M^{-1}s^{-1}$. To appreciate the meaning of K_m, suppose that $[S] = K_m$. The denominator in equation (25) then is equal to $2[S]$, which makes the velocity $v = V_{max}/2$. Thus, the K_m is the substrate concentration at which the velocity is half maximal (fig. 7.6).

K_m values for several enzyme–substrate pairs are given in table 7.1. The values vary over a wide range but typically lie between 10^{-6} and 10^{-1} M. With enzymes that can act on several different substrates, K_m can vary substantially from substrate to substrate. With the enzyme chymotrypsin, for example, the K_m for the substrate glycyltyrosinamide is about 50 times that for the substrate N-benzoyltyrosinamide.

Does the K_m indicate how tightly a particular substrate binds to the active site of an enzyme? Not necessarily. According to equation (23),

$$K_m = \frac{k_{-1} + k_2}{k_1} \qquad (23')$$

whereas the dissociation constant of the ES complex is

$$K_s = \frac{k_{-1}}{k_1} \qquad (27)$$

Comparing these two expressions, we see that K_m must always be larger than K_s. If $k_2 \gg k_{-1}$, then $K_m \approx k_2/k_1$, which is much greater than K_s. On the other hand, if $k_{-1} \gg k_2$, then K_m is approximately equal to K_s. In the latter circumstance, a smaller K_m means a smaller dissociation constant, which implies tighter binding to the enzyme. (This is the limiting case assumed in the Henri-Michaelis-Menten treatment, for when $k_{-1} \gg k_2$, the enzyme–substrate complex is in equilibrium with the free enzyme and substrate.) But whether or not this is a valid approximation depends on the particular enzyme and the substrate. For a multistep reaction the relationship between K_m and the rate constants can be considerably more complex.

Significance of the Turnover Number, k_{cat}. The turnover number of an enzyme, k_{cat}, is the maximum number of molecules of substrate that could be converted to product each second per active site. Because the maximum rate is obtained at high substrate concentrations, when all the active sites are occupied with substrate, the turnover number is a measure of how rapidly an enzyme can operate once the active site is filled. This is given simply by

$$k_{cat} = \frac{V_{max}}{[E_t]} \qquad (28)$$

Turnover numbers for some representative enzymes are listed in table 7.2. The enormous value of 4×10^7 molecules/s achieved by catalase is among the highest known; the low value for lysozyme is at the other end of the spectrum. As is the case with K_m, the relationship of k_{cat} to individual rate constants, such as k_2 and k_3, depends on the details of the reaction mechanism.

Significance of the Specificity Constant, k_{cat}/K_m. Under physiological conditions, enzymes usually do not operate at saturating substrate concentrations. More typically, the ratio of the substrate concentration to the K_m is in the range of 0.01–1.0. If [S] is much smaller than K_m, the denominator of the Briggs-Haldane equation [equation (25)] is approximately equal to K_m, so that the velocity of the reaction becomes

$$v \approx \frac{V_{max}}{K_m}[S] \qquad (when\ [S] \ll K_m)$$

$$= \frac{k_{cat}}{K_m}[E_t][S] \qquad (29)$$

The ratio k_{cat}/K_m is referred to as the specificity constant. Equation (29) indicates that the specificity constant provides a measure of how rapidly an enzyme can work at low [S]. Table 7.3 gives the values of the specificity constants for some particularly active enzymes.

The specificity constant k_{cat}/K_m is useful for comparing the relative abilities of different compounds to serve as a substrate for the same enzyme. If the concentrations of two substrates are the same, and are small relative to the K_m values, the ratio of the rates when the two substrates are present is equal to the ratio of the specificity constants.

Another use of the specificity constant is for comparing the rate of an enzyme-catalyzed reaction with the rate at which random diffusion brings the enzyme and substrate into contact. We mentioned previously that if every collision between a protein and a small molecule results in a reaction, the maximum value of the second-order rate constant is on the order of 10^8 to 10^9 $M^{-1}s^{-1}$. Some of the values of k_{cat}/K_m in table 7.3 are in this range. The reactions

Table 7.2

Values of k_{cat} for Some Enzymes

Enzyme	$k_{cat}\ (s^{-1})$
Catalase	40,000,000
Carbonic anhydrase	1,000,000
Acetylcholinesterase	14,000
Penicillinase	2,000
Lactate dehydrogenase	1,000
Chymotrypsin	100
DNA polymerase I	15
Lysozyme	0.5

these enzymes catalyze proceed at nearly the maximum possible speed, given a fixed, low concentration of substrate and given the restriction that the enzyme and substrate have to find each other by diffusion. The only practical way to go much faster is to have the substrate generated right on the enzyme or in its immediate vicinity, so that little diffusional motion is necessary.

Kinetics of Enzymatic Reactions Involving Two Substrates

Enzymes that catalyze reactions with two or more substrates work in a variety of ways. In some cases, the intermolecular reaction occurs when all the substrates are bound in a common enzyme–substrate complex; in others, the substrates bind and react one at a time. A frequent application of kinetic measurements is to distinguish between such alternatives.

Consider a reaction in which two substrates, S_1 and S_2, are converted to products P_1 and P_2. One way for the reaction to occur is for S_1 to bind to the enzyme first, forming the binary complex ES_1. Binding of S_2 can then form the ternary complex ES_1S_2, which gives rise to the products

$$E \xrightarrow{\ S_1\ } ES_1 \xrightarrow{\ S_2\ } ES_1S_2 \xrightarrow{\ P_1 + P_2\ } E \qquad (30)$$

This process is referred to as an ordered pathway. An alternative is a random-order pathway, in which the two substrates can bind to the enzyme in either order. Still another scheme is for S_1 to bind to the enzyme and be converted to

Table 7.3

Enzymes for Which k_{cat}/K_m Is Close to the Diffusion-Controlled Association Rate

Enzyme	Substrate	k_{cat} (s⁻¹)	K_m (M)	k_{cat}/K_m (M⁻¹s⁻¹)
Acetylcholinesterase	Acetylcholine	1.4×10^4	9×10^{-5}	1.6×10^8
Carbonic anhydrase	CO_2	1×10^6	0.012	8.3×10^7
	HCO_3^-	4×10^5	0.026	1.5×10^7
Catalase	H_2O_2	4×10^7	1.1	4×10^7
Crotonase	Crotonyl-CoA	5.7×10^3	2×10^{-5}	2.8×10^8
Fumarase	Fumarate	800	5×10^{-6}	1.6×10^8
	Malate	900	2.5×10^{-5}	3.6×10^7
Triosephosphate isomerase	Glyceraldehyde 3-phosphate	4.3×10^3	4.7×10^{-4}	2.4×10^8
β-Lactamase	Benzylpenicillin	2.0×10^3	2×10^{-5}	1×10^8

Source: From Alan Fersht, *Enzyme Structure and Mechanism,* 3d ed. Copyright © 1985 by W. H. Freeman & Company, New York. Reprinted with permission.

P_1, leaving the enzyme in an altered form, E′. S_2 then binds to E′ and is converted to P_2, returning the enzyme to its original form.

$$E \xrightarrow{\;S_1\;} ES_1 \xrightarrow{\;P_1\;} E' \xrightarrow{\;S_2\;} E'S_2 \xrightarrow{\;P_2\;} E \quad (31)$$

This process is called the Ping-Pong mechanism to emphasize the bouncing of the enzyme between two states, E and E′. Ping-Pong pathways are commonly observed with enzymes that contain bound coenzymes. Interconversion of the enzyme between the two forms usually involves modification of the coenzyme.

Kinetic equations for these and other mechanisms can be worked out just as we have done for reactions involving only one substrate. Techniques for doing this are described in the references at the end of the chapter; here we simply illustrate a few of the results. For the Ping-Pong mechanism [equation (31)], the double-reciprocal form of the final expression is

$$\frac{1}{v} = \frac{1}{V_{max}}\left(1 + \frac{K_{m2}}{[S_2]}\right) + \frac{K_{m1}}{V_{max}}\frac{1}{[S_1]} \quad (32)$$

where K_{m1} is the Michaelis constant for S_1, and K_{m2} is that for S_2. This expression is similar to that for a one-substrate reaction (equation 26), except that the first term on the right is multiplied by the factor $(1 + K_{m2}/[S_2])$. If we measure the rate as a function of $[S_1]$, keeping $[S_2]$ constant, a plot of $1/v$ versus $1/[S_1]$ is linear, but the intercept on the ordinate (the apparent V_{max}) depends on $[S_2]$. Increasing $[S_2]$ in-

Figure 7.10

Double-reciprocal plots ($1/v$ versus $1/[S_1]$) for the Ping-Pong mechanism. Measurements made at different values of $[S_2]$ give a set of parallel straight lines. The intercepts on the ordinate depend on $[S_2]$. K_{m1}, K_{m2}, and V_{max} can be obtained by replotting the intercepts as a function of $1/[S_2]$.

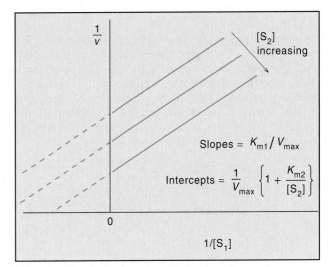

creases the apparent V_{max} (fig. 7.10). From a series of such plots, measured at different values of $[S_2]$, we can find the true V_{max} in addition to K_{m1} and K_{m2}.

An ordered pathway [equation (30)] results in the expression

$$\frac{1}{v} = \frac{1}{V_{max}}\left(1 + \frac{K_{m2}}{[S_2]} + \frac{K_{m1}}{[S_1]} + \frac{K_{m2}}{[S_2]}\frac{K_{s1}}{[S_1]}\right) \quad (33)$$

Figure 7.11

Double-reciprocal plots for an ordered pathway. Measurements made at different fixed values of $[S_2]$ give a set of lines that intersect to the left of the ordinate. The two values of K_m, V_{max} and K_{s1} can be obtained by replotting the slopes and intercepts of these lines as functions of $1/[S_2]$. A random pathway gives similar results, but can be distinguished by making such measurements for the reverse reaction $(P_1 + P_2 \rightarrow S_1 + S_2)$ in addition to the forward reaction.

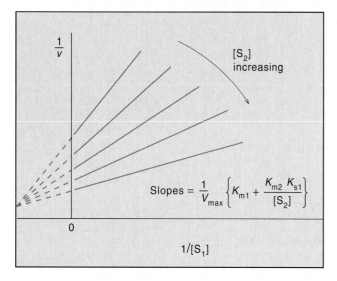

Figure 7.12

The activity of a typical enzyme as a function of temperature. The turnover number (k_{cat}) increases with temperature until a point is reached at which the enzyme is no longer stable. The temperature at which k_{cat} is greatest should not be interpreted as the "optimum temperature" for the enzyme. Because denaturation of the enzyme occurs continuously during the measurement, the position of the maximum depends on how quickly the experimenter is able to assay the enzyme activity.

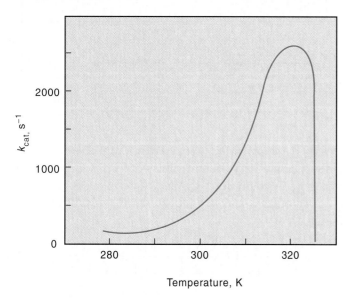

where K_{s1} is the dissociation constant for ES_1. A plot of $1/v$ versus $1/[S_1]$ at constant $[S_2]$ is still linear, but now both the slope and the intercept depend on $[S_2]$ (fig. 7.11). Again, all of the kinetic parameters can be obtained from a series of plots measured at different S_2 concentrations. However, additional measurements must be made to determine which substrate binds to the enzyme first. In some cases, the substrate that binds first can be shown to form a stable enzyme–substrate complex in the absence of the other substrate.

Effects of Temperature and pH on Enzymatic Activity

For most enzymes, the turnover number increases with temperature until a temperature is reached at which the enzyme is no longer stable (fig. 7.12). Above this point there is a precipitous, and usually irreversible, drop in activity. At lower temperatures, the temperature dependence of k_{cat} can be related to the activation energy of the slowest (rate-limiting) step in the catalytic pathway [see equation (12)]. With many enzymes a 10°C rise in temperature increases k_{cat} by about a factor of 2, which translates into an activation energy of about 12 kcal/mole.

Enzymes, like other proteins, are stable over only a limited range of pH. Outside this range, changes in the charges on ionizable amino acid residues result in modifications of the tertiary structure of the protein and eventually lead to denaturation. But within the range at which an enzyme is stable, both k_{cat} and K_m often depend on pH. The effects of pH can reflect the pK_a of ionizing groups on either the enzyme or the substrate. A substrate that has an amine group, for example, may bind to the enzyme best when this group is protonated. In many cases, however, the pH dependence reflects ionizable residues that constitute the active site on the enzyme or are essential for maintaining the structure of the active site, and the optimum pH is a characteristic more of the enzyme than of the substrate. Thus, the maximum activity of chymotrypsin always occurs around pH 8, the activity of pepsin peaks around pH 2, and acetylcholinesterase works best at pH 7 or higher (fig. 7.13). The activity of papain, on the other hand, is essentially independent of pH between 4 and 8.

Enzyme Inhibition

Most enzymes are sensitive to inhibition by specific agents that interfere with the binding of a substrate at the active site or with conversion of the enzyme–substrate complex into products. Two of the major applications of kinetic measurements are in distinguishing between

Figure 7.13

Enzyme activity (k_{cat}/K_m) as a function of pH for three different enzymes. The optimum pH usually is a characteristic of the enzyme and not the particular substrate. Often the pH sensitivity is an indication of an ionizable group at the active site.

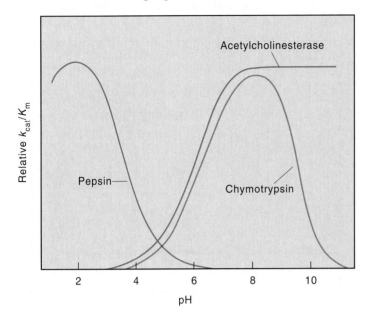

Figure 7.14

Types of enzyme inhibition. (*a*) A competitive inhibitor competes with the substrate for binding at the same site on the enzyme. (*b*) A noncompetitive inhibitor binds to a different site but blocks the conversion of the substrate to products. (*c*) An uncompetitive inhibitor binds only to the enzyme–substrate complex. (E = enzyme; S = substrate.)

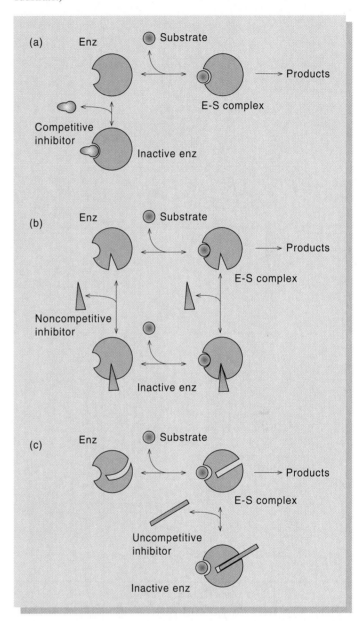

different types of inhibition and in providing quantitative information on the effectiveness of various inhibitors. Such information is essential for an understanding of how cells regulate their enzymatic activities. Comparisons of a series of inhibitors also can help to map the structure of an enzyme's active site and are a key step in the design of therapeutic drugs.

Competitive Inhibitors Bind at the Active Site

In many cases, an inhibitor resembles the substrate structurally and binds reversibly at the same site on the enzyme. This activity is called competitive inhibition because the inhibitor and the substrate compete for binding (fig. 7.14*a*). The inhibitor is prevented from binding if the active site is already occupied by the substrate. As an example consider the proteolytic enzyme trypsin, which cleaves polypeptide chains at peptide linkages adjacent to basic amino acid residues. Trypsin is inhibited competitively by benzamidine. The substrate-binding site on the enzyme consists of a pocket where a positively charged lysine or arginine side chain fits snugly and interacts with a negatively charged carboxylate group (fig. 7.15). When protonated, benzamidine is positively charged and has a flat, delocalized electronic structure resembling that of an arginine side chain. The binding pocket accepts and binds a benzamidine ion reversibly in place of its regular substrate.

An expression describing enzyme kinetics in the presence of a competitive inhibitor can be derived straightforwardly. Consider the reaction that we treated previously.

$$E + S \underset{k_{-1}}{\overset{k_1}{\rightleftharpoons}} ES \underset{k_{-2}}{\overset{k_2}{\rightleftharpoons}} E + P \qquad (19')$$

Figure 7.15

The specificity pocket of trypsin can accommodate an arginine side chain of a polypeptide substrate or a benzamidine ion, which acts as a competitive inhibitor. Asp = aspartic acid.

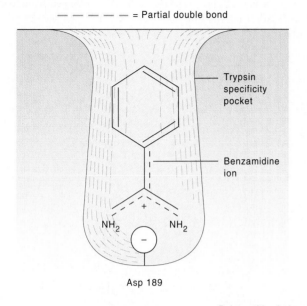

– – – – – = Partial double bond

Trypsin specificity pocket

Benzamidine ion

NH₂   NH₂

Asp 189

Polypeptide chain

C_a

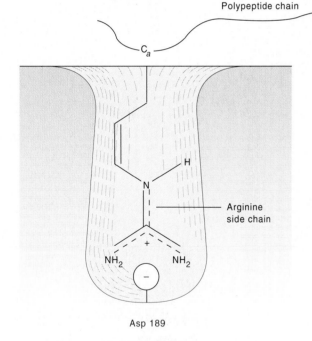

H

N

Arginine side chain

NH₂   NH₂

Asp 189

Arginine side chain

Figure 7.16

Competitive inhibition. A series of double-reciprocal plots ($1/v$ versus $1/[S]$) measured at different concentrations of the inhibitor (I) all intersect at the same point ($1/V_{max}$) on the ordinate. The slopes of the plots and the intercepts on the abscissa are simple, linear functions of $[I]/K_i$, where K_i is the dissociation constant of the inhibitor–enzyme complex.

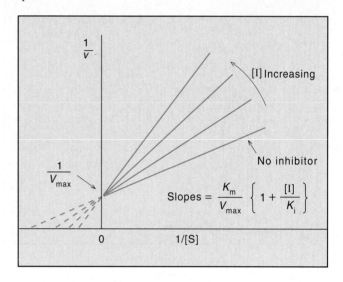

We now have the additional feature that the enzyme also reacts reversibly with the inhibitor (I) to give an inactive complex (EI).

$$E + I \rightleftharpoons EI \qquad (34)$$

The derivation proceeds just as the derivation that led to equations (25) and (26), except that the total enzyme concentration $[E_t]$ is now $[E] + [ES] + [EI]$, instead of just $[E] + [ES]$. As a result, in place of equation 26 we end up with

$$\frac{1}{v} = \frac{1}{V_{max}} + \frac{K_m}{V_{max}} \frac{1}{[S]}\left(1 + \frac{[I]}{K_i}\right) \qquad (35)$$

where K_i is the dissociation constant of the enzyme–inhibitor complex.

$$K_i = \frac{[E][I]}{[EI]} \qquad (36)$$

According to equation (35), a plot of $1/v$ versus $1/[S]$ is linear and passes through the same intercept on the ordinate as the plot obtained in the absence of the inhibitor ($1/V_{max}$). This is because the effect of the inhibitor disap-

Figure 7.17

Noncompetitive inhibition. The double-reciprocal plots pass through different points on the ordinate, but intersect at the same point ($-1/K_m$) on the abscissa. The slopes and the intercepts on the ordinate are linear functions of $[I]/K_i$.

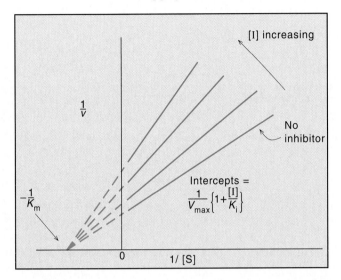

Table 7.4

Some Inhibitors of Enzymes that Form Covalent Linkages with Functional Groups on the Enzyme

Inhibitor	Enzyme Group that Combines with Inhibitor
Cyanide	Fe, Cu, Zn, other transition metals
p-Mercuribenzoate	Sulfhydryl
Diisopropylfluorophosphate	Serine hydroxyl
Iodoacetate	Sulfhydryl, imidazole, carboxyl, thioether

pears at infinite substrate concentration for a competitive inhibitor. The slope of the double-reciprocal plot, however, depends on the product

$$\frac{K_m}{V_{max}}\left(1 + \frac{[I]}{K_i}\right)$$

instead of simply K_m/V_{max} (fig. 7.16). As a result the apparent K_m is altered to K'_m, where

$$K'_m = K_m\left(1 + \frac{[I]}{K_i}\right)$$

By measuring the slope of the plot as a function of $[I]$ we can determine K_i.

Noncompetitive and Uncompetitive Inhibitors Do Not Compete Directly with Substrate Binding

Some inhibitors bind at sites other than the enzyme's active site and do not compete directly with binding of the substrate. Instead, they act by interfering with the reaction of the enzyme–substrate complex. An inhibitor that binds to

an enzyme whether or not the active site is occupied by the substrate is termed a noncompetitive inhibitor (fig. 7.14b). A noncompetitive inhibitor decreases the maximum velocity of an enzymatic reaction without affecting the K_m. The inhibitor removes a certain fraction of the enzyme from operation, no matter the concentration of the substrate. Plots of $1/v$ versus $1/[S]$ in the presence of different concentrations of a noncompetitive inhibitor intersect at the same point on the abscissa ($-1/K_m$) but pass through the ordinate at different points (fig. 7.17).

Still another possibility is that the inhibitor binds only to the enzyme–substrate complex and not to the free enzyme (fig. 7.14c). This reaction is called uncompetitive inhibition. Uncompetitive inhibition is rare in reactions that involve a single substrate but more common in reactions with multiple substrates. Plots of $1/v$ versus $1/[S]$ at different concentrations of an uncompetitive inhibitor give a series of parallel lines.

Irreversible Inhibitors Permanently Alter the Enzyme Structure

Competitive, noncompetitive, and uncompetitive inhibition are all reversible. If the inhibited enzyme is dialyzed to remove the inhibitor, the enzymatic activity recovers. There are, however, inhibitors that react essentially irreversibly, usually by forming a covalent bond to an amino acid side chain or a bound coenzyme. Examples of such inhibitors are given in table 7.4. Like reversible inhibitors, irreversible inhibitors can change either V_{max} or K_m or both.

Irreversible inhibitors often provide clues to the nature of the active site. Enzymes that are inhibited by iodo-acetamide, for example, frequently have a cysteine in the active site, and the cysteinyl sulfhydryl group often plays an essential role in the catalytic mechanism (fig. 7.18). An example is glyceraldehyde 3-phosphate dehydrogenase, in which the catalytic mechanism begins with a reaction of the cysteine with the aldehyde substrate (see fig. 12.21). As we discuss in chapter 8, trypsin and many related proteolytic enzymes are inhibited irreversibly by diisopropyl-fluorophosphate (fig. 7.18), which reacts with a critical serine residue in the active site.

Figure 7.18

Iodoacetamide is an irreversible inhibitor of many enzymes that contain a cysteine residue in the active site. Diisopropylfluoro-phosphate is an irreversible inhibitor of trypsin, chymotrypsin, and several related enzymes. It reacts with a serine residue at the active site.

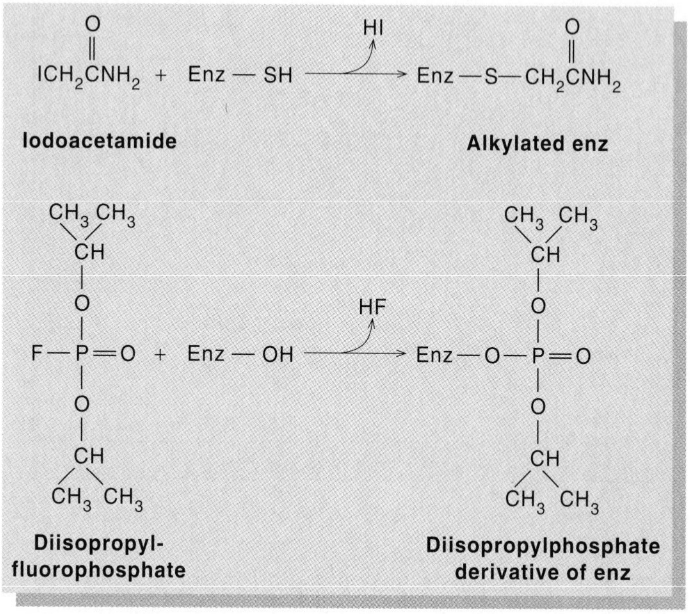

Summary

Enzymes are biological catalysts. Kinetic analysis is one of the most broadly used tools for characterizing enzymatic reactions.

1. The rate of a reaction depends on the frequency of collisions between the reacting species and on the fraction of the collisions that produce products. The former depends on the concentrations of the reactants; the latter depends on temperature and activation free energy $\Delta G^{\ddagger}$. $\Delta G^{\ddagger}$ can be interpreted as the free energy needed to convert the reactants to a transition state. A catalyst increases the reaction rate by lowering $\Delta G^{\ddagger}$.

2. Enzyme kinetics usually are studied by mixing the enzyme and substrates and measuring the initial rate of formation of product or the disappearance of a reactant. Special techniques are necessary to measure very fast reactions. It is common to measure the rate as a function of substrate concentration, pH, and temperature.

3. Enzymes have localized catalytic sites. The substrate (S) binds at the active site to form an enzyme–substrate complex (ES). Subsequent steps transform the bound substrate into product and regenerate the free enzyme. The overall speed of the reaction depends on the concentration of ES. Shortly after the enzyme and substrate are mixed, [ES] becomes approximately constant and remains so for a period of time termed the steady state. The rate (v) of the reaction in the steady state usually has a hyperbolic dependence on the substrate concentration. It is proportional to [S] at low concentrations but approaches a maximum (V_{max}) when the enzyme is fully charged with substrate. The Michaelis constant K_m is the substrate concentration at which the rate is half maximal. K_m and V_{max} often can be obtained from a plot of $1/v$ versus $1/[S]$. If ES is in equilibrium with the free enzyme and substrate, K_m is equal to the dissociation constant for the complex (K_s). More generally, K_m depends on at least three rate constants and is larger than K_s.

4. The turnover number k_{cat} is the maximum number of molecules of substrate converted to product per unit

time per active site and is V_{max} divided by the total enzyme concentration. The specificity constant k_{cat}/K_m is a measure of how rapidly an enzyme can work at low substrate concentrations. This is usually the best index of the effectiveness of an enzyme.

5. Enzymes that catalyze reactions of two or more substrates work in a variety of ways that can be distinguished by kinetic analysis. Some enzymes bind their substrates in a fixed order; others bind in random order. In some cases binding of one substrate gives a partial reaction before the second substrate binds.

6. Enzymes can be inhibited by agents that interfere with the binding of substrate or with conversion of the ES complex into products. Reversible inhibitors are classified as competitive, noncompetitive or uncompetitive. A competitive inhibitor competes with substrate for binding at the active site. Consequently, a sufficiently high concentration of substrate can eliminate the effect of a competitive inhibitor. Noncompetitive inhibitors bind at a separate site and block the reaction regardless of whether the active site is occupied by substrate. An uncompetitive inhibitor binds to the ES complex but not to the free enzyme. These three forms of inhibition are distinguishable by measuring the rate as a function of the concentrations of the substrate and inhibitor. Irreversible inhibitors often provide information on the active site by forming covalently linked complexes that can be characterized.

Selected Readings

Advances in Enzymology. New York: Academic Press. An annually published volume containing monographs on selected topics.

Boyer, P. D. (ed.), *The Enzymes.* New York: Academic Press. A continuing series of monographs on selected enzymes. See particularly the chapter entitled ''Steady State Kinetics'' by W. W. Cleland in vol. 2.

Fersht, A., *Enzyme Structure and Mechanism,* 2d ed. New York: W. H. Freeman & Co., 1985.

Frost, A. A., and R. G. Pearson, *Kinetics and Mechanism,* 2d ed. New York: Wiley, 1961. An excellent introduction to general chemical kinetics.

Purich, D. L., *Contemporary Enzyme Kinetics and Mechanism.* New York: Academic Press, 1983. Selected chapters from *Methods in Enzymology.* Detailed information on how to analyze kinetic data and on effects of temperature, pH, and inhibitors.

Segal, I. H., *Enzyme Kinetics.* New York: Wiley, 1975.

Problems

1. Explain what is meant by the order of a reaction, using the reaction below as an example. What is the reaction order for each reactant? For the overall reaction? (Consider the forward and reverse reaction.)

$$A + B \rightleftharpoons 2C$$

2. In a first-order reaction a substrate is converted to product so that 87% of the substrate is converted in 7 min. Calculate the first-order rate constant. In what time is 50% of the substrate converted to product?

3. Prove that the K_m equals the substrate concentration at one-half maximal velocity.

4. The Michaelis constant K_m is frequently equated with K_s, the [ES] dissociation constant. However, there is usually a disparity between those values. Why? Under what conditions are K_m and K_s equivalent?

5. When quantifying the activity of an enzyme, does it matter if you measure the appearance of a product or the disappearance of a reactant?

6. An enzyme was assayed with substrate concentration of twice the K_m value. The progress curve of the enzyme (product produced per minute) is shown here. Give two possible reasons why the progress curve becomes nonlinear.

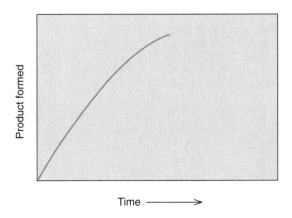

7. What is the steady-state approximation and under what conditions is it valid?

8. Assume that an enzyme-catalyzed reaction follows Michaelis-Menten kinetics with a K_m of 1 μM. The initial velocity is 0.1 μM/min at 10 mM substrate. Calculate the initial velocity at 1 mM, 10 μM, and 1 μM substrate. If the substrate concentration increased to 20 mM, would the initial velocity double? Why or why not?

9. If the K_m for an enzyme is 1.0×10^{-5} M and the K_i of a competitive inhibitor of the enzyme is 1.0×10^{-6} M, what concentration of inhibitor would be necessary to lower the reaction rate by a factor of 10 when the substrate concentration is 1.0×10^{-3} M? 1.0×10^{-5} M? 1.0×10^{-6} M?

10. Assume that an enzyme-catalyzed reaction follows the scheme shown:

$$E + S \underset{k_2}{\overset{k_1}{\rightleftharpoons}} ES \underset{k_4}{\overset{k_3}{\rightleftharpoons}} E + P$$

Where $k_1 = 10^9$ M^{-1} s^{-1}, $k_2 = 10^5$ s^{-1}, $k_3 = 10^2$ s^{-1}, $k_4 = 10^7$ M^{-1} s^{-1}, and [E_t] is 0.1 nM. Determine the value of each of the following.

(a) K_m
(b) V_{max}
(c) Turnover number
(d) Initial velocity when [S]$_\circ$ is 20 μM.

11. A colleague has measured the enzymatic activity as a function of reaction temperature and obtained the data shown in this graph. He insists on labeling point A as the "temperature optimum" for the enzyme. Try, tactfully, to point out the fallacy of that interpretation.

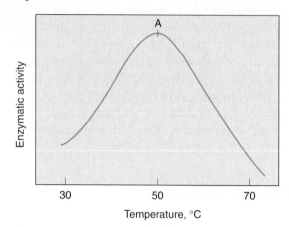

12. You have isolated a tetrameric NAD$^+$-dependent dehydrogenase. You incubate this enzyme with iodo-acetamide in the absence or presence of NADH (at 10 times the K_m concentration), and you periodically remove aliquots of the enzyme for activity measurements and amino acid composition analysis. The results of the analyses are shown in the table.

Table for Problem 12

	(No NADH Present)				(NADH Present)			
Time (min)	Activity (U/mg)	His	(Residues/mole)	Cys	Activity (U/mg)	His	(Residues/mole)	Cys
0	1,000	20		12	1,000	20		12
15	560	18.2		11.4	975	20		11.4
30	320	17.3		10.8	950	20		10.8
45	180	16.7		10.4	925	19.8		10.4
60	100	16.4		10.0	900	19.6		10.0

(a) What can you conclude about the reactivities of the cysteinyl and histidyl residues of the protein?

(b) Which residue can you implicate in the active site? On what do you base the choice? Are the data conclusive concerning the assignment of a residue to the active site? Why or why not?

(c) After 1 h you dilute the enzyme incubated with iodoacetamide but no NADH. Do you expect the enzyme activity to be restored? Explain.

13. The initial velocity data shown in the table were obtained for an enzyme.

[S] (mM)	Velocity (MS^{-1}) $\times 10^7$
0.10	0.96
0.125	1.12
0.167	1.35
0.250	1.66
0.50	2.22
1.0	2.63

Each assay at the indicated substrate concentration was initiated by adding enzyme to a final concentration of 0.01 nM. Derive K_m, V_{max}, k_{cat}, and the specificity constant.

14. You measured the initial velocity of an enzyme in the absence of inhibitor and with inhibitor A or inhibitor B. In each case, the inhibitor is present at 10 μM. The data are shown in the table.

[S] (mM)	Velocity (MS^{-1}) $\times 10^7$ Uninhibited	Velocity (MS^{-1}) $\times 10^7$ Inhibitor A	Velocity (MS^{-1}) $\times 10^7$ Inhibitor B
0.333	1.65	1.05	0.794
0.40	1.86	1.21	0.893
0.50	2.13	1.43	1.02
0.666	2.49	1.74	1.19
1.0	2.99	2.22	1.43
2.0	3.72	3.08	1.79

(a) Determine K_m and V_{max} of the enzyme.
(b) Determine the type of inhibition imposed by inhibitor A and calculate K_i(s).
(c) Determine the type of inhibition imposed by inhibitor B and calculate K_i(s).

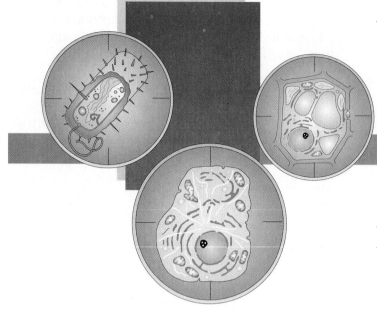

How Enzymes Work

Chapter Outline

General Themes in Enzymatic Mechanisms
- *The Proximity Effect: Enzymes Bring Reacting Species Close Together*
- *General-Base and General-Acid Catalysis Avoids the Need for Extremely High or Low pH*
- *Electrostatic Interactions Can Promote the Formation of the Transition State*
- *Enzymatic Functional Groups Provide Nucleophilic and Electrophilic Catalysis*
- *Structural Flexibility Can Increase the Specificity of Enzymes*

Detailed Mechanisms of Enzyme Catalysis
- *Serine Proteases: Enzymes that Use a Serine Residue for Nucleophilic Catalysis*
- *Ribonuclease A: An Example of Concerted Acid–Base Catalysis*
- *Triosephosphate Isomerase Has Approached Evolutionary Perfection*

The effectiveness of enzymes is due to the fact that they bring the reactants (substrates) together in a location where they are ideally oriented with respect to each other and the catalytic groups of the enzyme.

In chapter 7 we saw that enzymes can increase the rates of reactions by many orders of magnitude. We noted that enzymes are highly specific in the reactions they catalyze and in the particular substrates they accept. In this chapter we explore the mechanisms of several enzyme-catalyzed reactions in greater detail. Our goal is to relate the activity of each of these enzymes to the structure of the active site, where the functional groups of amino acid side chains, the polypeptide backbone, or bound cofactors must interact with the substrates in such a way as to favor the formation of the transition state. We explore enzyme catalytic mechanisms in many subsequent chapters as well but usually in less detail than here.

General Themes in Enzymatic Mechanisms

Several broad themes recur frequently in enzymatic reaction mechanisms. Among the most important of these are (1) proximity effects, (2) general-acid and general-base catalysis, (3) electrostatic effects, (4) nucleophilic or electrophilic catalysis by enzymatic functional groups, and

(5) structural flexibility. All known enzymes use at least one of these themes, and most use more than one. We start by discussing these five themes in general terms and then see how they apply to three enzymes for which the mechanisms have been studied in depth.

The Proximity Effect: Enzymes Bring Reacting Species Close Together

The idea of the proximity effect is that an enzyme can accelerate a reaction between two species simply by holding the two reactants close together in an appropriate orientation. It has long been known that intramolecular reactions involving two groups that are tied together in a single molecule are usually much faster than the corresponding intermolecular reactions between two independent molecules. The cyclization of succinic acid to form succinyl anhydride [equation (1)], for example, occurs much more rapidly than the formation of acetic anhydride from two molecules of acetic acid [equation (2)].

$$
\begin{array}{c}
\text{CH}_2\text{CO}_2\text{H} \\
| \\
\text{CH}_2 \\
| \\
\text{CH}_2 \\
\text{CO}_2\text{H}
\end{array}
\longrightarrow
\begin{array}{c}
\text{O} \\
\| \\
\text{C} \\
\text{CH}_2 \\
| \\
\text{CH}_2 \\
\text{C} \\
\| \\
\text{O}
\end{array}
\text{O} + \text{H}_2\text{O} \qquad (1)
$$

$$
\begin{array}{c}
\text{CH}_3\text{CO}_2\text{H} \\
\\
\text{CH}_3\text{CO}_2\text{H}
\end{array}
\longrightarrow
\begin{array}{c}
\text{O} \\
\| \\
\text{CH}_3\text{C} \\
\text{CH}_3\text{C} \\
\| \\
\text{O}
\end{array}
\text{O} + \text{H}_2\text{O} \qquad (2)
$$

We cannot compare the rate constants for these two reactions directly because they are expressed in different units. The intramolecular reaction [equation (1)] is kinetically first order, whereas the intermolecular reaction [equation (2)] is second order. But suppose the two molecules of acetic acid that enter into reaction (2) are labeled isotopically to make them distinguishable and that one type of molecule is present in great excess over the other. The process is then kinetically first order in the concentration of the limiting reactant. To make the rate constant the same as for reaction (1), the more abundant species has to be present at a concentration of 3×10^5 M! This is far above any concentration that can actually be obtained.

This example from organic chemistry shows that tying two reactants together can have an enormous effect on the rate of a reaction. The effect is due largely to differences between the entropy changes that accompany the inter- and intramolecular reactions. The formation of the transition state requires a larger decrease of translational and rotational entropy in the intermolecular reaction than it does in the intramolecular reaction. In the intramolecular reaction much of this entropy decrease has already occurred during the preparation of the reactant. Enzymes that catalyze intermolecular reactions take advantage of the proximity effect by binding the reactants close together at the active site so that the reactive groups are oriented appropriately for the reaction.

General-Base and General-Acid Catalysis Avoids the Need for Extremely High or Low pH

Chemical bonds are formed by electrons, and formation or breakage of bonds requires the migration of electrons. In broad terms, reactive chemical groups function either as electrophiles or as nucleophiles. Electrophiles are electron-deficient substances that react with electron-rich substances; nucleophiles are electron-rich substances that react with electron-deficient substances. The task of a catalyst often is to make a potentially reactive group more reactive by increasing its electrophilic or nucleophilic character. In many cases the simplest way to do this is to add or remove a proton.

As an example, consider the hydrolysis of an ester (fig. 8.1). Because the electronegativity of the oxygen atom in the C=O group is greater than that of the carbon, the oxygen has a fractional negative charge δ^-, and the carbon has a fractional positive charge δ^+. Hydrolysis of an ester in neutral aqueous solution can occur if the oxygen atom of H_2O, acting as a nucleophile, reacts with the carbonyl carbon. The initial product is an intermediate in which the carbon atom has four substituents in a tetrahedral arrangement. The reaction is completed by the breakdown of the tetrahedral intermediate to release the alcohol. However, water is a comparatively weak nucleophile, and its reaction with esters in the absence of a catalyst is very slow. Hydrolysis of esters occurs much more rapidly at high pH, when the negatively charged hydroxide ion replaces water as the reactive nucleophile (fig. 8.1a). The nucleophilic character of water itself also can be increased by interaction with a basic group other than OH^- (fig. 8.1b). By offering electrons to one of the protons of the water, the base increases the electron density on the oxygen.

The term "general base" is used to describe any substance that is capable of binding a proton in aqueous solution. Enzymes use a variety of functional groups in this role.

Figure 8.1

Several ways that the hydrolysis of an ester can occur. A red, curved arrow represents the movement of an electron pair from an electron donor to an acceptor. (*a*) Catalysis by free hydroxide ion. (*b*) General-base catalysis. (*c*) General-acid catalysis.

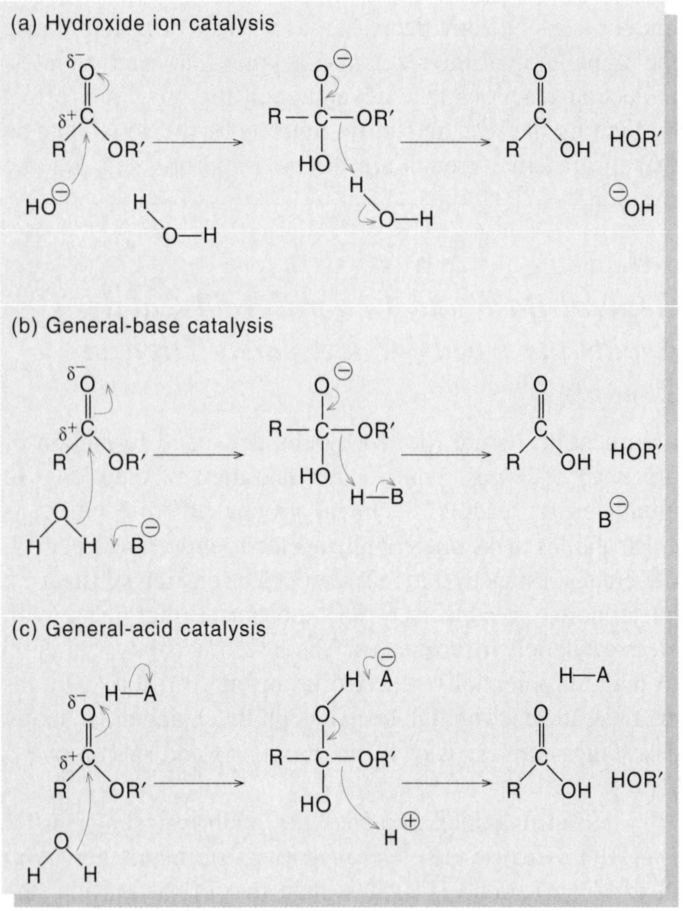

Two factors make free OH⁻ ions themselves unsuitable for enzymatic catalysis and thus favor the use of a general base. First, the low concentration of OH⁻ limits its availability at physiological pH. In contrast, proteins contain numerous functional groups that can serve as general bases at moderate pH or even under mildly acidic conditions. The only requirement is that the base can exist at least partly in its unprotonated form at the ambient pH. This condition can easily be met by selecting a basic group from among the ionizable or polar amino acid side chains from an amino-terminal —NH₂ group or a carboxy-terminal carboxylate (see table 3.3). The second advantage of using a general base instead of OH⁻ is that a basic group provided by the protein can be positioned precisely with respect to the substrate in the active site, allowing the proximity effect to come into play. Free OH⁻ ions are much more mobile. In exceptional cases in which OH⁻ does act as a nucleophile in

an enzymatic reaction, it usually is tightly bound to a metal cation.

The hydrolysis of an ester also can be catalyzed by an acid (fig. 8.1*c*). The acid donates a proton to the carbonyl oxygen, increasing the positive charge on the carbon. The term "general acid" is used to refer to any substance capable of releasing a proton in solution, and again enzymes almost always use such proton donors in preference to free H⁺ or H₃O⁺ ions, presumably because a general acid can operate at moderate pH and is easy to fix in position.

An important point to note in figure 8.1 is that the same general acid or base that catalyzes the formation of the tetrahedral intermediate also can participate in the decomposition of the intermediate. When a general acid (HA) donates a proton to the ester oxygen it becomes a base (A⁻), which can retrieve the proton as the intermediate breaks down. When a general base (B⁻) removes a proton from water it becomes an acid (BH), which can provide a proton to the alcohol. Note also that general-acid and general-base catalysis are not mutually exclusive; they can both occur in a concerted manner in the same step of a reaction.

Electrostatic Interactions Can Promote the Formation of the Transition State

The frequent use of general acids and general bases in enzymes illustrates the underlying principle that enzymes act by stabilizing the distribution of electrical charge in transition states. In the enzymatic hydrolysis of an ester, the key transition state probably resembles the tetrahedral intermediates shown in figure 8.1. To form such an intermediate, electrons must move from the nucleophile through the carbon atom of the C=O group to the oxygen. There is thus a net movement of negative charge from the nucleophile to the substrate. In the absence of a general acid or base, a charge approaching +1 appears on the nucleophile, and a charge approaching −1 appears on the C=O oxygen. A general base can stabilize this new distribution of charge by offering electrons to the nucleophile so that some of the positive charge moves to the base. By providing a proton to the C=O oxygen, a general acid can delocalize the negative charge here. But enzymes have other ways to achieve a similar stabilization. Suppose that the active site included a positively charged amino acid side chain, such as that of lysine or arginine, located near the oxygen atom of the C=O group. A positive charge in this region favors the formation of the tetrahedral intermediate. A negative charge in the region of the nucleophile has a similar effect. The interactions of such charges are termed electrostatic effects.

Electrostatic interactions can be significant even between groups whose net formal charge is zero. This is because charge distributions within molecular groups are not uniform but rather vary from atom to atom. We alluded to this previously when discussing the partial charges on the oxygen and carbon atoms of an ester (fig. 8.1). Similar considerations apply to other functional groups: The electron distributions around the nuclei leave each atom with a small net positive or negative charge. In an alcoholic —OH group, for example, the oxygen atom has a negative charge of approximately −0.4 atomic charge units, and the hydrogen has a charge of about +0.4.

As a reacting substrate is transformed into a transition state, the changing charges on its atoms interact with the charges on atoms of the surrounding protein and any nearby water molecules. The energy difference between the initial state and the transition state thus depends critically on the details of the protein structure. We see illustrations of this in the three enzymes discussed later on.

Enzymatic Functional Groups Provide Nucleophilic and Electrophilic Catalysts

Another strategy for catalyzing the hydrolysis of an ester or an amide is to replace water by a stronger nucleophilic group that is part of the enzyme's active site. The $HOCH_2$— group of a serine residue is often used in this way. In such cases, the reaction of the serine with the substrate splits the overall reaction into a two-step process. Instead of immediately yielding the free carboxylic acid, the breakdown of the initial tetrahedral intermediate yields an intermediate ester that is attached covalently to the enzyme.

$$R-\overset{\overset{O}{\|}}{C}-NHR' + HOCH_2-Enzyme \longrightarrow$$

$$R-\overset{\overset{O}{\|}}{C}-OCH_2-Enzyme + R'NH_2 \quad (3)$$

The *acyl-enzyme* ester intermediate must be hydrolyzed by a second reaction, in which water becomes the nucleophile.

$$R-\overset{\overset{O}{\|}}{C}-OCH_2-Enzyme + H_2O \longrightarrow$$

$$R-\overset{\overset{O}{\|}}{C}-OH + HOCH_2-Enzyme \quad (4)$$

The proteolytic enzymes trypsin, chymotrypsin, and elastase, discussed later on, all work in this way.

Nucleophilic groups on enzymes participate in a variety of other types of reactions in addition to hydrolytic reactions. An example is acetoacetic acid decarboxylase, which catalyzes the reaction

$$CH_3-\overset{\overset{O}{\|}}{C}-CH_2-CO_2H \longrightarrow CH_3-\overset{\overset{O}{\|}}{C}-CH_3 + CO_2 \quad (5)$$

Figure 8.2

In acetoacetic acid decarboxylase, the positive charge of a protonated Schiff base intermediate pulls electrons from a nearby carbon–carbon bond, thereby releasing CO_2.

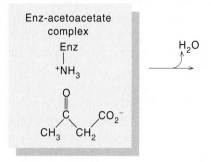

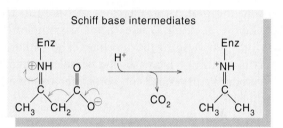

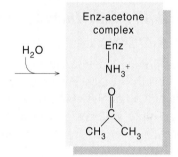

The reaction proceeds through an intermediate called a Schiff base, in which the substrate is covalently attached to the ϵ-amino group of a lysine residue at the enzyme's active site.

$$CH_3-\overset{\overset{O}{\|}}{C}-CH_2-CO_2H + Enzyme-NH_2 \longrightarrow$$

$$CH_3-\overset{\overset{N-Enzyme}{\|}}{C}-CH_2CO_2H + H_2O \quad (6)$$

Protonation of the nitrogen atom of the Schiff base introduces a positive charge that pulls electrons from the nearby carbon–carbon bond, causing decarboxylation (fig. 8.2).

The mechanisms outlined in equations 3–6 illustrate a basic feature: Nucleophilic catalysis by enzymes involves the formation of an intermediate state in which the substrate is covalently attached to a nucleophilic group of the enzyme. In addition to the —CH_2OH group of serine and the ϵ-amino of lysine, the —CH_2SH of cysteine is often used as a nucleophile. The carboxylate of aspartate or glutamate and the imidazole group of histidine can play a similar role. Some enzymes take advantage of bound coenzymes, such as thiamine, biotin, pyridoxamine, or tetrahydrofolate, to obtain additional nucleophilic reagents (see chapter 10).

There also are numerous enzymes that use bound metal ions to form complexes with substrates. In these enzymes, the metal ion usually serves as an electrophilic functional group rather than as a nucleophile. Carbonic anhydrase, for example, contains a Zn^{+2} ion that binds one of the substrates, hydroxide ion, as a ligand. The bound OH^- reacts with the other substrate, CO_2. In alcohol dehydrogenase, and in the proteolytic enzymes thermolysin and carboxypeptidase A, a Zn^{+2} ion forms a complex with a carbonyl oxygen atom of the substrate. The withdrawal of electrons by the Zn^{+2} increases the partial positive charge on the carbonyl carbon and thus promotes reaction with a nucleophile.

Structural Flexibility Can Increase the Specificity of Enzymes

Although precise positioning of the reactants is a fundamental aspect of enzyme catalysis, most enzymes undergo some change in their structure when they bind substrates. A particularly dramatic example is hexokinase, which catalyzes the transfer of a phosphate group from adenosine triphosphate (ATP) to glucose.

$$ATP + glucose \longrightarrow$$
$$ADP + glucose\text{—}6\text{—}phosphate \quad (7)$$

When hexokinase binds glucose, its structure changes in a way that brings together the elements of the active site (fig. 8.3). The enzyme literally closes like a set of jaws around the substrate! Such a structural change is often referred to as an induced fit.

Enfolding a substrate in this way can serve to maximize the favorable entropy change associated with removing a hydrophobic substrate molecule from water. It also allows the enzyme to control the electrostatic effects that promote formation of the transition state. The substrate is forced to respond to the directed electrostatic fields from the enzyme's functional groups, instead of the disordered fields from the solvent.

Structural changes also contribute to the high specificity of some enzymatic reactions. In hexokinase, the structural change induced by glucose promotes the binding of the other substrate, ATP. ATP does not bind to the enzyme

Figure 8.3

Models of the crystallographic structure of hexokinase in the "open" (*a*) and "closed" (*b*) conformations. The enzyme (shown in blue) adopts the open conformation in the absence of substrates, but switches to the closed conformation when it binds glucose (red). Hexokinase also has been crystallized with a bound analog of ATP. In the absence of glucose, the enzyme with the bound ATP analog remains in the open conformation. The structural change caused by glucose results in the formation of additional contacts between the enzyme and ATP. This can explain why the binding of glucose enhances the binding of ATP. (Courtesy of Dr. Thomas A. Steitz.)

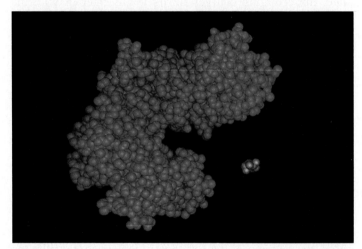

(a)

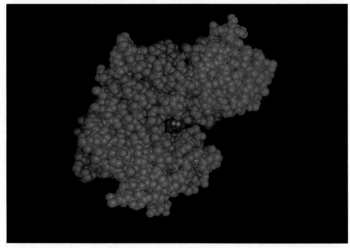

(b)

properly unless glucose is already present in the catalytic site. If ATP were to bind in the absence of glucose, the enzyme might have a tendency to catalyze the transfer of phosphate from ATP to water, resulting in a wasteful loss of ATP.

$$ATP + H_2O \longrightarrow ADP + HPO_4^{2-} + H^+ \quad (8)$$

Hexokinase does not catalyze this side reaction; it waits for glucose to bind first.

The structural changes that occur in hexokinase bring home the point that enzyme crystal structures give static snapshots of molecules that actually are highly flexible. In solution, the structure of an enzyme undergoes fluctuations that vary widely in amplitude and frequency from place to place in the protein. Vibrations and rotations involving only a few atoms occur on time scales of 10^{-13}–10^{-11} s. Somewhat larger motions, such as the flipping of the aromatic ring of a tyrosine, typically occur on scales of 10^{-9}–10^{-8} s. Major reorganizations may take 10^{-6}–10^{-3} s. All of these types of motions can be important in catalysis.

Detailed Mechanisms of Enzyme Catalysis

In the preceding sections, we discussed five themes that occur frequently in enzyme reaction mechanisms. We now examine several representative enzymes in finer detail. We focus on three enzymes for which crystal structures have been obtained, because the most decisive advances in our understanding of enzyme reaction mechanisms have come by inspecting such structures.

Serine Proteases: Enzymes that Use a Serine Residue for Nucleophilic Catalysis

The serine proteases are a large family of proteolytic enzymes that use the reaction mechanism for nucleophilic catalysis outlined in equations (3) and (4), with a serine residue as the reactive nucleophile. The best known members of the family are three closely related digestive enzymes: trypsin, chymotrypsin, and elastase. These enzymes are synthesized in the mammalian pancreas as inactive precursors termed zymogens. They are secreted into the small intestine, where they are activated by proteolytic cleavage in a manner discussed in chapter 9.

In the digestive system trypsin, chymotrypsin, and elastase work as a team. They are all endopeptidases, which means that they cleave protein chains at internal peptide bonds, but each preferentially hydrolyses bonds adjacent to a particular type of amino acid residue (fig. 8.4). Trypsin cuts just next to basic residues (lysine or arginine); chymotrypsin cuts next to aromatic residues (phenylalanine, tyrosine, or tryptophan); elastase is less discriminating but prefers small, hydrophobic residues such as alanine.

About half of the amino acid residues of trypsin are identical to the corresponding residues in chymotrypsin, and about a quarter of the residues are conserved in all three of the pancreatic endopeptidases (fig. 8.5). The structural simi-

Figure 8.4

Trypsin, chymotrypsin, and elastase—three members of the serine protease family—catalyze the hydrolysis of proteins at internal peptide bonds adjacent to different types of amino acids. Trypsin prefers lysine or arginine residues; chymotrypsin, aromatic side chains; and elastase, small, nonpolar residues. Carboxypeptidases A and B, which are not serine proteases, cut the peptide bond at the carboxyl-terminal end of the chain. Carboxypeptidase A preferentially removes aromatic residues; carboxypeptidase B, basic residues. (Illustration copyright by Irving Geis. Reprinted by permission.)

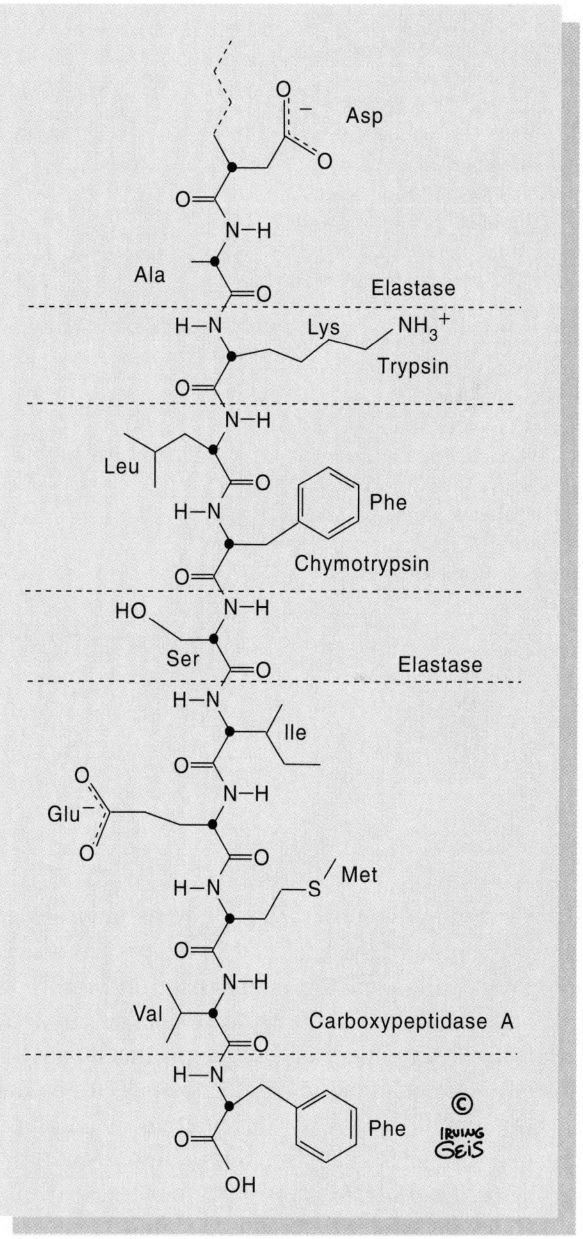

larities of trypsin, chymotrypsin, and elastase are even more evident in the crystal structures. As is shown in figure 8.6a the folding of the polypeptide chain is essentially the same in all three enzymes. These enzymes are classical illustra-

Figure 8.5

Schematic diagrams of the amino acid sequences of chymotrypsin, trypsin, and elastase. Each circle represents one amino acid. Amino acid residues that are identical in all three proteins are in solid color. The three proteins are of different lengths but have been aligned to maximize the correspondence of the amino acid sequences. All of the sequences are numbered according to the sequence in chymotrypsin. Long connections between nonadjacent residues represent disulfide bonds. Locations of the catalytically important histidine, aspartate, and serine residues are marked. The links that are cleaved to transform the inactive zymogens to the active enzymes are indicated by parenthesis marks. After chymotrypsinogen is cut between residues 15 and 16 by trypsin and is thus transformed into an active protease, it proceeds to digest itself at the additional sites that are indicated; these secondary cuts have only minor effects on the enzymes's catalytic activity. (Illustration copyright by Irving Geis. Reprinted by permission.)

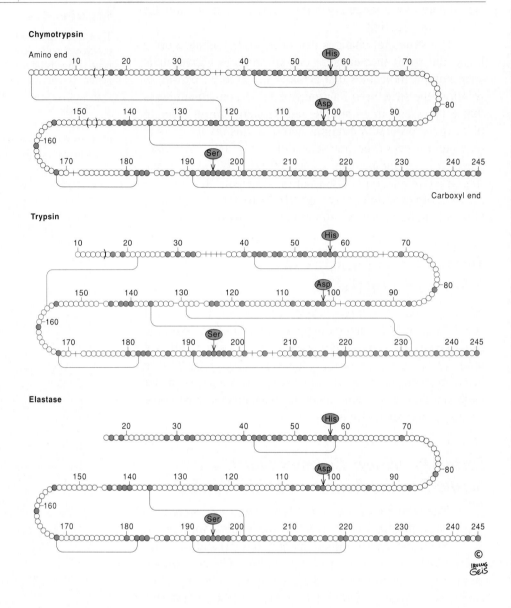

tions of diverging evolution from a common ancestor. The structures of some of the bacterial serine proteases also are homologous to those of the mammalian enzymes. On the other hand, subtilisin, a serine protease obtained from *Bacillus subtilis,* has an amino acid sequence that seems totally unrelated to the mammalian sequences. Its three-dimensional structure also is very different from those of the mammalian enzymes (see fig. 8.6b). Subtilisin probably joined the serine protease family by convergent evolution. Remarkably, a small set of critical amino acid residues comes together in the folded structure to form the essential elements of the active site in all of these proteases. Using the chymotrypsin numbering system, these are His 57, Asp 102, and Ser 195 (fig. 8.7).

That Ser 195 played an important role in the catalytic mechanism was known from early studies on the enzyme

inhibitor diisopropylfluorophosphate (see chapter 7). This inhibitor reacts irreversibly with chymotrypsin or trypsin to form an inactive derivative in which the diisopropylphosphate group is covalently attached to the serine residue (see fig. 7.18). A variety of other inhibitors react in a parallel manner with this same serine residue. As a rule, these inhibitors do not react with other serines in the enzyme or with serines in enzymes that are not part of the serine protease family. The reactivity of Ser 195 thus is not a property of serine residues in general but depends on the special surroundings of this residue in the protein. We will see shortly that it is the juxtaposition with His 57 and Asp 102 that makes the serine especially reactive.

Although His 57 is far removed from Ser 195 in the primary sequence, studies with affinity labels showed that it must be near the serine residue in the active site. A deriva-

Figure 8.6

(a) Superimposed computer-generated c-alpha tube structures for trypsin (green), chymotrypsin (yellow-gold) and elastase (pale pink). The side chains for the three key catalytic residues, Asp 102 (red), His 57 (lt. blue), and Ser 195 (orange) are shown for trypsin. The structure for trypsin also includes a bound inhibitor, benzamidine, in yellow. (b) Superimposed c-alpha tube structures for subtilisin (lavender) and trypsin (green), both with catalytic side chains showing. The three key catalytic residues, Asp 102 (red), His 57 (lt. blue), and Ser 195 (orange) for trypsin and Asp 32 (red), His 64 (lt. blue), and Ser 221 (orange) for subtilisin are shown. (Images (a) and (b): Copyright 1994 by the Scripps Research Institute/Molecular Graphics Images by Michael Pique using software by Yng Chen, Michael Connolly, Michael Carson, Alex Shah, and AVS, Inc. Visualization advice by Holly Miller, Wake Forest University Medical Center.)

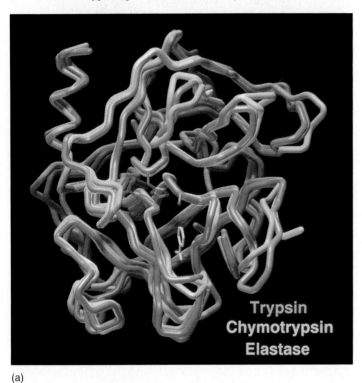

(a)

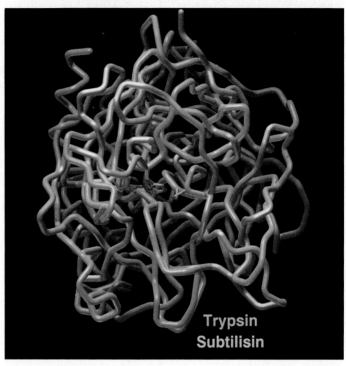

(b)

tive of phenylalanine containing a reactive chloromethylketone group was found to inhibit chymotrypsin irreversibly by reacting with the histidine. The inhibition can be prevented by the presence of other aromatic molecules that bind competitively at the active site.

The initial evidence for the formation of an acyl-enzyme ester intermediate came from studies of the kinetics with which chymotrypsin hydrolyzed analogs of its normal polypeptide substrates. The enzyme turned out to hydrolyze esters as well as peptides and simpler amides. Of particular interest was the reaction with the ester p-nitrophenyl acetate. This substrate is well suited for kinetic studies because one of the products of its hydrolysis, p-nitrophenol, has a yellow color in aqueous solution, whereas p-nitrophenyl acetate itself is colorless. The change in the absorption spectrum makes it easy to follow the progress of the reaction. When rapid-mixing techniques are used to add the substrate to the enzyme, an initial burst of p-nitrophenol is detected within the first few seconds, before the reaction settles down to a constant rate (fig. 8.8). The amount of p-nitrophe-nol released in the burst is approximately equal to the amount of enzyme present in the solution. These observations suggest that the overall enzymatic reaction occurs in two distinct steps, as shown in figure 8.9. In the first step, p-nitrophenol is released and the acetyl group is transferred to the enzyme, forming an acyl-enzyme intermediate. In the second step, the intermediate is hydrolyzed, and acetate is released. The finding that diisopropylfluorophosphate prevents the initial burst of p-nitrophenol suggested that the enzymatic group that forms the ester intermediate is Ser 195.

When the crystal structures of trypsin, chymotrypsin, and elastase with bound substrate analogs were determined, the substrate analogs were indeed located close to Ser 195 (see figs. 8.6 and 8.7). Histidine 57 sits nearby, in an orientation suggesting that the OH group of Ser 195 forms a hydrogen bond to the imidazole side chain of the histidine. Aspartic acid 102 sits on the opposite edge of the imidazole ring, where its negatively charged carboxylate group can interact with the proton on the other nitrogen of the ring.

Figure 8.7

Computer-generated c-alpha tube model of the crystal structure of trypsin, seen from the same perspective as in figure 8.6(*a*). The coloring of the amino acid side chains and the inhibitor (benzamidine) is as in figure 8.6(*a*). (*b*) A close-up view of the active site of trypsin with bound benzamidine. (Images (*a*) and (*b*): Copyright 1994 by the Scripps Research Institute/Molecular Graphics Images by Michael Pique using software by Yng Chen, Michael Connolly, Michael Carson, Alex Shah, and AVS, Inc. Visualization advice by Holly Miller, Wake Forest University Medical Center.)

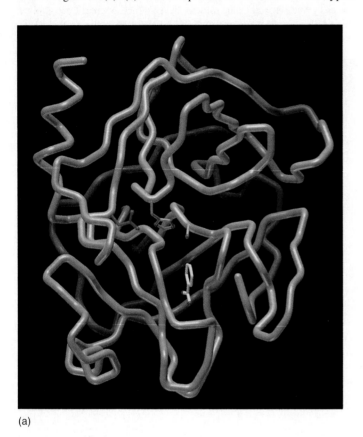

(a)

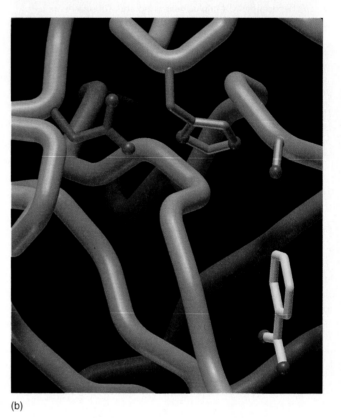

(b)

Figure 8.8

p-Nitrophenol formation as a function of time during the hydrolysis of *p*-nitrophenyl acetate by chymotrypsin. A rapid initial burst of *p*-nitrophenol is followed by a slower, steady-state reaction. The amount of *p*-nitrophenol released in the burst is approximately equal to the amount of enzyme present.

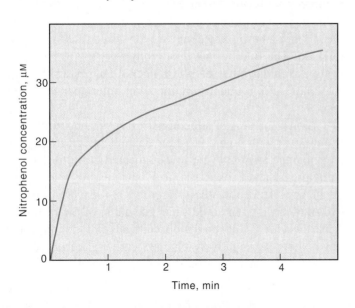

The side chains of the aspartate and histidine residues thus appear to be oriented so as to facilitate removal of the proton from the serine's OH group.

$$-\overset{\displaystyle O}{\overset{\|}{C}}-O^{-} \cdots HN \diagup N \cdots HOCH_2-$$

Because withdrawing the proton would increase the nucleophilic character of the oxygen, this arrangement explains why Ser 195 is exceptionally reactive. The dependence of the enzyme kinetics on pH agrees with this picture. Enzymatic activity depends on the presence of a basic group with a pK_a of about 6.8, which is in the range consistent with a histidine side chain. The k_{cat} decreases abruptly if this group is protonated (fig. 8.10). Protonating His 57 prevents the histidine from forming a hydrogen bond to Ser 195, thereby decreasing the nucleophilic reactivity of the serine.

The crystal structures also provided an explanation for the different substrate specificities of trypsin, chymotrypsin,

Figure 8.9

Steps in the hydrolysis of *p*-nitrophenyl acetate by chymotrypsin. In the hydrolysis of this and most other esters, the breakdown of the acyl-enzyme intermediate is the rate-determining step. In the hydrolysis of peptides and amides, the rate-determining step usually is the formation of the acyl-enzyme intermediate. This makes the transient formation of the intermediate more difficult to study because the intermediate breaks down as rapidly as it forms.

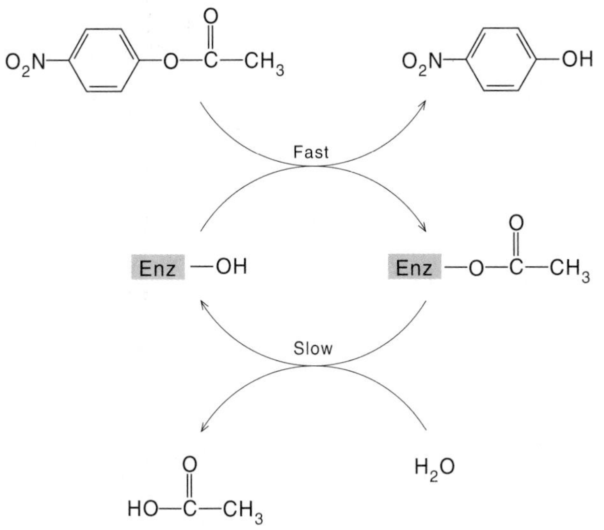

Figure 8.10

The turnover number (k_{cat}) and the Michaelis constant (K_m) as a function of pH for the hydrolysis of *N*-acetyl-L-tryptophanamide by chymotrypsin at 25°C. The decrease in k_{cat} as the pH is lowered between 8 and 6 probably reflects the protonation of His 57. The increase in K_m above pH 9 probably reflects the deprotonation of Ile 16, which results in the rotation of Gly 193 out of the substrate-binding site.

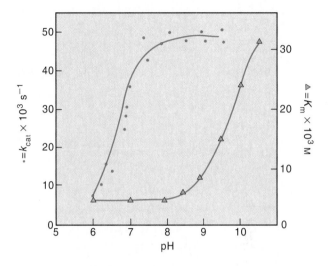

and elastase. In both trypsin and chymotrypsin, the side chain of the substrate fits snugly into a pocket (see figs. 8.6 and 8.7). At the far end of the pocket in trypsin is an aspartic acid residue. The negative charge of the aspartate carboxylate favors the binding of the positively charged side chain of lysine or arginine (see fig. 7.15). In chymotrypsin the aspartic acid residue is replaced by serine, creating a less polar environment that suits the side chain of tyrosine, phenylalanine, or tryptophan. In elastase, the binding pocket is padded with the bulky side chains of valine and threonine residues, so that it can accommodate only small substrates such as alanine.

Another important feature of the substrate-binding site in serine proteases is that the carbonyl oxygen atom of the peptide bond that is cleaved is hydrogen-bonded to one or more NH groups. In trypsin, chymotrypsin, and elastase, these are the amide NH groups of Ser 195 and Gly 193. The interaction with the two protons favors an increase in the negative charge on the oxygen, facilitating formation of a tetrahedral intermediate state, as we discussed previously in connection with general-acid catalysis and electrostatic effects. Figure 8.7 shows this part of the active site in trypsin.

The overall reaction mechanism of chymotrypsin is sketched in figure 8.11. Part *a* shows the enzyme–substrate complex, with an aromatic side chain of the substrate seated in the binding pocket and the carbonyl oxygen atom hy-

drogen-bonded to the amide NH hydrogens of Gly 193 and Ser 195. Aspartate 102 and His 57 are aligned as described above, with Ser 195 forming a hydrogen bond with the histidine. Part *b* shows a tetrahedral intermediate state. Serine 195 has released its proton to His 57 and launched a nucleophilic attack on the carbonyl carbon of the substrate. The histidine residue thus acts as a general-base catalyst in the formation of the intermediate. The movement of negative charge to the carbonyl oxygen creates an "oxyanion," which is stabilized largely by electrostatic interactions with the amide protons of Ser 195 and Gly 193. The tetrahedral oxyanion intermediate is not stable enough to be isolated, and the evidence for its existence is inferential. In part *c*, this intermediate has decomposed to form the more stable acyl-enzyme intermediate, in which the serine is linked as an ester to the carboxylic part of the substrate and the amine product has been released. Histidine 57 probably facilitates this decomposition by acting as a general acid.

To complete the reaction, the breakdown of the acyl-enzyme probably occurs as shown in parts *d*, *e*, and *f* of figure 8.11. The steps here are essentially a reversal of the steps through parts *a*, *b*, and *c*, except that water replaces the amino part of the substrate. The reaction probably proceeds by way of a tetrahedral intermediate similar to one shown in part *b* except for this substitution (part *e*).

In figure 8.11, note that Asp 102 does not actually remove a proton from His 57 in the formation of either of the tetrahedral intermediates. The effect of the aspartate's negatively charged carboxyl group probably is best viewed

Figure 8.11

The probable mechanism of action of chymotrypsin. The six panels show the initial enzyme-substrate complex (*a*), the first tetrahedral (oxyanion) intermediate (*b*), the acyl-enzyme (ester) intermediate with the amine product departing (*c*), the same acyl-enzyme intermediate with water entering (*d*), the second tetrahedral (oxyanion) intermediate (*e*), and the final enzyme-product complex (*f*). In the transition states between these intermediates, there probably is a more even distribution of negative charge between the different oxygen atoms attached to the substrate's central carbon atom.

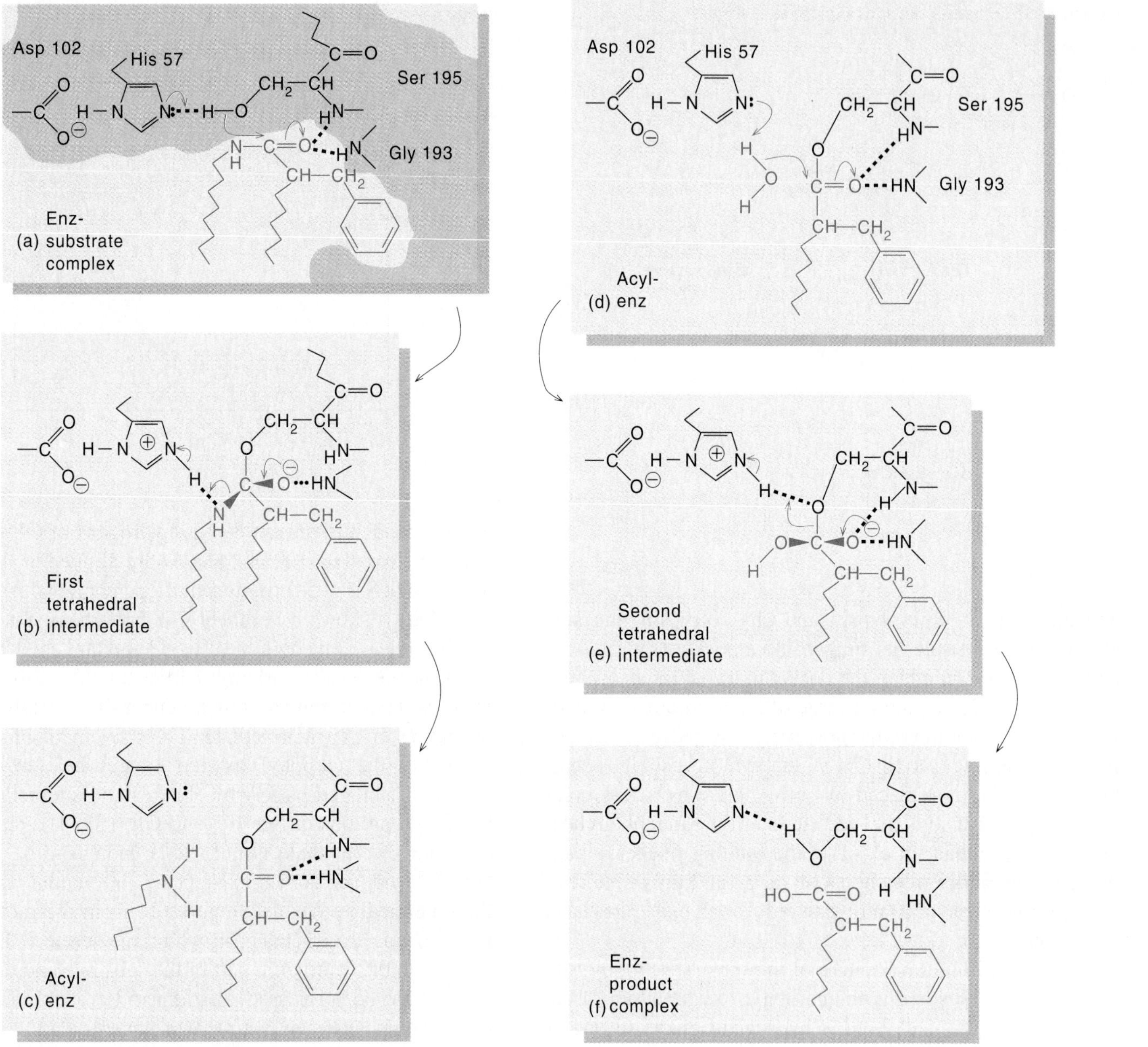

as an electrostatic effect that favors movement of positive charge from Ser 195 to His 57. The stabilization of the oxyanion intermediate by the amide protons of Gly 193 and Ser 195 can be described similarly as a favorable electrostatic interaction of the negatively charged oxygen with nearby atoms that have positive charges, rather than as an actual transfer of a proton to the oxygen. The carboxyl group of Asp 102 also helps to align the imidazole ring of His 57 in the proper orientation for removing the proton from Ser 195.

One way to test a scheme like that shown in figure 8.11 is to investigate the effects of modifying the amino acid residues that play important roles. The inhibitory effect of the reaction of Ser 195 with diisopropylfluorophosphate was mentioned earlier (also see fig. 7.18). To examine the importance of His 57, this residue can be modified by meth-

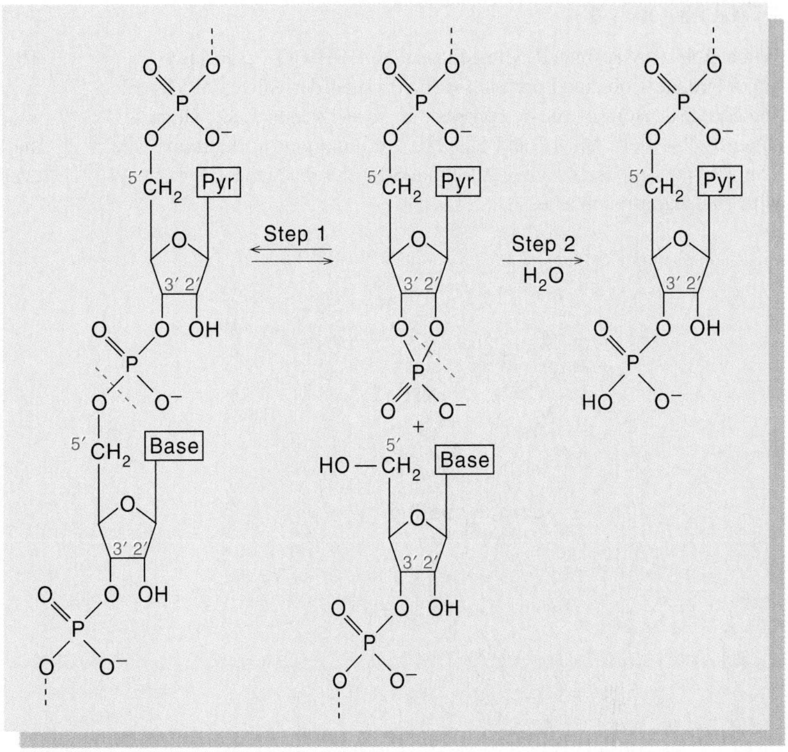

Figure 8.12

A portion of an RNA chain, indicating points of cleavage by pancreatic ribonuclease. "Pyr" refers to a pyrimidine; "Base" can be either a purine or a pyrimidine. The 2', 3', and 5' carbon atoms are labeled. The enzymatic reaction proceeds in two steps, with a cyclic 2',3'-phosphate diester as an intermediate.

ylation, which disrupts the interaction with Asp 102. This process decreases the activity of chymotrypsin by a factor of more than 10^3. The importance of Asp 102 was tested by site-directed mutagenesis of trypsin and subtilisin. Replacing the aspartate by asparagine reduces k_{cat} by a factor of about 10^4.

Ribonuclease A: An Example of Concerted Acid–Base Catalysis

Ribonucleases are a widely distributed family of enzymes that hydrolyze RNA by cutting the P—O ester bond attached to a ribose 5' carbon (fig. 8.12). A good representative of the family is the pancreatic enzyme ribonuclease A (RNase A), which is specific for a pyrimidine base (uracil or cytosine) on the 3' side of the phosphate bond that is cleaved. When the amino acid sequence of bovine RNase A was determined in 1960 by Stanford Moore and William Stein, it was the first enzyme and only the second protein to be sequenced. RNase A thus played an important role in the development of ideas about enzymatic catalysis. It was one of the first enzymes to have its three-dimensional structure elucidated by x-ray diffraction and was also the first to be synthesized completely from its amino acids. The synthetic protein proved to be enzymatically indistinguishable from the native enzyme.

RNase A is a relatively small protein, with a single polypeptide chain of 124 amino acid residues (see fig.

4.15). An early discovery by Frederick Richards that turned out to be useful was that the protein could be cleaved between residues 20 and 21 by the bacterial serine protease, subtilisin. The resulting two polypeptides were separated and purified. They were enzymatically inactive individually, but regained the activity of the native enzyme when they were recombined. This work shows that strong, noncovalent interactions occur that can hold protein chains together even when one of the peptide links is cut. It also makes it possible to modify specific amino acid residues of the two polypeptide chains independently and to explore how each residue contributes to the reassembly of the protein and the recovery of enzymatic activity.

The hydrolysis of RNA catalyzed by RNase A occurs in two distinct steps, with a 2',3'-phosphate cyclic diester intermediate (see fig. 8.12). The intermediate can be identified relatively easily, because its breakdown is much slower than its formation. Ribonucleases do not hydrolyze DNA, which lacks the 2'-hydroxyl group needed for the formation of the cyclic intermediate.

RNase A is completely inhibited if either of two histidine residues (His 12 or His 119) is modified by carboxymethylation with iodoacetate (fig. 8.13) suggesting that these histidines play important roles in the active site. In support of this conclusion, the reaction of iodoacetate with His 12 or His 119 is inhibited by cytidine-3'-phosphate and other small molecules that bind at the active site. Lysine 41 has been implicated similarly in the active site by the observation that enzymatic activity is destroyed by the reaction of

Figure 8.13

When RNase A is treated with iodoacetate ($ICH_2CO_2^-$), the two major products obtained are carboxymethylated derivatives of His 12 and His 119. Both of these enzymes are severely inhibited, which suggests that both His 12 and His 119 are important in the active site. The enzyme also is completely inhibited by the reaction of Lys 41 with fluorodinitrobenzene.

1-carboxymethyl-His 119

3-carboxymethyl-His 12

Dinitrobenzyl-Lys 41

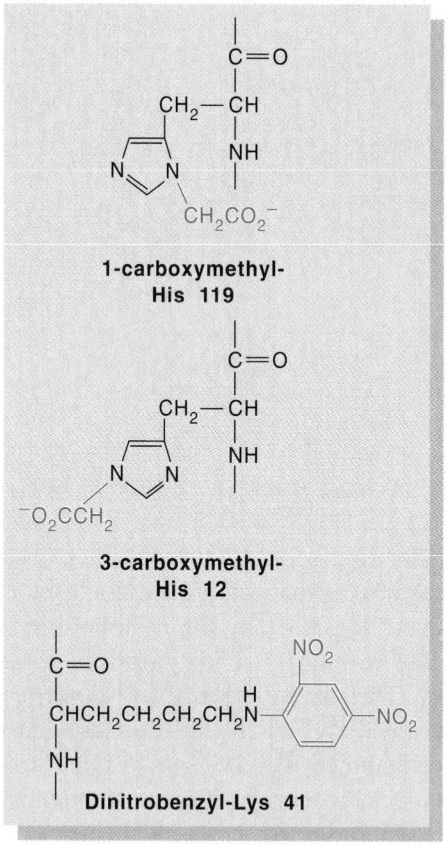

Figure 8.14

The dependence of k_{cat} of RNase A on pH. The bell-shaped curve suggests that one histidine residue must be in the protonated state and another must be unprotonated. Similar pH dependences are found for the hydrolysis of either RNA or pyrimidine nucleoside-2′,3′-phosphate cyclic diesters.

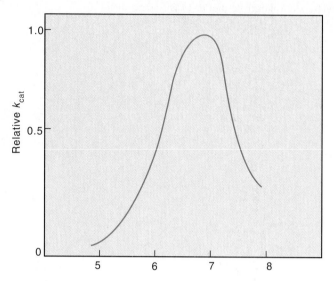

the ϵ-amino group of this residue with fluorodinitrobenzene (see fig. 8.13). A second lysine (Lys 7) also appears to be important for substrate binding, although the enzyme retains some activity when this residue is modified.

The enzymatic activity of RNase A shows a bell-shaped dependence on pH, with an optimum near pH 7 (fig. 8.14). The operation of the enzyme appears to require that a dissociable group with a pK_a of about 6.3 be in the protonated form, and that a group with a pK_a of about 8 be unprotonated. These groups have been identified as His 12 and His 119 by measuring the NMR spectrum of the enzyme as a function of pH. Spectral peaks due to histidine residues undergo characteristic shifts as the imidazole ring is protonated.

The pH dependence of the kinetics, taken with the information that histidines 12 and 119 are essential for enzymatic activity, suggests that the two histidines participate in the reaction by concerted general-base and general-acid catalysis. This supposition is supported by the crystal structure

of the enzyme. Figure 8.15 shows a model of the crystal structure of RNase S (the active enzyme obtained by cutting ribonuclease with subtilisin and recombining the two polypeptides) with a bound competitive inhibitor, "UpCH$_2$A." UpCH$_2$A resembles the dinucleotide substrate UpA but cannot be hydrolyzed because it has a methylene carbon in place of the oxygen between the P and the 5′—CH$_2$ of the ribose of adenosine (fig. 8.16). Similar crystal structures have been obtained with other substrate analogs. In all of the structures, His 12 and His 119 sit on opposite sides of the substrate's phosphate group, with His 12 positioned close to the 2′ OH group. Lysines 7, 14, and 66 also are located nearby, where their positively charged amino groups have favorable electrostatic interactions with the negatively charged phosphate. Near His 119 is the carboxyl group of an acidic residue, Asp 121. The electrostatic effect of Asp 121 favors the transfer of a proton from a molecule of water to His 119, much as Asp 102 of chymotrypsin induces a proton to move from Ser 195 to His 57.

The pyrimidine ring on the 3′ side of the substrate fits snugly into a groove on RNase A. Valine 43 is located on one side of the pyrimidine ring, and Phe 120 on the other. The specificity of the enzyme for pyrimidine nucleotides appears to result largely from hydrogen bonding to Thr 45. The threonine side chain can form a pair of complementary hydrogen bonds with either uracil or cytosine (fig. 8.17). Crystallographic studies have shown that if the pyrimidine is replaced by a purine, the compound can still bind to the enzyme, but the distance between His 12 and the 2′ OH of

Figure 8.15

A model of the binding of the substrate analog UpCH$_2$A (blue) to RNase S. (The structure of UpCH$_2$A is shown in figure 8.16.) RNase S is an active enzyme, although it differs from the native enzyme because of a peptide bond cleavage between residues 20 and 21 (light dashed line). The two histidines and the lysine that are crucial to the active site (His 12, His 119, and Lys 41) are indicated in gray. The dotted lines indicate some of the hydrogen bonds that maintain the protein structure.
(Illustration copyright by Irving Geis. Reprinted by permission.)

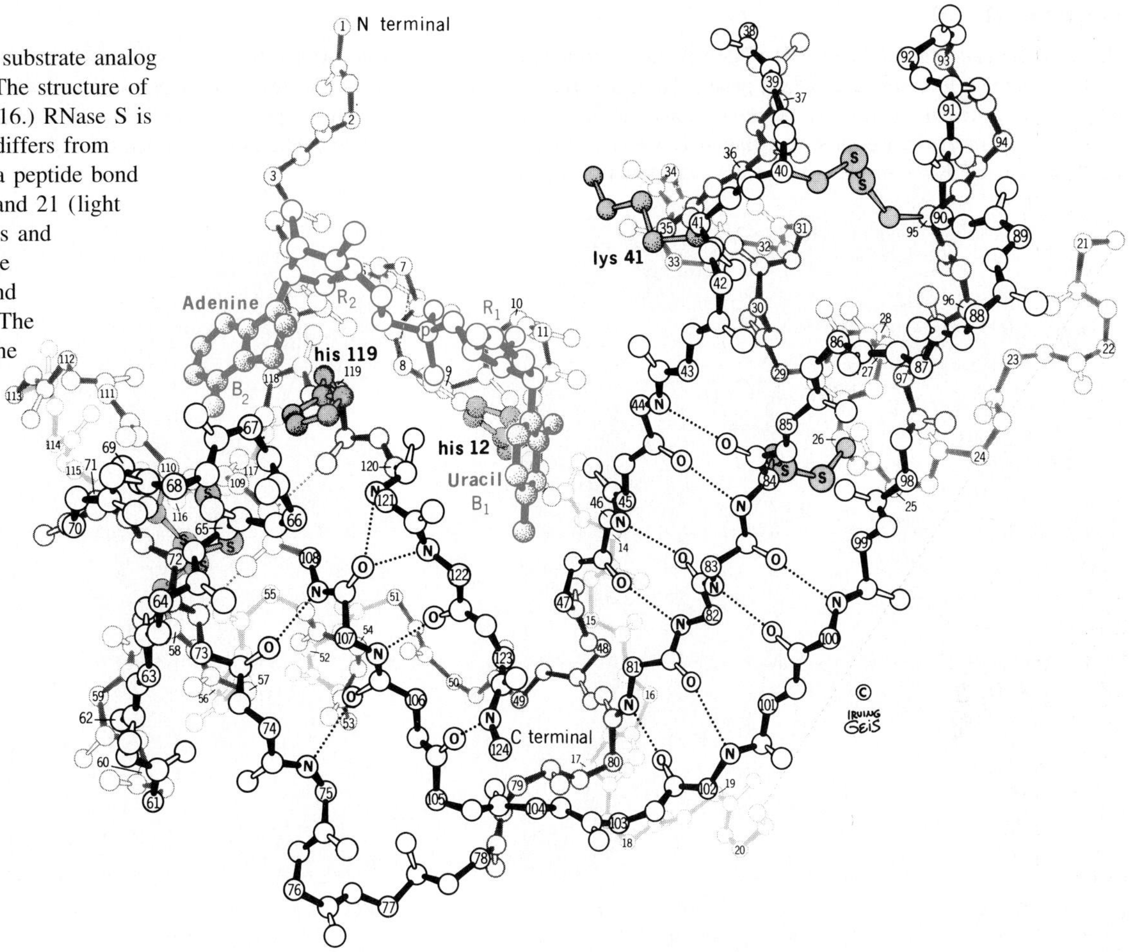

Figure 8.16

Structure of UpCH$_2$A, a competitive inhibitor of RNase A. UpCH$_2$A differs from the dinucleotide substrate UpA in having a methylene group instead of oxygen between the phosphorus and C-5′ of the adenosine.

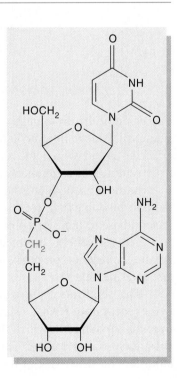

Figure 8.17

The side chain and amide NH group of Thr 45 in RNase A can form a pair of complementary hydrogen bonds with either uracil or cytosine. These probably account for the specificity of RNase A for pyrimidine nucleotides.

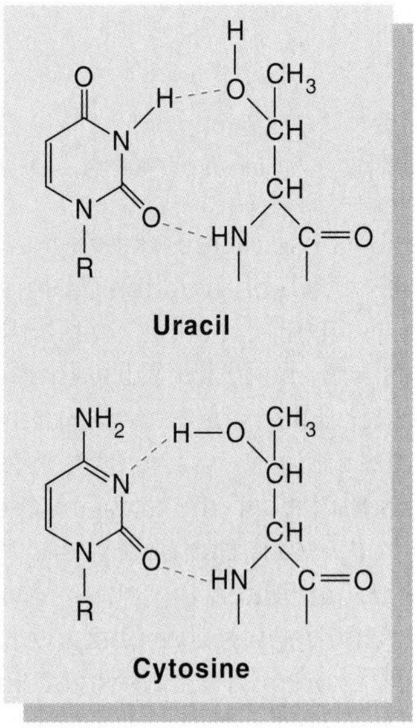

Uracil

Cytosine

Figure 8.18

The probable catalytic mechanism of RNase A. In the formation of the 2′,3′-cyclic ester, His 12 acts as a general base, and His 119 acts as a general acid. The increase in negative charge on the phosphate's oxygen atoms in the transition state is favored by electrostatic interactions with Lys 7 and Lys 41. In the breakdown of the cyclic ester, His 119 acts as a general base and His 12 acts as a general acid. The reaction proceeds through two short-lived intermediates in which the phosphorus atom is pentavalent, in addition to the more stable cyclic ester intermediate. Py = pyrimidine.

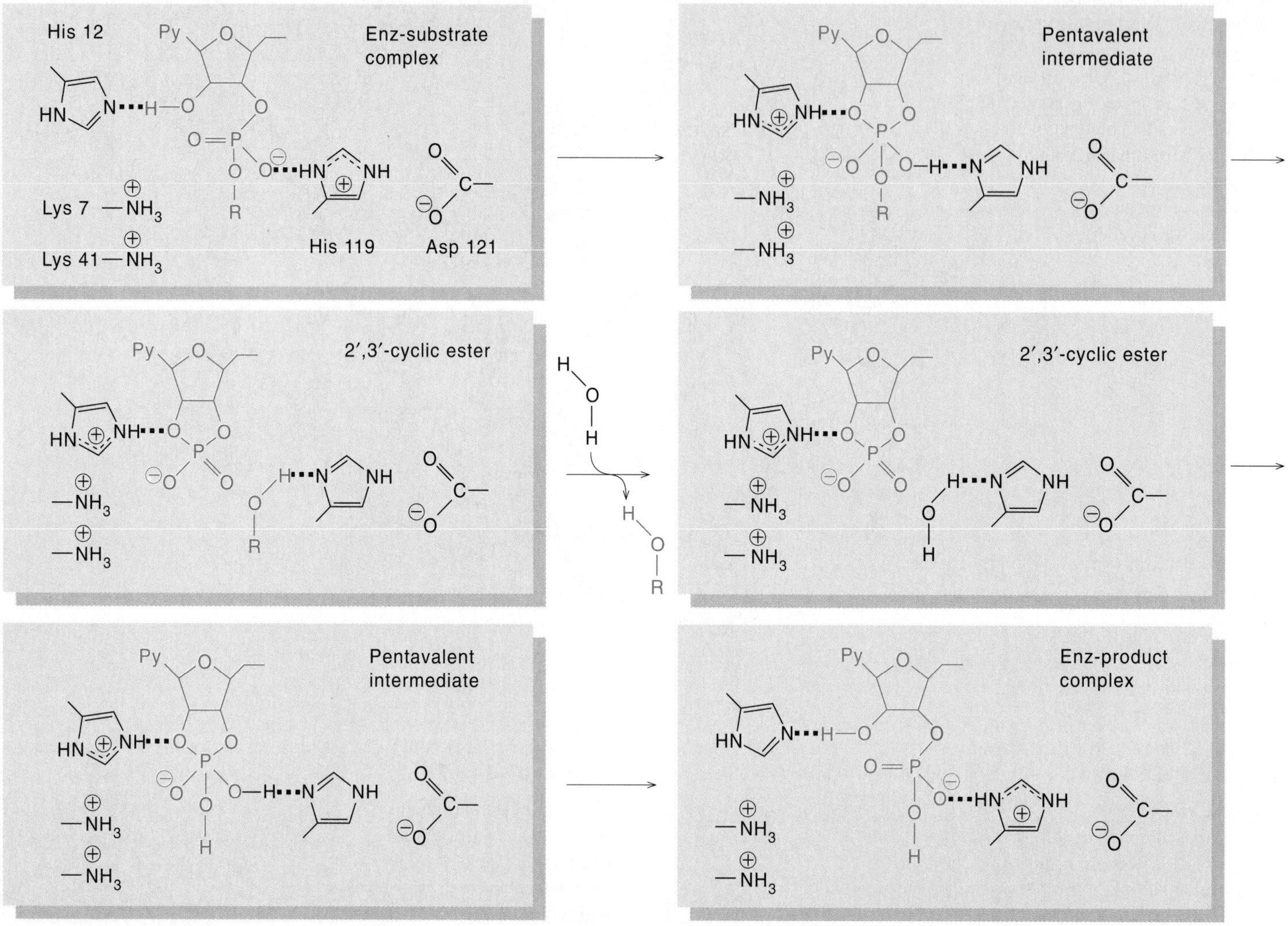

the ribose increases by about 1.5 Å. This evidently prevents the catalytic reaction from occurring.

The probable catalytic mechanism of RNase A is shown in figure 8.18. In the formation of the 2′,3′-cyclic ester, His 12 acts as a general base to remove the proton of the 2′ hydroxyl group, which increases the nucleophilic character of the oxygen atom. Reaction of the hydroxyl group with the phosphate creates a transient intermediate state in which the phosphorus atom is pentavalent. The increase in negative charge on the phosphate's oxygen atoms is favored by electrostatic interactions with Lys 7 and Lys 41. The formation of the pentavalent intermediate also is promoted by His 119, which acts as a general acid to protonate one of the phosphate oxygens. His 119 then probably participates in transferring a proton to the 5′ oxygen atom.

Expulsion of the protonated 5′ OH group converts the transient pentavalent intermediate into the more stable 2′,3′-cyclic ester, in which the substituents of the phosphorus have returned to a tetrahedral geometry.

In the breakdown of the 2′,3′-cyclic ester, the roles of the two histidines are reversed. Histidine 119 now acts as general base to remove a proton from H_2O, and His 12 acts as a general acid to protonate the substrate's 2′ oxygen. Again, the reaction probably proceeds by way of a pentavalent intermediate.

The pentavalent phosphoryl intermediates that figure in the reactions of RNase A probably have the geometry of a trigonal bipyramid in which three of the phosphorus atom's substituents lie in a plane. The entering and leaving oxygen atoms are on either side of the plane, forming a straight line

Figure 8.19

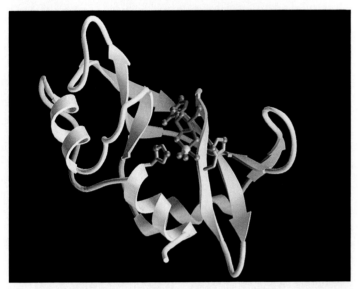

(a)

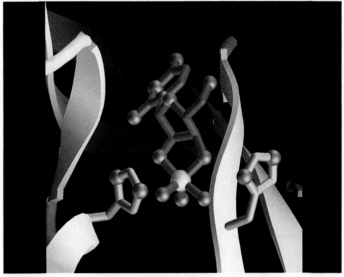

(b)

with the phosphorus. One observation that supports the formation of such pentavalent intermediates is that RNase A is inhibited strongly by uridine vanadate, in which the vanadium atom is surrounded by five oxygens with a similar geometry. Figure 8.19 shows the structure of the complex of the enzyme with the inhibitor. The disposition of histidines 12 and 119 on either side of the vanadate group in the crystal structure is consistent with the linear arrangements of the entering and leaving oxygen atoms on either side of the phosphorus atom in figure 8.18.

Triosephosphate Isomerase Has Approached Evolutionary Perfection

 Triosephosphate isomerase catalyzes the interconversion of dihydroxyacetone phosphate and 3-phosphoglyceraldehyde.

$$
\begin{array}{ccc}
CH_2OH & & H{-}C{=}O \\
| & & | \\
C{=}O & \rightleftharpoons & H{-}C{-}OH \qquad (10) \\
| & & | \\
CH_2OPO_3{}^{2-} & & CH_2OPO_3{}^{2-}
\end{array}
$$

Dihydroxyacetone phosphate   Glyceraldehyde-3-phosphate

The interconversion of the two triosephosphates is an essential step in the catabolism of carbohydrates (see chapter 12). Examining the catalytic mechanism of triosephosphate isomerase is instructive. It is among the enzymes that ap-proach evolutionary perfection in the sense that its specificity constant k_{cat}/K_m is greater than $10^8 \, s^{-1} M^{-1}$, which is near the limit set by the rates of diffusion of the substrate and protein (see table 7.3). The enzyme achieves this high value for the specificity constant by having both a relatively high k_{cat} and a relatively low K_m.

Crystal structures have been obtained for triosephosphate isomerase purified both from yeast and from chicken muscle. The proteins from the two organisms have identical residues at about half the positions in their amino acid sequences, and their three-dimensional structures are very similar. The overall structure is a β barrel composed of eight parallel β sheets linked by eight α helices (see fig. 4.17*b*). The active site is near the center of the molecule and contains a glutamic acid (Glu 165) with nearby histidine and lysine residues (fig. 8.20). Glu 165 was first implicated in the active site by the observation that it reacts irreversibly with chloroacetone phosphate, an inhibitor that is a close structural analog of the substrate dihydroxyacetone phosphate. (In chloroacetone phosphate, a chlorine atom replaces the hydroxyl group on C-1.) In crystal structures of the enzyme with bound dihydroxyacetone phosphate, the carboxyl group of Glu 165 is located close to C-1 of the substrate. Changing Glu 165 to aspartate by site-directed mutagenesis decreases the rate of the isomerization reaction catalyzed by the enzyme approximately 1,000-fold. This is a dramatic effect, considering that the substitution of Asp for Glu only entails shortening the amino acid side chain by one —CH_2-group.

Figure 8.20

(a) Ribbon diagram of the crystal structure of triosephosphate isomerase from chicken muscle. This figure shows one of the two subunits of the enzyme. The active site amino acid side chains Glu 165 (red), His 95 (light blue) and Lys 13 (dark blue) are shown. (b) Close-up view of the active site of triosephosphate isomerase.

(Copyright 1994 by the Scripps Research Institute/Molecular Graphics Images by Michael Pique using software by Yng Chen, Michael Connolly, Michael Carson, Alex Shah, and AVS, Inc. Visualization advice by Holly Miller, Wake Forest University Medical Center.)

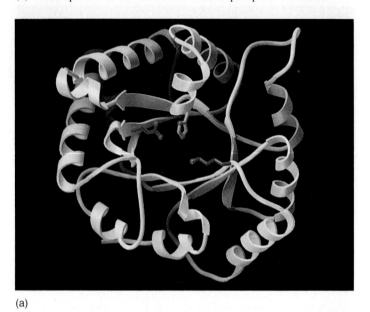

(a)

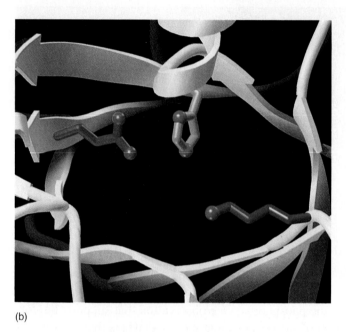

(b)

Figure 8.21

The probable reaction mechanism of triosephosphate isomerase. The γ-carboxylate group of Glu 165 acts as a general base to remove a proton from C-1 of the substrate, dihydroxyacetone phosphate (DHAP). This generates a planar ene-diolate intermediate that has two tautomeric forms. After an interconversion of ene-diolate tautomers, the protonated glutamic acid residue returns a proton to C-2. The electrostatic effects of His 95 and Lys 12 facilitate the formation of the negatively charged intermediate. GAP = glyceraldehyde-3-P.

The rate of the isomerase reaction decreases if a basic group is protonated by lowering the pH below 6.5. It seems clear that this is the carboxyl group of Glu 165 because lowering the pH also prevents the reaction of this residue with chloroacetone phosphate. This suggests that, in the enzymatic reaction, Glu 165 acts as a general base to remove a proton from C-1 of dihydroxyacetone phosphate, converting the substrate into an ene-diolate intermediate as shown in figure 8.21. To complete the reaction, the protonated glutamic acid residue can return a proton to C-2 of the ene-diolate instead of to C-1.

Additional support for this mechanism comes from experiments in which the enzymatic reaction is run in D_2O. Deuterium is incorporated stereospecifically onto carbon 2 of the product.

$$\begin{array}{ccc}
CH_2OH & & H\!-\!C\!=\!O \\
| & \xrightarrow{D_2O} & | \\
C\!=\!O & & D\!-\!C\!-\!OH \\
| & & | \\
CH_2OPO_3^{2-} & & CH_2OPO_3^{2-}
\end{array} \qquad (11)$$

In the absence of the enzyme, the carbon-bound hydrogen atom on 3-phosphoglyceraldehyde does not exchange with deuterium atoms of the solvent at any significant rate. (We are not concerned here with an exchange of the proton bound to the alcohol oxygen in the reactant or product. This proton does exchange rapidly with the solvent in either the presence or the absence of an enzyme, but the deuteron that is incorporated here can be removed immediately by putting the product back in H_2O.) The incorporation of deuterium can be explained as follows: After a proton is transferred from C-1 of the substrate to the carboxyl group of Glu 165, it has an opportunity to escape into the solution and to be replaced by a deuteron. A proton attached to a carboxylic acid oxygen usually exchanges rapidly with the solvent. The deuteron on Glu 165 then can be transferred to C-2 of the product.

The hydrogen-bonding pattern in the crystal and studies of the pH dependence of the reaction suggest that the histidine and lysine residues in the active site do not act as general acids in the catalytic pathway but rather serve to stabilize the negative charge on an ene-diolate intermediate by electrostatic effects (see fig. 8.21).

Assuming that the mechanism shown in figure 8.21 is generally correct, how can we determine whether the formation or the breakdown of the ene-diolate intermediate is the rate-limiting step in the overall reaction? One approach is to study the kinetics using substrates that are specifically labeled with deuterium. With [1(R)-^{2}H]-dihydroxyacetone phosphate (fig. 8.22a), k_{cat} is about three times smaller than the k_{cat} obtained with the ^{1}H analog. An isotope effect of this magnitude is consistent with the view that the bond holding the H or D atom to C-1 must be broken in the

Figure 8.22

The substrates of triosephosphate isomerase can be labeled specifically with deuterium. The rate of the isomerase reaction with [1(R)-^{2}H]-dihydroxyacetone phosphate (a) as substrate is slowed by about a factor of 3 relative to the rate with ordinary dihydroxyacetone phosphate. This isotope effect indicates that breaking the H—C bond to form the ene-diolate intermediate is the rate-limiting step in conversion of dihydroxyacetone phosphate to 3-phosphoglyceraldehyde. Little or no isotope effect is obtained when the reaction is run in the opposite direction starting with [2-^{2}H]-3-phosphoglyceraldehyde (b), indicating that the breakdown of the ene-diolate must be rate-limiting in this direction. Note that although the two hydrogen atoms on C-1 of dihydroxyacetone phosphate are chemically identical, they are stereochemically distinguishable. The enzyme always removes the one that has been replaced by deuterium in (a).

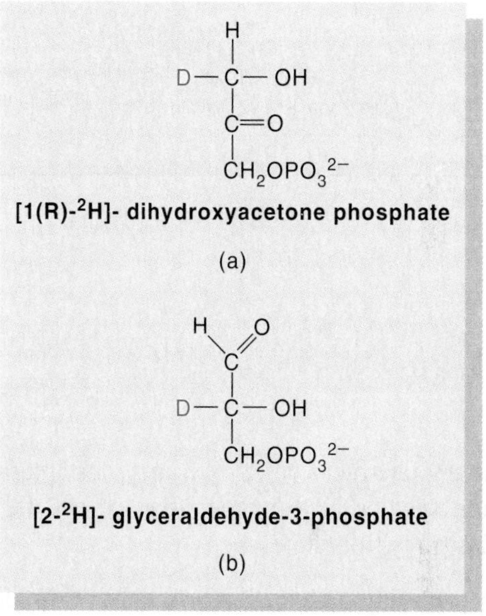

[1(R)-^{2}H]- dihydroxyacetone phosphate

(a)

[2-^{2}H]- glyceraldehyde-3-phosphate

(b)

rate-limiting step. This is consistent with the view that the formation of the ene-diolate is rate-limiting. Similar measurements with [2-^{2}H]-3-phosphoglyceraldehyde (fig. 8.22b) show that there is little or no isotope effect on k_{cat} for the reverse direction, from 3-phosphoglyceraldehyde to dihydroxyacetone phosphate. In this direction, the rate-limiting step must be the breakdown of the ene-diolate intermediate.

By measuring the kinetics of the isotopic exchange reactions catalyzed by the enzyme, along with the overall kinetics of the isomerization reactions in both directions, Jeremy Knowles and his colleagues obtained rate constants for the individual steps of the catalytic mechanism. Figure 8.23 shows the pathway in the form of a free energy profile. The free energy changes for binding of the reactant or product to the enzyme are calculated on the basis of typical concentrations of dihydroxyacetone phosphate and 3-phosphoglyceraldehyde found in muscle cells (40 μM). At this low concentration of substrate, when the overall rate of the

Figure 8.23

The calculated free energy profile of the reaction catalyzed by triosephosphate isomerase. (Enz = enzyme; DHAP = dihydroxyacetone phosphate; GAP = glyceraldehyde-3-phosphate.) The free energy changes associated with binding of DHAP and GAP to the enzyme are calculated on the assumption that DHAP and GAP are present at concentrations of 40 μM.

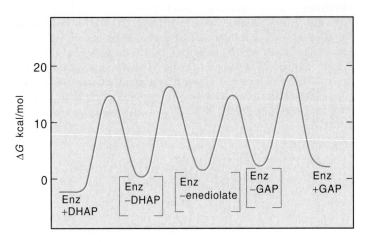

Reaction coordinate

reaction is described by the specificity constant k_{cat}/K_m, the rate-limiting step appears to be the formation of the enzyme–substrate complex. This step occurs with a rate constant of about 10^8 $M^{-1}s^{-1}$ and evidently is set by the rate of diffusion of the enzyme and substrate. It can be slowed down by increasing the viscosity of the solution. Because the activation free energies for the additional steps in the interconversion of the enzyme–substrate and enzyme–product complexes are low enough so that these steps do not limit the overall rate of the reaction, there is no evolutionary pressure to increase the rate constants for these steps any further. Evolutionary changes that led to tighter binding of the substrate might actually be harmful, because lowering the free energy of the enzyme–substrate complex can increase the activation free energy for one or more of the steps in the conversion of this complex into the product. Given the fixed, low concentration of its substrate, triosephosphate isomerase appears to have reached perfection!

Summary

Like ordinary chemical catalysts, enzymes interact with the reacting species in a manner that lowers the free energy of the transition state.

1. All known enzymatic reaction mechanisms depend on one or more of the following five themes:
   a. proximity effects (enzymes hold the reactants close together in an appropriate orientation)
   b. general-acid or general-base catalysis (acidic or basic groups of the enzyme donate or remove protons and often do first one and then the other)
   c. electrostatic effects (charged or polar groups of the enzyme favor the redistribution of electric charges that must occur to convert the substrate into the transition state)
   d. nucleophilic or electrophilic catalysis (nucleophilic or electrophilic functional groups of the enzyme or a cofactor react with complementary groups of the substrate to form covalently linked intermediates
   e. structural flexibility (changes in the protein structure can increase the specificity of enzymatic reactions by ensuring that substrates bind or react in an obligatory order and by sequestering bound substrates in pockets that are protected from the solvent).

2. The serine proteases trypsin, chymotrypsin, and elastase are very similar in structure but have substrate-binding pockets that are tailored for different amino acid side chains. The active site of each enzyme contains three critical residues: serine, histidine, and aspartate, which are positioned so that the serine hydroxyl group becomes a strong nucleophilic reagent for reaction with the substrate's peptide carbon atom. This reaction generates an acyl-enzyme intermediate. Formation and hydrolysis of the acyl-enzyme intermediate probably are promoted by electrostatic interactions that stabilize tetrahedral transition states and by general-acid and general-base catalysis.

3. Ribonuclease A hydrolyzes RNA adjacent to pyrimidine bases. The reaction proceeds through a 2′,3′-phosphate cyclic diester intermediate. Formation and breakdown of the cyclic diester appear to be promoted by concerted general-base and general-acid catalysis by two histidine residues, and by electrostatic interactions with two lysines. These reactions proceed through pentavalent phosphoryl intermediates. The geometry of these intermediates resembles the geometry of vanadate compounds that act as inhibitors of the enzyme.

4. Triosephosphate isomerase interconverts dihydroxyace-tone phosphate and 3-phosphoglyceraldehyde. A glutamic acid residue probably acts as a general base to remove a proton from the substrate, forming an enediolate intermediate. Histidine and lysine side chains stabilize this intermediate electrostatically. Triosephosphate isomerase appears to have reached evolutionary perfection in the sense that it catalyzes its reaction at the maximum possible rate given the concentration of the substrate in the cell.

Selected Readings

Albery, W. J., and J. R. Knowles, Free-energy profile of the reaction catalyzed by triosephosphate isomerase. *Biochem.* 15:5588, 5627, 1976.

Blackburn, P., and S. Moore, Pancreatic ribonuclease. *The Enzymes* 15:317, 1982.

Blow, D., Structure and mechanism of chymotrypsin. *Acc. Chem. Res.* 9:145, 1976.

Branden, C. -I., H. Jornvall, H. Eklund, and B. Furugren, Alcohol dehydrogenases. *The Enzymes* 11:104, 1975.

Breslow, R., How do imidazole groups catalyze the cleavage of RNA in enzyme models and enzymes? Evidence from "negative catalysis." *Acc. Chem. Res.* 24:317, 1991.

Eklund, H., B. V. Plapp, J. -P. Samama, and C. -I. Branden, Binding of substrate in a ternary complex of horse liver alcohol dehydrogenase. *J. Biol. Chem.* 257:14349, 1982.

Fersht, A., *Enzyme Structure and Mechanism,* 2d ed. New York: Freeman, 1985.

Fersht, A. R., D. M. Blow, and J. Fastrez, Leaving group specificity in chymotrypsin-catalyzed hydrolysis of peptides: A stereochemical interpretation. *Biochem.* 12:2035, 1973.

Findlay, D., D. G. Herries, A. P. Mathias, B. R. Rabin, and C. A. Ross, The active site and mechanism of pancreatic ribonuclease. *Nature* 190:781, 1961.

Holbrook, J. J., A. Liljas, S. J. Steindel, and M. G. Rossmann, Lactate dehydrogenase. *The Enzymes* 11:191, 1975.

Holmes, M. A., D. E. Tronrud, and B. W. Matthews, Structural analysis of the inhibition of thermolysin by an active-site-directed irreversible inhibitor. *Biochem.* 22:236, 1983.

Imoto, T., L. N. Johnson, A. C. T. North, D. C. Phillips, and J. A. Rupley, Vertebrate lysozymes. *The Enzymes* 7:665, 1972.

Kelly, J. A., A. R. Sielecki, B. D. Sykes, M. N. G. James, and D. C. Phillips, X-ray crystallography of the binding of the bacterial cell wall trisaccharide NAM-NAG-NAM to lysozyme. *Nature* 282:875, 1979.

Kraut, J., How do enzymes work? *Science* 242:533, 1988.

Lolis, E., T. Alber, R. C. Davenport, D. Rose, F. C. Hartman, and G. A. Petsko, Structure of yeast triosephosphate isomerase at 1.9-Å resolution. *Biochem.* 29:6609, 1990.

Markley, J. L., Correlation proton magnetic resonance studies at 250 MHz of bovine pancreatic ribonuclease. I. Reinvestigation of histidine peak assignments. *Biochem.* 14:3546, 1975.

Page, M. I. (ed.), *Enzyme Mechanisms.* London: Royal Society of Chemistry, 1987. See particularly the chapters by M. I. Page (Theories of Enzyme Catalysis), A. L. Fink (Acyl Group Transfer—The Serine Proteinases), D. S. Auld (Acyl Group Transfer—Metalloproteinases), P. M. Cullis (Acyl Group Transfer—Phosphoryl Transfer), M. L. Sinnott (Glycosyl Group Transfer), and J. P. Richard (Isomerization Mechanisms through Hydrogen and Carbon Transfer).

Reeke, G. N., J. A. Hartsuck, M. L. Ludwig, F. A. Quiocho, T. A. Steitz, and W. N. Lipscomb, The structure of carboxypeptidase A: Some results at 2.0-Å resolution, and the complex with glycyl tyrosine at 2.8-Å resolution. *Proc. Natl. Acad. Sci. USA* 58:2220, 1967.

Rose, I. A., Mechanism of the aldose-ketose isomerase reactions. *Adv. Enzymol.* 43:491, 1975.

Shoham, M., and T. Steitz, Crystallographic studies and model building of ATP at the active site of hexokinase. *J. Mol. Biol.* 140:1, 1980.

Warshel, A., G. Naray-Szabo, F. Sussman, and J. -K. Hwang, How do serine proteases really work? *Biochem.* 28:3629, 1989.

Problems

1. Hexokinase promotes the following reaction:

   Glucose + ATP $\longrightarrow$ ADP + Glucose-6-phosphate

   Hexokinase initially binds glucose, then the hexokinase–glucose complex binds ATP. How do you explain the purpose of this substrate-binding pattern to a fellow student without using the induced-fit concept?

2. (a) In what ways are the mechanistic features of chymotrypsin, trypsin, and elastase similar?

   (b) If the mechanisms of these enzymes are similar, what features of the enzyme active site dictate substrate specificity?

3. If a lysine were substituted for the aspartate in the trypsin side-chain-binding crevice, would you expect the enzyme to be functional? If it were functional, what effect would you predict the substitution to have on substrate specificity?

4. Plant proteolytic enzymes are cysteine proteases, that is, they have a cysteine that is critical for the catalytic activity. Plant proteases are thought to function by a mechanism reminiscent of that shown for chymotrypsin (fig. 8.11). Propose a structure for the acyl-enzyme intermediate that would exist for plant proteases.

5. How could the inhibitors presented in table 8.4 be used to indicate that an unknown protease was of plant versus nonplant origin?

6. Notice how the active site of chymotrypsin contains Asp 102, His 57, and Ser 195 (fig. 8.11), whereas the active site of RNase A contains Asp 121, Lys 41, Lys 7, His 12, and His 119 (fig. 8.18). What fundamental principle concerning protein structure is emphasized when the residue number in the active site of enzymes is considered?

7. RNase A utilizes a 2′,3′-cyclic phosphate ester as an intermediate (fig. 8.18) but yields only a 3′-phosphate product. Within the mechanism, what controls the final position of the phosphate (i.e., why isn't the 2′-phosphate a product)?

8. For many enzymes, V_{max} is dependent on pH. At what pH do you expect V_{max} of RNase to be optimal? Why?

9. RNase can be completely denatured by boiling or by treatment with chaotropic agents (e.g., urea), yet can refold to its fully active form on cooling or removal of the denaturant. By contrast, when enzymes of the trypsin family and carboxypeptidase A are denatured, they do not regain full activity on renaturation. What aspects of trypsin and carboxypeptidase A structure preclude their renaturation to the fully active form?

10. In figures 8.18 and 8.21 lysyl residues are shown in the lower right parts of the figures. An initial examination of the mechanisms indicates these lysyl groups are only observers. Why are they shown in the "mechanism," and what do they do?

11. Why do structural analogs of the transition-state intermediate of an enzyme inhibit the enzyme competitively and with low K_i values?

12. Transition-state analogs of a specific chemical reaction have been used to elicit antibodies with catalytic activity. These catalytic antibodies have great promise as experimental tools as well as having commercial value. Why is it reasonable to assume that the binding site for the transition-state analog on the antibody mimics the enzyme active site? What difficulties might be encountered if a catalytic antibody is sought for a reaction requiring a cofactor (coenzyme)?

13. Using site-directed mutagenesis techniques, you isolate a series of recombinant enzymes in which specific lysine residues are replaced with aspartate residues. The enzymatic assay results are shown in the table.

Enzyme Form	Activity (U/mg)
Native enzyme	1,000
Recombinant Lys 21 ⟶ Asp 21	970
Recombinant Lys 86 ⟶ Asp 86	100
Recombinant Lys 101 ⟶ Asp 101	970

(a) What do you infer about the role(s) of Lys 21, 86, and 101 in the catalytic mechanism of the native enzyme?

(b) Speculate on the location of Lys 21 and Lys 101. Would you expect these residues to be conserved in an evolutionary sense?

(c) Would you expect Lys 86 to be evolutionarily conserved? Why or why not?

CHAPTER

9

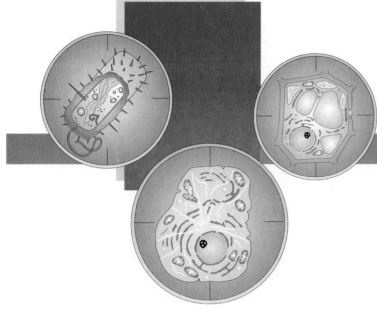

Regulation of Enzyme Activities

Chapter Outline

Partial Proteolysis Results in Irreversible Covalent
   Modifications

Phosphorylation, Adenylylation, and Disulfide Reduction
   Lead to Reversible Covalent Modifications

Allosteric Regulation Allows an Enzyme to Be
   Controlled Rapidly by Materials that Are Structurally
   Unrelated to the Substrate

   *Allosteric Enzymes Typically Exhibit a Sigmoidal
      Dependence on Substrate Concentration*

   *The Symmetry Model Provides a Useful Framework
      for Relating Conformational Transitions to
      Allosteric Activation or Inhibition*

Phosphofructokinase: Allosteric Control of Glycolysis Is
   Consistent with the Symmetry Model

Aspartate Carbamoyl Transferase: Allosteric Control of
   Pyrimidine Biosynthesis

Glycogen Phosphorylase: Combined Control by
   Allosteric Effectors and Phosphorylation

A select group of enzymes are regulated by one of two strategies: modification of the covalent structure of the enzyme or modification of their structure by the reversible binding of effector molecules.

The living cell resembles a complex factory with many different assembly lines of worker enzymes performing specific tasks. The activities of these workers have to be regulated precisely so that resources are used efficiently and a steady flow of parts keeps all the assembly lines moving smoothly. Cells use two basic strategies for regulating their enzyme activities. The first strategy is to adjust the amount and location of key enzymes. This requires mechanisms for the control of synthesis, degradation, and transport of proteins, which we discuss in Chapters 29, 30 and 31. The second strategy is to regulate the activities of the enzymes that are on hand. We deal exclusively with the second strategy in this chapter.

Enzymes subject to regulation are a select few of the total enzymes in a cell and are so subject for two reasons. First, it is not necessary to regulate the activity of every enzyme to achieve the desired level of control. In many cases, an entire metabolic pathway can be controlled by regulating only the enzyme that catalyzes the first step in the pathway (see chapter 11). Second, elaborate structural properties are required to create an enzyme that can be regulated.

In the first part of this chapter we survey the different methods that cells use to regulate enzyme activities. Then we consider in detail how these methods apply to three of the most thoroughly characterized regulatory enzymes.

In principle, the activities of many enzymes can be altered by changes in pH. Cells do take advantage of this possibility in a few cases. Lysozyme, for example, is most active in the pH 5 region, which is characteristic of some extracellular secretions, and is much less active in the intracellular pH region near 7. The activity of lysozyme thus remains low until the enzyme is secreted. But this is not a very practical solution to the problem of regulating the activities of intracellular enzymes, because most cells must hold their pH within narrow limits. Two strategies are much more widely applicable. The first is to modify the covalent structure of the enzyme in such a way as to alter either K_m or k_{cat}. The second is to use an inhibitor or activator, an *effector*, that binds reversibly to the enzyme and, again, alters either the K_m or k_{cat}. Such an effector may bind either at the active site itself or at some more distant site on the enzyme. In the latter case, it is termed an allosteric effector (from the Greek words *allos,* meaning ''other,'' and *stereos,* meaning ''space,'' and enzymes that are regulated by such effectors are called allosteric enzymes. Although allosteric effectors usually are small molecules such as ATP, some proteins are inhibited or activated when they bind to other proteins.

Partial Proteolysis Results in Irreversible Covalent Modifications

In chapter 8, we mentioned that the pancreas secretes trypsin, chymotrypsin, and elastase as inactive zymogens, which are activated by extracellular proteases. Trypsin is activated when the intestinal enzyme enteropeptidase cuts off an N-terminal hexapeptide. Trypsin in turn activates chymotrypsin by cutting it at the N-terminal end between Arg 15 and Ile 16. This type of change in the covalent structure of an enzyme is termed partial proteolysis. Delaying the activation prevents the digestive enzymes from destroying the pancreatic cells in which they are synthesized.

The crystal structures of chymotrypsin and its zymogen precursor *chymotrypsinogen* suggest an explanation for the increase in catalytic activity that results from trimming off the N-terminal end of the zymogen. Although the catalytic triad of Asp 102, His 57, and Ser 195 has a similar structure in chymotrypsinogen and chymotrypsin, the substrate-binding pocket is not properly formed in the zymogen (fig. 9.1a). The NH group of Gly 193 is not in position to form a hydrogen bond with the carbonyl oxygen of the sub-

Figure 9.1

A schematic drawing of the structural changes that occur when chymotrypsinogen (a) is converted to chymotrypsin (b). In chymotrypsinogen, the carboxylate group of Asp 194 forms a salt bridge to His 40; in chymotrypsin, the bridge goes to the new N terminal, the —NH3+ group of Ile 16. This change evidently allows Gly 193 to swing around so that its amide NH comes closer to the NH of Ser 195. The two amide groups form essential hydrogen bonds to the substrate in the enzyme–substrate complex (see figs. 8.7 and 8.11).

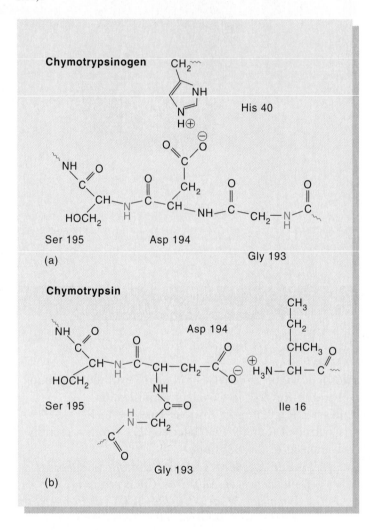

strate. The importance of this bond for the activity of the enzyme was discussed in chapter 8 (see figs. 8.7 and 8.11). The major constraint preventing the completion of the binding pocket appears to be a hydrogen bond between the carboxylate group of the neighboring residue, Asp 194, and His 40. When the zymogen is cleaved, the new N-terminal —NH3+ group of Ile 16 forms a salt bridge with Asp 194, thus allowing Gly 193 to rotate into the correct orientation (fig. 9.1b). A similar bridge between the N-terminal group and Asp 194 occurs in trypsin.

Figure 9.2

The blood coagulation cascade. Each of the curved red arrows represents a proteolytic reaction, in which a protein is cleaved at one or more specific sites. With the exception of fibrinogen, the substrate in each reaction is an inactive zymogen; except for fibrin, each product is an active protease that proceeds to cleave another member in the series. Many of the steps also depend on interactions of the proteins with Ca^{2+} ions and phospholipids. The cascade starts when factor XII and prekallikrein come into contact with materials that are released or exposed in injured tissue. (The exact nature of these materials is still not fully clear.) When thrombin cleaves fibrinogen at several points, the trimmed protein (fibrin) polymerizes to form a clot.

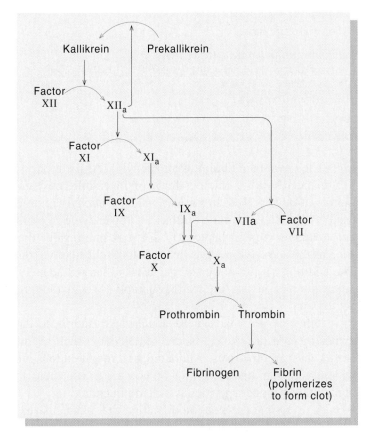

The enzymes that participate in blood clotting also are activated by partial proteolysis, which again serves to keep them in check until they are needed. The blood coagulation system involves a cascade of at least seven serine proteases, each of which activates the subsequent enzyme in the series (fig. 9.2). Because each molecule of activated enzyme can, in turn, activate many molecules of the next enzyme, initiation of the process by factors that are exposed in damaged tissue leads explosively to the conversion of prothrombin to thrombin, the final serine protease in the series. Thrombin then cuts another protein, fibrin, into peptides that stick together to form a clot.

Table 9.1

Some Enzymes Regulated by Partial Proteolysis

Digestive Enzymes
Trypsin, chymotrypsin, carboxypeptidase A and B, elastase, pepsin, phospholipase

Blood Coagulation Enzymes
Factors VII, IX, X, XI, and XIII, kallikrein, thrombin

Enzymes Involved in Dissolving Blood Clots
Plasminogen, plasminogen activator

Enzymes Involved in Programmed Development
Chitin synthetase, cocoonase, collagenase

Table 9.1 lists some of the other enzymes activated by partial proteolysis. A common pattern is for proteolysis to occur in a loop that connects two different domains of the protein, relieving a constraint that interferes with the formation of the active site.

The activation of an enzyme by partial proteolysis is an irreversible process. Once the enzyme is cut, it remains active until it is degraded or inhibited by some other means. This is fine for the digestive enzymes, but it clearly raises a problem in blood coagulation: A mechanism must exist to prevent the clot from spreading away from the site of the injury and taking over the entire blood supply. Several mechanisms work to this end. The activated enzymes are diluted by the flow of the blood, degraded in the liver, and inhibited by other blood proteins that bind tightly to the active enzymes and occlude their active sites. Blood coagulation thus depends on a delicate balance between a rapid, irreversible mechanism of enzyme activation, and rapid, irreversible mechanisms for disposing of the active enzymes.

Phosphorylation, Adenylylation, and Disulfide Reduction Lead to Reversible Covalent Modifications

A type of covalent modification used more widely than partial proteolysis is phosphorylation of the side chains of serine, threonine, or tyrosine residues. Phosphorylation differs from partial proteolysis in being reversible. The introduction and removal of the phosphate group are catalyzed by separate enzymes (phosphorylation by a protein kinase, and dephosphorylation by a phosphatase), which are themselves usually under metabolic regulation (fig. 9.3).

Figure 9.3

Phosphorylations of serine side chains in enzymes are catalyzed by kinases, and dephosphorylation by phosphatases. Threonine and tyrosine side chains undergo similar reactions, but these are less common than the phosphorylation of serine.

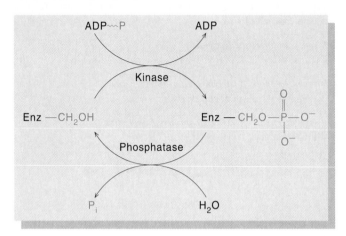

Table 9.2

Some Enzymes Regulated by Phosphorylation

Enzymes of Carbohydrate Metabolism
Glycogen phosphorylase
Phosphorylase kinase
Glycogen synthase
Phosphofructokinase-2
Pyruvate kinase
Pyruvate dehydrogenase

Enzymes of Lipid Metabolism
Hydroxymethylglutaryl-CoA reductase
Acetyl-CoA carboxylase
Triacylglycerol lipase

Enzymes of Amino Acid Metabolism
Branched-chain ketoacid dehydrogenase
Phenylalanine hydroxylase
Tyrosine hydroxylase

In eukaryotic organisms, phosphorylation is used to control the activities of literally hundreds of enzymes, a few of which are listed in Table 9.2. These enzymes are phosphorylated or dephosphorylated in response to extracellular signals, such as hormones or growth factors. The adrenal hormone epinephrine, for example, is transmitted through the blood to muscle and adipose tissue when there is a need for muscular exertion. On reaching the target tissue, epinephrine initiates a chain of events that leads to the activation of a protein kinase. The kinase [cyclic adenosine monophosphate (cAMP)-dependent protein kinase] catalyzes the phosphorylation of one or more specific enzymes, depending on the tissue. Some enzymes are activated when they are phosphorylated and inactivated when the phosphate is removed; others are inactivated by phosphorylation. In adipose tissue, phosphorylation activates triacylglycerol lipase, an enzyme that breaks down esters of fatty acids. In muscle, phosphorylation activates glycogen phosphorylase, an enzyme that breaks down glycogen, and it stops the synthesis of glycogen by switching off glycogen synthase.

Phosphorylation also can modify an enzyme's sensitivity to allosteric effectors. Phosphorylation of glycogen phosphorylase reduces its sensitivity to the allosteric activator adenosine monophosphate (AMP). Thus, a covalent modification triggered by an extracellular signal can override the influence of intracellular allosteric regulators. In other cases, variations in the concentrations of intracellular effectors can modify the response to the covalent modification, depending on the metabolic state of affairs in the cell.

The covalent addition of an adenylyl (AMP) group to a tyrosine residue is another form of reversible, covalent modification (fig. 9.4). In *Escherichia coli,* adenylylation is used to regulate glutamine synthase, a key enzyme in nitrogen metabolism (see chapter 21). The tyrosine residue that accepts the adenylyl group is located close to the active site. The addition of the bulky and negatively charged adenylyl group inhibits the enzyme, perhaps simply by occluding the active site.

Other groups that can be attached covalently to enzymes include fatty acids, isoprenoid alcohols such as farnesol, and carbohydrates. Although such modifications are widespread, our understanding of how cells use them to regulate enzymatic activities is still fragmentary.

A reversible covalent modification that plants use extensively is the reduction of cystine disulfide bridges to sulfhydryls. Many of the enzymes of photosynthetic carbohydrate synthesis are activated in this way (table 9.3). Some of the enzymes of carbohydrate breakdown are inactivated by the same mechanism. The reductant is a small protein called "thioredoxin," which undergoes a complementary oxidation of cysteine residues to cystine (fig. 9.5). Thioredoxin itself is reduced by electron-transfer reactions driven by sunlight, which serves as a signal to switch carbohydrate metabolism from carbohydrate breakdown to synthesis. In one of the regulated enzymes, phosphoribulokinase, one of the freed cysteines probably forms part of the catalytic active site. In nicotinamide-adenine dinucleotide phosphate (NADP)-malate dehydrogenase and fructose-1,6-bis-

Figure 9.4

The transfer of the adenylyl group from ATP to a tyrosine residue is used to regulate some enzymes. The other product of the adenylylation reaction *(top)* is inorganic pyrophosphate. Hydrolysis of the adenylyl-tyrosine ester bond releases AMP *(bottom).*

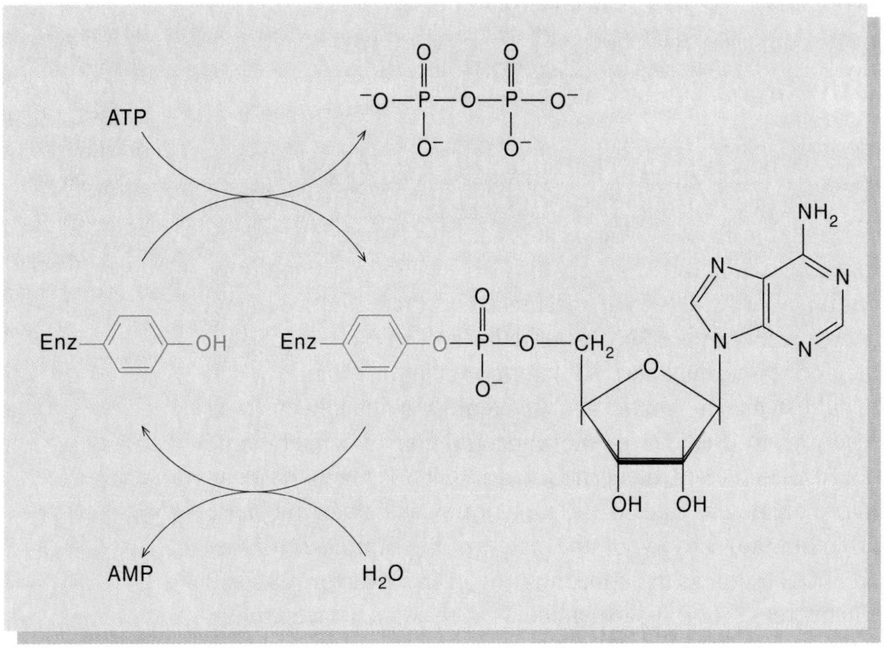

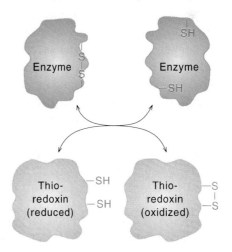

Table 9.3

Some Enzymes Regulated by Disulfide Reduction in Plants

Activated by Reduction
Fructose-1,6-bisphosphatase
Sedoheptulose-1,7-bisphosphatase
Glyceraldehyde-3-phosphate dehydrogenase
NADP-malate dehydrogenase
Phosphoribulokinase
Thylakoid ATP-synthase

Inhibited by Reduction
Phosphofructokinase

Figure 9.5

Reduction of the disulfide bond of cystine is used to activate enzymes of photosynthetic carbohydrate biosynthesis in plants. The reductant is a small protein called thioredoxin. Thioredoxin also serves as a reductant for the biosynthesis of deoxynucleotides in animals and microorganisms as well as in plants.

phosphatase, the cysteines both appear to be some distance from the catalytic site. How the structural changes that result from the reduction are transmitted to the active sites of these enzymes is not yet known.

Covalent modifications of enzymes allow a cell to regulate its metabolic activities more rapidly and in much more intricate ways than is possible by changing the absolute concentrations of the same enzymes. They still do not provide truly instantaneous responses to changes in conditions, however, because each modification requires the action of

another enzyme, which must itself be regulated. A lag also occurs in responding to the removal or inhibition of the enzyme that causes the modification, because reversing the modification requires still another enzyme.

Allosteric Regulation Allows an Enzyme to Be Controlled Rapidly by Materials that Are Structurally Unrelated to the Substrate

Whereas eukaryotic organisms use phosphorylation to handle responses to extracellular signals, both prokaryotic and eukaryotic organisms commonly use allosteric regulation in responding to changes in conditions within a cell. A typical circumstance that might demand such a response is a surplus or deficit of ATP or some other metabolic intermediate. Allosteric regulation enables a cell to adjust an enzymatic activity almost instantaneously in response to changes in the concentration of a metabolite because, unlike covalent modification, it does not require an intermediate enzyme. If the metabolite acts as an allosteric effector, the activity of the enzyme can increase (or decrease) as soon as the concentration of the effector rises and can decrease (or increase) again as soon as the concentration falls.

Regulation of enzymes by allosteric effectors is considerably more common than regulation by compounds that bind at the active site for two reasons: First, whereas an agent that binds at the active site usually acts as an inhibitor, a compound that binds at an allosteric site can serve as either an inhibitor or an activator, depending on the structure of the enzyme. Second, a substance that binds to an allosteric site does not need to have any structural relationship to the substrate. Consider, for example, the metabolic pathway of histidine biosynthesis in plants and bacteria. The pathway requires nine enzymes that work one after another. If histidine is already present in abundance, it is advantageous for a cell to cut off the entire pathway at the first step to avoid wasting energy or accumulating the products of the intermediate steps. The first step in the pathway is the reaction between phosphoribosylpyrophosphate and adenosine triphosphate (ATP), neither of which even vaguely resembles histidine, and yet the enzyme that catalyzes this step is strongly inhibited by histidine. Binding of histidine to an allosteric site causes a structural change that is transmitted to the active site.

Inhibition of the initial step of a biosynthetic pathway by an end product of the pathway is a recurrent theme in metabolic regulation. In addition, many key enzymes are regulated by ATP, adenosine diphosphate (ADP), AMP, or inorganic phosphate ion (P_i). The concentrations of these materials provide a cell with an index of whether energy is abundant or in short supply. Because ATP, ADP, AMP, or P_i often are chemically unrelated to the substrate of the enzyme that must be regulated, they usually bind to an allosteric site rather than to the active site.

Figure 9.6

Kinetics of the reaction catalyzed by phosphofructokinase. In the presence of 1.5 mM ATP, the rate has a sigmoidal dependence on the concentration of the substrate fructose-6-phosphate. Although ATP also is a substrate for the reaction, the sigmoidal kinetics seen under these conditions are associated with the binding of ATP to an inhibitory allosteric site. The kinetics become hyperbolic if a low concentration of AMP is added.

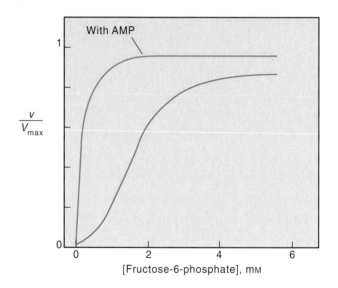

Allosteric Enzymes Typically Exhibit a Sigmoidal Dependence on Substrate Concentration

In chapter 5 we saw that the binding of O_2 to any one of the four subunits of hemoglobin increases the affinity of the other subunits for O_2. This effect reflects cooperative changes in the tertiary and quaternary structure of the protein. Binding of O_2 alters the interactions among the subunits in such a way that the entire protein tends to flip into a state with increased O_2 affinity; binding of glycerate-2,3-bisphosphate favors a transition in the opposite direction. When Jacques Monod, Jeffreys Wyman, and Jean-Pierre Changeux first advanced the idea of allosteric enzymes in 1963, they suggested that these enzymes might contain multiple subunits and that the changes in catalytic activity caused by allosteric effectors can reflect alterations in quaternary structure. This suggestion turned out to be remarkably accurate. Although there is no reason why an enzyme consisting of a single subunit cannot be sensitive to allosteric effectors, most of the enzymes regulated in this way do have multiple subunits, and the changes in activity often can be related to interactions among the subunits.

One indication of the importance of intersubunit interactions in allosteric enzymes is that many such enzymes do not obey the classical Michaelis-Menten kinetic equation. A plot of the rate of reaction as a function of substrate concentration is not hyperbolic, as described by the Michaelis-Menten equation, but rather sigmoidal, resembling the curve for the binding of O_2 to hemoglobin (see fig. 5.2). Furthermore, allosteric effectors often cause the kinetics to change from one of these forms to the other, much as glycerate-2,3-bisphosphate affects the degree of cooperativity in the binding of O_2 to hemoglobin. Figure 9.6 shows an illustration of these effects for phosphofructokinase, which catalyzes the formation of fructose-1,6-bisphosphate from fructose-6-phosphate and ATP.

$$\text{Fructose-6-phosphate} + \text{ATP} \xrightarrow{\text{phosphofructokinase}}$$
$$\text{Fructose-1,6-bisphosphate} + \text{ADP} \quad \textbf{(1)}$$

In the presence of 1.5 mM ATP, the kinetics have a sigmoidal dependence on the concentration of fructose-6-phosphate; at very low concentrations of ATP, or in the presence of AMP, the kinetics become hyperbolic.

Phosphofructokinase has four identical subunits. To explore how sigmoidal kinetics can arise in such an enzyme, consider an enzyme that has just two such subunits, each with its own catalytically active site. We can schematize the binding of substrate to the enzyme as follows:

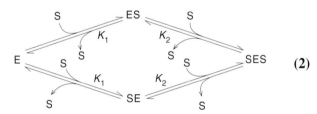

$$\textbf{(2)}$$

Here ES and SE represent complexes of the enzyme (E) with substrate (S) on the two different subunits, and SES represents the complex in which both binding sites are occupied. The two dissociation constants K_1 and K_2 are not necessarily identical because the binding of substrate to one subunit can affect the dissociation constant for the other subunit. In fact, this is just the point we want to explore.

Let's assume that the rate constant k_{cat} for the formation of products on either subunit is the same, whether only that site or both catalytic sites are occupied. Suppose also that ES, SE, and SES are in equilibrium with the free enzyme and substrate. By following the same procedure that led to the Henri-Michaelis-Menten equation in chapter 7, we can derive an expression for the rate of the enzymatic reaction in terms of [S], K_1, and K_2. Here we just give the result.

$$v = \frac{V_{max} \dfrac{[S]}{K_1}\left(1 + \dfrac{[S]}{K_2}\right)}{\left(1 + \dfrac{[S]}{K_1}\right) + \dfrac{[S]}{K_1}\left(1 + \dfrac{[S]}{K_2}\right)} \quad \textbf{(3)}$$

Figure 9.7

Kinetics of an enzyme with two identical subunits. The curves were calculated with equation (3). The abscissa is the ratio of the substrate concentration [S] to K_d, where $K_d = \sqrt{K_1 K_2}$ and K_1 and K_2 are the dissociation constants for the first and second molecule of substrate. K_d is taken to be the same for all three curves. For the *solid curve*, K_1 and K_2 are assumed to be identical ($K_1 = K_2 = K_d$); equation (3) then reduces to equation (4). For the *dashed curve*, $K_2 = K_1/25$ ($K_1 = 5K_d$ and $K_2 = K_d/5$). For the *dotted curve*, $K_2 = 10^{-4}K_1$ ($K_1 = 100K_d$ and $K_2 = K_d/100$); equation (3) then reduces to equation (5).

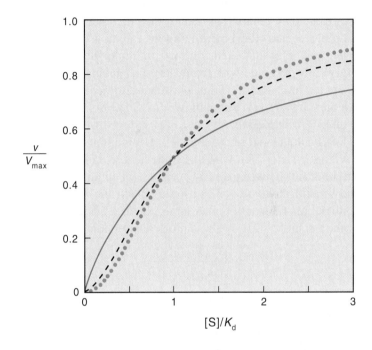

If the binding of S to one subunit does *not* affect binding to the other, then $K_2 = K_1$, and equation (3) reduces to

$$v = \frac{V_{max}[S]}{K_s + [S]} \quad \textbf{(4)}$$

where $K_s = K_1 = K_2$. This is simply the Henri-Michaelis-Menten equation [equation (18) in chapter 7]. A plot of v versus (S) according to equation (4) is hyperbolic. Such a plot is shown as the solid curve in figure 9.7. The dashed curve in figure 9.7 is a plot of equation (3) when $K_2 = K_1/25$. To facilitate comparison with the solid curve, K_1 is taken to be $5K_d$ and K_2 to be $K_d/5$, where K_d is the same for the two curves. This means that the overall $\Delta G°$ for the formation of SES from E + 2S is the same. (The overall equilibrium constant is $(1/K_1)(1/K_2)$, or $1/K_d^2$.) Note that the dashed curve is sigmoidal in shape. It starts out rising more slowly than the solid curve, rises steeply in the region of $[S] \approx K_d$, where $v/V_{max} \approx 0.5$, and then continues rising more slowly.

A ratio of 25 between K_1 and K_2 means that the $\Delta G°$ for binding the second molecule of substrate is only

1.9 kcal/mol more favorable than the $\Delta G°$ for binding the first molecule. This is the order of magnitude of the $\Delta G°$ for forming a single hydrogen bond. Evidently, sigmoidal kinetics might be obtained if binding the substrate causes even a relatively minor change in the conformation of the enzyme. In chapter 8, we noted that binding of glucose to hexokinase results in a pronounced conformational change that can be seen in the crystal structure. But to yield sigmoidal kinetics, it is essential that events occurring at two different binding sites be linked: Binding at one site must decrease the dissociation constant at the other site. If no such coupling occurs, the kinetics follow equation (4) and are hyperbolic even if the enzyme has multiple subunits. This situation is observed with many enzymes, including one we discussed in chapter 8. Triosephosphate isomerase exists as a dimer, but its two subunits work independently, and its kinetics are hyperbolic.

The dotted curve in figure 9.7 shows a plot of equation 6 with $K_2 = 10^{-4} K_1$, which is equivalent to a difference of 5.5 kcal/mol between the $\Delta G°$ values for binding the first and second molecules of substrate. When the ratio of the dissociation constants is this large, the kinetics can be described equally well by the simpler expression

$$v = \frac{V_{max}[S]^2}{K_1 K_2 + [S]^2} \qquad (5)$$

This is the limiting form of equation (3) when $K_2 \ll [S] \ll K_1$.

It is not uncommon for enzymes that are regulated allosterically to have four or even more subunits. The active form of acetyl-coenzyme A (acetyl-CoA) carboxylase consists of linear strings of 10 or more identical subunits (see fig. 18.11). In general, the larger the number of subunits that interact such that the binding of substrate to one subunit promotes binding to others, the more steeply the enzyme's kinetics depend on [S] in the region where $v/V_{max} \approx 0.5$. To derive a more general form of equation (5), consider a protein E with n identical subunits, each of which has a binding site for a substrate S. Suppose that binding of the first molecule of S strongly favors binding to all n subunits, giving the overall reaction

$$E + nS \rightleftharpoons ES_n \qquad (6)$$

The concentration of ES_n then is

$$[ES_n] \approx \frac{[E][S]^n}{K_h} \qquad (7)$$

where K_h is the product of the individual dissociation constants for all n steps leading to ES_n. The fraction of the protein that has taken up the substrate is

$$y \approx \frac{[ES_n]}{[E] + [ES_n]} = \frac{[S]^n}{K_h + [S]^n} \qquad (8)$$

This is called the Hill equation, and the exponent n is the Hill coefficient.

Equation (8) is an approximation because it ignores intermediate species that have some, but not all, of the binding sites occupied. Even so, the Hill coefficient provides a useful measure of cooperativity. The binding of O_2 to hemoglobin is described well by the Hill equation with $n \approx 2.8$. In the case of phosphofructokinase, which has four subunits, the dependence of the rate on the fructose-6-phosphate concentration at a fixed, relatively high concentration of ATP is described well with $n \approx 3.8$.

The Symmetry Model Provides a Useful Framework for Relating Conformational Transitions to Allosteric Activation or Inhibition

Several theoretical models have been developed for relating changes in dissociation constants to conformational changes in oligomeric proteins. We discussed two such models in connection with the oxygenation of hemoglobin (see fig. 5.14), and one of them is shown again in a slightly different form in figure 9.8. The basic idea is that the protein's subunits can exist in either of two conformations: R ("relaxed") and T ("tight," or "tense"). The substrate is assumed to bind more tightly to the R form than to the T form, which implies that binding of the substrate favors the transition from T to R. Conformational transitions of the individual subunits are assumed to be tightly linked, so that if one subunit flips from T to R, the others must do the same. Binding of the first molecule of substrate thus promotes the binding of a second. Because the concerted transition of all of the subunits from T to R or back preserves the overall symmetry of the protein, this model is called the "symmetry" model. The symmetry model can account for the behavior of allosteric activators if these compounds also bind preferentially to the R state and thus stabilize the conformation that is more effective at binding the substrate. Allosteric inhibitors act by stabilizing the T state.

Using the symmetry model, the fraction of the binding sites occupied at any given substrate concentration can be described with an expression that includes the substrate dissociation constants for the two conformations (K_R and K_T) and the equilibrium constant between the T and R conformations in the absence of substrate, $L = [T]/[R]$. Thus, the symmetry model attempts to explain the difference between K_1 and K_2 in equation (3) by introducing a third independent parameter. Considering that equation (3) can fit the experimental data for a dimeric enzyme with only two pa-

Figure 9.8

Symmetry model for allosteric transitions of a dimeric enzyme. The model assumes that the enzyme can exist in either of two different conformations (T and R), which have different dissociation constants for the substrate (K_T and K_R). Structural transitions of the two subunits are assumed to be tightly coupled, so that both subunits must be in the same state. L is the equilibrium constant (T)/(R) in the absence of substrate. If the substrate binds much more tightly to R than to T ($K_R \ll K_T$), the binding of a molecule of substrate to either subunit pulls the equilibrium between T and R in the direction of R. Because both subunits must go to the R state, the binding of the second molecule of substrate is promoted.

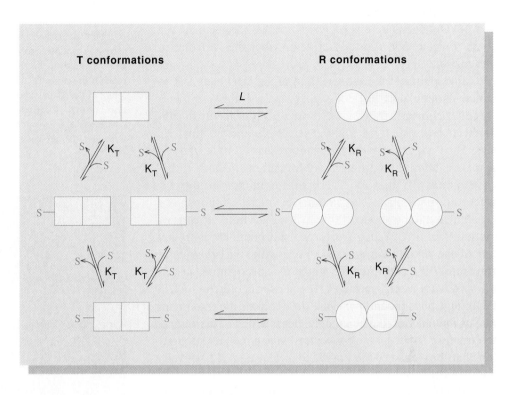

rameters, what do we gain by using a more complicated equation? The usefulness of the symmetry model is that it can be generalized to enzymes that have a larger number of subunits. The general expression still contains only four independent parameters: K_R, K_T, L, and the total number of subunits (n). But this simplicity comes with a cost: We have to assume that the entire oligomeric protein has just two different conformational states. Is it not more realistic to assume that the protein also can adopt a variety of intermediate conformations? The answer depends in part on the particular protein and in part on our goals in using any type of model. Models that allow intermediate conformations have been developed, but they of course require more independent parameters. The ''sequential'' model shown in figure 5.14 is one of these more elaborate models. Although the sequential model is more general than the symmetry model and may be more realistic, the available experimental data in most cases do not justify a distinction between the two.

The symmetry model is useful even if it does oversimplify the situation, because it provides a conceptual framework for discussing the relationships between conformational transitions and the effects of allosteric activators and inhibitors. In the following sections we consider three oligomeric enzymes that are under metabolic control and see that substrates and allosteric effectors do tend to stabilize each of these enzymes in one or the other of two distinctly different conformations.

Phosphofructokinase: Allosteric Control of Glycolysis Is Consistent with the Symmetry Model

Phosphofructokinase catalyzes the transfer of a phosphate group from ATP to fructose-6-phosphate (equation 1). This is a major site of regulation of glycolysis, the metabolic pathway by which glucose breaks down to pyruvate (see chapter 12). As we saw in figure 9.6, the kinetics of phosphofructokinase are strongly cooperative with respect to fructose-6-phosphate. The kinetics are noncooperative with respect to the other substrate, ATP, at low concentrations, but at concentrations above 0.5 mM ATP acts as an inhibitor. The inhibitory effect results from binding to an allosteric site that is distinct from the substrate-binding site for ATP. Phosphofructokinase also is inhibited by phosphoenolpyruvate and by citrate, an intermediate in two other metabolic pathways that embark from pyruvate. On the other hand, it is stimulated by ADP, AMP, guanosine diphosphate (GDP), cAMP, fructose-2,6-bisphosphate, and a variety of other compounds. Most if not all of the activators bind to the same allosteric site as ATP and may work simply by preventing the inhibitory effect of ATP. The effects of ATP, ADP, and AMP are such that phosphofructokinase is restrained when the cell's needs for energy have been satisfied, as reflected in a high ratio of [ATP] to [ADP] and [AMP], and is unleashed when the cell needs additional energy. The effects of cAMP and fructose-2,6-bisphosphate relate to hormonal control of carbohydrate metabolism in higher organisms, and is discussed further in chapters 12 and 24. Our focus here is on how fructose-6-phosphate and some of the major allosteric effectors alter the structure of the enzyme.

Phosphofructokinase was one of the first enzymes to which Monod and his colleagues applied the symmetry model of allosteric transitions. It contains four identical subunits, each of which has both an active site and an allosteric site. The cooperativity of the kinetics suggests that the enzyme can adopt two different conformations (T and R) that have similar affinities for ATP but differ in their affinity for fructose-6-phosphate. The binding for fructose-6-phosphate is calculated to be about 2,000 times tighter in the R conformation than in T. When fructose-6-phosphate binds to any one of the subunits, it appears to cause all four subunits to flip from the T conformation to the R conformation, just as the symmetry model specifies. The allosteric effectors ADP, GDP, and phosphoenolpyruvate do not alter the maximum rate of the reaction but change the dependence of the rate on the fructose-6-phosphate concentration in a manner suggesting that they change the equilibrium constant (L) between the T and R conformations.

Philip Evans and his coworkers have determined the crystal structures of phosphofructokinase from two species of bacteria, *E. coli* and *Bacillus stearothermophilus*. By crystallizing the enzyme in the presence and absence of the substrate and several allosteric effectors, they obtained detailed views of both the T and R conformations. This work has led to an explanation of why phosphofructokinase appears to be constrained largely to all-or-nothing transitions between these two states, rather than adopting a series of intermediate conformations.

Figure 9.9 shows the crystal structure of two of the subunits of phosphofructokinase from *B. stearothermophilus*. In the complete enzyme, the subunits are disposed symmetrically about three mutually perpendicular axes. Each of these axes is a twofold symmetry axis, which means that rotating the entire structure by half of a full circle (180°) around the symmetry axis results in an identical structure. This rotation is shown diagramatically in figure 9.10. Because ADP is a product of the enzymatic reaction as well as an allosteric activator, it binds at both the catalytic and allosteric sites (see figs. 9.9 and 9.10). The catalytic site for fructose-6-phosphate in each subunit is at the interface of the subunit with one of its neighbors, and the allosteric site is at the interface with a different neighbor.

In the transition between the T and R conformations, the four subunits rotate by about 7° with respect to each other (see fig. 9.10). This rotation is associated with coupled rearrangements of the structures at the interfaces between adjacent subunits. Figure 9.11 shows how these rearrange-

Figure 9.9

Computer-generated structure of two of the four subunits of phosphofructokinase from *Bacillus stearothermophilus*. The enzyme, shown as yellow and light blue tubes, was crystallized in the R conformation in the presence of the substrate fructose-6-phosphate (dark blue) and the allosteric activator ADP (pink). The magnesium ions (white/silver spheres), Mg^{2+}, bound to the ADP molecules are also shown. (Copyright 1994 by the Scripps Research Institute/Molecular Graphics Images by Michael Pique using software by Yng Chen, Michael Connolly, Michael Carson, Alex Shah, and AVS, Inc. Visualization advice by Holly Miller, Wake Forest University Medical Center.)

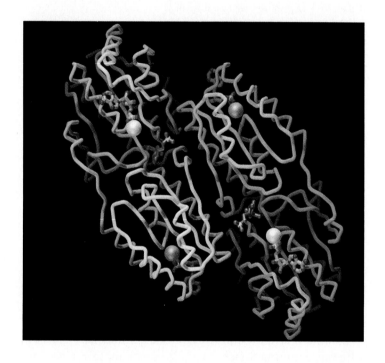

ments affect the binding site for fructose-6-phosphate. The most significant structural change in this region is an inversion of the orientation of the side chains of Glu 161 and Arg 162. In the R conformation, Arg 162 forms a hydrogen bond to the phosphate group of fructose-6-phosphate, whereas Glu 161 points in the opposite direction. In the T conformation, Arg 162 points away from the binding site, and Glu 161 inserts a negative charge into the site, where it forms a hydrogen bond with Arg 243. The change in the orientations of the negatively charged Glu 161 and the positively charged Arg 162 probably accounts for most of the difference between the dissociation constants for fructose-6-phosphate in the two states. Note that although figure 9.11*b* shows the molecule of fructose-6-phosphate bound on subunit A, Glu 161 and Arg 162 are residues of subunit D. Arginines 252 and 243 also contribute to the binding site from opposite sides of the boundary. Structural changes that occur on one of the subunits thus are intricately linked to changes on the other. The substrate-binding site for ATP, on

Figure 9.10

Outlines of phosphofructokinase in the T (solid lines) and R (dashed lines) conformations. The enzyme contains four identical subunits (A, B, C, and D). The locations of the catalytic and allosteric sites are indicated in the two subunits closest to the viewer (A and D). The binding sites for fructose-6-phosphate (F6P) are at the interface of these subunits; the allosteric sites are at the interfaces of A with B, and of D with C. Two of the three perpendicular symmetry axes are labeled p and q. A 180° rotation about axis q interchanges the positions of subunits A and C and also interchanges B and D. A similar rotation about p interchanges A with B, and C with D. (Source: From T. Schirmer and P. R. Evans, Structural basis of the allosteric behaviour of phospho-fructokinase, *Nature* 343:140, 1990.)

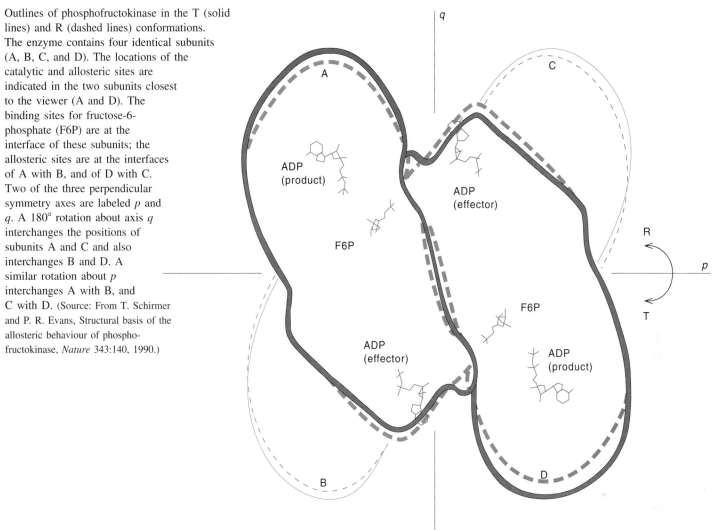

the other hand, is made up of residues from only one subunit (subunit A in figure 9.11). This probably explains why the binding of fructose-6-phosphate to the enzyme is strongly cooperative, whereas the binding of ATP as a substrate is not cooperative.

In the T conformation, Glu 161 and Arg 162 are located at the end of a stretch of polypeptide that winds up into a helical turn in the transition to the R structure (see fig. 9.11). This coiling is linked to a major structural change in an adjacent region of the interface between subunits A and D. The interface here includes a pair of antiparallel β strands, each of which is hydrogen-bonded to a parallel strand in its own subunit, as shown in figure 9.12. In the T conformation (fig. 9.12a), the β strands from the different subunits are hydrogen-bonded together directly across the interface. In the R conformation (fig. 9.12b), the strands have moved apart, and the region between them is filled by a row of hydrogen-bonded water molecules.

The insertion of water at the interface between subunits A and D is an essential component of the rotation of the subunits with respect to each other, and it appears to be an all-or-nothing effect. Intermediate conformations in which only some of the water molecules are present would have a less extensive network of hydrogen bonds and thus are probably less stable than either the R or the T conformation. The same might be said of the winding of the helical turn between residues 155 and 161; intermediates in which the helical turn is partially unwound probably would be destabilized by steric crowding or force the structure to expand in a way that leaves empty spaces in other regions. These considerations, taken with the close coupling between the individual subunits, seem to explain why all of the subunits undergo concerted transitions from the R to the T state or back, without giving appreciable concentrations of intermediate states.

Although the crystal structures offer a very plausible

Figure 9.11

Interface between subunits A and D of phosphofructokinase near the catalytic site in (*a*) the T and (*b*) the R structures. Crystals of the enzyme in the R state were obtained in the presence of fructose-6-phosphate and ADP (see fig. 9.9); crystals in the T state were obtained in the presence of a nonphysiological allosteric inhibitor, 2-phosphoglycolate. The wavy green line represents part of the boundary between subunits A and D. The heavy green line indicates the polypeptide backbone. The side chains of Glu 161 and Arg 162 are shown in red. Note the inversion of the positions of these side chains in the two structures. (Source: From T. Schirmer and P. R. Evans, Structural basis of the allosteric behaviour of phosphofructokinase, *Nature* 343:140, 1990.)

(a) T state

(b) R state

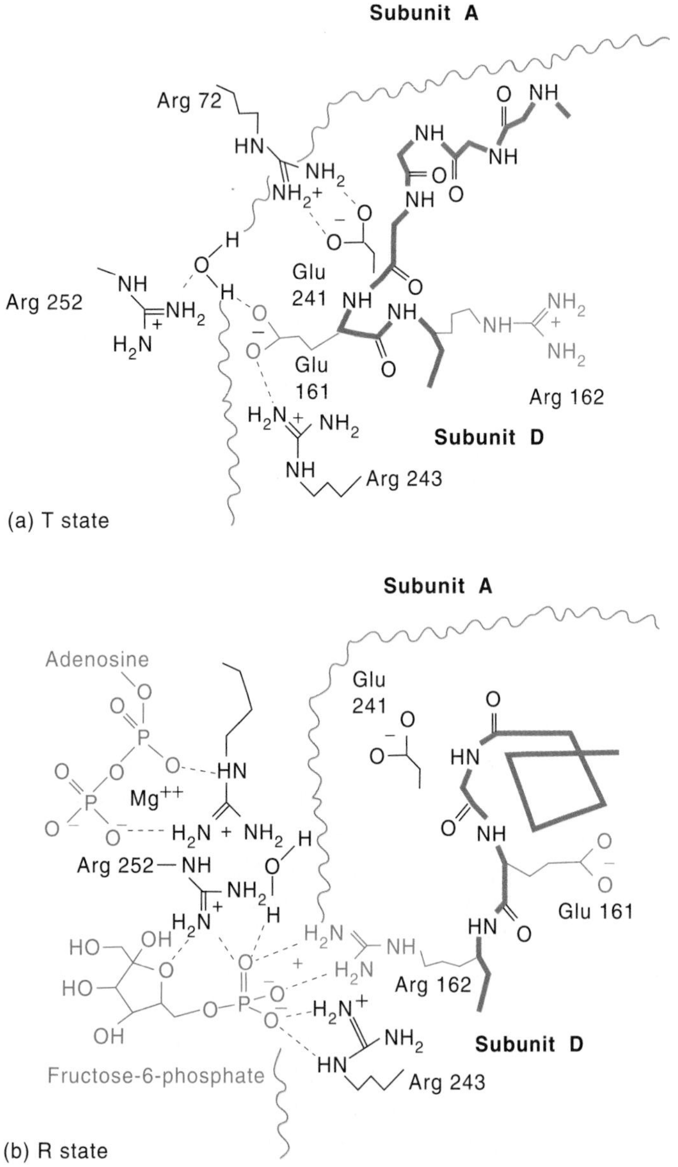

Figure 9.12

Hydrogen bonds of the peptide backbone and the side chain of threonine 245 at the interface between subunits A and D of phosphofructokinase, in (*a*) the T and (*b*) the R structures. Note the additional molecules of water (red) between the two subunits in the R structure. (Source: From T. Schirmer and P. R. Evans, Structural basis of the allosteric behaviour of phosphofructokinase, *Nature* 343:140, 1990.)

(a) T state

(b) R state

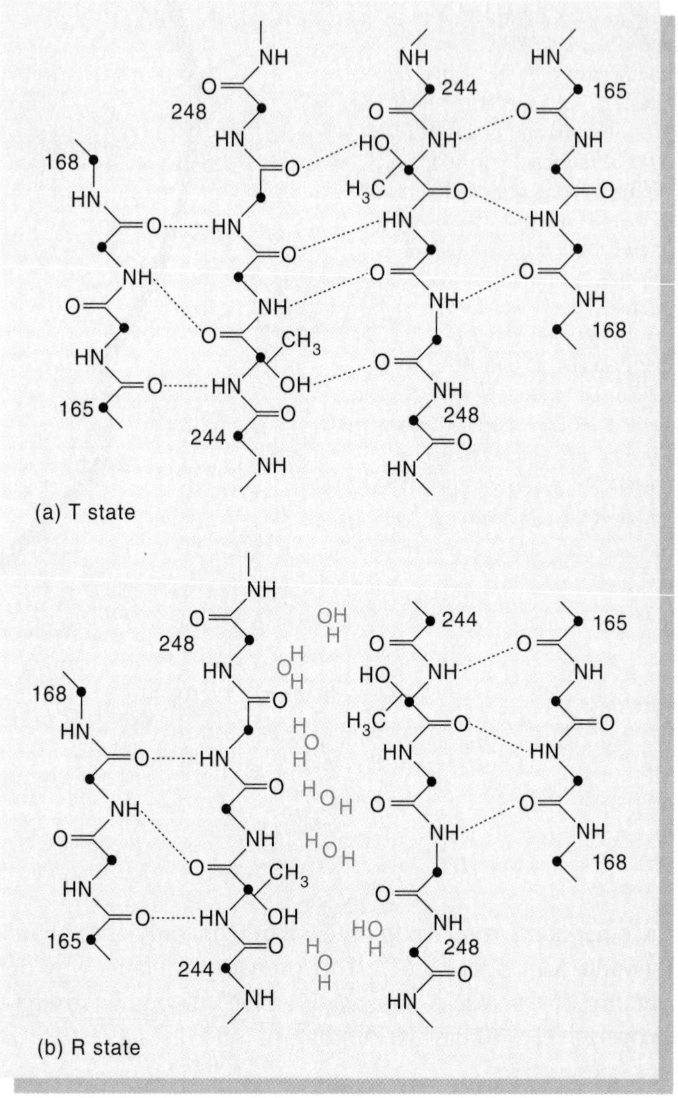

explanation for the cooperative binding of fructose-6-phosphate, it is still not clear why various allosteric effectors stabilize the protein in different conformational states. However, it is significant that the allosteric binding site lies at the interface of different subunits. The equilibrium between the R and T conformations appears to be sensitive to subtle structural changes in this region.

Figure 9.13

The reaction catalyzed by aspartate carbamoyl transferase, and the feedback inhibition of this enzyme in *E. coli* by the end product of the pathway, CTP. The series of small arrows represents additional reaction steps in the pathway from carbamoyl aspartate to CTP. These steps are discussed in chapter 1. The upward arrow with the negative sign indicates the feedback inhibition.

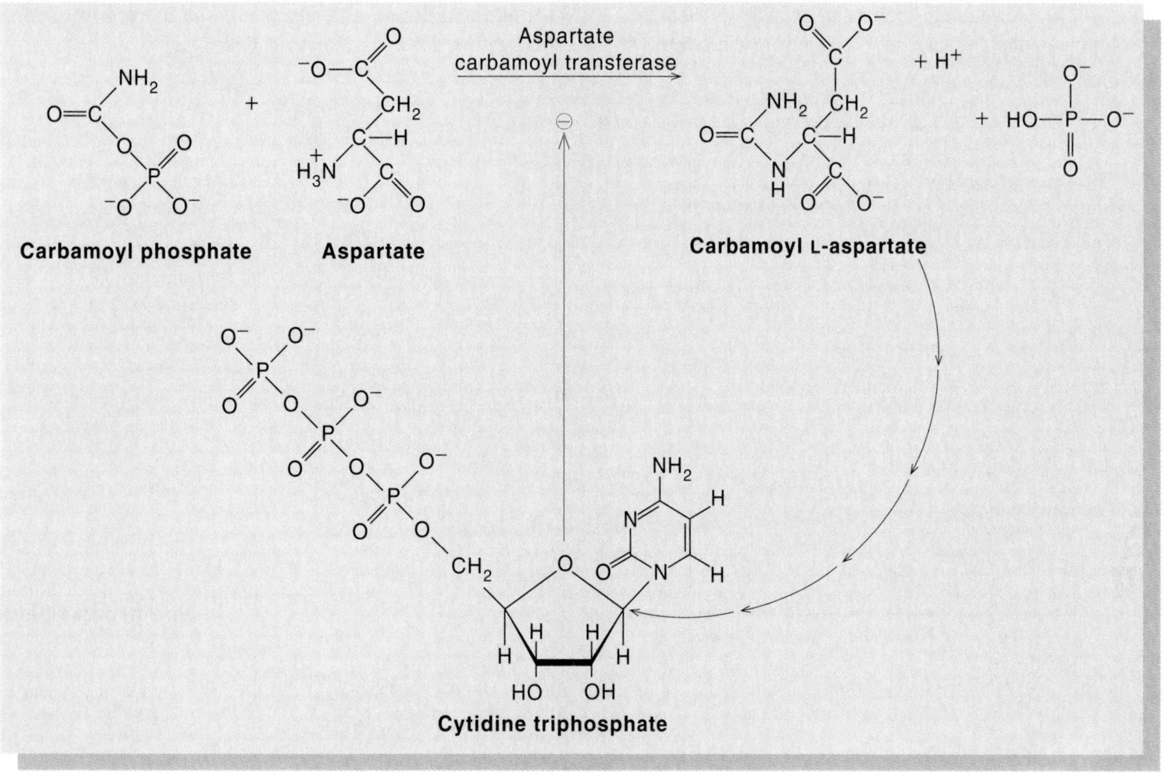

Aspartate Carbamoyl Transferase: Allosteric Control of Pyrimidine Biosynthesis

Aspartate carbamoyl transferase, or aspartate trans-carbamylase, catalyzes the transfer of a carbamoyl group

$$(H_2N-\overset{\overset{\displaystyle O}{\|}}{C}-)$$

from carbamoyl phosphate to aspartic acid to form carbamoyl aspartate (fig. 9.13). This step commits aspartate to the biosynthetic pathway for pyrimidines (see figure 23.13). Aspartate carbamoyl transferase is inhibited by cytidine triphosphate (CTP) and uridine triphosphate (UTP), the end products of the pathway, and it is stimulated by a purine, ATP. The opposing effects of CTP, UTP, and ATP serve to keep the biosynthesis of pyrimidines in balance with that of purines. This balance is important because cells need pu-

rines and pyrimidines in approximately equal amounts for the synthesis of nucleic acids.

The kinetics and physical properties of aspartate carbamoyl transferase from *E. coli* were studied in considerable detail by Howard Schachman and his colleagues. The kinetics have a sigmoidal dependence on the concentration of aspartate, as shown in figure 9.14. CTP shifts the kinetic curve to the right, and thus inhibits the enzyme strongly at low concentrations of aspartate but not at high concentrations. ATP reverses the effect of CTP, or in the absence of CTP eliminates the cooperativity altogether, making the kinetics hyperbolic instead of sigmoidal.

The behavior of aspartate carbamoylase changed dramatically when the enzyme was treated with the organic mercurial compound *p*-hydroxymercuribenzoate (see fig. 9.14). The binding of aspartate no longer showed positive cooperativity, and ATP or CTP were without effect. Exposure to mercurials was found to cause the enzyme to dissociate into two types of fragments, one of which retained the enzymatic activity but was no longer affected by CTP or

Figure 9.14

Effects of CTP, ATP, and mercurials on the rate of the reaction catalyzed by aspartate carbamoyl transferase. In the absence of CTP and ATP, the sigmoidal kinetics show positive cooperativity with respect to aspartate. CTP augments the positive cooperativity; ATP reverses the effect of CTP. An organic mercurial, or ATP in the absence of CTP, eliminates the cooperativity, converting the curve from sigmoidal to hyperbolic.

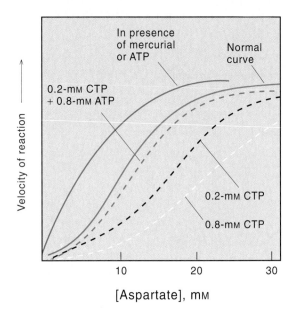

Figure 9.15

Subunit structure of aspartate carbamoyl transferase and the fragments produced by treating the enzyme with mercurials. In the complete enzyme *(top)*, the three sets of regulator dimers are sandwiched between two trimers of catalytic subunits (see fig. 9.17). The approximate location of the active site in each *c* subunit of the trimer facing the viewer is indicated with a *c*.

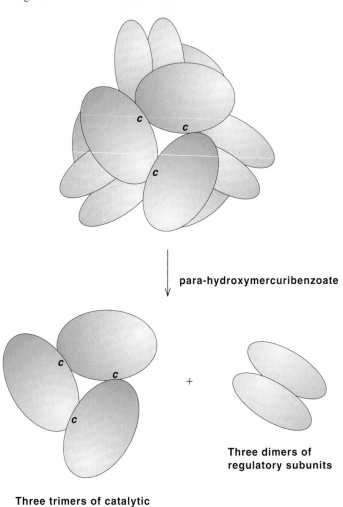

ATP. The other fragment bound CTP and ATP but had no enzymatic activity. In the native enzyme, each molecule is a complex of six catalytic (*c*) subunits and six regulatory (*r*) subunits (fig. 9.15). The fragments resulting from treatment with mercurials consist of trimers of the *c* subunit, and dimers of *r*. When the c_6r_6 complex was reconstituted from the fragments, the enzyme regained its sigmoidal kinetics and its sensitivity to CTP and ATP. These observations provided a convincing demonstration that the regulatory agents bind at an allosteric site and not at the active site. The two sites are on totally different subunits!

A substrate analog that proved particularly useful for studying aspartate carbamoyl transferase is *N*-phosphonacetyl-L-aspartate (PALA). PALA is structurally similar to a covalently linked adduct of carbamoyl phosphate and aspartate and thus resembles a likely intermediate in the enzymatic reaction (fig. 9.16). It binds to the enzyme in a highly cooperative manner, and its binding is promoted by ATP and opposed by CTP. The binding of PALA decreases the sedimentation and diffusion coefficients of the enzyme, indicating that the protein expands or changes shape. It also decreases the chemical reactivity of a cysteine residue in

each *c* subunit and increases the reactivity of several cysteines in each *r* subunit. CTP opposes these effects.

The changes in sedimentation coefficient and chemical reactivity caused by PALA fit the model that the enzyme exists in two distinct conformations (T and R) and that the binding of PALA to only one or two of the *c* subunits causes the entire c_6r_6 complex to flip from the T to the R state. The dissociation constant for PALA is higher in the T state than in the R state. The equilibrium constant $L = [T]/[R]$ has been calculated to be 250 in the absence of substrates and

Figure 9.16

N-phosphonacetyl-L-aspartate (PALA) is structurally similar to a likely intermediate in the reaction catalyzed by aspartate carbamoyl transferase. The binding of PALA to the enzyme is prevented competitively by carbamoyl phosphate, and PALA prevents the binding of aspartate. These observations support the view that PALA binds at the catalytic site.

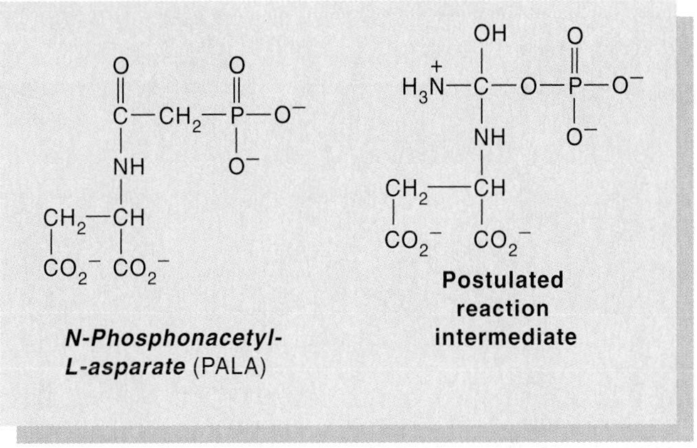

N-Phosphonacetyl-L-asparate (PALA)

Postulated reaction intermediate

allosteric effectors, 70 in the presence of ATP alone, and 1,250 in the presence of CTP alone. ATP thus shifts the equilibrium toward the conformational state that favors binding of the substrate, and CTP shifts the equilibrium in the direction of weaker binding.

Crystallization of aspartate carbamoyl transferase with and without bound PALA or CTP, by William Lipscomb and his colleagues, led to detailed pictures of the R and T conformations. Figures 9.17 and 9.18 show the crystal structures from two different perspectives. In both the R and the T conformations, the six c subunits are arranged in two equilateral trimers, one of which is inverted and stacked on top of the other (see figs. 9.15 and 9.17). The three c units in each trimer are related to each other by a threefold axis of symmetry. (Rotating the structure by one-third of a circle about this axis results in an identical structure.) The substrate-binding site on each c subunit is located in a pocket between two domains of the polypeptide. At the end of one of these domains the c subunit interacts with another c subunit in the same trimer, close to *its* substrate-binding site. In the other domain, it interacts more extensively with a c subunit in the other c_3 trimer. The six r subunits are arranged in three sets of dimers that form another equilateral triangle about the same axis of symmetry. Like the c subunits, each r subunit is folded into two domains: a peripheral domain where it interacts with its companion r subunit in the dimer, and a smaller domain that interacts with two adjacent c sub-

units. In the latter region, each r subunit binds an atom of Zn^{2+} that evidently plays a purely structural role. The binding site for the allosteric effectors CTP and ATP is located in the peripheral domain of the r subunit, at a considerable distance from the active sites.

In the transition from the T to the R conformation, the two c trimers rotate slightly with respect to each other about the threefold symmetry axis, so that they come into a more eclipsed alignment (see fig. 9.17). The r dimers rotate with respect to each other about a perpendicular axis and appear to act as a lever that moves the two c trimers apart by about 12 Å and opens up a cavity at the center of the entire structure (see fig. 9.18).

Figure 9.19 shows some of the details of the substrate-binding site in the R structure. As we mentioned above, the binding site consists of a pocket between two domains of a c subunit. Arginines 167 and 229 from one domain interact with the two carboxylate groups of PALA. Arginine 105 of the other domain interacts with the phosphonate group, and presumably does the same with the phosphate of carbamoyl phosphate. Histidine 134 appears to provide a hydrogen bond to the peptide oxygen atom of PALA. If it does the same to the corresponding oxygen of carbamoyl phosphate, it can serve as a general acid in the catalytic mechanism. In addition to these residues, the active site also includes two residues from a different c subunit, Ser 80 and Lys 84 (see fig. 9.19). As with phosphofructokinase, the location of the active site at the interface between two subunits provides a clue to how binding of substrate to one subunit can affect the binding at another.

The structural transition to the T state disrupts the active site in two major ways. First, the domain of the c subunit that includes Arg 105 and His 134, which interact with carbamoyl phosphate, is pulled away from the domain that interacts with aspartate, because some of the residues in both domains are tied up in an alternative set of hydrogen bonds. Arginine 105 is hydrogen-bonded to Glu 50 in the same domain, instead of to the substrate; His 134 interacts with a residue in the other c trimer. In addition, the loop of the c subunit that contains Ser 80 and Lys 84 is pulled out of the active site by hydrogen bonds to still another c subunit. A baroque net of interrelationships thus links the catalytic sites of all the c subunits in the complex.

The transition between the R and T states also involves large changes in the conformation of the r subunits. These changes include both the peripheral domain where CTP or ATP binds and the domain that interfaces with the c subunits (see figs. 9.17 and 9.18). However, it still is not clear how the binding of CTP to the peripheral domain tips the conformational equilibrium in favor of T, whereas ATP,

Figure 9.17

Structures of aspartate carbamoyl transferase in the T conformation (*a*) and the R conformation (*b*) viewed along the threefold symmetry axis. The enzyme contains two c_3 clusters and three r_3 clusters. The α-carbon chains of one of the c_3 groups are shown in aqua; those of the other c_3 group are in blue. One of the r subunits in each of the r_2 groups is shown in orange; the other, in red. The enzyme was crystallized in the T form in the absence of substrate or allosteric effectors; the R structure was obtained with bound PALA. The PALA molecules are seen in yellow in (*b*). Zinc ions bound to the r subunits are shown in white.

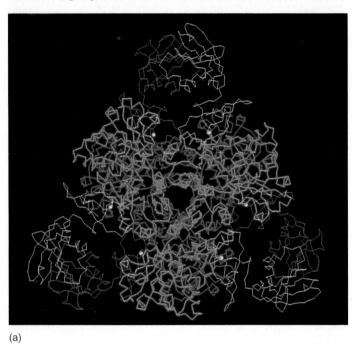

(a)

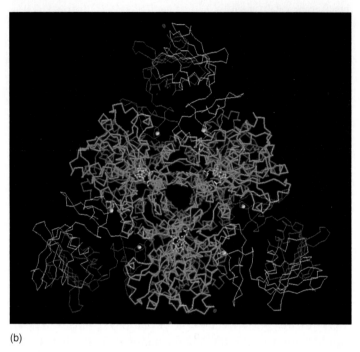

(b)

Figure 9.18

Structures of aspartate carbamoyl transferase in the T conformation (*a*) and the R conformation (*b*) viewed along an axis perpendicular to the threefold symmetry axis. The structures and the color coding are the same as in figure 9.17. Note the expansion of the cavity between the upper and lower c_3 groups in the R structure.

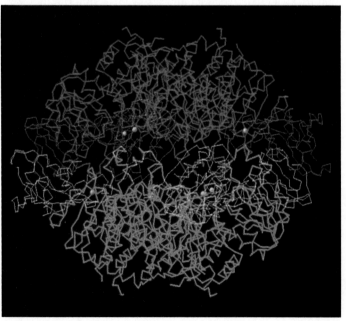

(a)

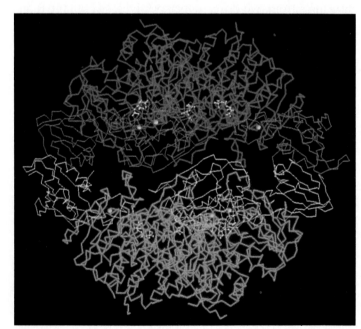

(b)

Figure 9.19

The binding of PALA to the R conformation of aspartate carbamoyl transferase. Arg 105 and His 134 are provided by one domain of a *c* subunit, and Arg 167 and Arg 229 by the other domain. Ser 80 and Lys 84 are part of a loop of protein from a different *c* subunit. The PALA is indicated by red. The wavy green lines indicate polypeptide backbone structure.

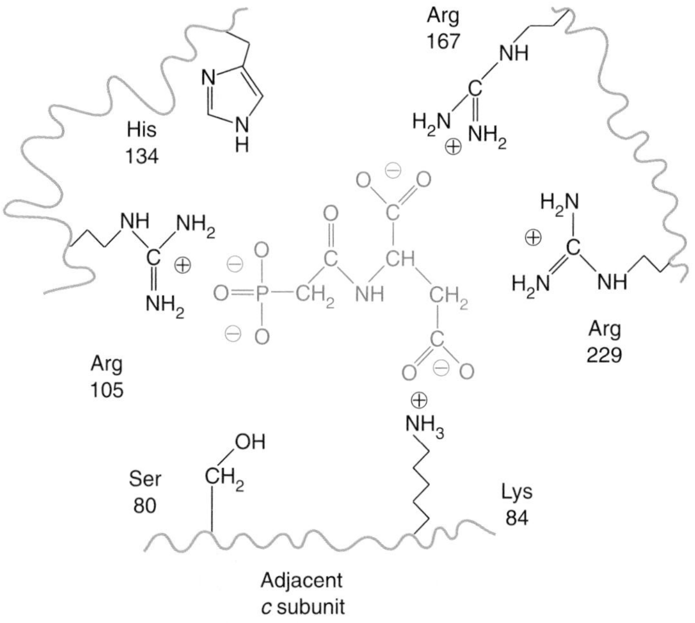

Figure 9.20

The rate of the reaction catalyzed by glycogen phosphorylase, as a function of the concentration of its main allosteric activator, AMP. The curves shown in color were obtained in the presence of ATP. Phosphorylase *b* (lower two curves) is almost completely inactive in the absence of AMP. Its activity is half maximal at an AMP concentration of about 40 μM. ATP greatly increases the concentration of AMP required for activity. Phosphorylase *a* (upper two curves) has about 80% of its maximal activity in the absence of AMP and reaches full activity at very low AMP concentrations; it also is relatively insensitive to inhibition by ATP.

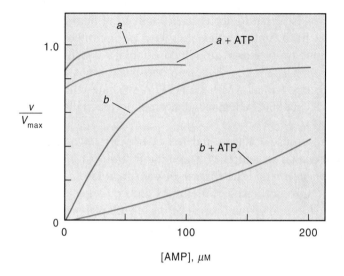

which binds to the same site as CTP, favors the formation of R. The crystal structures of the enzyme with bound ATP or CTP are very similar, both at the catalytic site and at the allosteric site.

Glycogen Phosphorylase: Combined Control by Allosteric Effectors and Phosphorylation

Glycogen phosphorylase catalyzes the removal of a terminal glucose residue from glycogen. (The structure of glycogen is shown in figs. 12.9 and 12.11.) The glycosidic bond is cleaved by a reaction with inorganic phosphate, so that the product is glucose-1-phosphate instead of free glucose.

$$(\text{glucose})_n + P_i \longrightarrow$$

$$(\text{glucose})_{n-1} + \text{glucose-1-phosphate} \quad \textbf{(9)}$$

This is the first step in the metabolic breakdown of glycogen to pyruvate.

In the early 1940s, Carl Cori and Gerty Radnitz Cori discovered that phosphorylase exists in two forms, *a* and *b*,

which differ greatly in their catalytic activities. As shown in figure 9.20, phosphorylase *b* has virtually no activity in the absence of AMP. It is activated by low concentrations of AMP, but the activation is inhibited competitively by ATP. At the concentrations of AMP and ATP that prevail in resting muscle tissue, phosphorylase *b* is essentially inactive. Phosphorylase *a*, on the other hand, has about 80% of its maximal activity in the absence of AMP and becomes fully active at very low concentrations of AMP. It also is relatively insensitive to inhibition by ATP (see fig. 9.20).

The Coris found that the interconversion of phosphorylases *a* and *b* is catalyzed by another enzyme, and subsequent work by Earl Sutherland showed that this process is under hormonal control. In muscle, conversion of phosphorylase *b* to *a* is stimulated by epinephrine; in liver, it is stimulated by both epinephrine and the pancreatic hormone glucagon. The structural basis for the difference between the two forms of phosphorylase remained unknown until the late 1950s, when Edwin Krebs and Edmund Fischer showed that phosphorylase *a* has a phosphate on serine 14. This phosphate is absent in the *b* form of the enzyme. Krebs and Fischer also showed that the kinase that catalyzes the addition of the phosphate is itself regulated by a phosphorylation catalyzed by another enzyme, the cAMP-dependent protein kinase!

The main effect of AMP on either phosphorylase *b* or phosphorylase *a* is to decrease the K_m for P_i. This change can be interpreted as we have interpreted the actions of allosteric effectors on phosphofructokinase and aspartate carbamoyl transferase, on the model that the enzyme can exist in two conformational states (R and T) with different affinities for the substrate. However, phosphorylase presents the additional complexity that the equilibrium constant (*L*) between the two conformational states can be altered by a covalent modification of the enzyme. In the absence of substrates, [T]/[R] appears to be greater than 3,000 in phosphorylase *b* but to decrease to about 10 in phosphorylase *a*.

Both forms of phosphorylase are inhibited by glucose or glucose-6-phosphate. Glucose inhibits by binding at the catalytic site while glucose-6-phosphate binds at the same allosteric site as AMP and ATP. A separate inhibitory allosteric site binds adenine, adenosine, or (much more weakly) AMP.

Louise Johnson and her coworkers have determined the crystal structures of T and the R forms of muscle phosphorylase *b* and the R form of phosphorylase *a*. In parallel with this work, Robert Fletterick and coworkers determined the structure of the T form of muscle phosphorylase *a*. The crystal structures provide an incisive look at the structural changes that accompany the transitions from the T to the R conformation and from the nonphosphorylated form of the enzyme to the phosphorylated form.

In keeping with the complexity of its allosteric and covalent regulation, phosphorylase is a large enzyme. It consists of a dimer of two identical subunits, each with a molecular weight of about 97,400 (fig. 9.21). The catalytic site is buried near the center of each subunit, at the end of a tunnel that opens to a concave surface. A binding site for glycogen on the surface can be recognized by the location of a small oligosaccharide in the crystal structure. Because the glycogen attachment site is about 30 Å from the catalytic site, the enzyme evidently clings to one branch of a glycogen particle while it chews on another branch. Near the catalytic site is a covalently bound molecule of the coenzyme pyridoxal phosphate (fig. 9.22), which probably participates as a general acid in the catalytic mechanism. The binding site for the allosteric effectors AMP, ATP, and glucose-6-phosphate is about 30 Å from the catalytic site, at one of the interfaces of the two subunits (see fig. 9.21). Serine 14, the locus of the covalent modification that converts phosphorylase *b* to *a*, is in the same region.

When phosphorylase *b* undergoes the transition from the T to the R conformation, major structural changes occur in a loop between residues 282 to 286, which connects two α-helical chains (fig. 9.23). In the T form, this loop obstructs the substrate-binding site for P_i; in the R form, the loop is pulled out of the P_i site. Some of the structural

Figure 9.21

(*a*) Ribbon diagram of the crystal structure of phosphorylase *a* in the R state. The view is along the twofold rotational symmetry axis of the dimer, with the allosteric sites and the phosphoserine (Ser-P) on each subunit facing forward. Regions where the positions of the C_d carbons differ by more than 1 Å between the R and T states are shown in orange for subunit 1 (*bottom*) and in pink for subunit 2 (*top*). Less mobile regions of the polypeptide chain are in green for subunit 1 and in blue for subunit 2. The N-terminal residues (10–23) and the C-terminal residues (837–842) are in white (residues 1–9 are disordered and cannot be seen in the crystal structure). The bound pyridoxal phosphate (PLP) indicates the location of the catalytic site. The allosteric effector site is occupied by AMP. The first two helices at the N-terminal end (labeled α_1 and α_2 in subunit 1) are connected by a loop (Cap) that forms one of the interfaces between the two subunits. (*b*) Ribbon diagram of phosphorylase *b* in the T state. The orientation of the dimer is the same as in (*a*). Regions where the C_d positions differ by more than 1 Å between the R and T states are represented in red for subunit 1 (*bottom*) and in yellow for subunit 2 (*top*); more fixed regions are in cyan and purple. The N-terminal residues and the C-terminal residues are in white. PLP is at the catalytic site and AMP at the allosteric effector site, as in (*a*). Maltopentaose is bound at the glycogen storage site. (From D. Barford, S. -H. Hu, and L. N. Johnson, *J. Mol. Biol.* 218:233, 1991. © 1991 Academic Press LTD., London England.)

(a)

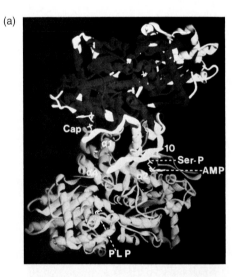

(b)

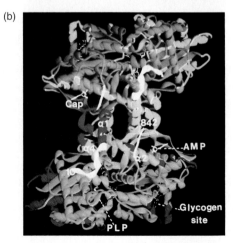

Figure 9.22

Pyridoxal phosphate, a prosthetic group in glycogen phosphorylase, is covalently attached to a lysine side chain of the enzyme. The phosphate group of the pyridoxal phosphate probably acts as a general acid to transfer a proton to inorganic phosphate in the enzymatic mechanism.

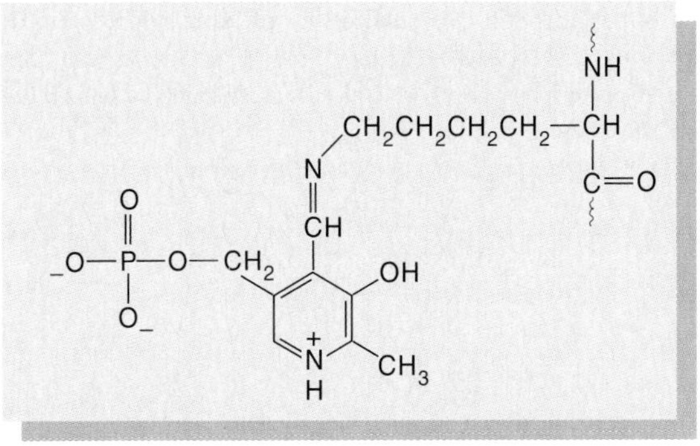

Figure 9.23

(a) In the T state of phosphorylase b, a loop of the polypeptide chain between residues 282 and 286 obstructs the active site. Asp 283, near the middle of this loop, inserts its negatively charged side chain into the binding site for phosphate. The locations of the aspartate and the pyridoxal phosphate prosthetic group (PLP) are indicated in red.

(b) In the R state, the subunits rotate with respect to one another. The loop from residues 282 to 286 is disordered, and Asp 283 leaves the phosphate-binding site. The green lines indicate polypeptide backbone structures. (Source: From D. Barford and L. N. Johnson, The allosteric transition of glycogen phosphorylase, *Nature* 340:609, 1989.)

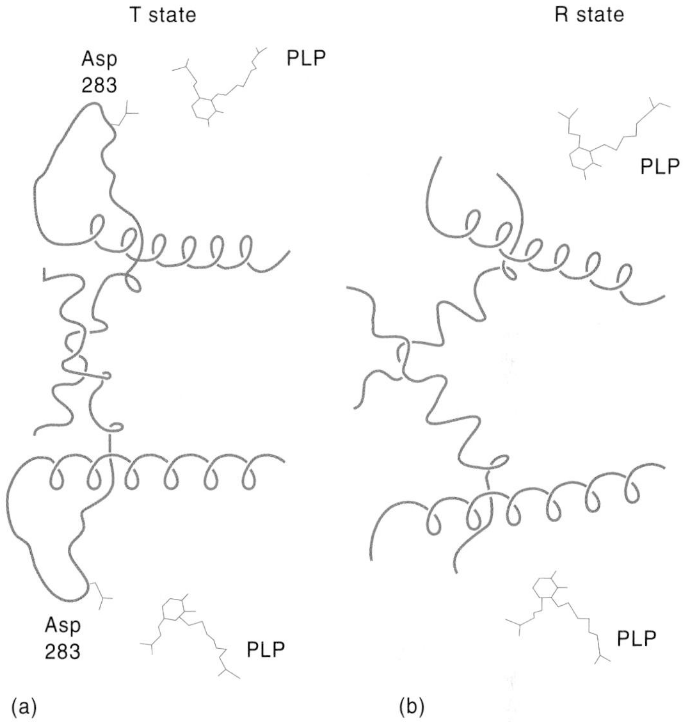

changes are shown in figure 9.24. In the R form, the substrate P_i is bound to Arg 569, Gly 135, Lys 574 (not shown in the figure), and the pyridoxal phosphate. In the T form, the side chain of Asp 283 sits in this region, and Arg 569 is pulled away by hydrogen bonding to Pro 281. The replacement of the positively charged arginine side chain by the negatively charged aspartate explains why the affinity for P_i is much lower in the T conformation.

The transformation from phosphorylase b to phosphorylase a includes structural changes that tighten the interactions between the two subunits. Figures 9.21 and 9.25 show some of these changes. In phosphorylase a, the phosphate group attached to each serine 14 is hydrogen-bonded to Arg 69 of its own subunit but also to Arg 43 of the other subunit. Hydrogen bonds also exist between Arg 10 and Leu 115 of different subunits and between Gln 72 and Asp 42. All of these intersubunit hydrogen bonds are missing in phosphorylase b. Instead, a bond exists between Arg 43 and a leucine in the same subunit, and one intersubunit hydrogen bond exists between His 36 and Asp 838. The N-terminal portion of the chain is ejected from this region in phosphorylase b and is replaced by residues from the C-terminal end, including Asp 838. At the nearby allosteric effector site, the pulling together of the two subunits in phosphorylase a enhances the binding of AMP but disfavors the binding of the inhibitor glucose-6-phosphate.

Figure 9.24

Residues in the region of the substrate-binding site for phosphate in phosphorylase *b* in the T state (*a*) and the R state (*b*). The side chain of Asp 283 leaves the binding site in the R structure, and the side chain of Arg 569 becomes available to interact with the phosphate. Key side chains and bound phosphate anion are shown in red. Portions of the polypeptide backbone are drawn in heavy black. (Source: From D. Barford and L. N. Johnson, The allosteric transition of glycogen phosphorylase, *Nature* 340:609, 1989.)

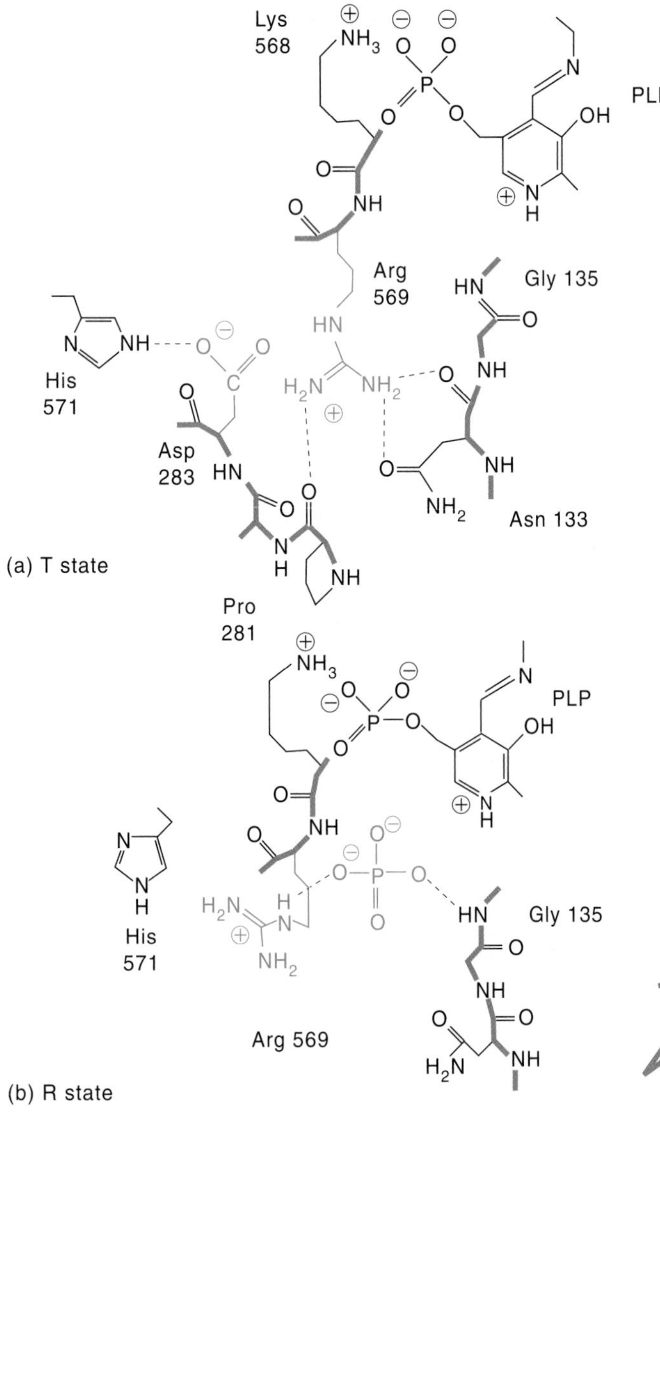

Figure 9.25

Structural changes that accompany the conversion of phosphorylase *b* to phosphorylase *a*. This figure shows the interface between the two subunits in the region of Ser 14, the residue that gains a phosphate group in phosphorylase *a*. Portions of the polypeptide backbone and the side chains of some residues from one of the subunits are drawn in red; the other subunit is drawn in black. A larger number of hydrogen bonds link the two subunits in phosphorylase *a (top)* than in phosphorylase *b (bottom)*. (Source: From S. R. Sprang et al., Structural changes in glycogen phosphorylase induced by phosphorylation, *Nature* 336:215, 1988.)

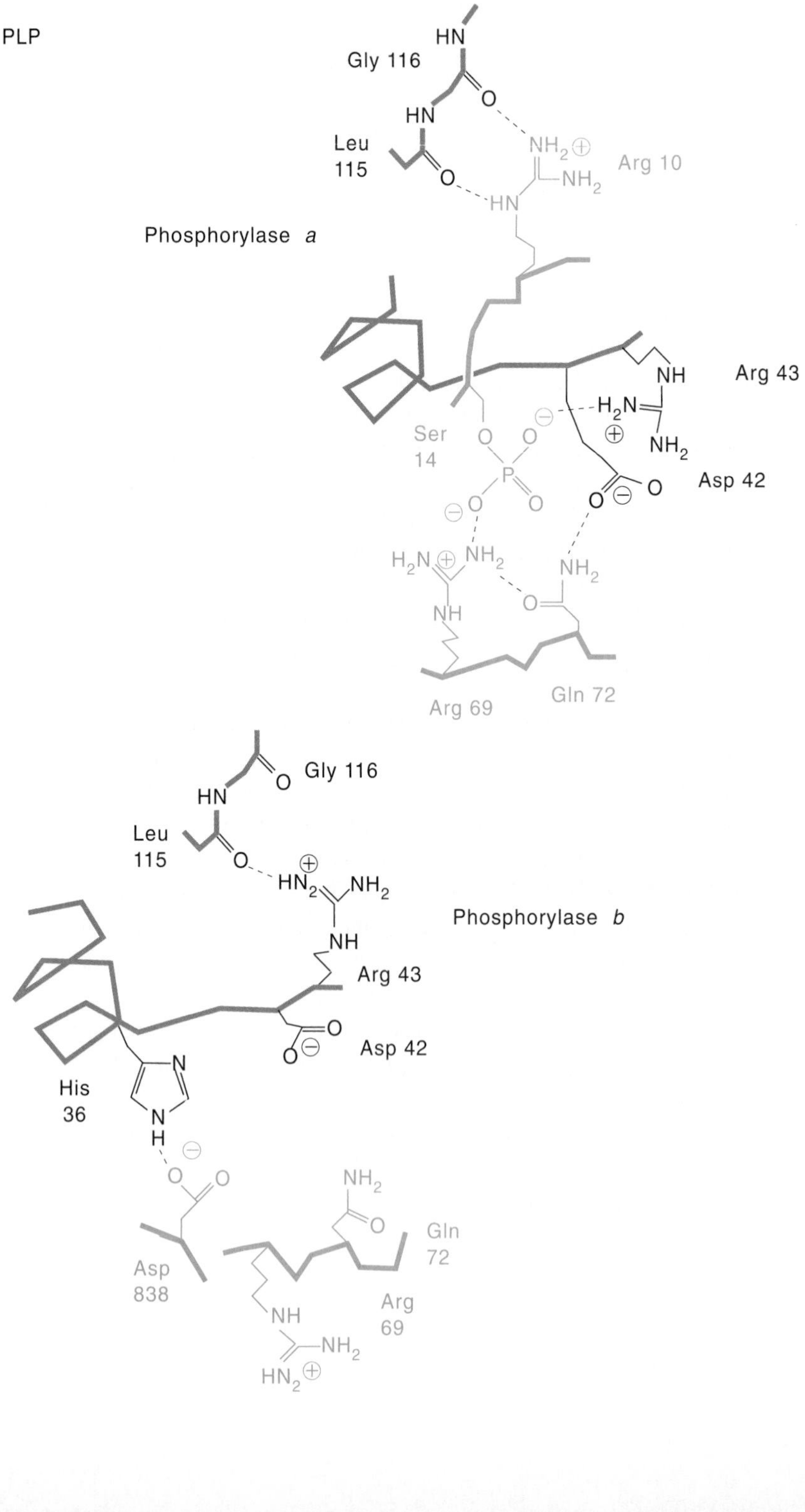

Summary

Cells regulate their metabolic activities by controlling rates of enzyme synthesis and degradation and by adjusting the activities of specific enzymes. Enzyme activities vary in response to changes in pH, temperature, and the concentrations of substrates or products, but also can be controlled by covalent modifications of the protein or by interactions with activators or inhibitors.

1. Partial proteolysis, an irreversible process, is used to activate proteases and other digestive enzymes after their secretion and to switch on enzymes that cause blood coagulation. Common types of reversible covalent modification include phosphorylation, adenylylation, and disulfide reduction.

2. Allosteric effectors are inhibitors or activators that bind to enzymes at sites distinct from the active sites. Allosteric regulation allows cells to adjust enzyme activities rapidly and reversibly in response to changes in the concentrations of substances that are structurally unrelated to the substrates or products. The initial steps in a biosynthetic pathway commonly are inhibited by the end products of the pathway, and numerous enzymes are regulated by ATP, ADP, or AMP.

3. The kinetics of allosteric enzymes typically show a sigmoidal dependence on substrate concentration rather than the more common hyperbolic saturation curves. These enzymes usually have multiple subunits, and the sigmoidal kinetics can be ascribed to cooperative interactions of the subunits. Binding of substrate to one subunit changes the dissociation constant for substrate on another subunit. The extent of the cooperativity can be described by the Hill equation or by the "symmetry" model or the more general "sequential" model. The symmetry model postulates that the enzyme can exist in two conformations, T and R. It is assumed that the substrate binds more tightly to the R conformation than to the T conformation, that binding of the substrate or an allosteric effector changes the equilibrium between these conformations, and that cooperative interactions make all of the subunits switch from one conformation to the other in a concerted manner. The symmetry model provides a useful conceptual framework that is consistent with the behavior of many allosteric enzymes.

4. Phosphofructokinase, the key regulatory enzyme of glycolysis, has four identical subunits. It exhibits sigmoidal kinetics with respect to fructose-6-phosphate and is inhibited by ATP and stimulated by ADP. In the transition between the R and T conformations, the subunits rotate with respect to each other and there is a rearrangement of the binding site for fructose-6-phosphate, which is located at an interface between subunits. At another interface, antiparallel β strands of two subunits are hydrogen-bonded together in the T structure but are separated by a row of water molecules in the R structure. This structural feature explains the cooperative nature of the conformational transition in all of the subunits: Intermediate structures probably would be less stable than either the R or the T structure.

5. Aspartate carbamoyl transferase, the first enzyme in the biosynthesis of pyrimidines, is inhibited allosterically by CTP, an end product of the pathway, and is stimulated by ATP. It has six identical catalytic (c) subunits and six regulatory (r) subunits. The c and r subunits can be separated by treating the enzyme with mercurials. Binding of a substrate analog to one of the c subunits causes the entire c_6r_6 complex to flip to the R conformation; binding of CTP to an r subunit favors the T conformation. Again, the substrate-binding sites are located at interfaces between subunits, and the T to R transition results in a rotation of the subunits and brings together components of the active site.

6. Glycogen phosphorylase breaks down glycogen to glucose-1-phosphate. It exists in two forms that differ by a covalent modification. Phosphorylase a is phosphorylated on a serine residue and is the more active form under cellular conditions. Phosphorylase b, which lacks the phosphate, is stimulated allosterically by AMP but inhibited by ATP. The enzyme has two identical subunits, with the binding site for allosteric effectors at the interface. In the T to R transition there is a repositioning of arginine and aspartate residues at the binding site for P_i. Conversion of phosphorylase b to phosphorylase a tightens the interactions between the subunits and favors the transition to the R form.

Suggested Reading

Barford, D., S. -H. Hu, and L. N. Johnson, Structural mechanism for glycogen phosphorylase control by phosphorylation and AMP. *J. Mol. Biol.* 218:233, 1991.

Barford, D., and L. N. Johnson, The allosteric transition of glycogen phosphorylase. *Nature* 340:609, 1989.

Cséke, C., and B. B. Buchanan, Regulation of the formation and utilization of photosynthate in leaves. *Biochim. Biophys. Acta.* 853:43, 1986.

Furie, B., and B. C. Furie, The molecular basis of blood coagulation. *Cell.* 53:505, 1988.

Lipscomb, W. N., Structure and function of allosteric enzymes. *Chemtracts-Biochem. Mol. Biol.* 2:1, 1991.

Schachman, H. R., Can a simple model account for the allosteric transition of aspartate transcarbamylase? *J. Biol. Chem.* 263, 18583, 1988.

Schirmer, T., and P. R. Evans, Structural basis of the allosteric behaviour of phosphofructokinase. *Nature* 343:140, 1990.

Sprang, S. R., et al., Structural changes in glycogen phosphorylase induced by phosphorylation. *Nature* 336:215, 1988.

Stevens, R. C., J. E. Gouaux, and W. N. Lipscomb, Structural consequences of effector binding to the T state of aspartate carbamoyltransferase: Crystal structures of the unligated and ATP- and CTP-complexed enzymes at 2.6 Å resolution. *Biochem.* 29:7691, 1990.

Problems

1. Notice the salt bond in figure 9.1b between the beta carboxyl group of Asp 194 and the N terminus (Ile 16) in the structure of chymotrypsin. Why is the N-terminus amino acid number 16?

2. Many hemophiliacs are sustained by regular injections of one of the cascade components, for example, factor VIII (fig. 9.2). Why doesn't the addition of factor VIII cause an uncontrolled clotting of the blood in these individuals?

3. While reading this chapter you should have noticed that protein regulation mechanisms fall into two major categories: Covalent modifications and allosteric regulation. Does one approach seem to be inherently less complicated? Why?

4. What similarities and differences can you find in the covalent modification of proteins shown in figures 9.3 and 9.4?

5. The cAMP-dependent protein kinases phosphorylate specific Ser (Thr) residues on target proteins. Given the availability of serine and threonine residues on the surface of globular proteins, how might a protein kinase select the "correct" residues to phosphorylate?

6. Is the interaction between an allosteric effector and an allosteric enzyme always an equilibrium?

7. The substrate concentration yielding half-maximal velocity is equal to K_m for hyperbolically responding enzymes. Is this relationship true for allosterically or sigmoidally responding enzymes? How do positive or negative allosteric effectors change the substrate concentration required for half-maximal velocity?

8. ATP is both a substrate and an inhibitor of the enzyme phosphofructokinase (PFK). Although the substrate fructose-6-phosphate binds cooperatively to the active site, ATP does not bind cooperatively. Explain how ATP may be both a substrate and an inhibitor of PFK.

9. Aspartate carbamoyltransferase is an allosteric enzyme in which the active sites and the allosteric effector binding sites are on different subunits. Explain how it might be possible for an allosteric enzyme to have both kinds of sites on the same subunit.

10. Examine the relationship of aspartate car-
    bamoyltransferase (ACTase) activity to aspartate
    concentration shown in the figure. Estimate the
    $(S_{0.9}/S_{0.1})$ ratio for the reaction under the following
    conditions: (a) normal curve, (b) plus 0.2 mM CTP,
    (c) plus 0.8 mM ATP. Do these ratios differ signifi-
    cantly? Explain.

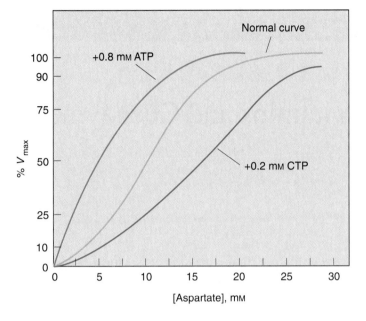

11. Why do you think the researchers who initially pre-
    pared PALA (fig. 9.16) utilized a methylene between
    the carbonyl and phosphoryl group? Why didn't they
    "just" use a "regular" phosphate group. *Hint:*
    Compare the postulated reaction intermediate (fig.
    9.16) with the reaction shown in figure 9.13.

12. In some instances, protein kinases are activated or
    inhibited by low-molecular-weight modifiers. Explain
    how a metabolite might more effectively regulate an
    enzyme by modifying a protein kinase rather than
    directly inhibiting the target enzyme.

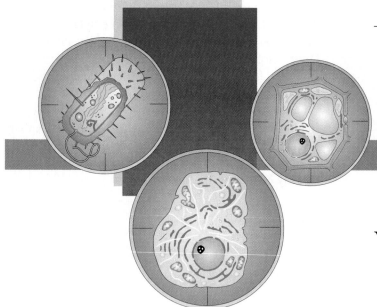

Vitamins and Coenzymes

Chapter Outline

Water-Soluble Vitamins and Their Coenzymes
- *Thiamine Pyrophosphate Is Involved in C—C and C—X Bond Cleavage*
- *Pyridoxal-5'-Phosphate Is Required for a Variety of Reactions with α-Amino Acids*
- *Nicotinamide Coenzymes Are Used in Reactions Involving Hydride Transfers*
- *Flavins Are Used in Reactions Involving One or Two Electron Transfers*
- *Reactions Requiring Acyl Activation Frequently Use Phosphopantetheine Coenzymes*
- *α-Lipoic Acid Is the Coenzyme of Choice for Reactions Requiring Acyl-Group Transfers Linked to Oxidation–Reduction*
- *Biotin Mediates Carboxylations*
- *Folate Coenzymes Are Used in Reactions for One-Carbon Transfers*
- *Ascorbic Acid Is Required to Maintain the Enzyme that Forms Hydroxyproline Residues in Collagen*
- *Vitamin B₁₂ Coenzymes Are Associated with Rearrangements on Adjacent Carbon Atoms*
- *Iron-Containing Coenzymes Are Frequently Involved in Redox Reactions*

Metal Cofactors

Lipid-Soluble Vitamins

Coenzymes are small organic molecules that act in concert with the enzyme to catalyze biochemical reactions.

The types of chemical reactions that can be catalyzed by proteins alone are limited by the chemical properties of the functional groups found in the side chains of nine amino acids: the imidazole ring of histidine; the carboxyl groups of glutamate and aspartate; the hydroxyl groups of serine, threonine, and tyrosine; the amino group of lysine; the guanidinium group of arginine; and the sulfhydryl group of cysteine. These groups can act as general acids and bases in catalyzing proton transfers and as nucleophilic catalysts in group transfer reactions.

Many metabolic reactions involve chemical changes that cannot be brought about by the structures of the amino acid side chain functional groups in enzymes acting by themselves. In catalyzing these reactions, enzymes act in cooperation with other smaller organic molecules or metallic cations, which possess special chemical reactivities or structural properties that are useful for catalyzing reactions. In this chapter we introduce these small molecules and survey the range of reactions they catalyze.

It has been known since the middle of the nineteenth century that small amounts of certain substances are very important to healthy nutrition. Vitamins are organic molecules essential in small quantities for healthy nutrition in rats or humans. The list of such molecules grew as they were purified from foodstuffs and shown to cure various disorders in animals maintained on deficient diets. The name "vitamine" was given in 1911 to the first vitamin to be isolated, thiamine. When it became clear that a number of essential organic micronutrients were not amines, the -*e* was dropped.

Vitamins are divided into water-soluble and lipid-soluble groups. In addition to vitamins, vitaminlike nutrients are required in small amounts by the organism and frequently function in similar capacities to vitamins. These vitaminlike compounds are not classified as vitamins because rats and humans have a limited capacity to synthesize them, provided that the diet contains the essential precursors. Table 10.1 lists both vitamins and vitaminlike nutrients.

In this chapter we are concerned, not primarily with vitamins *per se,* but with coenzymes. Many coenzymes are modified forms of vitamins. The modifications take place in the organism after ingestion of the vitamins. Coenzymes act in concert with enzymes to catalyze biochemical reactions. Tightly bound coenzymes are sometimes referred to as prosthetic groups. A coenzyme usually functions as a major component of the active site on the enzyme, which means that understanding the mechanism of coenzyme action usually requires a complete understanding of the catalytic process.

Water-Soluble Vitamins and Their Coenzymes

As we have noted, some vitamins are soluble in water, and others are soluble in lipids. In the following sections we survey the range of biochemical reactions in which water-soluble coenzymes participate.

Thiamine Pyrophosphate Is Involved in C—C and C—X Bond Cleavage

The structure of thiamine pyrophosphate (TPP) is given in figure 10.1. The vitamin, thiamine, or vitamin B_1, lacks the pyrophosphoryl group. Thiamine pyrophosphate is the essential coenzyme involved in the actions of enzymes that catalyze cleavages of the bonds indicated in red in figure 10.1. The bond scission in figure 10.1*b* is representative of those in many α-keto acid decarboxylations, nearly all of which require the action of TPP. The phosphoketolase reaction involves both cleavages shown in figure 10.1*c*, whereas the transketolase reaction (see fig. 12.33) involves the cleavage of the carbon–carbon bond but not the elimination of —OH. Acetolactate and acetoin arise by the formation of

Table 10.1

Vitamins and Vitaminlike Nutrients

Vitamin	Function
Water-Soluble Vitamins	
Thiamine (B_1)	Precursor of the coenzyme thiamine pyrophosphate. Deficiency can cause beriberi.
Riboflavin (B_2)	Precursor of the coenzymes flavin mononucleotide and flavin adenine dinucleotide. Deficiency leads to growth retardation.
Pyridoxine (B_6)	Precursor of the coenzyme pyridoxal phosphate. Deficiency causes dermatitis in rats.
Nicotinic acid (niacin)	Precursor of the coenzymes nicotinamide adenine dinucleotide and nicotinamide adenine dinucleotide phosphate. Deficiency leads to pellagra.
Pantothenic acid	Precursor of coenzyme A (CoA). Deficiency leads to dermatitis in chickens.
Biotin	Precursor of the coenzyme biocytin. Deficiency leads to dermatitis in humans.
Folic acid	Precursor of the coenzyme tetrahydrofolic acid. Deficiency causes anemias.
Vitamin B_{12}	Precursor of the coenzyme deoxyadenosyl cobalamin. Deficiency leads to pernicious anemia.
Vitamin C	Cosubstrate in the hydroxylation of proline in collagen. Deficiency leads to scurvy.
Lipid-Soluble Vitamins	
Vitamin A	Vision, growth, and reproduction (see supplement 2).
Vitamin D	Regulation of calcium and phosphate metabolism (see chapter 24).
Vitamin E	Antisterility factor in rats.
Vitamin K	Important for blood coagulation.
Vitaminlike Nutrients	
Inositol	Mediator of hormone action (see chapters 13, 19, and 24).
Choline	Important for integrity of cell membranes and lipid transport (see chapter 19).
Carnitine	Essential for transfer of fatty acids to mitochondria (see chapter 18).
α-Lipoic acid	Coenzyme in the oxidative decarboxylation of keto acids (this chapter).
p-Aminobenzoate (PABA)	Component of folic acid (this chapter).
Coenzyme Q (ubiquinones)	Important for electron transport in mitochondria (see chapter 14).

Figure 10.1

(a) Structure of thiamine pyrophosphate and (b, c) the bonds it cleaves or forms. Reactive part of the coenzyme and the bonds subject to cleavage in (b) and (c) are indicated in red.

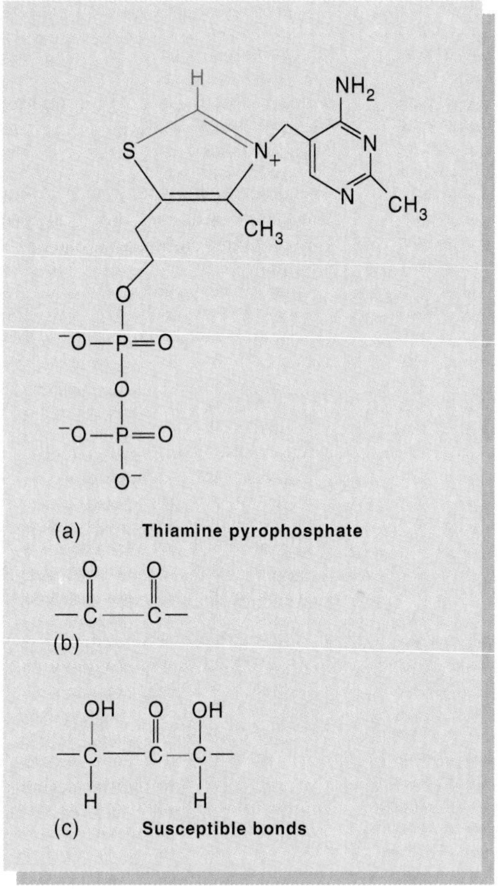

(a)   **Thiamine pyrophosphate**

(b)

(c)   **Susceptible bonds**

The active intermediate shown in equation (1) undergoes nucleophilic addition to the bond of polar carbonyl groups in substrates to produce intermediates, such as

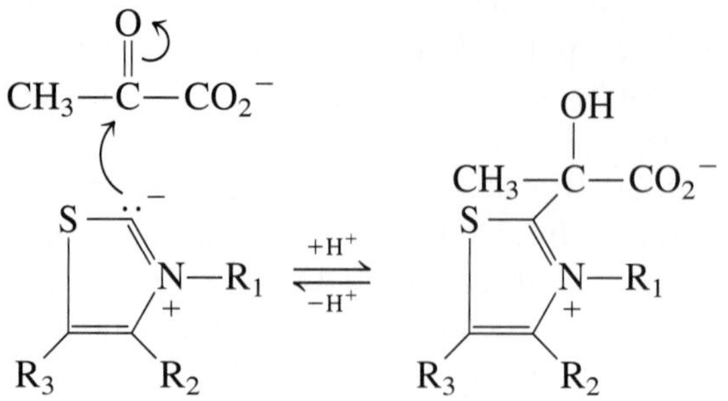

$$(2)$$

Intermediates of this type have the necessary chemical reactivity for cleaving the bonds indicated in figure 10.1b and c. The decarboxylated product of the pyruvate adduct shown in equation (2) is resonance-stabilized by the thiazolium ring (fig. 10.2a). This intermediate may be protonated to α-hydroxyethyl thiamine pyrophosphate (fig. 10.2d); alternatively, it may react with other electrophiles, such as the carbonyl groups of acetaldehyde or pyruvate, to form the species in figure 10.2b and c; or it may be oxidized to acetyl-thiamine pyrophosphate (fig. 10.2e). The fate of the intermediate depends on the reaction specificity of the enzyme with which the coenzyme is associated.

Pyridoxal-5'-Phosphate Is Required for a Variety of Reactions with α-Amino Acids

Pyridoxal-5'-phosphate is the coenzyme form of vitamin B_6, and has the structure shown in figure 10.3. The name vitamin B_6 is applied to any of a group of related compounds lacking the phosphoryl group, including pyridoxal, pyridoxamine, and pyridoxine.

Pyridoxal-5'-phosphate participates in many reactions with α-amino acids, including transaminations, α decarboxylations, racemizations, α,β eliminations, β,γ eliminations, aldolizations, and the β decarboxylation of aspartic acid.

the carbon–carbon bond in figure 10.1c (the structures of these two compounds are shown in figure 10.2).

The mechanism for the bond cleavages indicated in figure 10.1b was clarified by Ronald Breslow. In one of the earliest applications of nuclear magnetic resonance to biochemical mechanisms, he demonstrated that the proton bonded to C-2 in the thiazolium ring is readily exchangeable with the protons of H_2O and deuterons of D_2O in a base-catalyzed reaction

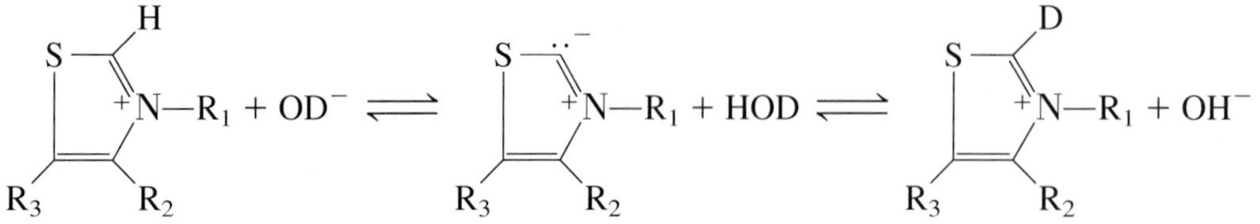

$$(1)$$

Figure 10.2

Mechanism of thiamine pyrophosphate action. Intermediate (*a*) is represented as a resonance-stabilized species. It arises from the decarboxylation of the pyruvate-thiamine pyrophosphate addition compound shown at the left of (*a*) and in equation (2). It can react as a carbanion with acetaldehyde, pyruvate, or H$^+$ to form (*b*), (*c*), or (*d*), depending on the specificity of the enzyme. It can also be oxidized to acetyl-thiamine pyrophosphate (TPP) (*e*) by other enzymes, such as pyruvate oxidase. The intermediates (*b*) through (*e*) are further transformed to the products shown by the actions of specific enzymes.

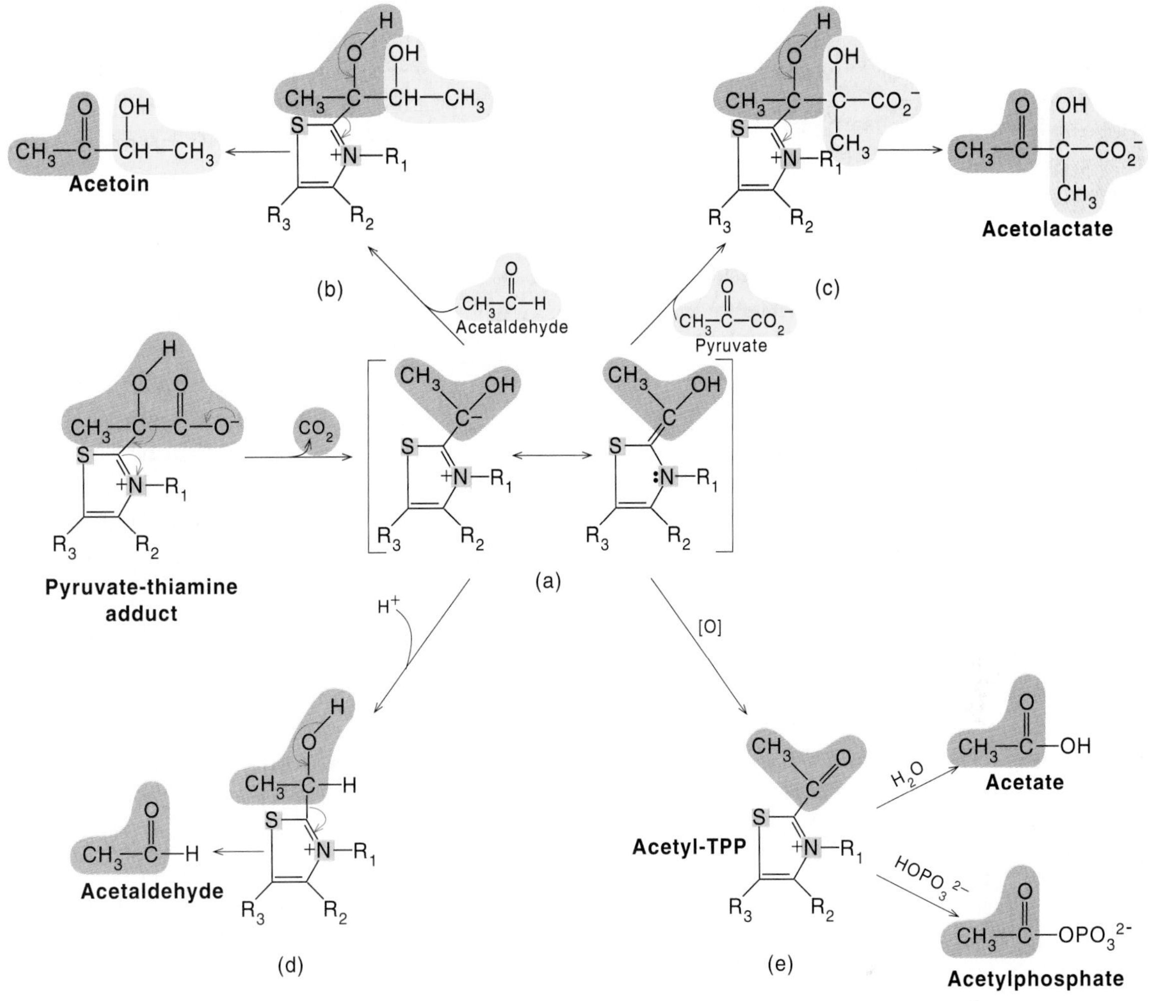

Figure 10.3

Structures of vitamin B$_6$ derivatives and the bonds cleaved or formed by the action of pyridoxal phosphate (*a*). The reactive part of the coenzyme is shown in red in (*a*). The bonds shown in red in (*d*) are the types of bonds in substrates that are subject to cleavage.

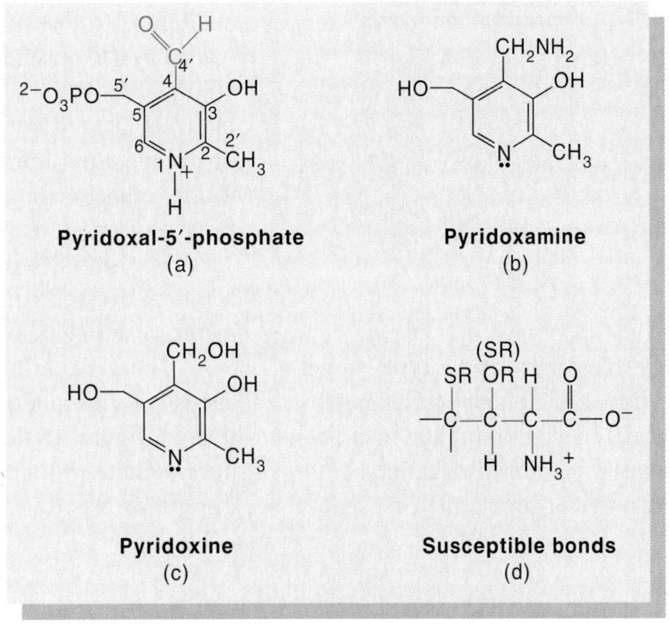

The following equations illustrate several reactions in which pyridoxal-5′-phosphate acts as a coenzyme.

$$R_1{-}\overset{\displaystyle H}{\underset{\displaystyle NH_3^+}{C}}{-}CO_2^- + R_2{-}\overset{\displaystyle O}{C}{-}CO_2^- \underset{}{\overset{Transaminase}{\rightleftharpoons}} \tag{3}$$

$$R_1{-}\overset{\displaystyle O}{C}{-}CO_2^- + R_2{-}\overset{\displaystyle H}{\underset{\displaystyle NH_3^+}{C}}{-}CO_2^-$$

$$R{-}\overset{\displaystyle H}{\underset{\displaystyle NH_3^+}{C}}{-}CO_2^- + H^+ \underset{}{\overset{Decarboxylase}{\rightleftharpoons}} \tag{4}$$

$$CO_2 + R{-}CH_2{-}NH_3^+$$

$$R{-}\overset{\displaystyle H}{\underset{\displaystyle NH_3^+}{C}}{-}CO_2^- \underset{}{\overset{Racemase}{\rightleftharpoons}} R{-}\overset{\displaystyle NH_3^+}{\underset{\displaystyle H}{C}}{-}CO_2^- \tag{5}$$

$$R{-}\overset{\displaystyle OH}{C}H{-}\overset{\displaystyle H}{\underset{\displaystyle NH_3^+}{C}}{-}CO_2^- \underset{}{\overset{Aldolase}{\rightleftharpoons}} \tag{6}$$

$$R{-}CHO + \underset{\displaystyle NH_3^+}{C}H_2{-}CO_2^-$$

$$^-O_2C{-}CH_2{-}\overset{\displaystyle H}{\underset{\displaystyle NH_3^+}{C}}{-}CO_2^- + H^+ \underset{}{\overset{Aspartate\text{-}\beta\text{-}decarboxylase}{\rightleftharpoons}} \tag{7}$$

$$CO_2 + CH_3{-}\overset{\displaystyle H}{\underset{\displaystyle NH_3^+}{C}}{-}CO_2^-$$

These equations involve bond cleavages of the type shown in color in figure 10.3d. Pyridoxal-5′-phosphate promotes these heterolytic bond cleavages by stabilizing the resulting electron pairs at the α- or β-carbon atoms of α-amino acids. To do this, the aldehyde group of the coenzyme first reacts with the α-amino group of an amino acid to produce an aldimine (fig. 10.4a) or Schiff's base, which is internally stabilized by H bonding. Loss of the α hydrogen as H^+ produces a resonance-stabilized species (fig. 10.4b) in which the electron pair is delocalized into the pyridinium system. This active intermediate may undergo further reactions at the carbon to form products determined by the reaction specificity of the enzyme. If, for example, the enzyme is a racemase, the species resulting from the loss of the proton from the α carbon may accept a proton from the opposite side to produce, ultimately, the enantiomer of the amino acid.

When the substrate is substituted at the β carbon with a potential leaving group, such as —OH, —SH, $-OPO_3^{3-}$ (see fig. 10.3d), the corresponding α-carbanion intermediate (see fig. 10.4b) can eliminate the group. This is an essential step in α,β eliminations. Upon hydrolysis, the elimination intermediate produces pyridoxal-5′-phosphate and the substrate-derived enamine, which spontaneously hydrolyzes to ammonia and an α-keto acid.

The full series of intermediates in a transamination is shown in figure 10.5a. After protonation at the aldimine carbon of pyridoxal-5′-phosphate (step 3), hydrolysis (step 4) forms an α-keto acid and pyridoxamine-5′-phosphate. The reverse of this sequence with a second α-keto acid (steps 5 through 8) completes the transamination reaction.

An intermediate analogous to that in figure 10.4b but generated from glycine and so lacking the β and γ carbons, can react as a carbanion with an aldehyde to produce a β-hydroxy-α-amino acid. These reactions are catalyzed by aldolases, such as threonine aldolase or serine hydroxymethyl transferase.

β-Decarboxylases (fig. 10.5b) generate intermediates analogous to those in figure 10.4b by catalyzing the elimination of CO_2 instead of H^+ from the intermediate in figure 10.4a (step 4, fig. 10.5b). Protonation of the α-carbanionic intermediates by protons from H_2O, followed by hydrolysis of the resulting imines, produces the amines corresponding to the replacement of the carboxylate group in the substrate by a proton (steps 5 through 8, fig. 10.5b).

The stability of the resonance hybrid (see fig. 10.4b) accounts for the catalytic action of pyridoxal-5′-phosphate in the reactions shown in equations (3) through (6).

The β decarboxylation of aspartate (equation 7) proceeds by elimination of a β-carbanionic intermediate like that in figure 10.4d from the ketimine, analogous to the intermediate produced by loss of the α proton from the aldimine of aspartate with pyridoxal-5′-phosphate.

Figure 10.4

Structures of catalytic intermediates in pyridoxal-phosphate–dependent reactions. The initial aldimine intermediate resulting from Schiff's base formation between the coenzyme and the α-amino group of an amino acid (*a*). This aldimine is converted to the resonance-stabilized intermediate (*b*) by loss of a proton at the α carbon. Further enzyme-catalyzed proton transfers to intermediates (*c*) and (*d*) may occur, depending on the specificity of a given enzyme. The enzymes use their general acids and bases to catalyze these proton transfers.

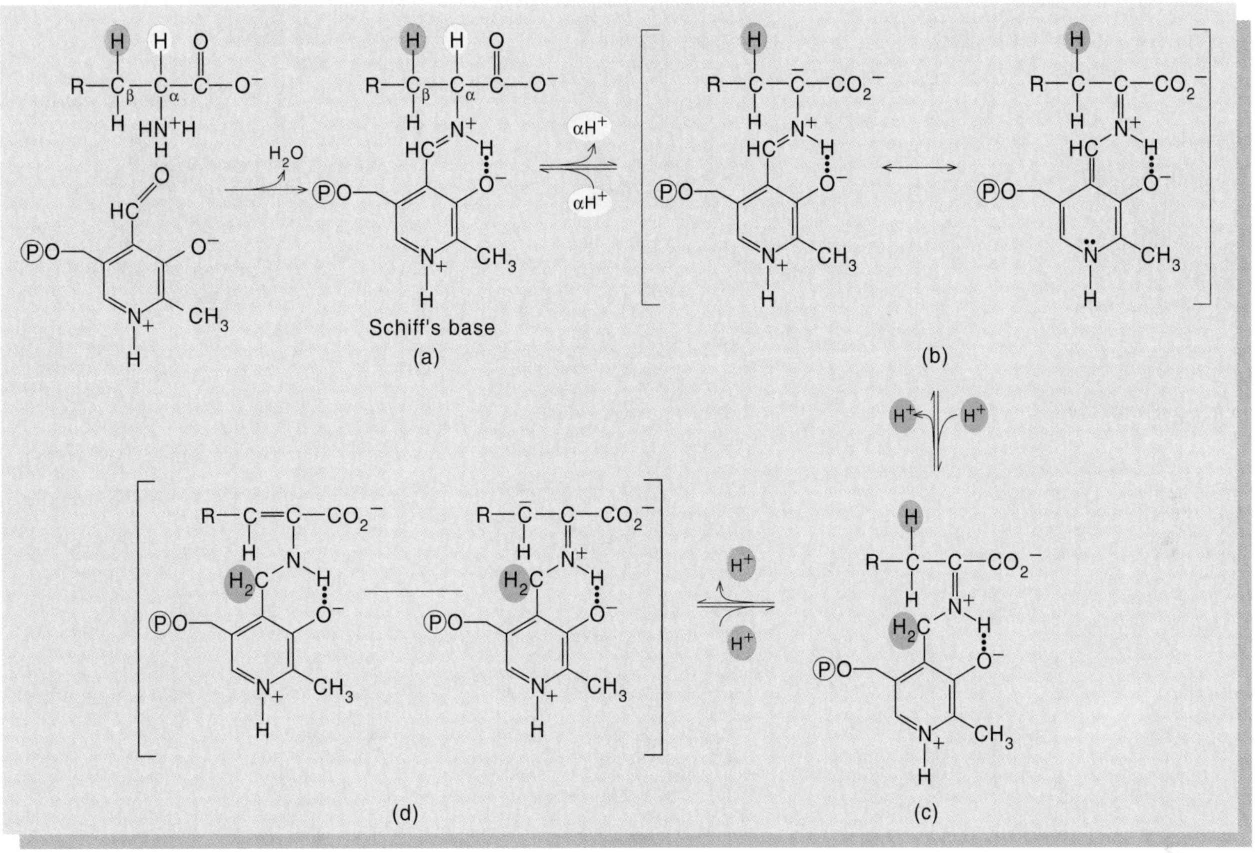

Nicotinamide Coenzymes Are Used in Reactions Involving Hydride Transfers

Nicotinamide adenine dinucleotide (NAD$^+$) is one of the two coenzymatic forms of nicotinamide (fig. 10.6). The other is nicotinamide adenine dinucleotide phosphate (NADP$^+$), which differs from NAD$^+$ by the presence of a phosphate group at C-2$'$ of the adenosyl moiety.

The fundamental biochemical function of pyridoxal-5$'$-phosphate is the formation of aldimines with α-amino acids that stabilize the development of carbanionic character at the α and β carbons of α-amino acids in intermediates, such as those in figures 10.4*b* and *c*. Enzymes acting alone cannot stabilize these carbanions and so cannot, by themselves, catalyze reactions requiring their formation as intermediates.

The nicotinamide coenzymes are biological carriers of reducing equivalents (electrons). The most common function of NAD$^+$ is to accept two electrons and a proton (H$^-$ equivalent) from a substrate undergoing metabolic oxidation to produce NADH, the reduced form of the coenzyme. This then diffuses or is transported to the terminal-electron transfer sites of the cell and reoxidized by terminal-electron acceptors, O$_2$ in aerobic organisms, with the concomitant formation of ATP (chapter 14). Equations (8), (9), and (10) are typical reactions in which NAD$^+$ acts as such an acceptor.

Figure 10.5

Mechanisms of action of pyridoxal phosphate: (a) in glutamate-oxaloacetate transaminase, and (b) in aspartate β-decarboxylase.

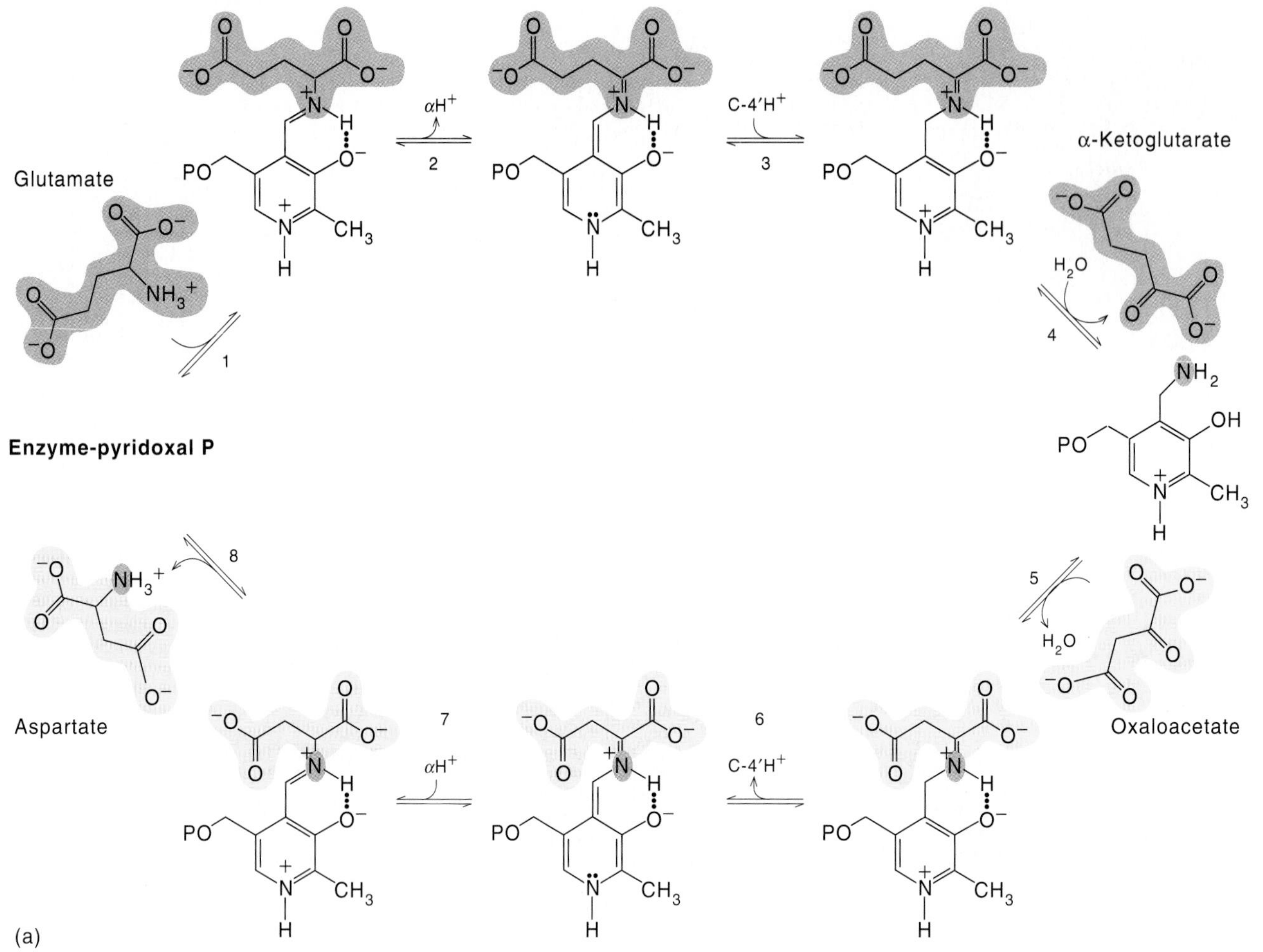

(a)

$$NAD^+ + CH_3CH_2OH \xrightleftharpoons[\text{dehydrogenase}]{\text{Alcohol}} CH_3{-}\overset{\overset{\displaystyle O}{\|}}{C}H + NADH + H^+ \quad \textbf{(8)}$$

$$^-O_2C(CH_2)_2\overset{\overset{\displaystyle NH_3{}^+}{|}}{C}HCO_2{}^- + NAD^+ + H_2O \xrightleftharpoons[\text{dehydrogenase}]{\text{Glutamate}} {}^-O_2C(CH_2)_2\overset{\overset{\displaystyle O}{\|}}{C}CO_2{}^- + NADH + NH_4{}^+ + H^+ \quad \textbf{(9)}$$

$$HPO_4{}^{2-} + {}^{2-}O_3POCH_2\overset{\overset{\displaystyle OH}{|}}{C}H{-}\overset{\overset{\displaystyle O}{\|}}{C}H + NAD^+ \xrightleftharpoons[\text{dehydrogenase}]{\text{Glyceraldehyde-3P}} {}^{2-}O_3POCH_2\overset{\overset{\displaystyle OH}{|}}{C}H\overset{\overset{\displaystyle O}{\|}}{C}OPO_3{}^{2-} + NADH + H^+ \quad \textbf{(10)}$$

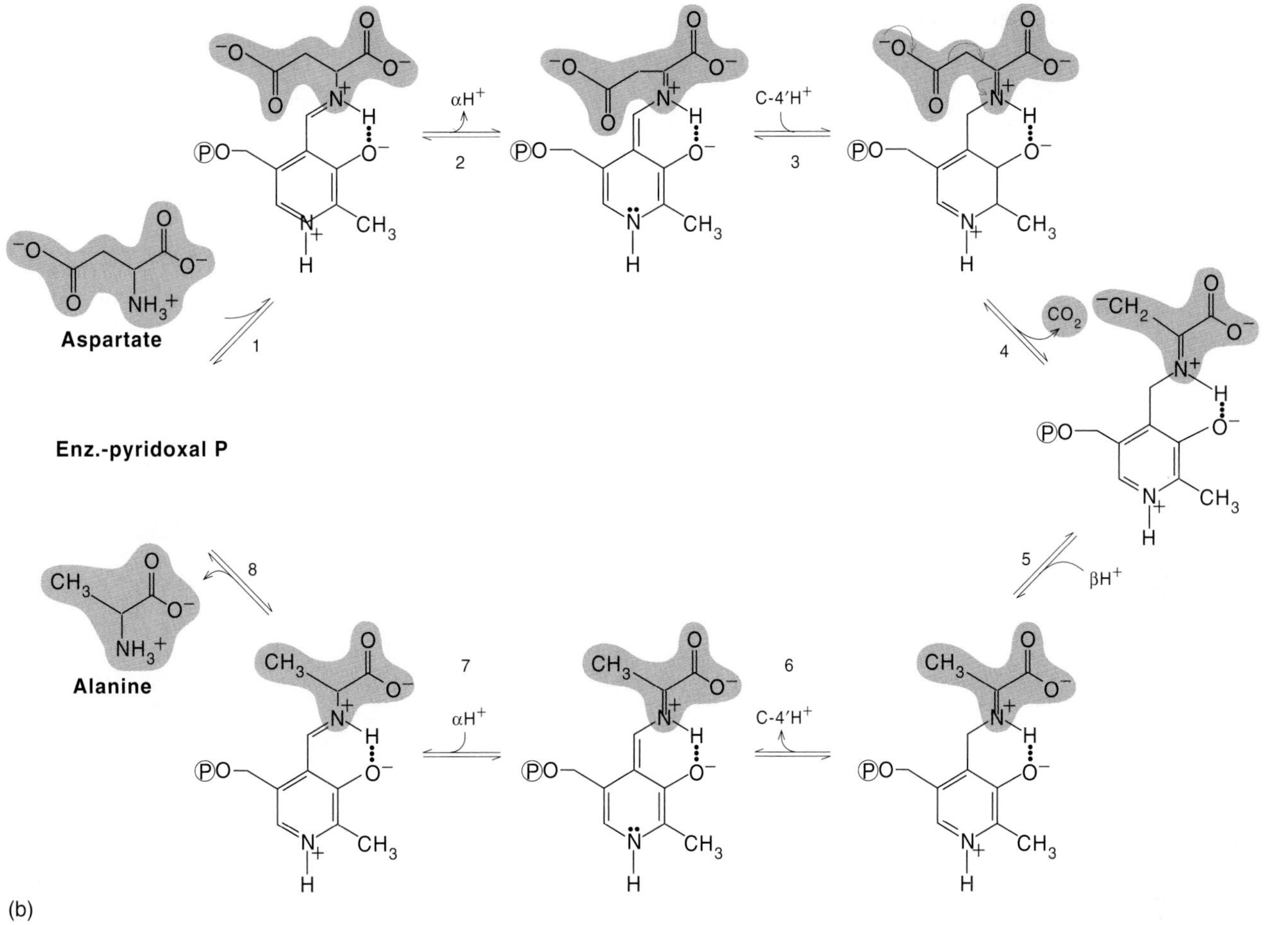

(b)

Figure 10.6

Structures of nicotinamide and nicotinamide coenzymes. The reactive
sites of the coenzymes are shown in red.

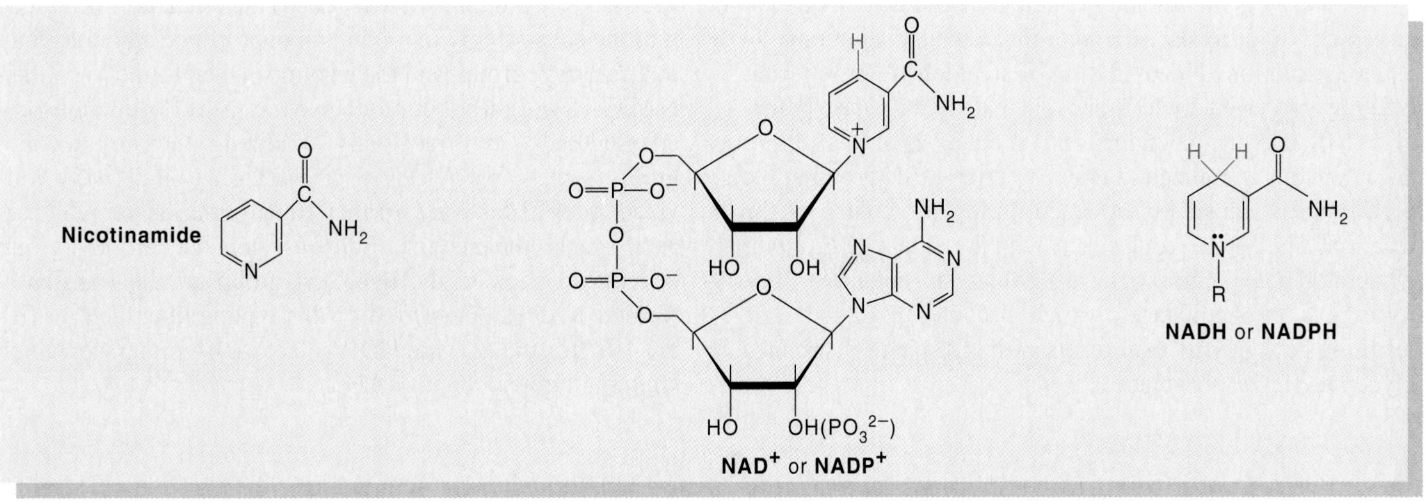

Figure 10.7

Mechanism of NAD^+ action in UDPgalactose-4-epimerase. No net oxidation or reduction occurs. Only the intermediate is oxidized.

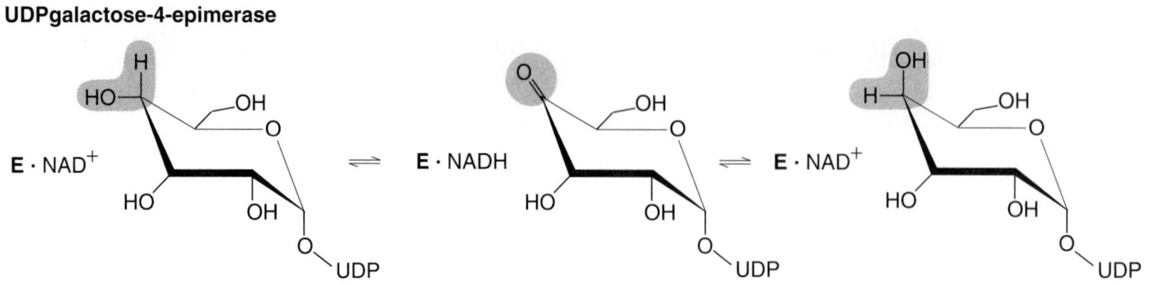

UDPgalactose-4-epimerase

The chemical mechanisms by which NAD^+ is reduced to NADH in equations (8) through (10) are probably similar, as represented in generalized forms in equation (11).

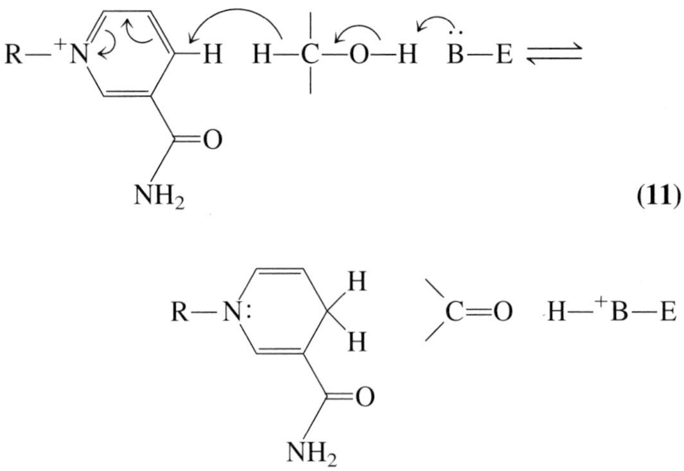

$$(11)$$

$$(12)$$

According to this formulation, the immediate oxidation product in equation (9), where $-NH_2$ replaces $-OH$ in equation (11), is the imine of α-ketoglutarate, which quickly undergoes hydrolysis to α-ketoglutarate and ammonia in aqueous solution. The oxidation of an aldehyde group catalyzed by glyceraldehyde-3-phosphate dehydrogenase [equation (10)] also can be understood on the basis of this formulation once it is realized that an essential $-SH$ group at the active site is transiently acylated during the course of the reaction. The $-SH$ group reacts with the aldehyde group of glyceraldehyde-3-phosphate according to equation (12), forming a thiohemiacetal, which becomes oxidized. The resulting acylenzyme then reacts with phosphate to produce glycerate-1,3-bisphosphate.

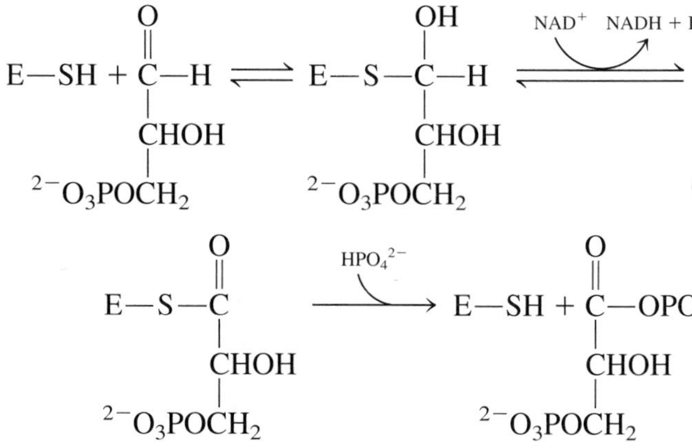

In addition to acting as a cellular electron carrier, NAD^+ also acts as a true coenzyme with certain enzymes. Enzymes are sometimes confronted with the problem of catalyzing such reactions as epimerizations, aldolizations, and eliminations on substrates lacking the intrinsic chemical reactivities required for these reactions to occur at significant rates. Sometimes such reactivities can be introduced into the substrate by oxidizing an appropriate alcohol group to a carbonyl group, and the enzyme is then found to contain NAD^+ as a tightly bound coenzyme. NAD^+ functions coenzymatically by transiently oxidizing the key alcohol group to the carbonyl level, producing an oxidatively activated intermediate the further transformation of which is catalyzed by the enzyme. In the last step, the carbonyl group is reduced back to the hydroxyl group by the transiently formed NADH. A reaction of this type is illustrated in figure 10.7 for the enzyme UDP-galactose-4-epimerase, which contains tightly bound NAD^+.

Figure 10.8

Structures of the vitamin riboflavin (*a*) and the derived flavin coenzymes (*b*). Like NAD^+ and $NADP^+$, the coenzyme pair FMN and FAD are functionally equivalent coenzymes, and the coenzyme involved with a given enzyme appears to be a matter of enzymatic binding specificity. The catalytically functional portion of the coenzymes is shown in red.

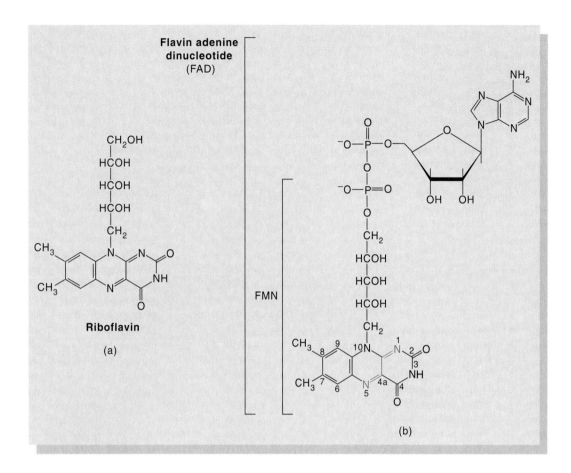

Flavins Are Used in Reactions Involving One or Two Electron Transfers

Flavin adenine dinucleotide (FAD) (fig. 10.8) and flavin mononucleotide (FMN) are the coenzymatically active forms of vitamin B_2, riboflavin. Riboflavin is the N^{10}-ribityl isoalloxazine portion of FAD, which is enzymatically converted into its coenzymatic forms first by phosphorylation of the ribityl C-5′ hydroxy group to FMN and then by adenylylation to FAD. FMN and FAD are functionally equivalent coenzymes, and the one that is involved with a given enzyme appears to be a matter of enzymatic binding specificity.

The catalytically functional portion of the coenzymes is the isoalloxazine ring, specifically N-5 and C-4a (see fig. 10.8*b*), which is thought to be the immediate locus of catalytic function, although the entire chromophoric system extending over N-5, C-4a, C-10a, N-1, and C-2 should be

regarded as an indivisible catalytic entity, as are the nicotinamide, pyridinium, and thioazolium rings of NAD^+, pyridoxal phosphate, and thiamine pyrophosphate, respectively.

Flavin-containing enzymes are known as flavoproteins and, when purified, normally contain their full complements of FAD or FMN. The bright yellow color of flavoproteins is due to the isoalloxazine chromophore in its oxidized form. In a few flavoproteins, the coenzyme is known to be covalently bonded to the protein by means of a sulfhydryl or imidazole group at the C-8 methyl group and in at least one case at C-6. In most flavoproteins, the coenzymes are tightly but noncovalently bound, and many can be resolved into apoenzymes that can be reconstituted to holoenzymes by readdition of FAD or FMN.

Flavin coenzymes exist in three spectrally distinguishable oxidation states that account in part for their catalytic functions; the yellow oxidized form, the red or blue one-electron reduced form, and the colorless two electron re-

Figure 10.9

Oxidation states of flavin coenzymes. The flavin coenzymes exist in three spectrally distinguishable oxidation states that account in part for their catalytic functions. They are the yellow oxidized form, the red or blue one-electron reduced form, and the colorless two-electron reduced form. Groups in red are those which are centrally involved in oxidation–reduction reactions.

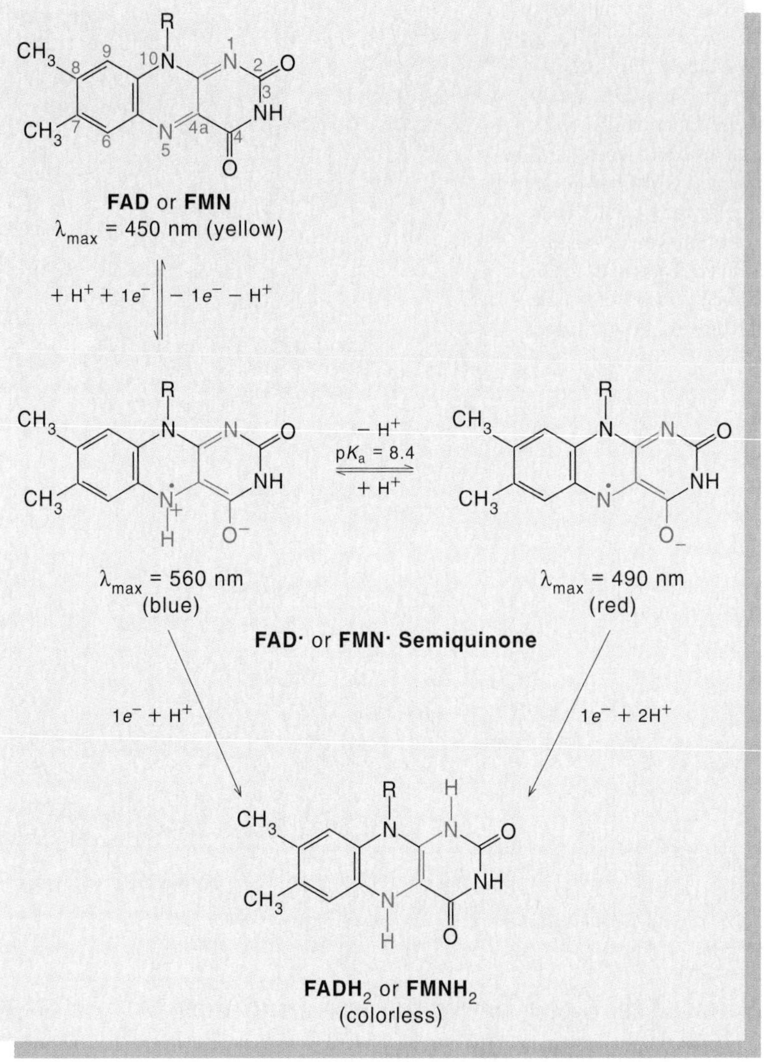

duced form. Their structures are depicted in figure 10.9. These and other less well defined forms often have been detected spectrally as intermediates in flavoprotein catalysis.

Flavins are very versatile redox coenzymes. Flavoproteins are dehydrogenases, oxidases, and oxygenases that catalyze a variety of reactions on an equal variety of substrate types. Since these classes of enzymes do not consist exclusively of flavoproteins, it is difficult to define catalytic specificity for flavins. Biological electron acceptors and donors in flavin-mediated reactions can be two-electron acceptors, such as NAD^+ or $NADP^+$, or a variety of one-electron acceptor systems, such as cytochromes (Fe^{2+}/Fe^{3+}) and quinones, and molecular oxygen is an electron acceptor for flavoprotein oxidases as well as the source of oxygen for oxygenases. The only obviously common aspect of flavin-dependent reactions is that all are redox reactions.

Typical reactions catalyzed by flavoproteins are listed in table 10.2, which groups flavoproteins into those that do not utilize molecular oxygen as a substrate and those that do. You can best appreciate the significance of this difference when you realize that the reduced form of FAD ($FADH_2$), a likely intermediate in many flavoprotein reactions, spontaneously reacts with O_2 to produce H_2O_2. In the case of the dehydrogenases, therefore, either $FADH_2$ is not an intermediate or is somehow prevented from reacting with O_2. Among the dehydrogenases are two that utilize the two-electron acceptor substrates NAD^+ or $NADP^+$, and it is reasonable to suppose that the two-electron reduction of NAD^+ by an intermediate $E \cdot FADH_2$ is involved. Also listed in table 10.2 are other dehydrogenases for which the electron acceptors from $E \cdot FADH_2$ are not given. These enzymes are membrane-bound and transfer electrons directly to membrane-bound acceptors, mainly one-electron

Table 10.2

Reactions Catalyzed by Flavoproteins

Flavoprotein	Reaction
Dehydrogenases	
Glutathione reductase	$H^+ + GSSG + NADPH \rightleftharpoons 2\,GSH + NADP^+$
Acyl-CoA dehydrogenases	$RCH_2CH_2COSCoA + NAD^+ \rightleftharpoons RCH{=}CHCOSCoA + NADH + H^+$
Succinate dehydrogenase	$^-O_2CCH_2CH_2CO_2{}^- + E \cdot FAD \rightleftharpoons {}^-O_2CCH{=}CHCO_2{}^- + E \cdot FADH_2$
D-Lactate dehydrogenase	$CH_3{-}CHOH{-}CO_2{}^- + E \cdot FAD \rightleftharpoons CH_3{-}CO{-}CO_2{}^- + E \cdot FADH_2$
Oxidases	
Amino acid oxidases	$R{-}\overset{\overset{\textstyle NH_3{}^+}{\mid}}{C}H{-}CO_2{}^- + O_2 + H_2O \longrightarrow R{-}CO{-}CO_2{}^- + H_2O_2 + NH_4{}^+$
Monoamine oxidase	$R{-}CH_2\overset{+}{N}H_3 + O_2 + H_2O \longrightarrow R{-}CHO + H_2O_2 + \overset{+}{N}H_4$
Monooxygenases	
Lactate oxidase	$CH_3{-}CHOH{-}CO_2{}^- + O_2 \longrightarrow CH_3{-}CO_2{}^- + CO_2 + H_2O$
Salicylate hydroxylase	$2H^+ + \text{(salicylate)}{-}CO_2{}^- + O_2 + NADH \longrightarrow \text{(catechol)}{-}OH + CO_2 + NAD^+ + H_2O$

acceptors, such as quinones and cytochromes (Fe^{2+}/Fe^{3+}). The stability of the flavin semiquinone, FAD $\cdot$ and FMN $\cdot$ in figure 10.9, gives flavins the capability of interacting with one-electron acceptors in electron-transport systems.

The other classes of flavoproteins in table 10.2 interact with molecular oxygen either as the electron-acceptor substrates in redox reactions catalyzed by oxidases or as the substrate sources of oxygen atoms for oxygenases. Molecular oxygen also serves as an electron acceptor and source of oxygen for metalloflavoproteins and dioxygenases, which are not listed in the table. These enzymes catalyze more complex reactions, involving catalytic redox components, such as metal ions and metal–sulfur clusters in addition to flavin coenzymes.

A recurrent theme in many flavoprotein reactions is the probable involvement of $FADH_2$ or the reduced form of FMN ($FMNH_2$) as transient intermediates. Figure 10.10 illustrates a reasonable catalytic pathway for the first enzyme listed in table 10.2; this reaction shows the likely involvement of $E \cdot FADH_2$ in each case. The mechanisms by which $E \cdot FAD$ is reduced to $E \cdot FADH_2$ by NADPH in the for-

ward direction and by glutathione in the reverse direction are undoubtedly different.

The biochemical importance of flavin coenzymes appears to be their versatility in mediating a variety of redox processes, including electron transfer and the activation of molecular oxygen for oxygenation reactions. An especially important manifestation of their redox versatility is their ability to serve as the switch point from the two-electron processes, which predominate in cytosolic carbon metabolism, to the one-electron transfer processes, which predominate in membrane-associated terminal electron-transfer pathways. In mammalian cells, for example, the end products of the aerobic metabolism of glucose are CO_2 and NADH (see chapter 13). The terminal electron-transfer pathway is a membrane-bound system of cytochromes, non-heme iron proteins, and copper-heme proteins—all one-electron acceptors that transfer electrons ultimately to O_2 to produce H_2O and NAD^+ with the concomitant production of ATP from ADP and P_i. The interaction of NADH with this pathway is mediated by NADH dehydrogenase, a flavoprotein that couples the two-electron oxidation of NADH with the one-electron reductive processes of the membrane.

Figure 10.10

Mechanism of the flavin-dependent glutathione reductase reaction. The first steps, not shown, involve the reduction of FAD to FADH$_2$ by NADPH and the binding of glutathione (glutathione is a sulfhydryl compound, see figure 22.15). The mechanism by which oxidized glutathione is reduced by the E · FADH$_2$ is shown.

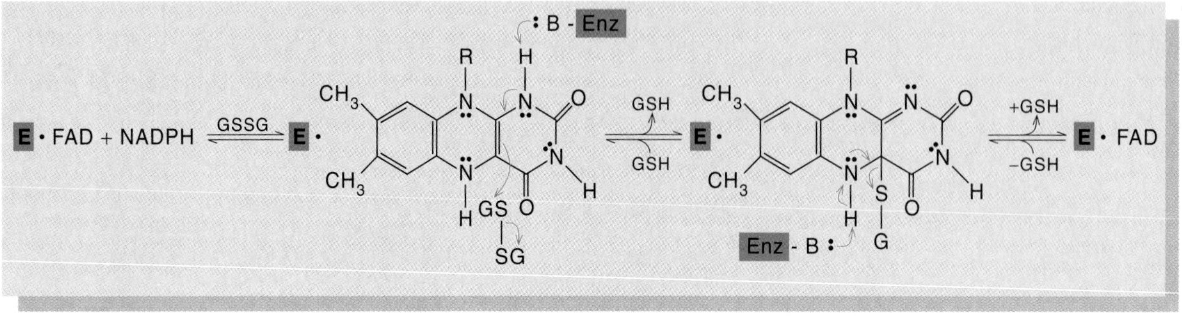

Reactions Requiring Acyl Activation Frequently Use Phosphopantetheine Coenzymes

4'-Phosphopantetheine coenzymes are the biochemically active forms of the vitamin pantothenic acid. In figure 10.11, 4'-phosphopantetheine is shown as covalently linked to an adenylyl group in coenzyme A; or it can also be linked to a protein such as a serine hydroxyl group in acyl carrier protein (ACP). It is also found bonded to proteins that catalyze the activation and polymerization of amino acids to polypeptide antibiotics. Coenzyme A was discovered, purified, and structurally characterized by Fritz Lipmann and colleagues in work for which Lipmann was awarded the Nobel Prize in 1953.

The sulfhydryl group of the β-mercaptoethylamine (or cysteamine) moiety of phosphopantetheine coenzymes is the functional group directly involved in the enzymatic reactions for which they serve as coenzymes. From the standpoint of the chemical mechanism of catalysis, it is the essential functional group, although it is now recognized that phosphopantetheine coenzymes have other functions as well. Many reactions in metabolism involve acyl-group transfer or enolization of carboxylic acids that exist as unactivated carboxylate anions at physiological pH. The predominant means by which these acids are activated for acyl transfer and enolization is esterification with the sulfhydryl group of pantetheine coenzymes.

The mechanistic importance of activation is exemplified by the condensation of two molecules of acetyl-coenzyme A to acetoacetyl-coenzyme A catalyzed by β-ketothiolase:

$$CH_3-\overset{\overset{\displaystyle O}{\|}}{C}-SCoA + CH_3-\overset{\overset{\displaystyle O}{\|}}{C}-SCoA \rightleftharpoons$$

$$CH_3-\overset{\overset{\displaystyle O}{\|}}{C}-CH_2-\overset{\overset{\displaystyle O}{\|}}{C}-SCoA + CoASH \quad (13)$$

The two important steps of the reaction depend on both acetyl groups being activated, one for enolization and the other for acyl-group transfer. In the first step, one of the molecules must be enolized by the intervention of a base to remove an α proton, forming an enolate:

$$B:H-CH_2-\overset{\overset{\displaystyle O}{\|}}{C}-SCoA \rightleftharpoons$$

$$\overset{+}{B}-H + CH_2\!=\!\overset{\overset{\displaystyle O}{\|}}{\underset{}{C}}-SCoA \quad (14)$$

The enolate is stabilized by delocalization of its negative charge between the α carbon and the acyl oxygen atom and by interactions with enzymatic groups, making it thermodynamically accessible as an intermediate. Moreover, this developing charge is also stabilized in the transition state preceding the enolate, so it is also kinetically accessible; that is, it is rapidly formed. If, by contrast, the same enolization reaction were carried out by the acetate anion, it would result in the generation of a second negative charge in the enolate, an energetically and kinetically unfavorable process.

Figure 10.11

Structures of the vitamin pantothenic acid (in red) and coenzyme A. The terminal —SH (in blue) is the reactive group in coenzyme A (CoASH).

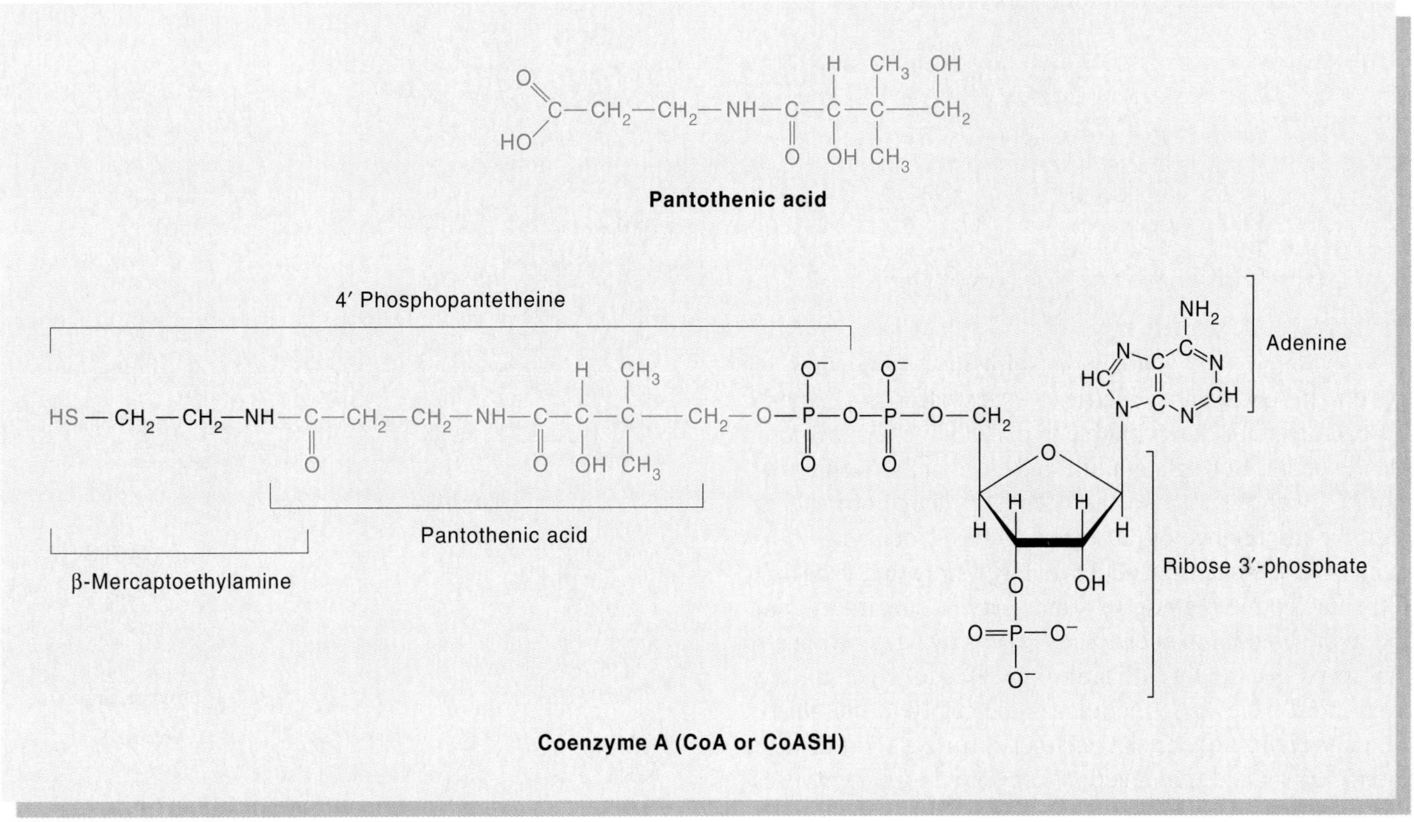

Pantothenic acid

Coenzyme A (CoA or CoASH)

The second stage of the condensation is the reaction of the enolate anion with the acyl group of a second molecule of acetyl-CoA:

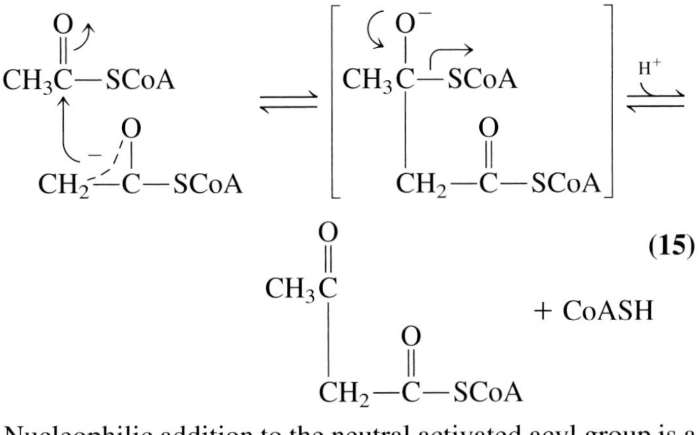

(15)

Nucleophilic addition to the neutral activated acyl group is a favored process, and coenzyme A is a good leaving group from the tetrahedral intermediate. The occurrence of this process with the acetate anion, that is, acetate reacting with an enolate anion, again provides a sharp contrast with the process of equation (15), for it would entail the nucleophilic addition of an anion to an anionic center, generating a dianionic transition state—an unfavorable process from both thermodynamic and kinetic standpoints. Moreover, the resulting intermediate would not have a very good leaving group other than the enolate anion itself, so the transition-state energy for acetoacetate formation would be high. Finally, the K_{eq} for the condensation of 2 mol of acetate to 1 mol of acetoacetate is not favorable in aqueous media, whereas the condensation of 2 mol of acetyl-CoA to produce acetoacetyl-CoA and coenzyme A is thermodynamically spontaneous. The maintenance of metabolic carboxylic acids involved in enolization and acyl-group transfer reactions as coenzyme A esters provides the ideal lift over the kinetic and thermodynamic barriers to these reactions.

The foregoing discussion, in emphasizing the purely electrostatic energy barriers, does not address the question of whether there is an activation advantage in thiol esters relative to oxygen esters. Why thiol esters in preference to oxygen esters? Thiol esters are more readily enolized than oxygen esters. They are more "ketonelike" because of their electronic structures, in which the degree of resonance-electron delocalization from the sulfur atom to the acyl group is less than that of oxygen esters. As a result the charge-separated resonance form is a smaller contributor to the electronic structure in thiol esters than in oxygen esters.

$$\left[\begin{matrix} O \\ \parallel \\ R_1-C-\ddot{S}-R_2 \end{matrix} \longleftrightarrow \begin{matrix} O^- \\ \mid \\ R_1-C=\overset{+}{\underset{..}{S}}-R_2 \end{matrix}\right] \quad (16)$$

$$\left[\begin{matrix} O \\ \parallel \\ R_1-C-\ddot{O}-R_2 \end{matrix} \longleftrightarrow \begin{matrix} O^- \\ \mid \\ R_1-C=\overset{+}{\underset{..}{O}}-R \end{matrix}\right]$$

Although the pantetheine sulfhydryl group has the appropriate chemical properties for activating acyl groups, this characteristic is not unique to pantetheine coenzymes in the biosphere. Both glutathione and cysteine, as well as cysteamine, would serve, so the chemistry does not itself explain the importance of these coenzymes. Coenzyme A has many binding determinants in its large structure, especially in the nucleotide moiety, so it may serve a specificity function in the binding of coenzyme A esters by enzymes. It also may serve as a binding "handle" in cases in which the acyl group must have some mobility in the catalytic site, that is, if it must enolize at one site and then diffuse a short distance to undergo an addition reaction to a ketonic group of a second substrate.

One system in which pantetheine almost certainly performs such a carrier role is the fatty acid synthase from *E. coli,* in which 4′-phosphopantetheine is a component of the acyl carrier protein (see chapter 18).

α-*Lipoic Acid Is the Coenzyme of Choice for Reactions Requiring Acyl-Group Transfers Linked to Oxidation–Reduction*

α-Lipoic acid is the internal disulfide of 6,8-dithiooctanoic acid, the structural formula of which is given in figure 10.12. It is the coupler of electron and group transfers catalyzed by α-keto acid dehydrogenase multienzyme complexes. The pyruvate and α-ketoglutarate dehydrogenase complexes are centrally involved in the metabolism of carbohydrates by the glycolytic pathway (chapter 12) and the tricarboxylic acid cycle (chapter 13). They catalyze two of the three decarboxylation steps in the complete oxidation of glucose, and they produce NADH and activated acyl compounds from the oxidation of the resulting ketoacids:

Figure 10.12

(*a*) Lipoic acid. (*b*) Reduced lipoid acid. (*c*) Lipoic acid bound to the ε-amino group of a lysine residue. Structures shown in red are those centrally involved in coenzyme reactions.

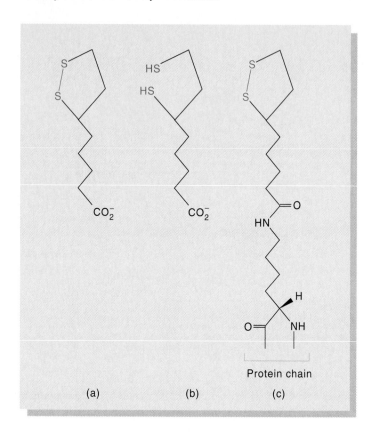

(a)   (b)   (c)

$$R-\overset{\displaystyle O}{\overset{\parallel}{C}}-CO_2^- + NAD^+ + CoASH \longrightarrow$$

$$CO_2 + NADH + R-\overset{\displaystyle O}{\overset{\parallel}{C}}-SCoA \quad (17)$$

The chemical aspect of the coenzymatic action of α-lipoic acid is to mediate the transfer of electrons and activated acyl groups resulting from the decarboxylation and oxidation of α-keto acids within the complexes. In this process, lipoic acid is itself transiently reduced to dihydrolipoic acid (see fig. 10.12), and this reduced form is the acceptor of the activated acyl groups. Its dual role of electron and acyl-group acceptor enables lipoic acid to couple the two processes.

The coenzymatic capabilities of α-lipoyl groups result from a fusion of its chemical and physical properties, the ability to act simultaneously as both electron and acyl-group acceptor, the ability to span long distances to interact with sites separated by up to 2.8 nm, and the ability to act cooperatively with other α-lipoyl groups by disulfide interchange to relay electrons and acyl groups through distances that exceed its reach. This reaction is discussed in chapter 13.

Biotin Mediates Carboxylations

The biotin structure shown in figure 10.13 is an imidazolone ring *cis*-fused to a tetrahydrothiophene ring substituted at position 2 by valeric acid. In carboxylase enzymes, biotin is covalently bonded to the proteins by an amide linkage between its carboxyl group and a lysyl-ϵ-NH_2 group in the polypeptide chain. This arrangement places the imidazolone ring at the end of a long flexible chain of atoms extending a maximum of about 1.4 nm from the α carbon of lysine.

Biotin is the essential coenzyme for carboxylation reactions involving bicarbonate as the carboxylating agent. Several reactions have been described in which ATP-dependent carboxylation occurs at carbon atoms activated for enolization by ketonic or activated acyl groups. One reaction is known in which a nitrogen atom of urea is carboxylated.

A general formulation of the ATP-dependent carboxylation of an α carbon by ^{18}O-enriched bicarbonate is

$$RCH_2\overset{\overset{\displaystyle O}{\|}}{C}-SCoA + ATP + HC^{18}O_3^{-} \xrightarrow{\text{Biotinyl carboxylase}}$$

$$H^+ + R-\underset{\underset{\displaystyle C^{18}O_2^-}{|}}{CH}-\overset{\overset{\displaystyle O}{\|}}{C}-SCoA + ADP + H^{18}OPO_3^{2-} \quad \textbf{(18)}$$

The appearance of ^{18}O in inorganic phosphate verifies that the function of ATP in the reaction is essentially the "dehydration" of bicarbonate.

The ATP-dependent carboxylation of biotin by bicarbonate is believed to control the transient formation of carbonic-phosphoric anhydride, or "carboxyphosphate," as an active carboxylation intermediate:

$$HO-\overset{\overset{\displaystyle O}{\|}}{C}-O^- \underset{ADP}{\overset{ATP}{\rightleftharpoons}} \; ^{-}O-\underset{\underset{\displaystyle OH}{|}}{\overset{\overset{\displaystyle O}{\|}}{P}}-O-\overset{\overset{\displaystyle O}{\|}}{C}-O^- \underset{P_i}{\rightarrow} \quad \textbf{(19)}$$

N^1-carboxybiotinyl-E

Figure 10.13

Structures of biotinyl enzyme and N^1-carboxybiotin. The reactive portions of the coenzyme and the active intermediate are shown in red. In carboxylase enzymes, biotin is covalently bonded to the proteins by an amide linkage between its carboxyl group and a lysyl-ϵ-NH_2 group in the polypeptide chain.

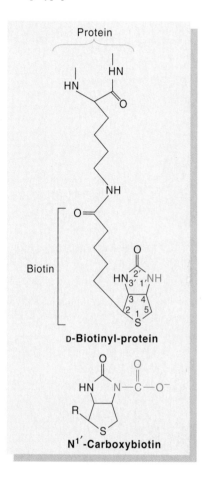

D-**Biotinyl-protein**

$N^{1'}$-**Carboxybiotin**

The coenzymatic function of biotin appears to be to mediate the carboxylation of substrates by accepting the ATP-activated carboxyl group and transferring it to the carboxyl acceptor substrate. There is good reason to believe that the enzymatic sites of ATP-dependent carboxylation of biotin are physically separated from the sites at which N^1-carboxybiotin transfers the carboxyl group to acceptor substrates, that is, the transcarboxylase sites. In fact, in the case of the acetyl-CoA carboxylase from *E. coli* (see chapter 18), these two sites reside on two different subunits, while the biotinyl group is bonded to a third, a small subunit designated biotin carboxyl carrier protein.

Figure 10.14

Structures and enzymatic interconversions of folate coenzymes. The reactive centers of the coenzymes are shown in red. The most active forms of the coenzyme contain oligo- or polyglutamyl groups.

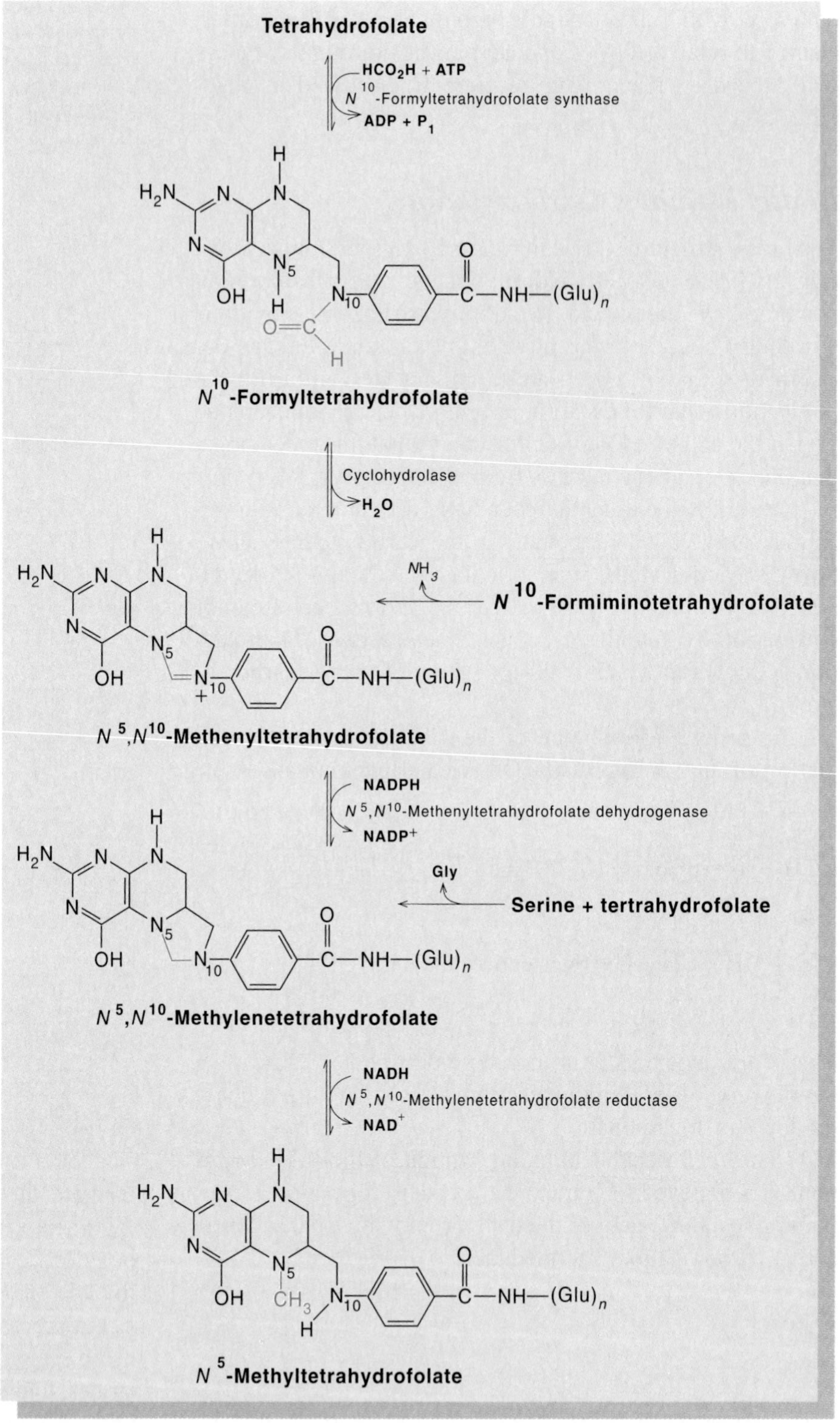

Biotin appears to have just the right chemical and structural properties to mediate carboxylation. It readily accepts activated carboxyl groups at N^1 and maintains them in an acceptably stable yet reactive form for transfer to acceptor substrates. Since biotin is bonded to a lysyl group, the N^1-carboxyl group is at the end of a 1.6-nm chain with bond rotational freedom about nine single bonds, giving it the capability to transport activated carboxyl groups through space from the carboxyl activation sites to the carboxylation sites.

Figure 10.15

Involvement of folate coenzymes in one-carbon metabolism. Shown in red are the one-carbon units of the end products that originate with the reactive one-carbon units of the folate coenzymes.

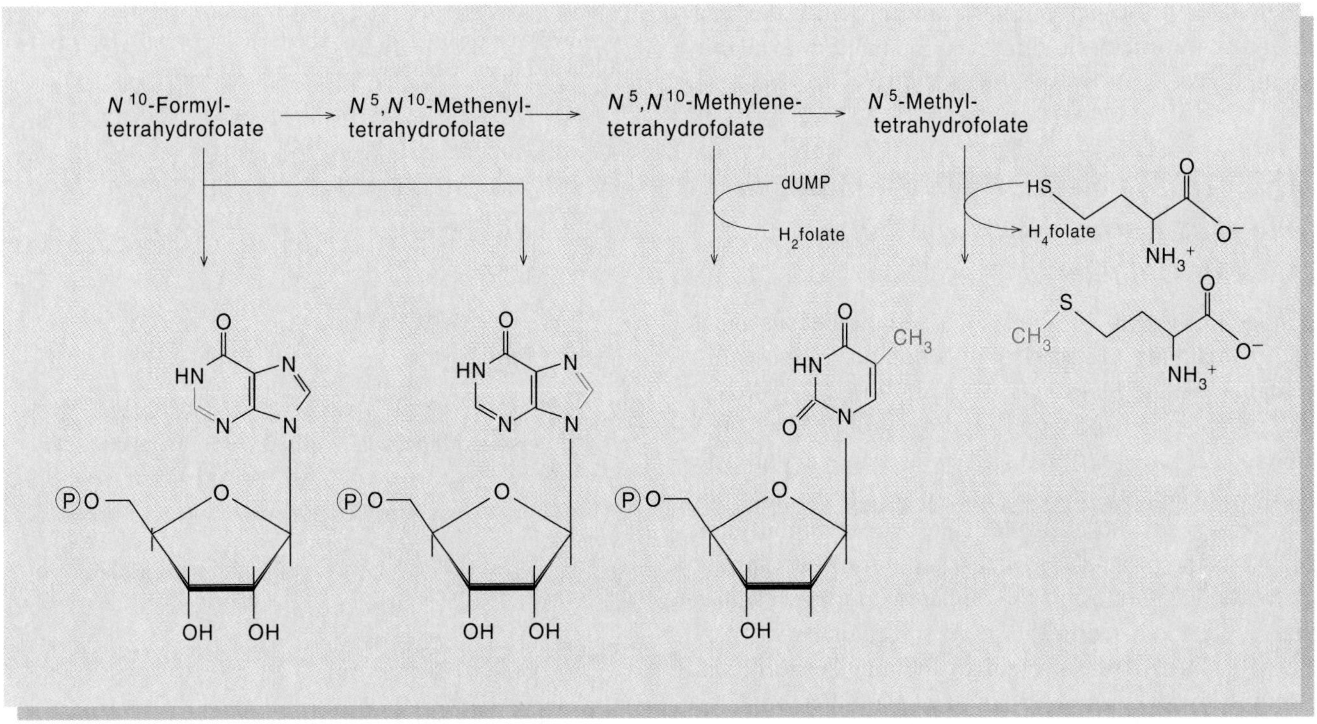

Folate Coenzymes Are Used in Reactions for One-Carbon Transfers

Tetrahydrofolate and its derivatives N^5,N^{10}-methylene-tetrahydrofolate, N^5,N^{10}-methenyltetrahydrofolate, N^{10}-formyltetrahydrofolate, and N^5-methyltetrahydrofolate are the biologically active forms of folic acid, a four-electron oxidized form of tetrahydrofolate. The structural formulas are given in figure 10.14, which also shows how they arise from tetrahydrofolate. The structures are shown glutamylated on the carboxyl group of the p-aminobenzoyl group; the most active forms contain oligo- or polyglutamyl groups, linked through the γ-carboxyl groups.

The tetrahydrofolates do not function as tightly enzyme-bound coenzymes. Rather, they function as cosubstrates for a variety of enzymes associated with one-carbon metabolism. N^{10}-formyltetrahydrofolate is produced enzymatically from tetrahydrofolate and formate in an ATP-linked process in which formate is activated by phosphorylation to formyl phosphate: the formyl group of formyl phosphate is then transferred to N^{10} of tetrahydrofolate. N^{10}-Formyltetrahydrofolate is a formyl donor substrate for some enzymes and is interconvertible with N^5,N^{10}-methenyltetrahydrofolate by the action of cyclohydrolase.

N^5,N^{10}-Methylenetetrahydrofolate is a hydroxymethyl-group donor substrate for several enzymes and a methyl-group donor substrate for thymidylate synthase (fig. 10.15). It arises in living cells from the reduction of N^5,N^{10}-methenyltetrahydrofolate by NADPH and also by the serine hydroxymethyltransferase-catalyzed reaction of serine with tetrahydrofolate.

N^5-Methyltetrahydrofolate is the methyl-group donor substrate for methionine synthase, which catalyzes the transfer of the five-methyl group to the sulfhydryl group of homocysteine. This and selected reactions of the other folate derivatives are outlined in figure 10.15, which emphasizes the important role tetrahydrofolate plays in nucleic acid biosynthesis by serving as the immediate source of one-carbon units in purine and pyrimidine biosynthesis.

Formaldehyde is a toxic substance that reacts spontaneously with amino groups of proteins and nucleic acids, hydroxymethylating them and forming methylene-bridge crosslinks between them. Free formaldehyde therefore wreaks havoc in living cells and could not serve as a useful hydroxymethylating agent. In the form of N^5,N^{10}-methylenetetrahydrofolate, however, its chemical reactivity is attenuated but retained in a potentially available form where needed for specific enzymatic action. Formate, however, is quite unreactive under physiological conditions and must be activated to serve as an efficient formylating agent. As N^{10}-formyltetrahydrofolate it is in a reactive state suitable for transfer to appropriate substrates. The fundamental biochemical importance of tetrahydrofolate is to maintain formaldehyde and formate in chemically poised states, not so reactive as to pose toxic threats to the cell but available for essential processes by specific enzymatic action.

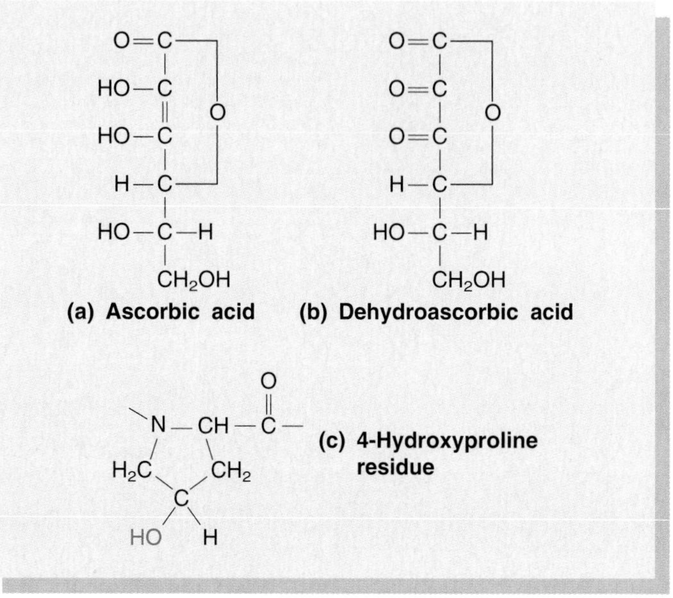

Ascorbic Acid Is Required to Maintain the Enzyme that Forms Hydroxyproline Residues in Collagen

Ascorbic acid (vitamin C; fig. 10.16) is the reducing agent required to maintain the activity of a number of enzymes, most notably proline hydroxylase, which forms 4-hydroxyproline residues in collagen. Hydroxyproline (see fig. 10.16c) is not synthesized biologically as a free amino acid but rather is created by modification of proline residues already incorporated into collagen. The hydroxylation reaction occurs as the protein is synthesized in the endoplasmic reticulum. At least a third of the numerous proline residues in collagen are modified in this way, substantially increasing the resistance of the protein to thermal denaturation.

Proline hydroxylase is a diooxygenase that requires ferrous iron as a cofactor and uses α-ketoglutarate as its second substrate. One oxygen atom from O_2 is incorporated into hydroxyproline; the other goes to the α-ketoglutarate, which decomposes to succinate and CO_2:

$$\text{prolyl-peptide} + {}^-O_2CCH_2CH_2\overset{\overset{\displaystyle O}{\|}}{C}\!-\!CO_2{}^- + O_2 \longrightarrow$$

$$\text{hydroxyprolyl-peptide} +$$

$$\qquad {}^-O_2CCH_2CH_2CO_2{}^- + CO_2 \quad (20)$$

Ascorbic acid is synthesized by plants and many animals but not by primates or guinea pigs. In scurvy, the disease associated with a severe deficiency of ascorbic acid, connective tissues throughout the body deteriorate. Weakening of the capillary walls results in hemorrhages, wounds heal poorly, and lesions occur in the bones.

Vitamin B_{12} Coenzymes Are Associated with Rearrangements on Adjacent Carbon Atoms

The principal coenzymatic form of vitamin B_{12} is 5′-deoxyadenosylcobalamin, the structural formula of which is given in figure 10.17. The structure includes a cobalt–

Figure 10.16

(a) Ascorbic acid is a reductant that appears to be required to keep some enzymes in their active states. Oxidation converts ascorbic acid to (b) dehydroascorbic acid, which decomposes irreversibly by hydrolysis of the lactone ring. One enzyme that requires ascorbic acid is proline hydroxylase, which converts proline residues in collagen to (c) 4-hydroxyproline (hydroxy group shown in red). Hydroxyproline residues help to stabilize the structure of collagen.

carbon bond between the 5′ carbon of the 5′-deoxyadenosyl moiety and the cobalt (III) ion of cobalamin. [Note: The metal oxidation state may be denoted either as Co^{3+} or as Co(III). The former stresses the free-ion character, while the latter stresses the bound character.] Vitamin B_{12} itself is cyanocobalamin, in which the cyano group is bonded to cobalt in place of the 5′-deoxyadenosyl moiety. Other forms of the vitamin have water (aquocobalamin) or the hydroxyl group (hydroxycobalamin) bonded to cobalt.

The vitamin was discovered in liver as the antipernicious anemia factor in 1926, but discovery of its complete structure had to await its purification, chemical characterization, and crystallization, which required more than 20 years. Even then the determination of such a complex structure proved to be an elusive goal by conventional approaches of that day and had to await the elegant x-ray crystallographic study of Lenhert and Hodgkin in 1961, for which Dorothy Hodgkin was awarded the Nobel Prize in 1964.

Most 5′-deoxyadenosylcobalamin-dependent enzymatic reactions are rearrangements that follow the pattern of equation (21), in which a hydrogen atom and another group

Figure 10.17

Structure of 5′-deoxyadenosylcobalamin coenzyme (vitamin B_{12}). The reactive groups are shown in red.

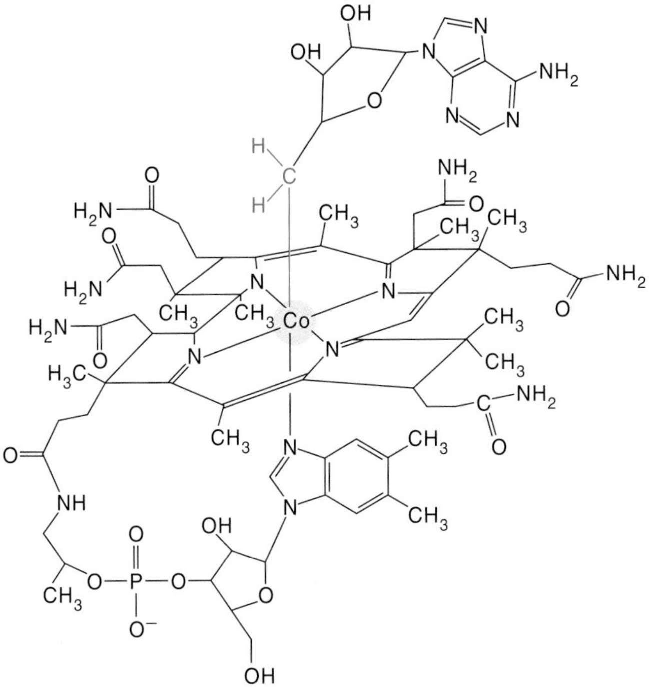

(designated X) bonded to an adjacent carbon atom exchange positions, with the group X migrating from C_α to C_β:

$$a-\underset{\underset{X}{|}}{\overset{\overset{b}{|}}{C}}_\alpha-\underset{\underset{H}{|}}{\overset{\overset{c}{|}}{C}}_\beta-d \rightleftharpoons a-\underset{\underset{H}{|}}{\overset{\overset{b}{|}}{C}}_\alpha-\underset{\underset{X}{|}}{\overset{\overset{c}{|}}{C}}_\beta-d \quad \textbf{(21)}$$

Three specific examples of rearrangement reactions are given in equations (22) through (24). It is interesting and significant that the migrating groups —OH, —COSCoA, and —CH $(NH_3{}^+)CO_2{}^-$ have little in common and that the hydrogen atoms migrating in the opposite direction are often chemically unreactive.

$$\overset{\text{Glutamate mutase}}{\underset{\underset{NH_3{}^+}{|}}{{}^-O_2C-CH_2CH_2-CH-CO_2{}^-}} \rightleftharpoons$$

$$\underset{\underset{NH_3{}^+}{|}}{{}^-O_2C-\underset{\overset{|}{CH_3}}{CH}-CH-CO_2{}^-} \quad \textbf{(22)}$$

$$\overset{\overset{\overset{O}{\|}}{\text{Methylmalonyl-CoA mutase}}}{{}^-O_2C-CH_2CH_2-\overset{\overset{O}{\|}}{C}-SCoA} \rightleftharpoons$$

$$\underset{\overset{|}{CH_3}}{{}^-O_2C-\overset{CH_3\ O}{\underset{\|}{CH}-\overset{O}{C}-SCoA}} \quad \textbf{(23)}$$

$$\underset{\overset{|}{OH}}{CH_3CHCH_2OH} \xrightarrow{\text{Dioldehydrase}} \underset{\overset{|}{OH}}{CH_3CH_2CH-OH} \xrightarrow{\text{Dioldehydrase}}$$

$$CH_3CH_2CHO + H_2O \quad \textbf{(24)}$$

Indeed, hydrogen migrations in all the B_{12} coenzyme-dependent rearrangements proceed without exchange with the protons of water; that is, isotopic hydrogen in substrates is conserved in the products. This fact plus spectroscopic evidence implicating Co(II) and organic radicals as catalytic intermediates in the reaction have led to the proposal of the mechanism illustrated in figure 10.18.

The reaction begins by homolytic cleavage of the Co—C bond (fig. 10.18), generating Co(II) and 5′-deoxyadenosyl free radical (step 1). The radical abstracts a hydrogen atom from the substrate, the migrating hydrogen in equation (21), generating 5′-deoxyadenosine and a sub-strate-derived free radical as intermediates (step 2). The substrate-radical undergoes rearrangement to a product-derived free radical, which abstracts a hydrogen atom to form the final product and regenerate the coenzyme (steps 3–5).

The most fundamental property of 5′-deoxyadenosyl-cobalamin leading to its unique action as a coenzyme is the weakness of the Co—C bond. This bond has a low dissociation energy, less than 30 kcal/mole, strong enough to be essentially stable in free solution but weak enough to be broken as a result of strain induced by multiple binding interactions between the enzyme binding sites and the adenosyl and cobalamin portions of the coenzyme. The radicals resulting from cleavage of this bond and abstraction of hydrogen from substrates undergo the rearrangements characteristic of B_{12}-dependent reactions.

Iron-Containing Coenzymes Are Frequently Involved in Redox Reactions

Iron as a cofactor in catalysis is receiving increasing attention. The most common oxidation states of iron are Fe^{2+} and Fe^{3+}. Iron complexes are nearly all octahedral, and practically all are paramagnetic (as a result of unpaired electrons in the $3d$ orbital). The most common form of iron in biological systems is heme. Heme groups (Fe^{2+}) and hematin (Fe^{3+}) most frequently involve a complex with protoporphyrin IX (fig. 10.19). They are the coenzymes (prosthetic

Figure 10.18

Hypothetical partial mechanism of vitamin B_{12}-dependent rearrangements. The designations Co(III) and Co(II) refer to species that are spectrally and magnetically similar to Co^{3+} and Co^{2+}, respectively. Co(III) is diamagnetic and red, and Co(II) is paramagnetic (unpaired electron) and yellow. The metal does not undergo a change in electrostatic charge when the cobalt–carbon bond breaks homolytically (i.e., without charge separation), because one electron remains with the metal and the other with 5'-deoxyadenosine.

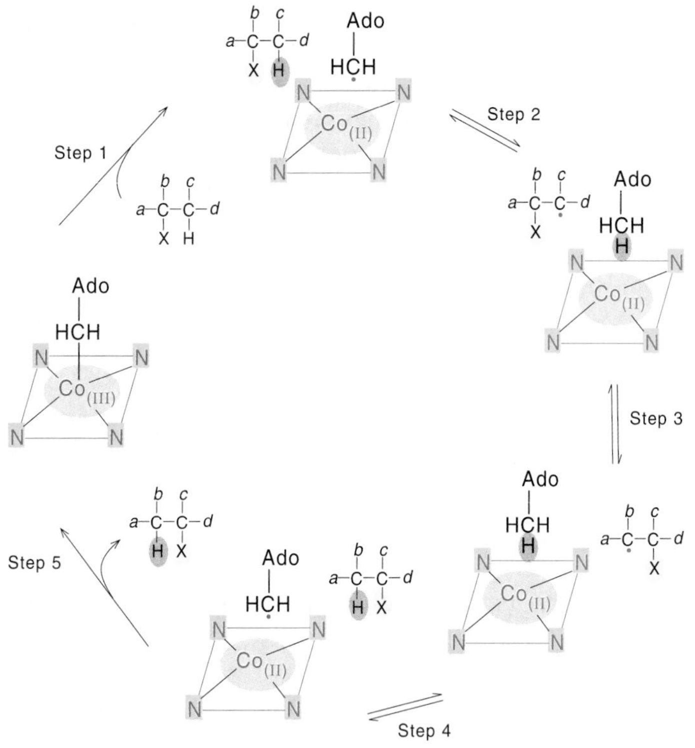

Figure 10.19

Structure of protoporphyrin IX. This coenzyme acts in conjunction with a number of different enzymes involved in oxidation and reduction reactions.

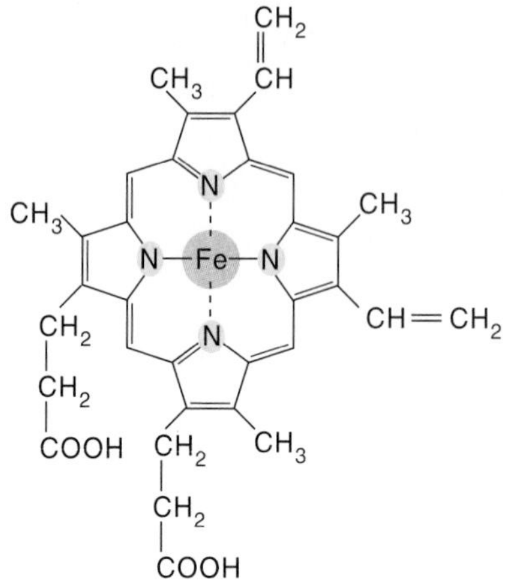

the identities of the upper axial ligands donated by the substrates. Spectral data show clearly that the heme coenzymes participate directly in catalysis; however, the mechanisms of action of hemes are not as well understood as those of other coenzymes.

Many redox enzymes contain iron–sulfur clusters that mediate one-electron transfer reactions. The clusters consist of two or four irons and an equal number of inorganic sulfide ions clustered together with the iron, which is also liganded to cysteinyl-sulfhydryl groups of the protein (fig. 10.20). The enzyme nitrogenase, which catalyzes the reduction of N_2 to $2 NH_3$ contains such clusters in which some of the iron has been replaced by molybdenum (see chapter 21). Electron-transferring proteins involved in one-electron transfer processes often contain iron–sulfur clusters. These proteins include the mitochondrial membrane enzymes NADH dehydrogenase and succinate dehydrogenase (chapter 13), which are flavoproteins, and the small-molecular-weight proteins ferredoxin, rubredoxin, adrenodoxin, and putidaredoxin (chapters 14, 15, and 20).

Heme coenzymes, iron–sulfur clusters, flavin coenzymes, and nicotinamide coenzymes cooperate in multienzyme systems to catalyze the chemically remarkable hydroxylations of hydrocarbons such as steroids (chapter 20). In these hydroxylation systems, the heme proteins constitute a family of proteins known as cytochrome P450, named for the wavelength corresponding to the most intense absorption band of the carbon monoxide-liganded heme, an inhib-

groups) for a number of redox enzymes, including catalase, which catalyzes dismutation of hydrogen peroxide (equation (25)), and peroxidases, which catalyze the reduction of alkyl hydroperoxides by such reducing agents as phenols, hydroquinones, and dihydroascorbate (represented as AH_2 in equation (26)).

$$2 H_2O_2 \rightleftharpoons 2 H_2O + O_2 \qquad (25)$$

$$R—O—O—H + AH_2 \longrightarrow A + R—O—H + H_2O \qquad (26)$$

Heme proteins exhibit characteristic visible absorption spectra as a result of protoporphyrin IX; their spectra differ depending on the identities of the lower axial ligand donated by the protein and the oxidation state of the iron as well as

Figure 10.20

Structures of iron–sulfur clusters. Many redox enzymes contain iron–sulfur clusters that mediate one-electron transfer reactions.

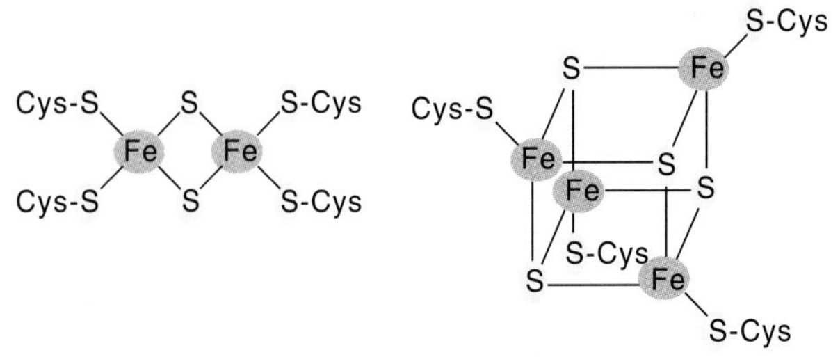

Figure 10.21

Steps in the formation of the oxygenating species of cytochrome P450. The oxygenating species of cytochrome P450 is generated by the transfer of two electrons in discrete steps from NADPH via the flavoprotein reductase and iron–sulfur clusters to the iron porphyrin complex, together with the reaction with oxygen. The Fe(III)-peroxide then undergoes a further expulsion of water to form the oxygenating species shown in brackets. The oxygenating species is not directly observed, since it reacts quickly, but it is thought to contain Fe(IV) and a delocalized radical-cation in the porphyrin ring.

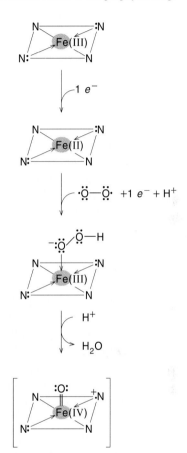

Oxygenating intermediate

ited form. The reactions catalyzed by these systems are represented in generalized form by equation (27), which also shows the fate of the two oxygens from $^{18}O_2$.

$$H^+ + NADPH + {}^{18}O_2 + R{-}\overset{\overset{\displaystyle H}{|}}{\underset{\underset{\displaystyle H}{|}}{C}}{-}R \longrightarrow$$

$$R{-}\overset{\overset{\displaystyle {}^{18}OH}{|}}{\underset{\underset{\displaystyle H}{|}}{C}}{-}R + NADP^+ + H_2{}^{18}O \quad (27)$$

One oxygen of O_2 is incorporated into the hydroxyl group of the product, whereas the other is incorporated into water. The enzymes usually include a cytochrome P450, an iron–sulfur cluster-containing protein, such as adrenodoxin or putidaredoxin, and a flavoprotein reductase.

In the mechanism of oxygenation by these enzymes, the flavoproteins and iron–sulfur proteins supply reducing equivalents in one-electron units from NADPH to cytochrome P450, and reduced cytochrome P450 reacts directly with O_2. These reactions generate a Fe(III)-peroxide complex, shown in figure 10.21, which undergoes a further dehydration to an oxygenating species of cytochrome P450. The oxygenating species is thought to be an oxo-complex of Fe(IV), in which the porphyrin ring has been oxidized to a radical-cation by loss of an electron. The positive charge and unpaired electron in figure 10.21 are stabilized by delocalization through the conjugated π-electron system of the porphyrin ring (see fig. 10.19).

The most widely accepted mechanism for cytochrome P450-catalyzed oxygenation of substrates is illustrated for a hydrocarbon substrate in figure 10.22. The oxygenating species, the oxo-Fe(IV) radical-cation, abstracts a hydrogen

atom from the hydrocarbon to form a hydrocarbon radical and a hydroxy-Fe complex. Hydrogen abstraction proceeds by the pairing of one electron from the oxo-Fe(IV) species and one electron from the carbon–hydrogen bond of the hydrocarbon. In the second step of oxygenation the hydrocarbon radical abstracts the hydroxy group from the hydroxy-Fe(III) complex by a pairing of the unpaired radical electron with one electron of the hydroxyl group, leaving cytochrome P450 in the +3 oxidation state, where it began in figure 10.21. The mechanism shown in figure 10.21 is known as the "rebound" mechanism of oxygenation. The

Figure 10.22

The rebound mechanism for oxygenation of hydrocarbons by cytochrome P450. The oxygenating species of cytochrome P450 in figure 10.21 is a highly reactive paramagnetic species that can abstract an unactivated and unreactive hydrogen from a hydrocarbon in the first step to form a hydrocarbon radical. In the rebound step the hydrocarbon radical abstracts a hydroxyl radical from the hydroxy-porphyrin intermediate.

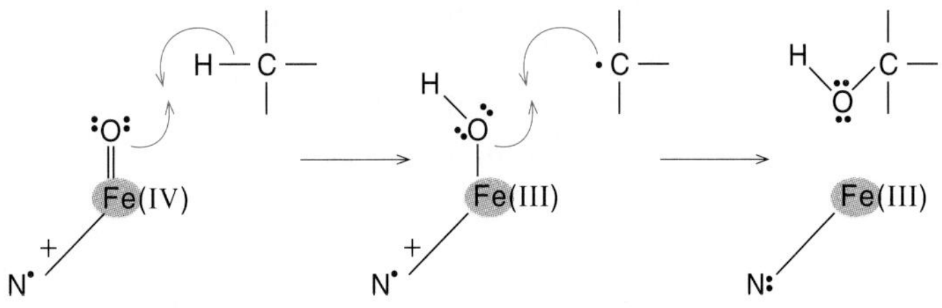

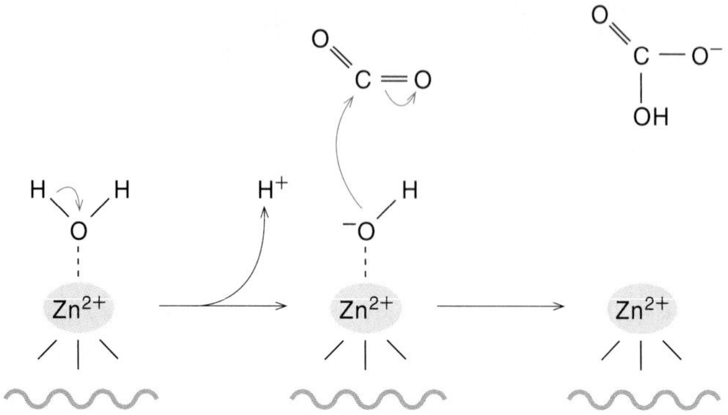

second step of the mechanism is the rebound step, and the rate constant for this step in a microsomal cytochrome P450 has been estimated to be greater than 10^{10} s^{-1}. The magnitude of this rate constant explains why the radical and hydroxy-Fe intermediates have not been detected by presently available spectroscopic methods.

Metal Cofactors

In addition to cobalt and iron (discussed above), other metals frequently function as cofactors in enzyme-catalyzed reactions. Like coenzymes, they are useful because they offer something not available in amino acid side chains. The most important of such features of metals are their high concentration of positive charge, their directed valences for interacting with two or more ligands, and their ability to exist in two or more valence states.

The alkali metal and alkaline-earth metal ions are spherically symmetrical with respect to charge distribution, so they do not usually show directed valences. There are notable exceptions, such as the case of Mg^{2+} in chlorophyll (see chapter 15), which adopts an approximately square planar distribution of orbitals. Alkali metals, Na^+ and K^+, with their single positive charges and lack of d-electronic orbitals for sharing, almost never make tight complexes with proteins. On rare occasions the alkaline-earth divalent cations, Mg^{2+} and Ca^{2+}, can make strong complexes. The alkali metal and alkaline-earth cations are asymmetrically distributed in the organism, with K^+ and Mg^{2+} being concentrated in the cytosol and Na^+ and Ca^{2+} being concentrated in the organelles and the blood.

Most of the remaining metals found in biological systems are from the first transition series, in which the $3d$

Figure 10.23

Proposed role of Zn^{2+} in carbonic anhydrase. Carbonic anhydrase catalyzes the reaction $H_2O + CO_2 \rightleftharpoons HCO_3^- + H^+$. In the enzyme, Zn^{2+} forms a complex with H_2O. Proton displacement generates an OH^- ion still bound to the Zn^{2+}. Nucleophilic attack of CO_2 by the OH^- ion generates bicarbonate.

orbitals are only partially filled. These metals all prefer structures with multiple coordination numbers, and, except for Zn^{2+}, which has a filled $3d$ orbital, they can exist in more than one oxidation state, a feature making them potentially useful in oxidation–reduction reactions. The stability of zinc's electronic state combined with its preference for a coordination number of 4–6 makes zinc singularly useful in enzyme reactions, in which it acts as a Lewis acid (e.g., see fig. 10.23), and in structural situations, in which it chelates nitrogen (usually in histidine side chains) and/or sulfur-containing amino acid side chains (usually cysteines) to rigidify the structural domain of a protein (see fig. 31.19).

The involvement of transition metals in biochemical reactions is discussed in later chapters.

Lipid-Soluble Vitamins

We have seen that most water-soluble vitamins are converted by single or multiple steps to coenzymes. Our understanding of the lipid-soluble vitamins (see table 10.1) and of how they are utilized by the organism is much less extensive. We briefly discuss the structure and functions of some lipid-soluble vitamins.

Vitamin D_3 (cholecalciferol) can be made in the skin from 7-dehydrocholesterol in the presence of ultraviolet light (see fig. 24.13). Vitamin D_3 is formed by the cleavage of ring 3 of 7-dehydrocholesterol. Vitamin D_3 made in skin or absorbed from the small intestine is transported to the liver and hydroxylated at C-25 by a microsomal mixed-

Figure 10.24

Structures of vitamins K_1 and K_2. K_1 is found in plants, K_2 in animals and bacteria.

Vitamin K₁
(phylloqinone)

Vitamin K₂
(menaqinone)

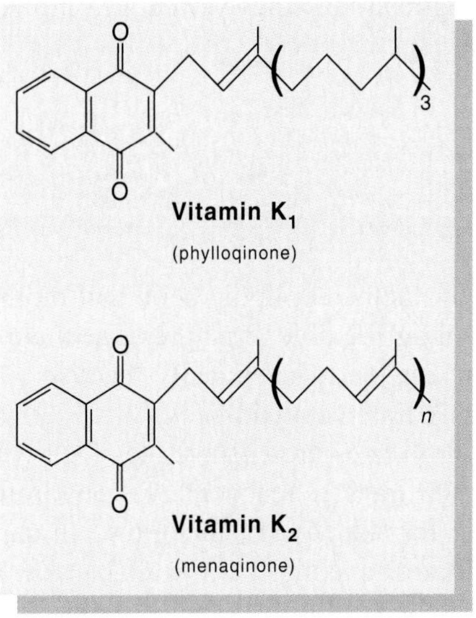

Figure 10.25

Vitamin-K-dependent carboxylation of a glutamic acid residue in a protein. This reaction is essential to blood clotting.

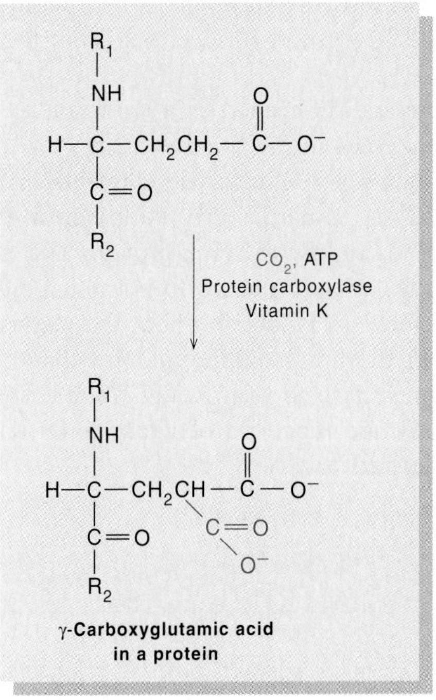

CO₂, ATP
Protein carboxylase
Vitamin K

γ-Carboxyglutamic acid in a protein

Figure 10.26

Structure of vitamin E (α-tocopherol).

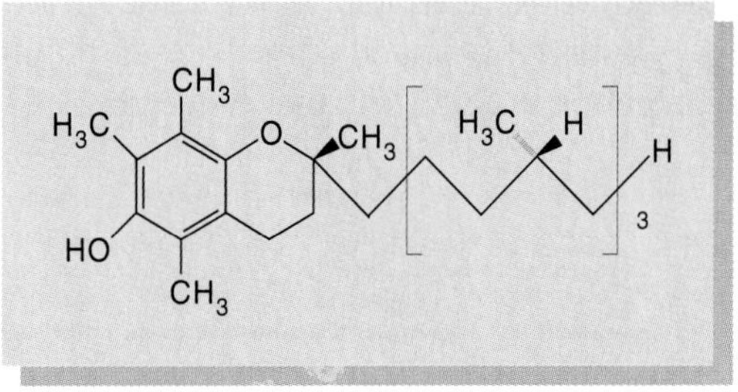

function oxidase. 25-Hydroxyvitamin D_3 appears to be biologically inactive until it is hydroxylated at C-1 by a mixed-function oxidase in kidney mitochondria. The 1,25-dihydroxyvitamin D_3 is delivered to target tissues for the regulation of calcium and phosphate metabolism. The structure and mode of action of 1,25-dihydroxyvitamin D_3 is analogous to that of the steroid hormones (see chapter 24).

Vitamin K was discovered by Henrik Dam in Denmark in the 1920s as a fat-soluble factor important in blood coagulation (K is for "koagulation"). The structures of vitamins K_1 and K_2 (fig. 10.24) were elucidated by Edward Doisy. Vitamin K_1 is found in plants, vitamin K_2 in animals and bacteria. How this vitamin functions in blood coagulation eluded scientists until 1974, when vitamin K was shown to be needed for the formation of γ-carboxyglutamic acid (fig. 10.25) in certain proteins. γ-Carboxyglutamic acid specifically binds calcium, which is important for blood coagulation. Such modified glutamic acid residues appear to be important in many other processes involving calcium transport and calcium-regulated metabolic sequences.

Vitamin E (α-tocopherol) (fig. 10.26) was recognized in 1926 as an organic-soluble compound that prevented sterility in rats. The function of this vitamin still has not been clearly established. A favorite theory is that it is an antioxidant that prevents peroxidation of polyunsaturated fatty acids. Tocopherol certainly prevents peroxidation *in vitro*, and it can be replaced by other antioxidants. However, other antioxidants do not relieve all the symptoms of vitamin E deficiency.

Vitamin A (*trans*-retinol) is called an isoprenoid alcohol because it consists, in part, of units of a single five-carbon compound called isoprene:

Isoprene is also a precursor of steroids and terpenes. This relationship becomes clear when we examine the biosynthesis of these compounds (see chapter 20). Vitamin A is either biosynthesized from β-carotene (see fig. S2.4) or absorbed in the diet. Vitamin A is stored in the liver predominantly as an ester of palmitic acid. For many decades, it has been known to be important for vision and for animal growth and reproduction. The form of vitamin A active in the visual process is 11-*cis*-retinal, which combines with the protein opsin to form rhodopsin. Rhodopsin is the primary light-gathering pigment in the vertebrate retina (see supplement S2).

Summary

Coenzymes are molecules that act in cooperation with enzymes to catalyze biochemical processes, performing functions that enzymes are otherwise chemically not equipped to carry out. Most coenzymes are derivatives of the water-soluble vitamins, but a few, such as hemes, lipoic acid, and iron–sulfur clusters, are biosynthesized in the body. Each coenzyme plays a unique chemical role in the enzymatic processes of living cells.

1. Thiamine pyrophosphate promotes the decarboxylation of α-keto acids and the cleavage of α-hydroxy ketones.
2. Pyridoxal-5′-phosphate promotes decarboxylations, racemizations, transaminations, aldol cleavages, and elimination reactions of amino acid substrates.
3. Nicotinamide coenzymes act as intracellular electron carriers to transport reducing equivalents between metabolic intermediates. They are cosubstrates in most of the biological redox reactions of alcohols and carbonyl compounds and also act as cocatalysts with some enzymes.
4. Flavin coenzymes act as cocatalysts with enzymes in a large number of redox reactions, many of which involve O_2.
5. Phosphopantetheine coenzymes form thioester linkages with acyl groups, which they activate for group transfer reactions.
6. Lipoic acid mediates electron transfer and acyl-group transfer in α-keto acid dehydrogenase complexes.
7. Biotin mediates carboxylation of activated methyl groups.
8. Phosphopantetheine, lipoic acid, and biotin, by virtue of their long, flexible structures, facilitate the physical translocation of chemically reactive species among separate catalytic sites.
9. Tetrahydrofolates are cosubstrates for a variety of one-carbon transfer reactions. Tetrahydrofolates maintain formaldehyde and formate in chemically poised states, making them available for essential processes by specific enzymatic action.
10. Heme coenzymes participate in a variety of electron-transfer reactions, including reactions of peroxides and O_2. Iron-sulfur clusters, composed of Fe and S in equal numbers with cysteinyl side chains of proteins, mediate other electron-transfer processes, including the reduction of N_2 to $2\ NH_3$. Nicotinamide, flavin, and heme coenzymes act cooperatively with iron–sulfur proteins in multienzyme systems that catalyze hydroxylations of hydrocarbons and also in the transport of electrons from foodstuffs to O_2.
11. Ascorbic acid is needed as a reductant to maintain some enzymes in their active forms.
12. Metal ions serve as catalytic elements in some enzymes; Zn^{2+} is particularly important in this regard. Ca^{2+} binds tightly to some proteins and acts to trigger intracellular responses to hormonal signals.
13. In general, less is known about the mechanisms of action of the lipid-soluble vitamins than about the coenzymes derived from water-soluble vitamins. The structures and functions of vitamins D, K, E, and A are discussed briefly.

Selected Readings

Bruice, T. C., and S. J. Benkovic, *Bioorganic Chemistry,* vols. 1 and 2. Menlo Park, Calif.: Benjamin, 1966. A detailed discussion of the mechanisms of bioorganic reactions, including those involving coenzymes.

DiMarco, A. A., T. A. Bobik, and R. S. Wolfe, Unusual coenzymes of methanogenesis. *Ann. Rev. Biochem.* 59:355 (1990). A review of the recently characterized coenzymes required for the biosynthesis of methane in methanogenic bacteria.

Dolphin, D. (ed.), B_{12}, vols. 1 and 2. New York: Wiley-Interscience, 1982. Chemistry and mechanism of action of vitamin B_{12}.

Dolphin, D., R. Poulson, and O. Avamovic (eds.), *Pyridoxal Phosphate,* part A & part B. New York: Wiley-Interscience, 1987. Chemistry and mechanism of action of pyridoxal phosphate.

Dolphin, D., R. Poulson, and O. Avamovic (eds.), *Pyridine Nucleotide Coenzymes,* part A & part B. New York: Wiley-Interscience, 1987. Chemical, biochemical, and medical aspects of pyridine nucleotide coenzymes.

Frausto de Silva, J. J. K., and R. J. P. Williams, *The Biological Chemistry of the Elements.* Oxford: Clarendon Press, 1991.

Frey, P. A., The importance of organic radicals in enzymatic cleavage of unactivated C—H bonds. *Chem. Rev.* 90:1343, 1990. A brief review of coenzymes required to cleave unreactive C—H bonds.

Jencks, W. P., *Catalysis in Chemistry and Enzymology.* New York: McGraw-Hill, 1969. A detailed analysis of mechanisms of enzymatic and nonenzymatic reactions, including those involving coenzymes.

Knowles, J. R., The mechanism of biotin-dependent enzymes. *Ann. Rev. Biochem.* 58:195, 1989. Review of the chemical mechanism of biotin-dependent carboxylation reactions.

Ortiz de Montellano, P. R. (ed.), *Cytochrome P-450: Structure, Mechanism and Biochemistry.* New York: Plenum, 1986. A treatise on the structure and function of cytochrome P450 monooxygenases.

Phipps, D. A., *Metals and Metabolism.* Oxford Chemistry Series. Oxford: Clarendon Press, 1976. Examines the importance of metal ions in metabolic processes.

Popják, G., Stereospecificity of enzymic reactions. In P. D. Boyer (ed.), *The Enzymes,* vol. 2. New York: Academic Press, 1970. p. 115.

Walsh, C. T., *Enzymatic Reaction Mechanisms.* San Francisco: Freeman, 1977. Provides discussion of the mechanisms of enzymatic reactions. An indepth treatment of coenzymes.

Problems

1. Which of the coenzymes listed in table 10.1 can be considered to be derivatives of AMP?

2. What structural features of biotin and lipoic acid allow these cofactors to be covalently bound to a specific protein in a multienzyme complex yet participate in reactions at active sites on other enzymes of the complex?

3. The following reactions are catalyzed by pyridoxal-5'-phosphate-dependent enzymes. Write a reaction mechanism for each, showing how pyridoxal-5'-phosphate is involved in catalysis.

(a)
$$CH_3-\underset{\underset{O}{\|}}{C}-COO^- + R-\underset{\underset{NH_3^+}{|}}{CH}-COO^- \longrightarrow$$

$$CH_3-\underset{\underset{NH_3^+}{|}}{CH}-COO^- + R-\underset{\underset{O}{\|}}{C}-COO^-$$

(b) $H_3^+N-(CH_2)_4-\underset{\underset{NH_3^+}{|}}{CH}-CO_2^- \longrightarrow$

$$CO_2 + H_3^+N-(CH_2)_5-NH_3^+$$

(c) $^-O_2C-CH_2-\underset{\underset{NH_3^+}{|}}{CH}-CO_2^- \longrightarrow$

$$CO_2 + CH_3-\underset{\underset{NH_3^+}{|}}{CH}-CO_2^-$$

4. $NADP^+$ differs from NAD^+ only by phosphorylation of the C-2' OH group on the adenosyl moiety. The redox potentials differ only by about 5 mV. Why do you suppose it is necessary for the cell to employ two such similar redox cofactors?

5. Thiamine-pyrophosphate-dependent enzymes catalyze the reactions shown below. Write a chemical mechanism that shows the catalytic role of the coenzyme.

(a)
$$CH_3-\underset{\underset{O}{\|}}{C}-CO_2^- \longrightarrow CO_2 + CH_3-\underset{\underset{O}{\|}}{C}-H$$

(b)
$$ⓅOCH_2-(CHOH)_2-\underset{\underset{OH}{|}}{CH}-\underset{\underset{O}{\|}}{C}-CH_2OH + HOPO_3^{2-} \longrightarrow$$

$$Ⓟ O-CH_2-(CHOH)_2-CHO$$

$$+ CH_3-\underset{\underset{O}{\|}}{C}-OPO_3^{2-} + H_2O$$

6. What chemical features allow flavins (FAD, FMN) to mediate electron transfer from NAD(P)H to cytochromes or iron–sulfur proteins?

7. What coenzymes would you expect to participate in the following reactions (specifically indicate which are enzyme-bound coenzymes or "substrate/product-like" coenzymes).

(a) 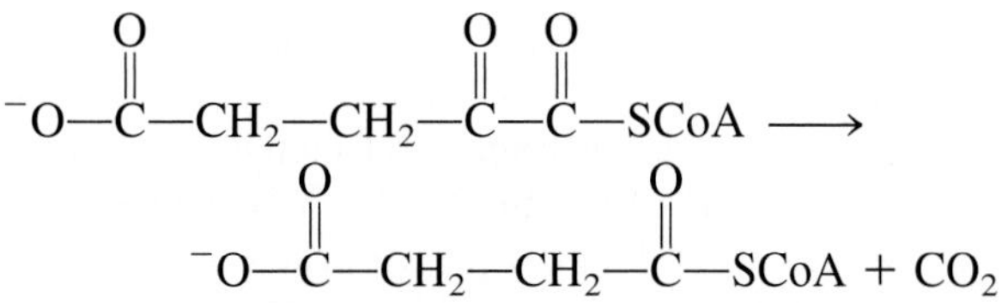 $^-O-\overset{O}{\overset{\|}{C}}-CH_2-CH_2-\overset{O}{\overset{\|}{C}}-\overset{O}{\overset{\|}{C}}-SCoA \longrightarrow$

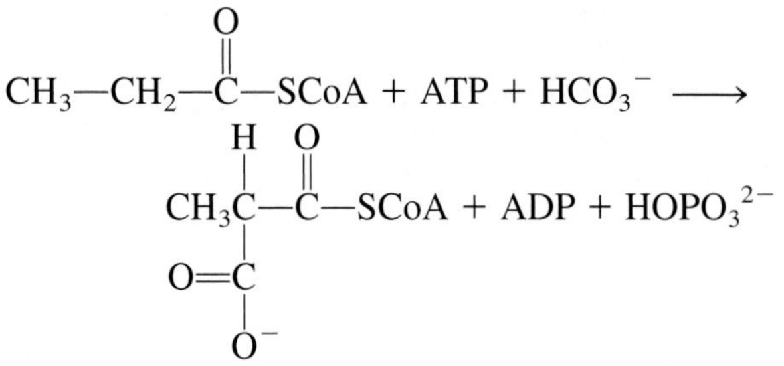

$$^-O-\overset{O}{\overset{\|}{C}}-CH_2-CH_2-\overset{O}{\overset{\|}{C}}-SCoA + CO_2$$

(b) $CH_3-CH_2-\overset{O}{\overset{\|}{C}}-SCoA + ATP + HCO_3^- \longrightarrow$

$$\overset{H}{\underset{|}{C}}H_3C-\overset{O}{\overset{\|}{C}}-SCoA + ADP + HOPO_3^{2-}$$

$$\underset{\underset{O^-}{|}}{O=C}$$

(c) $CH_3-CH_2-CH_2-\overset{O}{\overset{\|}{C}}-SCoA \longrightarrow$

$$CH_3-CH=CH-\overset{O}{\overset{\|}{C}}-SCoA$$

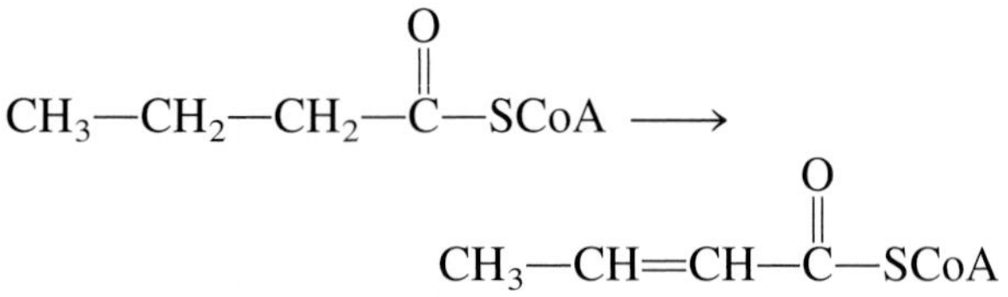

8. Write mechanisms that indicate the involvement of biotin in the following reactions:

(a)

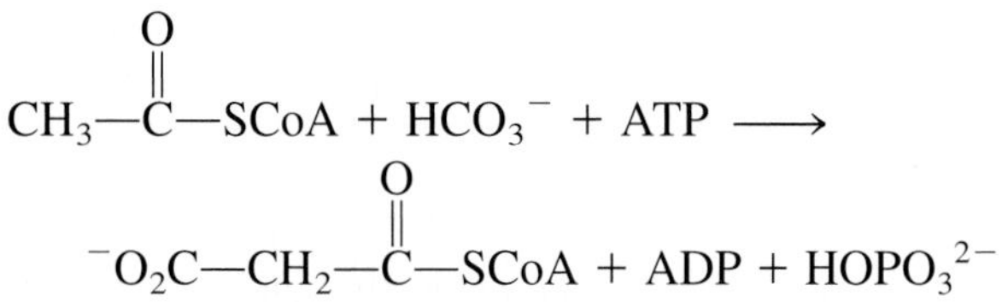

$$CH_3-\overset{O}{\overset{\|}{C}}-SCoA + HCO_3^- + ATP \longrightarrow$$

$$^-O_2C-CH_2-\overset{O}{\overset{\|}{C}}-SCoA + ADP + HOPO_3^{2-}$$

(b)

$$CH_3-CH_2-\overset{O}{\overset{\|}{C}}-SCoA$$

$$+ \ ^-O_2C-CH_2-\overset{O}{\overset{\|}{C}}-CO_2^- \ \rightleftharpoons$$

$$^-O_2C-\underset{\underset{CH_3}{|}}{CH}-\overset{O}{\overset{\|}{C}}-SCoA + CH_3-\overset{O}{\overset{\|}{C}}-CO_2^-$$

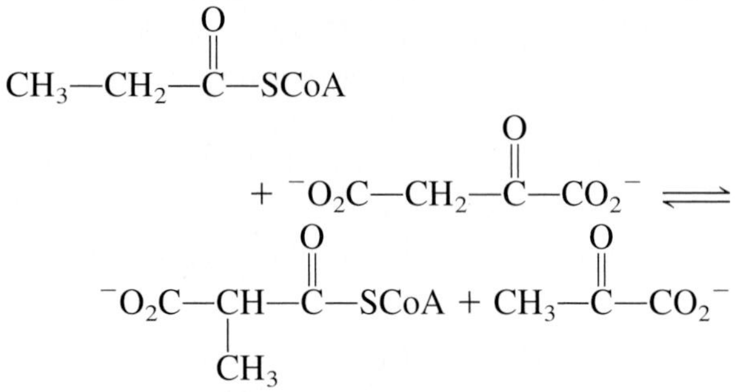

9. (a) What metabolic advantage is gained by having flavin cofactors covalently or tightly bound to the enzyme?

(b) Would covalently bound NAD^+ ($NADP^+$) be a metabolic advantage or disadvantage?

10. Given the amino acid of the general structure

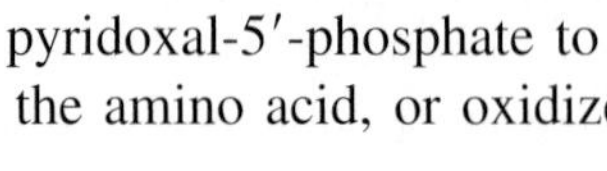

$$CH_3-\underset{\underset{X}{|}}{CH}-\underset{\underset{NH_3^+}{|}}{CH}-COO^-$$

we could use pyridoxal-5'-phosphate to eliminate X, decarboxylate the amino acid, or oxidize the α car-

bon to a carbonyl with formation of pyridoxamine-5'-phosphate. The metabolic diversity afforded by PLP, unchanneled, could wreak havoc in the cell. What other components are required to channel the PLP-dependent reaction along specific reaction pathways?

11. What coenzymes covered in this chapter are (a) biological redox agents, (b) acyl carriers, (c) both redox agents and acyl carriers.

12. For each of the following enzymatically catalyzed reactions, identify the coenzyme involved:

(a) 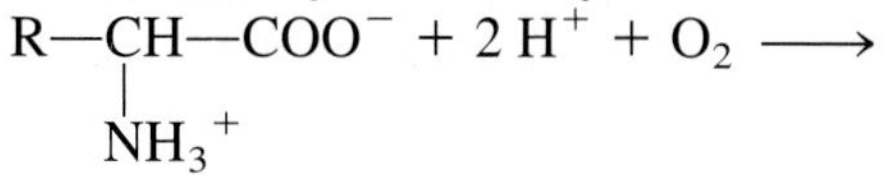 $R-\underset{\underset{NH_3^+}{|}}{CH}-COO^- + 2\,H^+ + O_2 \longrightarrow$

$$R-\overset{O}{\overset{\|}{C}}-COO^- + NH_4^+ + H_2O_2$$

(b) $HO-CH_2-\underset{\underset{NH_3^+}{|}}{CH}-CO_2^- \longrightarrow$

$$CH_3-\overset{O}{\overset{\|}{C}}-CO_2^- + NH_4^+$$

(c) $CH_3-CH_2-\overset{O}{\overset{\|}{C}}-SCoA + HCO_3^- + ATP \longrightarrow$

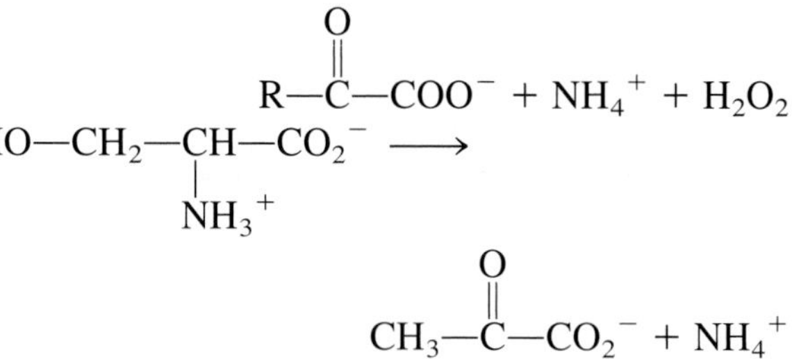

$$^-O_2C-\underset{\underset{CH_3}{|}}{CH}-\overset{O}{\overset{\|}{C}}-SCoA + ADP + P_i$$

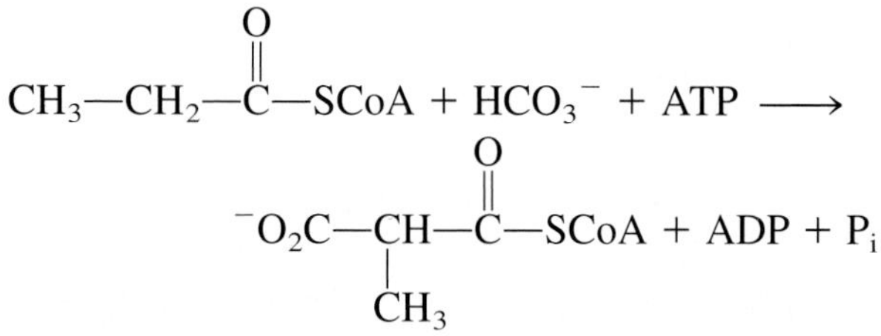

(d)

$$\underset{\underset{O\textcircled{P}}{|}}{CH_2}-\underset{\underset{OH}{|}}{CH}-\underset{\underset{OH}{|}}{CH}-\overset{O}{\overset{\|}{C}}H-\overset{OH}{\overset{|}{C}}-CH_2$$

$$+ CH_2-CH-CHO \rightleftharpoons$$
$$\underset{O\textcircled{P}}{|} \ \ \underset{OH}{|}$$

$$\underset{\underset{O\textcircled{P}}{|}}{CH_2}-\underset{\underset{OH}{|}}{CH}-\underset{\underset{OH}{|}}{CH}-CHO$$

$$+ \underset{\underset{O\textcircled{P}}{|}}{CH_2}-\underset{\underset{OH}{|}}{CH}-\overset{OH}{\overset{|}{C}}H-\overset{O}{\overset{\|}{C}}-\underset{\underset{OH}{|}}{CH_2}$$

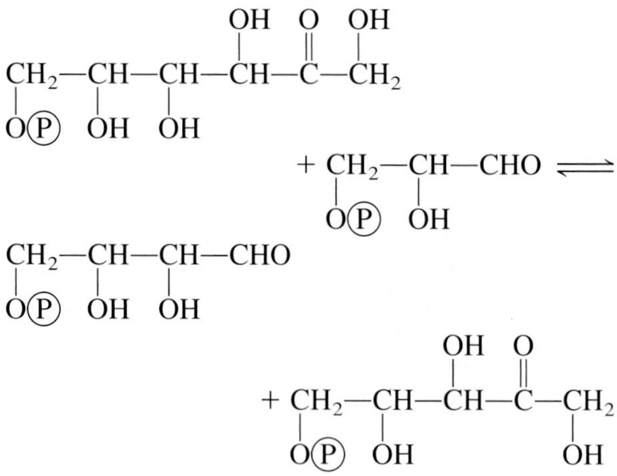

13. The structure of ascorbic acid is shown in figure 10.16. Which hydrogen is the most acidic? Why?

14. Some bacterial toxins use NAD^+ as a true substrate rather than as a coenzyme. The toxins catalyze the transfer of ADP-ribose to an acceptor protein. Examine the structure of NAD^+ and indicate which portion of the molecule is transferred to the protein. What is the other product of the reaction?

Appendices

Comparative Sizes of Biomolecules, Viruses and Cells

Frame 1

Water, Amino Acids, DNA and Protein Structure

Starting at the far left, we see a water molecule, two common amino acids, alanine and tryptophan, a segment of a DNA double helix, a segment of a protein single helix, and the folded polypeptide chain of the enzyme copper, zinc superoxide dismutase or SOD. With respect to the relative sizes of some of these molecules and structures, the water molecule is roughly half a nanometer (nm) across, the DNA and protein helices are about 2 nm and 1 nm in diameter, respectively, and the SOD, a small, globular protein of about 150 amino acids, is about 6 nm in width. SOD catalyzes the breakdown of harmful, negatively charged oxygen radicals, thereby protecting people against neurodegenerative diseases such as Lou Gehrig's disease.

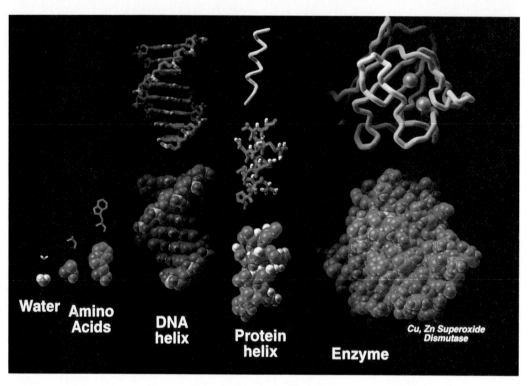

Frame 2

Proteins and Viruses

At the far left, we can see the nucleic acid and protein structures shown in frame 1. In addition, we show a much larger protein, the immunoglobulin G antibody molecule. Four separate polypeptide chains join to make up an antibody molecule: two heavy chains (blue) of about 400 amino acids and two light chains (purple) of about 200 amino acids. The antibody is about 16 nm in width. Finally, at the far right, we show the core particle from a small plant virus, the reovirus. Only the icosahedral protein coat of the virus can be seen. The reovirus particle is about 60 nm across. The nucleic acids of the virus are sequestered inside the virus core. The reovirus family is unusual in that its nucleic acids are all double-stranded RNA molecules.

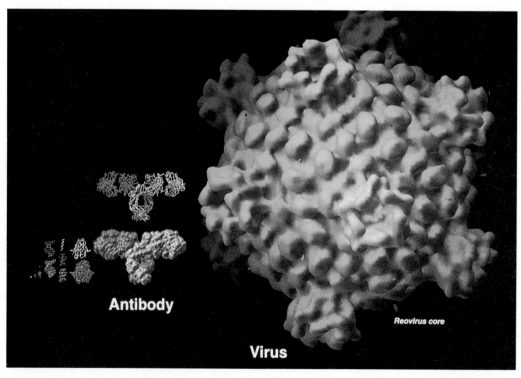

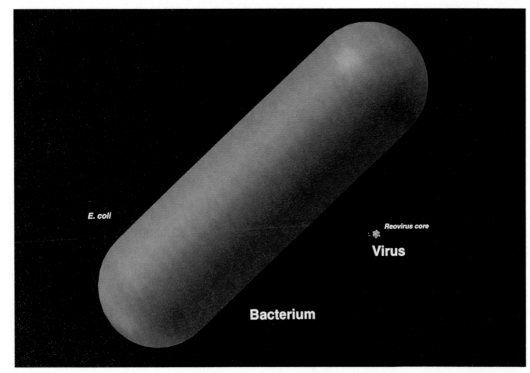

Frame 3

A Bacterial Cell

The common bacterium *E. coli,* shown here in a highly stylized form, is about 2 μm long. The vertebrate gut contains enormous numbers of *E. coli* cells which aid in digestive processes. To the right of the bacterial cell, the same reovirus core particle shown in frame 2 is displayed for size comparison.

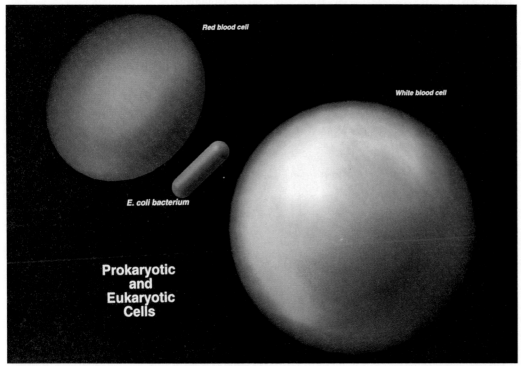

Frame 4

Two Human Cells

Two human blood cells, a white cell about 10 μm across and a red cell about 7 μm across, are shown here in highly stylized representations. The *E. coli* cell from frame 3 is shown between the two human blood cells for size comparison. (Frames 1–4: SOD enzyme by Parge, Hallewell, Getzoff, and Tainer. Antibody by Silverton, Navia, and Davies. Reovirus by Dryden and Yeager. Graphics by Michael Pique using AVS software. Concept by Arthur Olson. Visualization advice by David Goodsell. Images copyright/cW/1994 by The Scripps Research Institute.)

Common Abbreviations in Biochemistry

A	adenine
ACh	acetylcholine
ACTH	adrenocorticotropic hormone
ADP	adenosine-5′-diphosphate
AIDS	acquired immune deficiency syndrome
Ala (A)	alanine
AMP	adenosine-5′-monophosphate
Asn (N)	asparagine
Asp (D)	aspartic acid
ATP	adenosine-5′-triphosphate
BChl	bacteriochlorophyll
bp	base pair
BPG	D-2,3-bisphosphoglycerate
C	cytosine
Cys (C)	cysteine
CAP	catabolite gene activator protein
CDP	cytidine-5′-diphosphate
CMP	cytidine-5′-monophosphate
CTP	cytidine-5′-triphosphate
CoA or CoASH	coenzyme A
CoQ	coenzyme Q (ubiquinone)
cAMP	adenosine 3′,5′-cyclic monophosphate
cGMP	guanosine 3′,5′-cyclic monophosphate
cyt	cytochrome
d	2′-deoxy-
DHAP	dihydroxyacetone phosphate
DHF	dihydrofolate
DHFR	dihydrofolate reductase
DMS	dimethyl sulfate
DNA	deoxyribonucleic acid
cDNA	complementary DNA
DNase	deoxyribonuclease
DNP	2,4-dinitrophenol
ER	endoplasmic reticulum
FAD	flavin adenine dinucleotide (oxidized form)
$FADH_2$	flavin adenine dinucleotide (reduced form)
FBP	fructose-1,6-bisphosphate
fMet	N-formylmethionine
FMN	flavin mononucleotide (oxidized form)
$FMNH_2$	flavin mononucleotide (reduced form)
F1P	fructose-1-phosphate
F6P	fructose-6-phosphate
G	guanine
G protein	guanine-nucleotide binding protein
GDP	guanosine-5′-diphosphate
Gly (G)	glycine
GMP	guanosine-5′-monophosphate
Gln (Q)	glutamine
Glu (E)	glutamic acid
GSH	glutathione
GSSG	glutathionine disulfide
GTP	guanosine-5′-triphosphate
Hb	hemoglobin
HDL	high-density lipoprotein
HETPP	hydroxyethylthiamine pyrophosphate
HGPRT	hypoxanthine-guanosine phosphoribosyl transferase
His (H)	histidine
HIV	human immunodeficiency virus
HMG-CoA	β-hydroxy-β-methylglutaryl-CoA
HPLC	high-performance liquid chromatography
IDL	intermediate-density lipoprotein
IF	initiation factor
IgG	immunoglobulin G
Ile (I)	isoleucine
IMP	inosine-5′-monophosphate
IP_1	inositol-1-phosphate
IP_3	inositol-1,4,5-trisphosphate
IPTG	isopropylthiogalactoside
K_m	Michaelis constant

kb	kilobase pair	RFLP	restriction-fragment length polymorphism
kDa	kilodaltons	RNA	ribonucleic acid
LDL	low-density lipoprotein	hnRNA	heterogeneous nuclear RNA
Leu (L)	leucine	mRNA	messenger RNA
Lys (K)	lysine	rRNA	ribosomal RNA
Man	mannose	snRNA	small nuclear RNA
MHC	major histocompatibility complex	tRNA	transfer RNA
Met (M)	methionine	snRNP	small ribonucleoprotein
NAD$^+$	nicotinamide-adenine dinucleotide (oxidized form)	RNase	ribonuclease
		Ru1,5P	ribulose-1,5-bisphosphate
NADH	nicotinamide-adenine dinucleotide (reduced form)	Ru5P	ribulose-5-phosphate
		R5P	ribose-5′-phosphate
NADP$^+$	nicotinamide-adenine dinucleotide phosphate (oxidized form)	RSV	Rous sarcoma virus
		s	Svedberg constant
NADPH	nicotinamide-adenine dinucleotide phosphate (reduced form)	SAM	S-adenosylmethionine
		SDS	sodium dodecyl sulfate
NDP	nucleoside-5′-diphosphate	Ser (S)	serine
NAM	N-acetylmuramic acid	S7P	sedoheptulose-7-phosphate
NMR	nuclear magnetic resonance	SRP	signal recognition particle
NTP	nucleoside-5′-triphosphate	T	thymine
Phe (F)	phenylalanine	THF	tetrahydrofolate
P$_i$	inorganic orthophosphate	Thr (T)	threonine
PEP	phosphoenolpyruvate	TLC	thin-layer chromatography
PFK	phosphofructokinase	TMV	tobacco mosaic virus
PG	prostaglandin	TPP	thiamine pyrophosphate
2PG	2-phosphoglycerate	Trp (W)	tryptophan
3PG	3-phosphoglycerate	TTP	thymidine-5′-triphosphate
PIP$_2$	phosphatidylinositol-4,5-bisphosphate	Tyr (Y)	tyrosine
PK	pyruvate kinase	U	uracil
PLP	pyridoxal-5-phosphate	UDP	uridine-5′-diphosphate
PP$_i$	inorganic pyrophosphate	UDPG	UDP-glucose
Pro (P)	proline	UMP	uridine-5′-monophosphate
PRPP	phosphoribosylpyrophosphate	UQ	ubiquinone
PS	photosystem	Val (V)	valine
Q	ubiquinone or plastoquinone	VLDL	very-low-density lipoprotein
QH$_2$	ubiquinol or plastoquinol	XMP	xanthosine-5′-monophosphate
RER	rough endoplasmic reticulum	Xu5P	xylulose-5′-phosphate
RF	release factor or replicative form		

Organic Chemistry and Its Relationship to Biochemistry

By definition organic chemistry deals with the chemistry of carbon regardless of its origin. Since biochemistry deals with carbon chemistry only insofar as it concerns living processes, it comprises a distinct subdivision of organic chemistry. In addition to this major difference between organic chemistry and biochemistry, there are two others. The range of conditions used in organic chemistry far exceeds that of biochemistry. Organic chemistry is the study of carbon chemistry in both organic and aqueous solvents, whereas biochemistry is the study of reactions that take place in aqueous solvents. Organic chemistry often involves the study of reaction conditions that are devised by the chemist in the laboratory; biochemistry is concerned exclusively with reactions that occur in living systems. Finally, in organic chemistry the reactions and the reaction conditions often serve no purpose except the ones intended by the chemist. In biochemistry the reactions are functionally related to the needs of the organism.

Regardless of these differences, organic chemistry and biochemistry overlap considerably. As a result, the principles that govern organic chemistry provide a strong foundation for exploring biochemical phenomena. Here we review some of the regions of overlap between the two disciplines. We start with the fundamental properties that affect the chemistry of atoms and molecules. We then consider some of the molecules, their functional groups, their reactions, and the catalysts that accelerate their reactions.

Atoms Are Composed of Protons, Neutrons and Electrons

All atoms contain a centrally located nucleus composed of a mixture of protons and neutrons (except for the hydrogen nucleus, which contains only a single proton). Electrons surround the nucleus in a series of shells. Electrons in the innermost shell are the hardest ones to remove from the atom, and electrons in the outermost shell are the easiest to remove. The outermost shell is called the valence shell because, in most chemical reactions between atoms, electrons are either added to or removed from this shell.

Electrons rotate around the nucleus at high speeds, and determining their exact location at any given time is impossible. Nevertheless, for simplicity, electrons often are depicted as small spheres occupying discrete orbits (fig. 1). A more realistic depiction, the electron orbital model, indi-

Figure A.1

Bohr models of the six most common atoms found in living organisms.

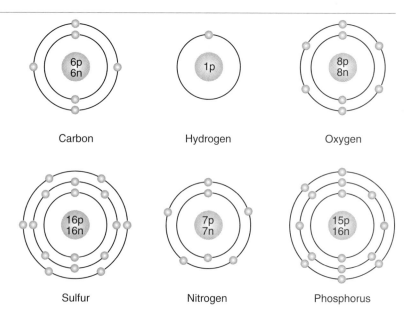

Carbon

Hydrogen

Oxygen

Sulfur

Nitrogen

Phosphorus

Figure A.2

Electron orbitals. Model depicting the volume of space in which an electron is likely to be found 90% of the time. (*a*) The first energy level consists of one spherical orbital containing up to two electrons. The second energy level has four orbitals, each describing the distribution of up to two electrons. One of the orbitals of the second energy level is spherical; the other three are dumbbell-shaped and arranged perpendicular to one another, as the axis lines indicate (*b*). The nucleus is at the center, where the axes intersect.

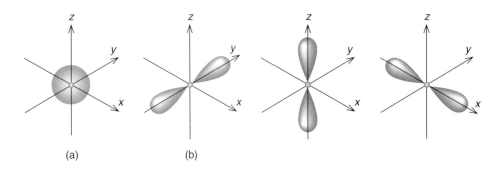

(a)   (b)

cates the volume of space in which a particular electron is likely to be found most of the time (fig. 2). The shell closest to the nucleus consists of a single orbital containing one or two electrons. Both the second and third shells consist of four orbitals, each containing one or two electrons.

Subatomic particles from which all atoms are composed are distinguished by mass and electrical charge. A proton has a mass of one unit and a positive electrical charge of one unit. A neutron also has a mass of one unit, but it has no net charge. An electron has a mass only about 1/1800 that of a nuclear particle, but it has a negative charge equivalent in magnitude to the positive charge of the proton. The net charge and mass of an atom are the sums of the charges and masses of its constituent particles. In calculating mass we can usually ignore the weight of the electrons. Thus an atom's weight approximately equals the total number of protons and neutrons. Another term, atomic number, equals the number of protons in an atom's nucleus.

Atoms Combine to Form Molecules

Except for the inert gases, atoms tend to interact with other atoms to form molecules. Hydrogen, oxygen, and nitrogen each readily form simple diatomic molecules. Invariably, molecules have properties that are quite different from those of the constituent elements. For example, a molecule of sodium chloride contains one atom of sodium (Na) and one atom of chlorine (Cl). Sodium is a highly reactive silvery metal, whereas chlorine is a corrosive yellow gas. When equal numbers of Na and Cl atoms interact, vigorous reaction occurs and white crystalline solid sodium chloride is formed.

Molecules are described by writing the symbols of the constituent elements and indicating the numbers of atoms of each element in the molecule as subscripts. For example, the sugar molecule glucose is represented as $C_6H_{12}O_6$, which indicates that it contains 6 atoms of carbon, 12 atoms of hydrogen, and 6 atoms of oxygen.

Atoms react with one another by gaining, losing, or sharing electrons to produce molecules. The type of chemi-cal bond that forms between atoms depends on the number of electrons the atoms have in their outermost valence shells.

Ionic Bonds Are Formed Between Oppositely Charged Atoms (Ions)

Atoms of the elements prevalent in living things (carbon, nitrogen, oxygen, phosphorus, and sulfur) are most stable chemically when they have eight electrons in their valence shells. This chemical tendency to fill a valence shell is called the octet rule. One way that this is done is for an atom with one, two, or three electrons in this shell to lose them to an atom with, correspondingly, seven, six, or five electrons in the valence shell—a kind of chemical give-and-take. A sodium atom, for example, has a single electron in its outermost shell. As a result, sodium has a strong tendency to lose this single electron because then it will have 8 electrons in its outermost shell. By contrast chlorine has a strong tendency to gain a single electron to fill its valence shell, which contains only seven electrons.

The attraction between oppositely charged ions results in an ionic bond, such as the one that holds NaCl together. The oppositely charged ions Na^+ and Cl^-, attract each other in such an ordered manner that a crystal results (fig. 3).

Covalent Bonds Are Formed Between Atoms That Share Electron Pairs

Atoms that have three, four, or five electrons in their valence shells are more likely to share electrons in a covalent bond than to swap them in the electronic give-and-take of an ionic bond. In both organic chemistry and biochemistry covalent bonds are much more common than ionic bonds. Carbon, with four electrons in its outer shell, can attain the stable eight-electron configuration in its outer shell by sharing electrons with four hydrogen atoms, each of which has one electron in its only shell. The resulting compound containing four single covalent bonds is methane, CH_4 (fig. 4).

Two or three electron pairs can also be shared in covalent bonds called double and triple bonds, respectively. A single atom of carbon forms a double bond with the oxygen

Figure A.3

An ionically bonded molecule (NaCl). (*a*) A sodium atom (Na) can donate the one electron in its valence shell to a chlorine atom (Cl), which has seven electrons in its outermost shell. The resulting ions (Na$^+$ and Cl$^-$) bond to form the compound sodium chloride (NaCl). The octet rule has been satisfied. (*b*) The ions that constitute NaCl form a regular crystalline structure in the solid state.

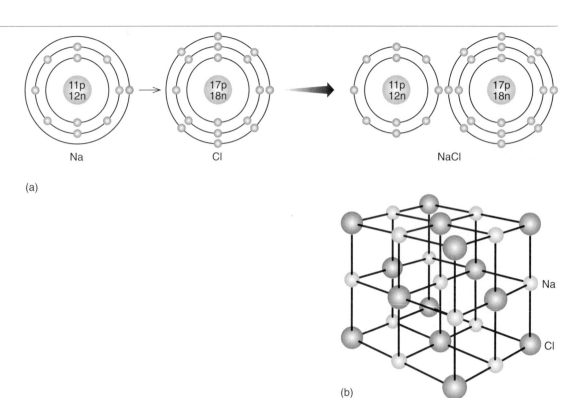

(a)

(b)

in formaldehyde and a triple bond with the nitrogen in hydrogen cyanide (fig. 5).

Weak Forces Lead to Intramolecular or Intermolecular Interactions

The strong interactions that result in either ionic or covalent bonds determine the <u>primary structure</u> of molecules. Molecules often interact, not by forming more bonds of this type,

but rather by forming weaker bonds by polar and apolar interactions. <u>Polar interactions</u> result in polar molecules when the electrons in a covalent bond are not equally shared. In a water molecule (H_2O), for example, the single oxygen atom has a greater affinity for the shared electrons than do the two hydrogen atoms. Thus most of the time the shared electrons are closer to the oxygen atom than to the hydrogen atoms. Because of this unequal sharing, the hy-

Figure A.4

A covalently bonded molecule. By sharing electrons, one carbon and four hydrogen atoms complete their outermost shells to form methane. In carbon, the outermost shell contains four electrons whereas in hydrogen the outermost shell contains only one electron.

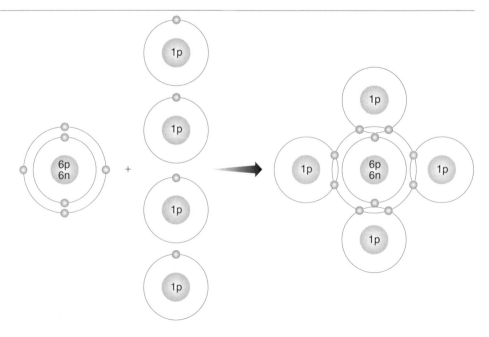

Figure A.5

Covalent linkages formed by carbon with hydrogen, oxygen and nitrogen. Each shared electron pair is represented by two dots or a single straight line.

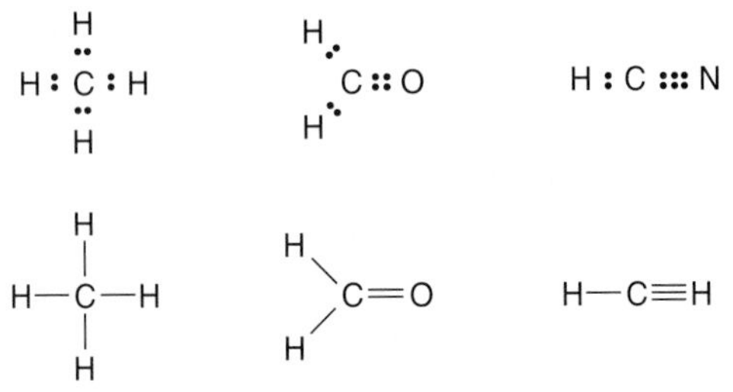

Figure A.6

Two commonly occurring fatty acids (*a*) and comparable hydrocarbons (*b*).

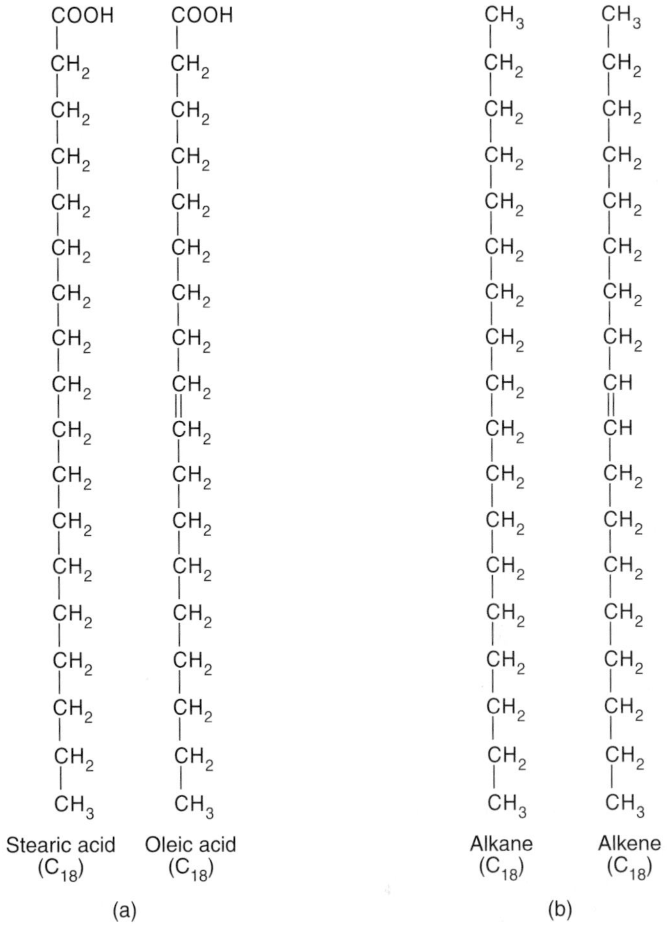

drogen atoms have a partial positive charge and the oxygen atoms have a partial negative charge. This polarity within individual water molecules results in electrostatic interactions between nonbonded H's and O's in liquid water. Apolar molecules are attracted to each other by so-called van der Waals forces. Whereas polar and apolar bonds are much weaker than ionic bonds and covalent bonds, they frequently outnumber the stronger bonds in the liquid state when molecules are nearly close packed. As a result, these weak forces exert a strong influence on the structure of water and the conformation of polymers in an aqueous solvent. Since water itself is polar, a polymer in water tends to fold in a way that exposes its polar groups to the aqueous solvent while burying its apolar groups so that they can interact with each other. In an apolar solvent the exact opposite occurs. The apolar groups on the polymer are exposed for interaction with the apolar organic solvent, whereas the polar groups are buried within the polymer structure so that they can interact with each other.

Four Types of Organic Compounds Are Most Commonly Found in Biochemical Systems

Chemically, the simplest organic compounds found in biochemical systems are the fatty acids (fig. 6). For the most part fatty acids consist of long zig-zag chains with carbon backbones and hydrogen substituents. At one end they usually contain a carboxyl group. In organic chemistry the same compound without the carboxyl group is called an alkane. If two hydrogen atoms are removed from adjacent carbons in a fatty acid, the carbons form a double bond with one another. The comparable organic compound without the carboxyl group is called an alkene. Alkanes and alkenes are highly insoluble in aqueous solvents because of their uniformly apolar character. Fatty acids are partially soluble in aqueous media because of their polar carboxyl group. The carboxyl group also is an important functional group that participates in chemical reactions.

Sugars also possess a carbon backbone but they differ in major respects from fatty acids in that one of the hydrogens attached to each carbon usually is replaced by a hydroxyl group. This change has a dramatic effect on the solubility properties of sugars, making them highly soluble in aqueous solvents. Most sugars are much shorter than fatty acids because of the tendency of sugars to form five- or six-membered rings (fig. 7).

Amino acids are more varied in composition than either fatty acids or sugars (fig. 8). Most amino acids important to biochemistry contain a central carbon with four different substituents: a hydrogen, an amino group, a carboxyl group, and an R group that varies for different types of amino acids. These R groups may be positively charged, contain sulfur, be negatively charged, or be neutral. Neutral R groups are further divided into polar and apolar types. The water solubility of amino acids varies by R group.

At the highest level of complexity in biochemical systems are the nucleotides (fig. 9). These molecules contain three components: a five-carbon sugar to which are attached

Figure A.7

Most sugars contain only 5 or 6 carbons. In aqueous solution, the circular hemiacetal form is strongly favored over the linear form.

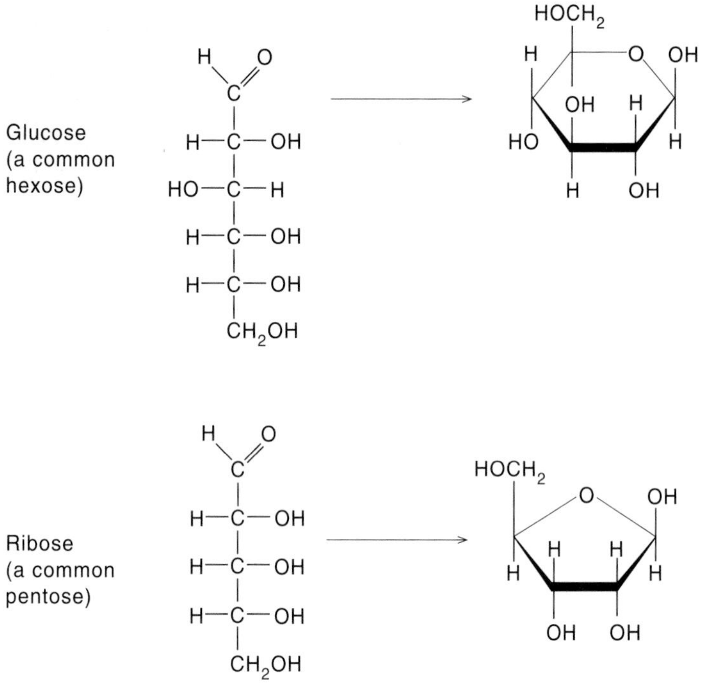

Glucose (a common hexose)

Ribose (a common pentose)

sents a component of a considerably larger structure in living cells. Fatty acids are major components of lipids and membranes. Sugars are major components of polysaccharides and cell walls. Amino acids are the substance from which proteins are synthesized. Nucleotides are the building blocks of nucleic acids and chromosomes.

Most Molecules Found in Cells Are Chiral

When you look into a mirror, you see a reflection that is very similar in appearance to yourself—the main difference being that your right and left sides are switched. No matter which way you turn, you cannot duplicate your mirror image. The same is true of most amino acids, which have four different substituents attached to their α carbon atoms. Amino acids exist in two chiral forms just like you and your mirror image. We can take this analogy a step further. Just as you know that you will never see anybody who looks just like your mirror image, so in biomolecules only one of two chiral forms is usually found. This characteristic is true of most of the biomolecules found in cells.

Functional Groups Are Important in Both Organic Chemistry and Biochemistry

Biomolecules contain a limited number of functional groups that are the reactive centers of the molecules (table 1). Biochemical reactions involving these functional groups are closely related to reactions studied in organic chemistry. Here we describe some of the best known functional groups and the reactions in which they participate.

Alcohols, which contain a hydroxyl functional group, can undergo dehydration reactions with either carboxylic or phosphoric acid to form esters. The double arrows used in this and subsequent equations indicate that the reaction can go in either direction.

a phosphate group on one side and a heterocyclic base on the other side. Because of the phosphate group and the hydroxyl groups on the sugar, nucleotides tend to be highly soluble in water and poorly soluble in apolar organic solvents. Whereas twenty different types of amino acid are commonly found in biochemical systems, only five types of heterocyclic nitrogen bases are commonly found in nucleotides.

Each of the molecules described in this section repre-

Figure A.8

The structure of α-amino acids found in proteins. A central carbon atom is linked to four different substituents (a). The R group substituents are of 20 different types.

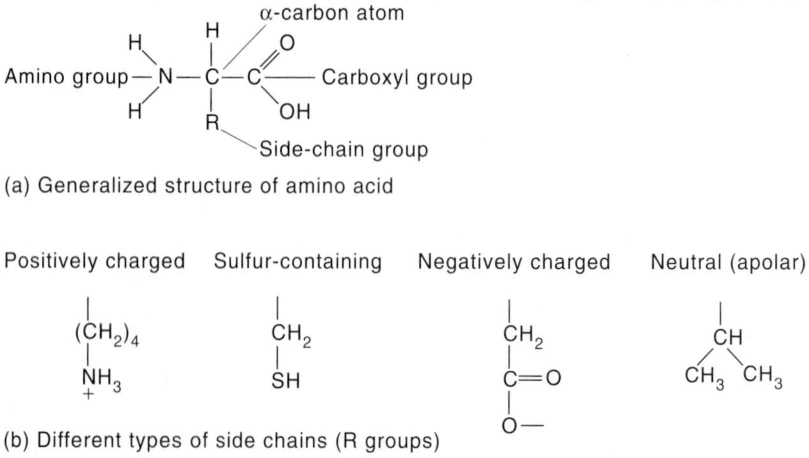

(a) Generalized structure of amino acid

| Positively charged | Sulfur-containing | Negatively charged | Neutral (apolar) |

(b) Different types of side chains (R groups)

Figure A.9

The nucleotide is a three component structure (*a*). There are five different bases commonly found in nucleotides (*b*).

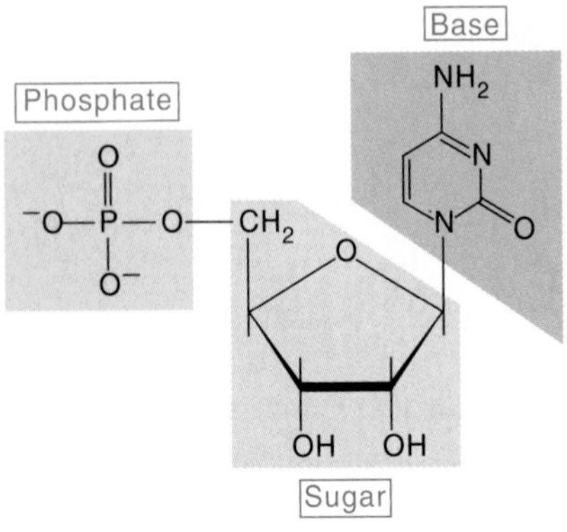

(a) Generalized structure of a nucleotide

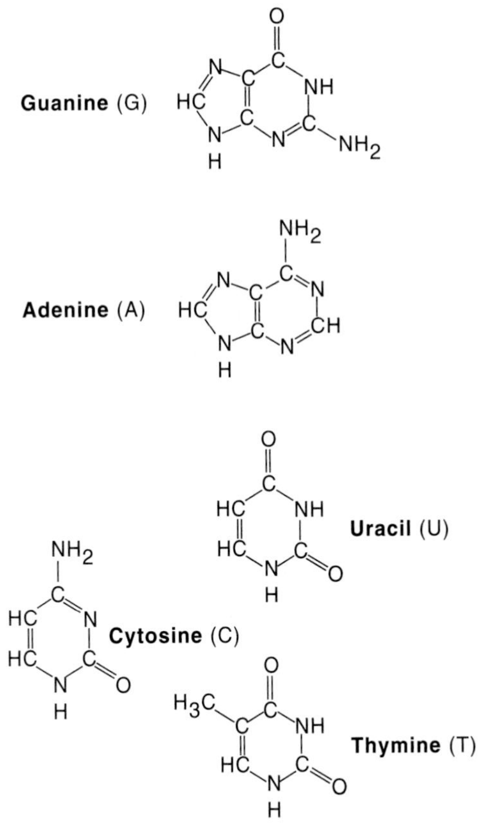

Guanine (G)

Adenine (A)

Uracil (U)

Cytosine (C)

Thymine (T)

(b) Different bases found in nucleotides

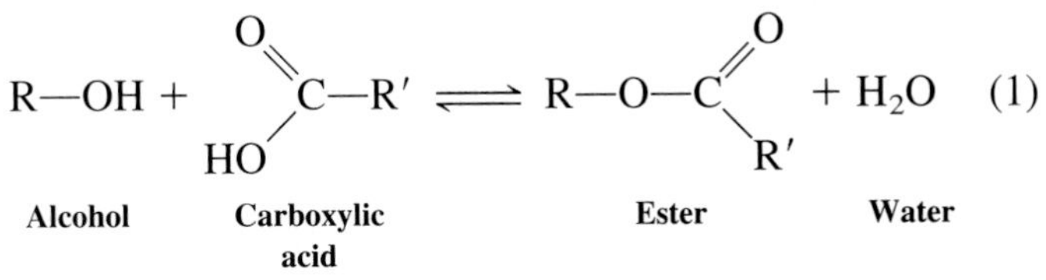

$$R\text{—}OH + \underset{HO}{\overset{O}{\underset{\|}{C}}}\text{—}R' \rightleftharpoons R\text{—}O\text{—}\overset{O}{\overset{\|}{C}}\text{—}R' + H_2O \quad (1)$$

Alcohol   **Carboxylic acid**   **Ester**   **Water**

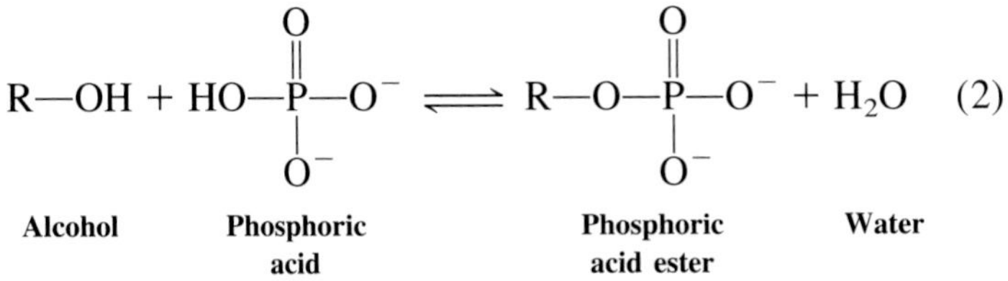

$$R\text{—}OH + HO\text{—}\overset{O}{\underset{O^-}{\overset{\|}{P}}}\text{—}O^- \rightleftharpoons R\text{—}O\text{—}\overset{O}{\underset{O^-}{\overset{\|}{P}}}\text{—}O^- + H_2O \quad (2)$$

Alcohol   **Phosphoric acid**   **Phosphoric acid ester**   **Water**

Thiols contain sulfhydryl groups (—SH) and can substitute for alcohols in some reactions, leading to the formation of thiol esters.

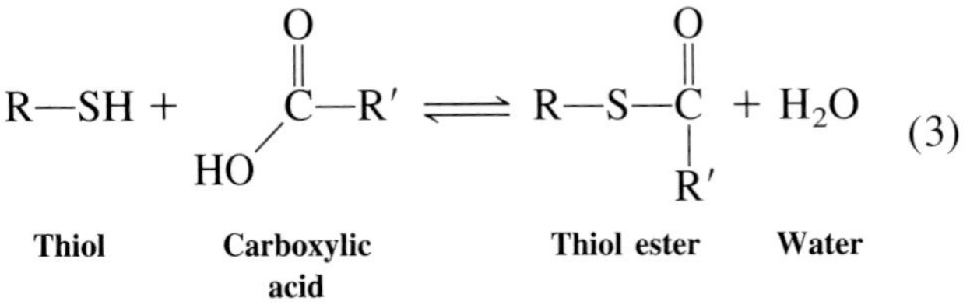

$$R\text{—}SH + \underset{HO}{\overset{O}{\overset{\|}{C}}}\text{—}R' \rightleftharpoons R\text{—}S\text{—}\overset{O}{\overset{\|}{C}}\text{—}R' + H_2O \quad (3)$$

Thiol   **Carboxylic acid**   **Thiol ester**   **Water**

Two alcohols can react with each other to form an ether.

$$R\text{—}OH + HO\text{—}R' \longrightarrow R\text{—}O\text{—}R' + H_2O \quad (4)$$

Alcohol$_1$ **Alcohol$_2$**   **Ether**

Alcohols may undergo a dehydrogenation reaction to form a carbonyl derivative (aldehyde or ketone).

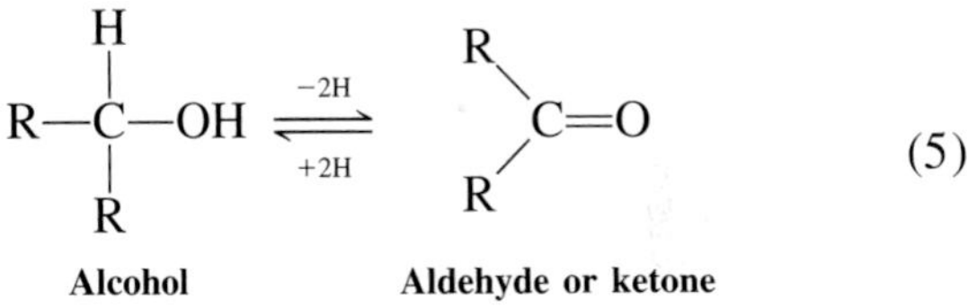

$$R\text{—}\overset{H}{\underset{R}{\overset{\|}{C}}}\text{—}OH \underset{+2H}{\overset{-2H}{\rightleftharpoons}} \overset{R}{\underset{R}{C}}\text{=}O \quad (5)$$

Alcohol   **Aldehyde or ketone**

Amines undergo reactions with carboxylic acids comparable to the formation of esters from alcohols. The product is known as an amide.

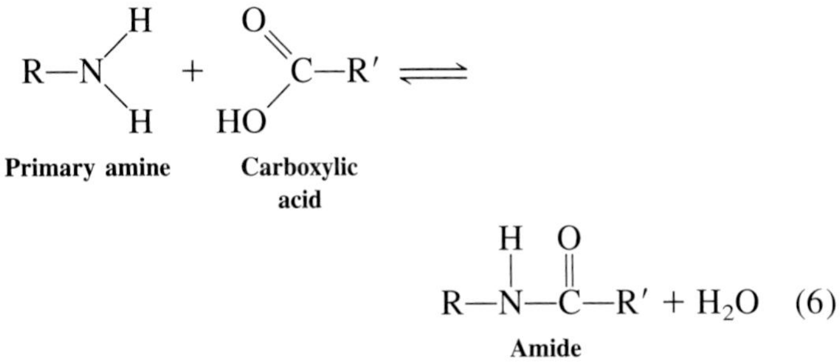

$$R\text{—}\overset{H}{\underset{H}{N}} + \underset{HO}{\overset{O}{\overset{\|}{C}}}\text{—}R' \rightleftharpoons$$

Primary amine   **Carboxylic acid**

$$R\text{—}\overset{H}{\underset{}{N}}\text{—}\overset{O}{\overset{\|}{C}}\text{—}R' + H_2O \quad (6)$$

Amide

Amines may undergo dehydrogenation reactions leading to the formation of imines, which are frequently unstable in water and hydrolyze to ketones, or, in cases where one of the R groups is an H, to aldehydes.

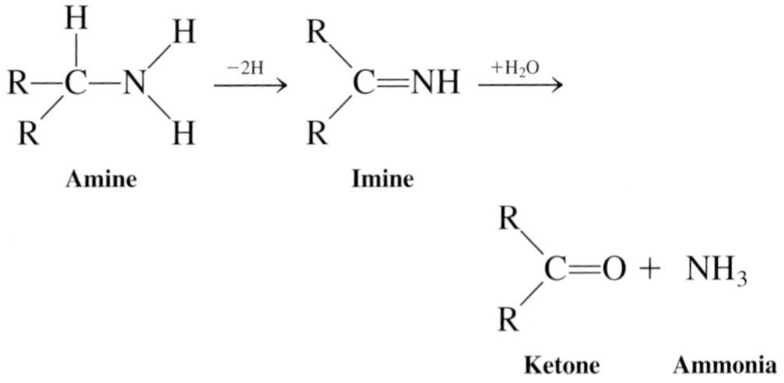

Amine → Imine → Ketone + Ammonia (equation 7):

$$R_2CH\text{-}NH_2 \xrightarrow{-2H} R_2C\text{=}NH \xrightarrow{+H_2O} R_2C\text{=}O + NH_3 \quad (7)$$

Aldehydes and ketones both may be reduced to alcohols by hydrogenation (see the alcohol dehydrogenation reaction, equation 5). Aldehydes may react with either water or alcohol to form aldehyde hydrates or hemiacetals, respectively (also see figure 7 for intramolecular hemiacetals formed by sugars). Reaction of an aldehyde with two molecules of alcohol leads to acetal formation.

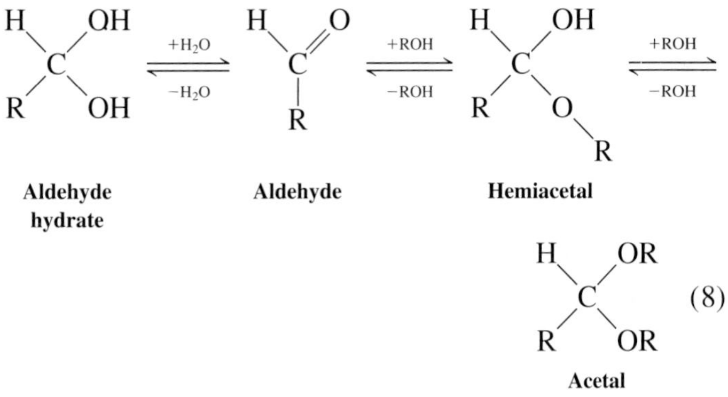

Aldehyde hydrate $\xrightarrow[-H_2O]{+H_2O}$ Aldehyde $\xrightarrow[-ROH]{+ROH}$ Hemiacetal $\xrightarrow[-ROH]{+ROH}$ Acetal (8)

Dehydrogenation of an aldehyde hydrate leads to carboxylic acid formation.

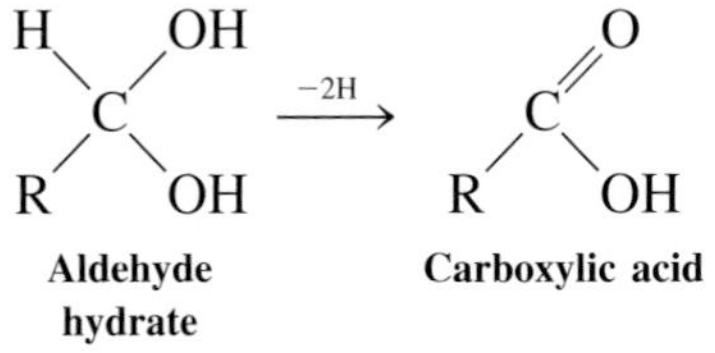

Aldehyde hydrate $\xrightarrow{-2H}$ Carboxylic acid (9)

Aldehydes and ketones may also isomerize to the enol form as long as the adjacent carbon atom is bonded to at least one hydrogen atom. In the reaction a hydrogen migrates and the double bond shifts. At equilibrium the keto form is strongly preferred.

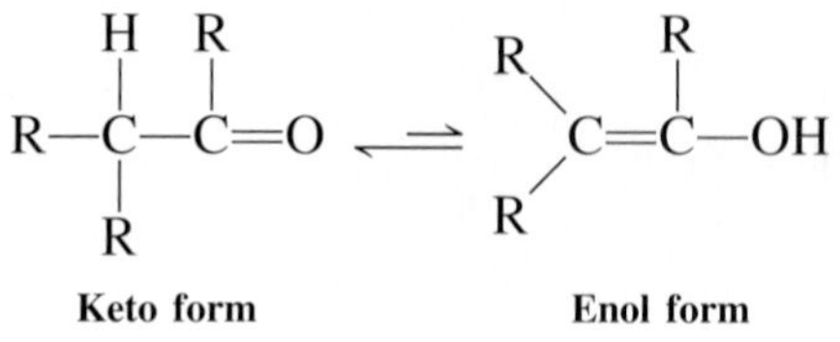

Keto form $\rightleftharpoons$ Enol form (10)

Pyrophosphates may hydrolyze to inorganic phosphoric acid (phosphate) and an organophosphoric acid.

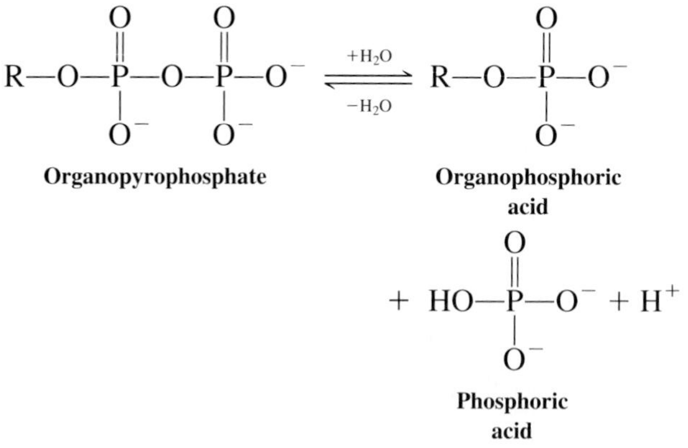

Organopyrophosphate $\xrightarrow[-H_2O]{+H_2O}$ Organophosphoric acid + Phosphoric acid $+ H^+$ (11)

Such hydrolysis reactions yield considerable energy, which can be utilized in biosynthesis.

All the functional groups described are electrostatically neutral in organic solvents. However, in water many of these functional groups either lose or gain protons to become charged species. Such ionization reactions are extremely important in biochemical systems because they frequently influence solubility and reactivity.

Carboxylic and phosphoric acids lose one or more protons in water to become negatively charged. The ionized forms are stabilized by resonance as shown:

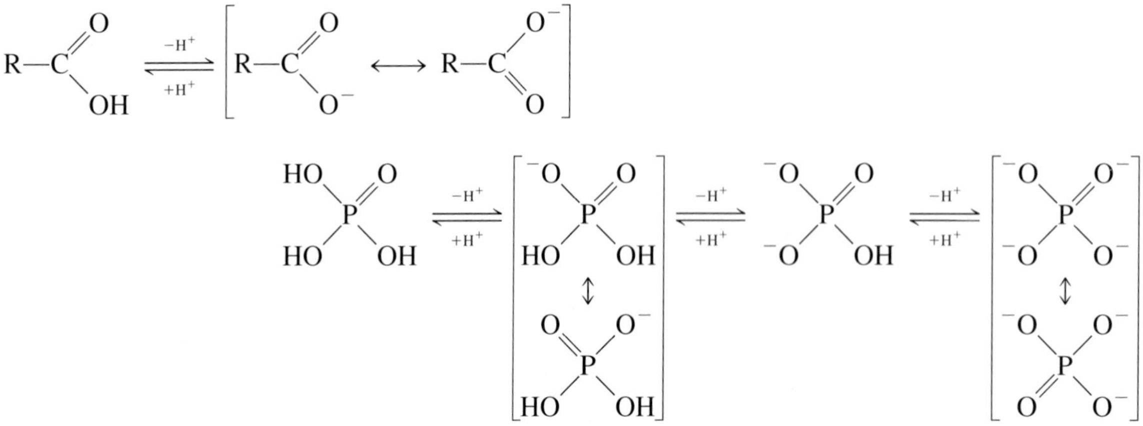

$$R\text{-}COOH \xrightleftharpoons[+H^+]{-H^+} \left[R\text{-}CO_2^- \longleftrightarrow R\text{-}CO_2^- \right]$$

Equation (12): phosphoric acid successive ionizations stabilized by resonance

Table 1

Organic Functional Groups of Biochemical Importance

Type of Compound	General Structural Formula	Characteristic Functional Group	Name of Functional Group	Example Biomolecule with Functional Group	Chapter Reference
Alcohols	R—O—H	—OH	Hydroxyl group	Glycerol (polyhydroxyl) H_2C—OH / H—C—OH / H_2C—OH	17
Ethers	R—O—R'	—O—	Ether	Thromboxane A_2 (cyclic ether)	19
Thiols	R—S—H	—SH	Sulfhydryl	Coenzyme A CoASH	10
Aldehydes	$\overset{O}{\overset{\|}{R-C-H}}$	$\overset{O}{\overset{\|}{-C-H}}$	Aldehyde	Glyceraldehyde-3-phosphate	12

Class			Functional group	Example
Ketones	$R-\overset{\overset{O}{\|}}{C}-R'$	$-\overset{\overset{O}{\|}}{C}-$	Keto	Fructose-6-phosphate 12
Carboxylic acids	$R-\overset{\overset{O}{\|}}{C}-OH$	$-\overset{\overset{O}{\|}}{C}-OH$	Carboxy	Stearic acid 17
Mixed acid anhydride	$R-\overset{\overset{O}{\|}}{C}-O-\overset{\overset{O}{\|}}{\underset{\overset{\|}{OH}}{P}}-OH$	$-\overset{\overset{O}{\|}}{C}-O-\overset{\overset{O}{\|}}{\underset{\overset{\|}{OH}}{P}}-OH$	Phosphoric acid anhydride	Glycerate-1,3-bisphosphate 12
Esters	$R-\overset{\overset{O}{\|}}{C}-O-R'$	$-\overset{\overset{O}{\|}}{C}-O-R'$	Ester	Cholesterol ester 20
Hemiacetals	$R-\overset{\overset{OH}{\|}}{\underset{\overset{\|}{H}}{C}}-O-R'$	$-\overset{\overset{OH}{\|}}{\underset{\overset{\|}{H}}{C}}-O-R'$	Hemiacetal	Glucose 12

Stearic acid: $CH_3(CH_2)_{15}CH_2-C(=O)-OH$

Table 1

Organic Functional Groups of Biochemical Importance (continued)

Type of Compound	General Structural Formula	Characteristic Functional Group	Name of Functional Group	Example Biomolecule with Functional Group	Chapter Reference
Acetals	$\begin{array}{c} O-R'' \\ \mid \\ R-C-O-R' \\ \mid \\ H \end{array}$	$\begin{array}{c} O-R'' \\ \mid \\ -C-O-R' \\ \mid \\ H \end{array}$	Acetal	Sucrose	16
Lactones	(cyclic ester structure)	(cyclic ester structure)	Lactone	6-Phospho-D-gluconolactone	12
Amines	$R-NH_2$	$-NH_2$	Amino	Lysine	3
Amides	$\begin{array}{c} O \quad R''' \\ \parallel \quad \mid \\ R-C-N-R''' \end{array}$	$\begin{array}{c} O \quad R''' \\ \parallel \quad \mid \\ -C-N-R''' \end{array}$	Amido	N-Acetyl-D-glucosamine	16
Alkenes	$\begin{array}{c} R'' \quad\quad R''' \\ \diagdown \quad\quad \diagup \\ C=C \\ \diagup \quad\quad \diagdown \\ R'' \quad\quad R''' \end{array}$	$\diagup C=C \diagdown$	Alkenyl	Palmitoleic acid	17

Note: R, R', R'' are abbreviations for any alkyl or aryl group.
R''' is the abbreviation for any alkyl group or hydrogen.

Amines usually add a proton to become positively charged.

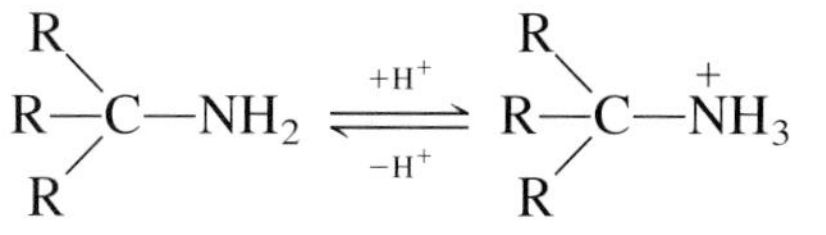

$$R{-}\underset{\underset{R}{|}}{\overset{\overset{R}{|}}{C}}{-}NH_2 \underset{-H^+}{\overset{+H^+}{\rightleftharpoons}} R{-}\underset{\underset{R}{|}}{\overset{\overset{R}{|}}{C}}{-}\overset{+}{N}H_3 \qquad (13)$$

Near neutrality (10^{-7} M H^+), where most biochemical systems function, the carboxyl group exists mainly in the negatively charged form, phosphoric acid exists mainly in the diionized form, and amino groups exist mainly in the positively charged form. These conditions have interesting consequences for amino acids which contain one amino group and one carboxyl group. The amino acids are usually neutral overall, even though they contain two charged groups, one resulting from the deprotonation of the carboxyl group and the other resulting from the protonation of the amino group. Amino acids existing as dipolar ions are called zwitterions.

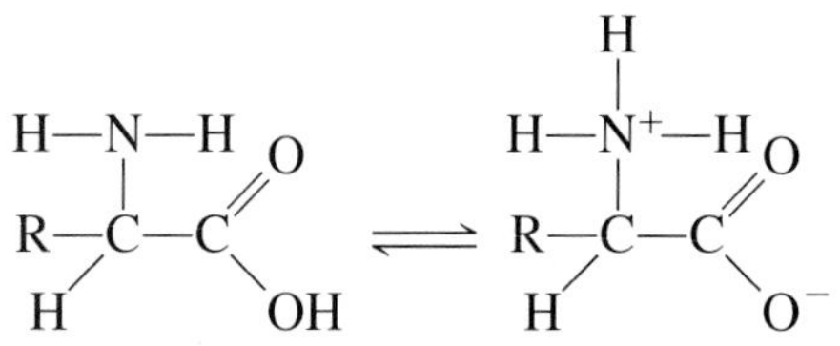

$$\underset{\textbf{Uncharged}}{H{-}\underset{\underset{H}{|}}{\overset{\overset{H-N-H}{|}}{C}}{-}\overset{O}{\underset{OH}{C}}} \rightleftharpoons \underset{\textbf{Zwitterion}}{R{-}\underset{\underset{H}{|}}{\overset{\overset{H-\overset{+}{N}-H}{|}}{C}}{-}\overset{O}{\underset{O^-}{C}}} \qquad (14)$$

The preceding are some of the more important reactions involving covalent bond breakage or formation in biochemistry. By now two things should be apparent about biochemical reactions: (1) the number of reactions in biochemistry is much more limited than in ordinary chemistry; and (2) as far as the reactants and products are concerned, biochemical reactions may be understood in the same terms as ordinary chemical reactions.

Reaction Mechanisms, Arrow Formalism, and Catalysis

So far we have described reactions strictly in terms of their reactants and products. That doesn't tell us much about how or why a reaction proceeds from reactants to products. The reaction mechanism sometimes is quite a complex process, and for many well-known reactions we still don't understand the mechanism. We confine this discussion to consideration of the relatively simple mechanism of ester hydrolysis. Ester hydrolysis entails the reaction of a water molecule with an ester, leading to the production of an alcohol and a carboxylic acid. It is depicted in equation 1, going from right to left.

In attempting to find a mechanism for this or any other reaction we first must recall that all covalent bonds are formed by electrons; the rearrangement or breakage of such bonds starts with the migration of electrons. In the most general sense, reactive groups function either as electrophiles or nucleophiles. The former are electron-deficient substances that are attacked by electron-rich substances. The latter are electron-rich substances that attack electron-deficient substances.

In the case of ester hydrolysis the O of the water is an electron-rich substance and the carbonyl carbon of the ester is an electron-deficient substance. Ester hydrolysis occurs in three stages: (1) the initial stage in which electrons flow from the water molecule to the ester; (2) the intermediate stage in which the ester carbonyl forms a tetrahedral complex involving the hydroxyl group originating from the water molecule that releases a proton; and (3) the final stage in which carboxylic acid and alcohol are formed.

This reaction is kinetically favored because the oxygen of water is a nucleophile, whereas the carbonyl carbon is an electrophile. In the initial step the curved, colored arrows that represent the flow of electron pairs suggest three electron pair migrations that happen in quick succession (figure 10a). An electron pair migrates from the O—H of the water molecule to the O. An electron pair from the O attacks the carbonyl carbon, and an electron pair between the C and the O in the carbonyl migrates to the O of the carbonyl. The intermediate step involves the results of these and two additional electron pair migrations from the relatively unstable intermediate products that lead to the final products.

Normally, ester hydrolysis at ambient temperatures in water at neutral pH will occur very slowly—over a period of many days—unless something is done to accelerate the reaction. Usually, reactions are accelerated by raising the temperature because molecules react more vigorously as the temperature rises. However, a better way to accelerate a reaction is to add a catalyst, that is, a molecule that accelerates the reaction without undergoing any net consumption. Frequently, the main job of the catalyst is to make a potentially reactive center more attractive for reaction, i.e., to potentiate the electrophilic or nucleophilic character of the reactive centers of the reacting species.

In aqueous solution, protons (actually present as hydronium ions) or hydroxide ions are the catalysts most commonly used for nonenzymatic reactions. The way in which acid or base catalysts work in ester hydrolysis is illustrated in Figure 10b and c. As a result of the electronegativity of the oxygen atom in the ester $\rangle C{=}O$ group, the oxygen has a fractional negative charge δ^- and the carbon has a fractional positive charge δ^+. Either acid or base catalysts may be used to accelerate hydrolysis of the ester. In acid cataly-

Figure A.10

Reaction mechanisms for ester hydrolysis in the absence (*a*) and presence of catalysts (*b–e*).

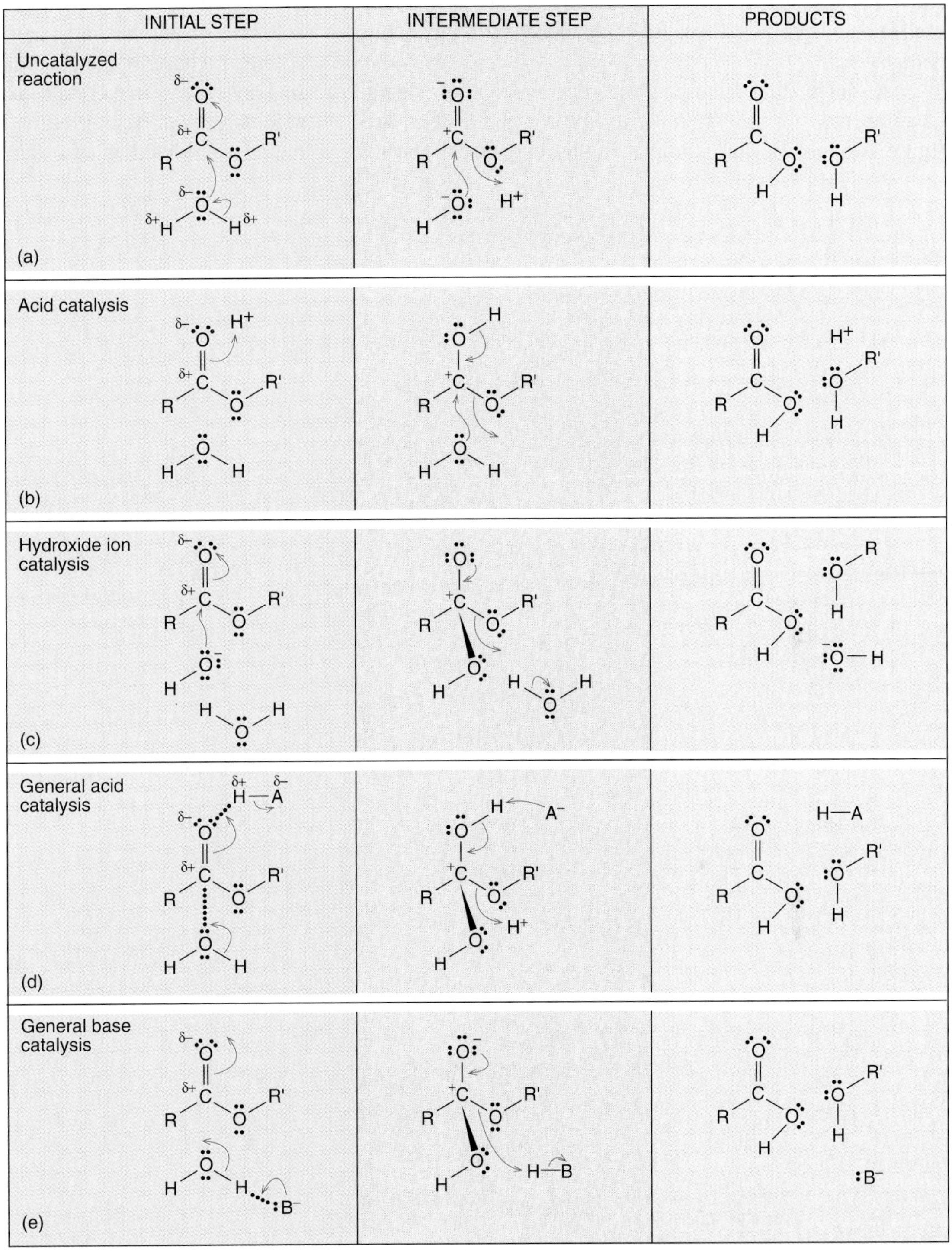

	INITIAL STEP	INTERMEDIATE STEP	PRODUCTS
Uncatalyzed reaction (a)			
Acid catalysis (b)			
Hydroxide ion catalysis (c)			
General acid catalysis (d)			
General base catalysis (e)			

sis, a proton acting as an electrophile is attracted to the oxygen. That leads to an intermediate stage that accentuates the positive charge on the carbon atom, making it a more attractive electrophile to an attacking nucleophile—in this case, water. Water is a poor nucleophile and would attack the carbon slowly without such an inducement, so this step is the key to the catalysis. The remaining reactions leading to ester hydrolysis and regeneration of the catalyst occur

rapidly and spontaneously. In hydroxide ion catalysis of the ester, the carbon atom is attacked directly by a stronger nucleophile, OH^-. Again after hydrolysis, the catalyst is regenerated.

To avoid the harshness of pH extremes, generalized acids or bases frequently are used to replace protons or hydroxide ions as catalysts. These compounds are capable of yielding protons or absorbing protons, respectively, during the course of a reaction. Enzymes almost always utilize generalized acids or bases because of the delicacy of biochemical systems.

An interesting feature of the general acid and general base mechanisms shown is the catalysis of both steps (see figure 10d and e). For example, in the first step of general acid catalysis, HA adds a proton and thus is acting as an acid, but in the second step A$^-$ removes a proton and is acting as a base. In the general base catalysis mechanism the sequence is reversed. Such sequential catalysis by a general acid or general base group is much more common in enzymatic reactions than in ordinary chemical reactions.

Appendix A
Some Landmark Discoveries in Biochemistry

In this appendix we list, in chronological order, some of the most important discoveries made in biochemistry during the past two centuries. It is impossible, for reasons of space, to give credit to every worker who has made a significant contribution, but it is possible to identify certain events as milestones, and thus to show how progress in this field has accelerated with the passage of time.

1770–1774
Priestly showed that oxygen is produced by plants and consumed by animals.

1773
Rouelle isolated urea from urine.

1828
Wohler synthesized the first organic compound, urea, from inorganic components.

1838
Schleiden and Schwann proposed that all living things are composed of cells.

1854–1864
Pasteur proved that fermentation is caused by microorganisms.

1864
Hoppe-Seyler crystallized hemoglobin.

1866
Mendel demonstrated the segregation and independent assortment of alleles in pea plants.

1893
Ostwald showed that enzymes are catalysts.

1898
Camillio Golgi described the Golgi apparatus.

1905
Knoop deduced the β oxidation mechanisms for fatty acid degradation.

1907
Fletcher and Hopkins showed that lactic acid is formed quantitatively from glucose during anaerobic muscle contraction.

1910
Morgan discovered sex-limited inheritance in *Drosophila.*

1912
Warburg postulated a respiratory enzyme for the activation of oxygen.

1913
Michaelis and Menten developed a kinetic theory of enzyme action.

1922
McCollum showed that lack of vitamin D causes rickets.

1926
Sumner crystallized the first enzyme urease.

1926
Jansen and Donath isolated vitamin B_1 (thiamine) from rice polishings.

1926–1930
Svedberg invented the ultracentrifuge and used it to demonstrate the existence of macromolecules.

1928
Levene showed that nucleotides are the building blocks of nucleic acids.

1928
Szent-Gyorgyi isolated ascorbic acid (Vitamin C).

1928–1933
Warburg deduced the iron-prophyrin presence in the respiratory enzyme.

1929
Burr and Burr discovered that linoleic acid is an essential fatty acid for animals.

1931
Englehardt discovered that phosphorylation is coupled to respiration.

1932
Warburg and Christian discovered the ''yellow enzyme,'' a flavoprotein.

1933
Krebs and Henseleit discovered the urea cycle.

1933

Embden and Meyerhof demonstrated the intermediates in the glycolytic pathway.

1935

Schoenheimer and Rittenberg first used isotopes as tracers in the study of intermediary metabolism.

1935

Stanley first crystallized a virus, tobacco mosaic virus.

1937

Krebs discovered the citric acid cycle.

1937

Warburg showed how ATP formation is coupled to the dehydrogenation of glyceraldehyde-3-phosphate.

1938

Hill found that cell-free suspensions of chloroplasts yield oxygen when illuminated in the presence of an electron acceptor.

1939

C. Cori and G. Cori demonstrated the reversible action of glycogen phosphorylase.

1939

Lipmann postulated the central role of ATP in the energy-transfer cycle.

1939–1946

Szent-Gyorgyi discovered actin and the actin-myosin complex.

1940

Beadle and Tatum deduced the one gene–one enzyme relationship.

1942

Bloch and Rittenberg discovered that acetate is the precursor of cholesterol.

1943

Chance applied spectrophotometric methods to the study of enzyme–substrate interactions.

1943

Martin and Synge developed partition chromatography.

1944

Avery, MacLeod, and McCarty demonstrated that bacterial transformation is caused by DNA.

1947–1950

Lipmann and Kaplan isolated and characterized coenzyme A.

1948

Leloir discovered the role of uridine nucleotides in carbohydrate metabolism.

1948

Hogeboom, Schneider, and Palade refined the differential centrifugation method for fractionation of cell parts.

1948

Kennedy and Lehninger discovered that the tricarboxylic acid cycle, fatty acid oxidation, and oxidative phosphorylation all take place in mitochondria.

1949

Christian deDuve discovered lysosomes.

1950–1953

Chargaff discovered the base equivalences in DNA.

1951

Pauling and Corey proposed the α-helix structure for α-keratins.

1951

Lynen postulated the role of coenzyme A in fatty acid oxidation.

1952

Palade, Porter, and Sjostrand perfected thin sectioning and fixation methods for electron microscopy of intracellular structures.

1952–1954

Zamecnik and his colleagues developed the first cell-free systems for the study of protein synthesis.

1953

Vincent du Vigneaud synthesized the first biologically active peptide hormone, ocytocin.

1953

Woodward and Bloch postulated a cyclization scheme for squalene, leading to cholesterol.

1953

Sanger and Thompson determined the complete amino acid sequence of insulin.

1953

Hokin and Hokin showed that acetylcholine induces the rapid biosynthesis of phosphatidylinositol in pigeon pancreas.

1953

Horecker, Dickens, and Racker elucidated the 6-phosphogluconate pathway of glucose catabolism.

1953

Watson and Crick and Wilkins determined the double-helix structure of DNA.

1954

Hugh Huxley proposed the sliding filament model for muscular contraction.

1955

Ochoa and Grunberg-Manago discovered polynucleotide phosphorylase.

1955

Kennedy and Weiss described the role of CTP in the biosynthesis of phosphatidylcholine.

1956

Kornberg discovered the first DNA polymerase.

1956

Umbarger reported that the end product isoleucine inhibits the first enzyme in its biosynthesis from threonine.

1956

Dorothy Crawfoot Hodgkin determined the structure of coenzyme B12.

1956

Ingram showed that normal and sickle-cell hemoglobin differ in a single amino acid residue.

1956

Anfinsen and White concluded that the three-dimensional conformation of proteins is specified by their amino acid sequence.

1956

Leloir determined the pathway to uridine diphosphate glucose (UDPG).

1957

Hoagland, Zamecnik, and Stephenson isolated tRNA and determined its function.

1957

Sutherland discovered cyclic AMP.

1958

Weiss, Hurwitz, and Stevens discovered DNA-directed RNA polymerase.

1958

Meselson and Stahl demonstrated that DNA is replicated by a semiconservative mechanism.

1959

Wakil and Ganguly reported that malonyl-CoA is a key intermediate in fatty acid biosynthesis.

1960

Kendrew reported the x-ray analysis of the structure of myoglobin.

1961

Jacob and Monod proposed the operon hypothesis.

1961

Jacob, Monod, and Changeux proposed a theory of the function and action of allosteric enzymes.

1961

Mitchell postulated the chemiosmotic hypothesis for the mechanism of oxidative phosphorylation.

1961

Nirenberg and Matthaei reported that polyuridylic acid codes for polyphenylalanine.

1961

Marmur and Doty discovered DNA renaturation.

1962

Racker isolated F_1 ATPase from mitochondria and reconstituted oxidative phosphorylation in submitochondrial vesicles.

1966

Maizel introduced the use of sodium dodecylsulfate (SDS) for high-resolution electrophoresis of protein mixtures.

1966

Crick proposed the wobble hypothesis.

1966

Gilbert and Muller-Hill isolated the lac repressor.

1968

Glomset proposed the theory of reverse cholesterol transport in which HDL is involved in the return of cholesterol to the liver.

1968

Meselson and Yuan discovered the first DNA restriction enzyme. Shortly thereafter Smith and Wilcox discovered the first restriction enzyme that cuts DNA at a specific sequence.

1969

Zubay and Lederman developed the first cell-free system for studying the regulation of gene expression.

1970

Howard Temin and David Baltimore discovered reverse transcriptase.

1971

Vane discovered that aspirin blocks the biosynthesis of prostaglandins.

1972

Jon Singer and Garth Nicolson proposed the fluid mosaic model for membrane structure.

1973

Cohen, Chang, Boyer, and Helling reported the first DNA cloning experiments.

1975

Brown and Goldstein described the low-density lipoprotein receptor pathway.

1975

Sanger and Barrell developed rapid DNA-sequencing methods.

1976

Michael Bishop and Harold Varmus discovered the c-src gene in uninfected cells, which is homologous to the v-src gene in the Rous sarcoma virus.

1977

Starlinger discovered the first DNA insertion element.

1977

McGarry, Mannaerts, and Foster discovered that malonyl-CoA is a potent inhibitor of β oxidation.

1977

Splicing of RNA simultaneously discovered in Broker's and Sharp's laboratories.

1977

Nishizuka and coworkers reported the existence of protein kinase C.

1978

Shortles and Nathans did the first experiments in directed mutagenesis.

1978

Tonegawa demonstrated DNA splicing for an immunoglobulin gene.

1981

Cech discovered RNA self-splicing.

1981

Steitz determined the structure of CAP protein.

1981–1982

Palmiter and Brinster produced transgenic mice.

1983

Mullis amplified DNA by the polymerase chain reaction (PCR) method.

1984

Schwartz and Cantor developed pulsed field gel electrophoresis for the separation of very large DNA molecules.

1984

Michel, Deisenhofer, and Huber determined the structure of the photosynthetic reaction center.

1984

Blobel discovered the mechanism for protein translocation across the endoplasmic reticulum membrane—the signal hypothesis.

1988

Elion and Hitchings shared the Nobel Prize for design and synthesis of therapeutic purines and pyrimidines.

1989

Synder and colleagues purified and reconstituted the inositol-1,3,4-P_3 receptor.

Appendix B
A Guide to Career Paths in the Biological Sciences

After You Receive Your Bachelor's Degree, What's Next?

When you receive your bachelor's degree in the biological sciences (e.g., biology, biochemistry, microbiology, etc.), you may choose to continue your education by attending graduate school, medical school, or some other professional school, or you may want to look for a job. If you are interested in going on to graduate school for a masters (MS) or doctorate (Ph.D.) degree in the biological sciences or to medical school for an MD or MD/Ph.D. degree, you can usually obtain information on specific schools and programs from your faculty advisor, on-campus career center, or library.

Job opportunities are available for bachelor's level biological sciences graduates if you are interested in finding employment directly after graduation. Because majors and job titles don't always match and colleges and universities don't typically have employment services, finding employment most often requires quite a bit of research on your part. Extensive job descriptions for virtually every type of position available are provided in *The Dictionary of Occupational Titles,* published by the U.S. Department of Labor. Also, research and reference firms such as Peterson's Guides, Inc., publish guides to job opportunities for science graduates.

Job Opportunities for Bachelor's Level Biological Science Graduates

As indicated in *Peterson's Job Opportunities for Engineering, Science, and Computer Graduates 1993,* * job titles listed in employment advertisements can be categorized by functional occupational area. They are accounting/finance, administration, information systems/processing, marketing/sales, production/operations, research/development, and technical/professional services. All these functional areas are not directly applicable to people with degrees in the biological sciences, but thinking about jobs in terms of these areas can help broaden your career options.

The most applicable functional occupational areas for people with a degree in the biological sciences are mar-

keting/sales, production/operations, research/development, and technical/professional services. A large number of companies and other organizations have entry-level positions available for bachelor's level biological sciences graduates, including the following.

Marketing / Sales

American Cyanamid Company
1 Cyanamid Plaza
Wayne, NJ 07470
Contact: Office of Technical Recruiting

Eli Lilly and Company
Lilly Corporate Center
Indianapolis, IN 46285
Contact: Office of Corporate Recruiting

Johnson & Johnson
1 Johnson & Johnson Plaza
New Brunswick, NJ 08933
Contact: College Relations Coordinators in the Personnel Office of each Johnson & Johnson Division

Marion Merrell Dow, Inc.
10236 Marion Park Drive
Kansas City, MO 64137
Contact: Staffing Manager

Parke-Davis, Warner-Lambert Company
201 Tabor Road
Morris Plains, NJ 07950
Contact: Director, Human Resources, Sales & Marketing

Production / Operations

Calgon Vestal Laboratories
Calgon Corporation
St. Louis, MO 63166
Contact: Personnel Department

Ciba-Geigy Corporation
444 Saw Mill River Road
Ardsley, NY 10502
Contact: Manager, College Relations and Staffing

* *Peterson's Job Opportunities for Engineering, Science, and Computer Graduates 1993,* Copyright © 1992 Peterson's Guides, Inc., Princeton, New Jersey.

Genentech, Inc.
460 Point San Bruno Boulevard
South San Francisco, CA 94080
Contact: Human Resources Department

Merck & Co., Inc.
126 East Lincoln Avenue
Rahway, NJ 07065
Contact: Office of College Relations

Millipore Corporation
80 Ashby Road
Bedford, MA 01730
Contact: Manager of Employment

Research / Development

Argonne National Laboratory, University of Chicago
9700 South Cass Avenue
Argonne, IL 60439
Contact: Recruiting Coordinator

Coca-Cola Foods
2000 St. James Place
Houston, TX 77056
Contact: Manager, Professional Staffing

General Mills, Inc.
One General Mills Boulevard
Minneapolis, MN 55446
Contact: Director of Recruitment and College Relations

Memorial Sloan-Kettering Cancer Center
1275 York Avenue
New York, NY 10021
Contact: Administrator, College Relations

Centers for Disease Control
Department of Health and Human Services
1600 Clifton Road, NE
Atlanta, GA 30333
Contact: Employment Office

National Cancer Institute, National Institutes of Health
9000 Rockville Pike
Bethesda, MD 20892
Contact: Personnel Staffing Specialist

Technical / Professional

American Software USA, Inc.
470 East Paces Ferry Road, NE
Atlanta, GA 30305
Contact: Manager, Corporate Recruiting

Energy and Environment Analysis, Inc.
1655 North Fort Meyer Drive
Arlington, VA 22209
Contact: Recruiting Coordinator

General Electric Company
3135 Easton Turnpike
Fairfield, CT 06431
Contact: Recruiting Support Services

Patent and Trademark Office
U.S. Department of Commerce
2011 Crystal Drive, One Crystal Park
Arlington, VA 20231
Contact: Office of Personnel

Consumer Product Safety Commission
5401 Westband Avenue
Bethesda, MD 20816
Contact: Director, Division of Personnel Management

Appendix C
Answers to Selected Problems

Chapter 2

1. Chemists can often make a thermodynamically unfavorable reaction proceed to some extent by manipulating experimental reaction parameters such as pressure, temperature, or concentrations of reactants. Biochemists generally cannot alter such reaction parameters because organisms function within limited concentration ranges and at essentially constant pressure and temperature.

3. A state function describes the thermodynamic parameters of the system under consideration at a particular moment. Only the difference between initial and final states, not the path taken to achieve these states, is important in most thermodynamic considerations. Enthalpic contributions defining the thermodynamic state are considered only at the initial and final states. Enthalpy is independent of pathway and is therefore a state function.

5. In thermodynamics the total system is the universe, consisting of a particular system and its environment. If the particular system under study is a closed system, it can exchange internal energy as heat and work with its surroundings, but no exchange of matter can occur. An open system can also exchange matter with its surroundings. Therefore, although any energy lost by a particular system is gained by its environment (and vice versa), the energy of the universe remains constant. This is the first law of thermodynamics. Distinguishing between the system and the universe is important to differentiating between a situation in which energy can change and a situation in which energy is constant.

7. Entropy decreases because of hydration effects. The ionized species will order much of the water during hydration, decreasing the total number of free molecules.

9. The reaction will proceed toward oxaloacetate formation in the cell if low product concentration is maintained. Oxidation of NADH by the mitochondrial electron transport system and utilization of oxaloacetate in the formation of citrate shifts the malate-oxaloacetate reaction toward oxaloacetate production.

11. (a) For reaction (P1), $\Delta G^{\circ\prime} = -2.4$ kcal/mole;
    (b) $\Delta G^{\circ\prime}$ of ATP hydrolysis is -7.9 kcal/mole.

13. (a) $\Delta G^{\circ\prime} - 1.5$ kcal/mole; thermodynamically favorable as written.
    (b) $\Delta G^{\circ\prime} = +1.7$ kcal/mole; thermodynamically unfavorable as written.
    (c) $\Delta G^{\circ\prime} = -7.3$ kcal/mole; thermodynamically favorable as written.

15. In the first case, where the repressor protein is cut in half, the binding enthalpy for each part would be essentially half the enthalpy value for the intact repressor. The entropy would be less favorable (less positive) because of the chelation effect. As a result, the free energy would be less favorable (less negative).

    In the second case, where one of the binding sites on the DNA is eliminated, the binding enthalpy for the repressor would be approximately half that for repressor binding to the unmodified DNA sequence. The entropy would depend on the extent of hydration and the extent of mobility of the unbound portion of the repressor. Again, we would expect the free energy for binding to be less favorable.

Chapter 3

1. Berzelius's proposal of $C_{40}H_{62}N_{10}O_{12}$ has a molecular mass of 874 g/mole, of which 140 g/mole is nitrogen (or 16.0% nitrogen by mass).

3. $\text{pH} = 6.39 + \log \dfrac{0.0133 \text{ mole}/0.25 \text{ L}}{0.0060 \text{ mole}/0.25 \text{ L}} = 6.39 + 0.35$

$$= 6.74$$

$[\text{H}^+] = 10^{-6.74} = 1.82 \times 10^{-7}$ M

5. pI histidine $= [(\text{p}K_{amino}) + (\text{p}K_{imidazole})]/2$
   pI $= 7.69$
   pI aspartic acid $= [(\text{p}K_{carboxyl}) + (\text{p}K_R)]/2$
   pI $= 2.95$
   pI arginine $= (\text{p}K_{amino} + \text{p}K_R)/2$
   pI $= 10.74$

The sum of positive and negative charge contributions is

α-amino group	α-carboxyl	β-carboxyl	
(+1)	(−0.9)	(0.1)	= 0

These data demonstrate that the net charge on aspartic acid is zero at pH 2.95, which verifies 2.95 as the isoelectric pH.

7. Aspartic acid: pH 2, 4, 10. (The pH range 1 to 5 will be buffered by the α-carboxyl and the β-carboxyl groups.)
Histidine: pH 2, 9, 6 (imidazole side chain).
Serine: pH 2, 9. (The pK_a of the alcohol is outside the range of pH normally considered for buffers.)

9. Threonine and isoleucine

11. NH_2-Ala-Ala-Lys-Ala-Ala-Phe-Ala

Chapter 4

1. Consider the entropic effect of decreasing water organization by moving the hydrophobic residue side chains from an aqueous to a nonaqueous environment.

3. The α helix is a rodlike element that cannot easily change direction. Loops, β bends, and "random" structure break the helical structure and allow these directional changes.

5. An α helix broken at the Pro-Asn-Ala region with the hydrophobic residues on the exterior should insert into the membrane.

7. The right- or left-handedness of a helix is the same as a conventional screw or bolt. When turned clockwise, a right-handed screw advances. The same is true of a helix or a helical spring.

9. These substances can cause the loss of a protein's shape by disrupting hydrogen bonding and electrostatic interactions.

11. The detergent replaces the membrane, producing a soluble enzyme, and allows the enzymes to be purified free of the membrane.

Chapter 5

1. The early chemists had no way of knowing that each hemoglobin contains four Fe^{2+}.

$$4.9 \text{ mg of } Fe_2O_3$$

3. (a) $CO_2 + H_2N$-hemoglobin $\longrightarrow$

$$\overset{O}{\overset{\|}{^-OCHN}}\text{-hemoglobin} + H^+$$

(b) $HOCO_2^- + H_2N$-hemoglobin $\longrightarrow$

$$\overset{O}{\overset{\|}{^-OCHN}}\text{-hemoglobin} + H_2O$$

5. Red blood cells pass single file through the capillaries. In a sickle cell crisis the red blood cells "jam together," clogging the capillaries. The associated tissues become starved for oxygen, producing the pain of a sickle cell crisis.

7. Individuals with sickle cell anemia have a functional gene for the gamma chain. If the production of fetal hemoglobin could be "turned back on" the affected individuals could function normally except during pregnancy.

9. Unprotonated. The protonated form of histidyl F8 is positively charged and is less likely to interact favorably with the Fe^{2+} of the heme.

11. 457 Å. Because the two 457-Å-long strands are twisted into a helix, the resulting TM is shorter than 457 Å.

13. Rigor is probably caused by the depletion of ATP and a considerable discharge of calcium from the sarcoplasmic reticulum.

Chapter 6

1. It is often a rapid effective way to reduce the volume of crude extracts and at the same time eliminate a major portion of the total protein.

3. Difference in charge allows the separation of phosphorylase a from phosphorylase b with the use of DEAE-cellulose. Phosphorylase a and phosphorylase b should elute as a single peak upon gel filtration.

5. Heat treatment of protein solutions denatures and precipitates some of the proteins, while others remain both soluble and stable. Thermal lability is determined empirically for each enzyme or protein of interest.

7. The protein is positively charged at a pH more acidic than the pI. Therefore the protein will probably adhere to (bind to) the CM-cellulose if the pH is between 4 and 6.

9. The student's "pure" protein contains at least two components separable by the criterion of mass/charge but which share a common subunit molecular weight. The multiple protein bands appearing in the nondenaturing gel possibly arose through deamidation of glutamine or asparagine residue side chains.

11. (a) Proteins 1 and 3 should elute in the initial wash buffer, but proteins 2 and 4 are predicted to bind to the column. Based solely on isoelectric point, we might predict that protein 4, then protein 2, would be eluted in the salt gradient.

(b) Proteins 2 and 4 would be eluted in the initial wash buffer from the column, whereas proteins 1 and 3 would be predicted to adhere to the column. Based solely on pI values, protein 3 would be predicted to elute prior to protein 1 in the KCl gradient.

(c) Proteins with molecular weight greater than the limit (protein 62,000 M_r) are excluded from entry into the gel and elute in the void volume (V_o). The other proteins in the solution will elute in the order protein 3, protein 1, and protein 4.

Chapter 7

1. Reaction order is the power to which a reactant concentration is raised in defining the rate equation. The example is first order in A and B, second order overall, and second order in C.

3. Let $v = V_{max}/2$ and substitute into equation 25:

$$\frac{V_{max}}{2} = \frac{V_{max}[S]}{[S] + K_M}$$

Solving yields $K_M = [S]$.

5. No

7. Steady-state approximation is based on the concept that the formation of [ES] complex by binding of substrate to free enzyme and breakdown of [ES] to form product plus free enzyme occur at equal rates. A graphical representation of the relative concentrations of free enzyme, substrate, enzyme-substrate complex, and product is shown in figure 7.8 in the text. Derivation of the Michaelis-Menten expression is based on the steady-state assumption. Steady-state approximation may be assumed until the substrate concentration is depleted, with a concomitant decrease in the concentration of [ES].

9. $[S] = 1.0 \times 10^{-3}$ M, $[I] = 9.1 \times 10^{-4}$ M
   $[S] = 1.0 \times 10^{-5}$ M, $[I] = 1.8 \times 10^{-5}$ M
   $[S] = 1.0 \times 10^{-6}$ M, $[I] = 9.9 \times 10^{-6}$ M

11. An enzyme is a catalyst for a chemical reaction, so the rate of an enzyme catalyzed reaction increases with increased temperature. However, the catalyst, a protein, is structurally labile and is inactivated (denatured) at elevated temperatures. The precise temperature at which the enzyme is inactivated varies with the specific enzyme. There is no "temperature optimum" for a catalyst (enzyme).

13. $V_{max} = 3.3 \times 10^{-7}$ Ms^{-1}; $K_m = 2.4 \times 10^{-4}$ M; $k_{cat} = 3.3 \times 10^4$ s^{-1}; specificity constant $= 1.4 \times 10^8$ M^{-1}s^{-1}

Chapter 8

1. The substrate recognition site is a pocket on hexokinase into which glucose, then ATP, bind. Glucose and hexokinase bind together, changing the shape of both and forming a binding site for ATP.

3. Specificity of bond cleavage would most certainly change. The favorable electrostatic interaction between the Lys or Arg side chain of the peptide substrate and the aspartate in the binding pocket would be replaced by electrostatic repulsion upon substitution of the lysine residue. Thus selective binding of Arg or Lys to the substrate pocket would be precluded. The modified trypsin might therefore (a) cleave peptide bonds randomly, but exclude bonds on the carboxyl side of Lys or Arg residues, or (b) exhibit specificity for peptide cleavage on the carboxyl side of acidic amino acids (Asp, Glu). Side chains of these amino acids may fit well into the substrate binding pocket and have favorable electrostatic interaction with the lysine therein.

5. We expect iodoacetate and para-mercuribenzoate to inhibit plant proteases and diisopropylfluorophosphate to inhibit serine proteases.

7. Histidyl 12 accepts the 2'-OH hydrogen to start the reaction. The protonated histidyl 12 ultimately provides the hydrogen to the cyclic ester to produce the 3'-phosphate product.

9. Renaturation of denatured protein is dictated by the primary structure of the protein. The trypsin family of enzymes and carboxypeptidase A are synthesized as proenzymes that are proteolytically activated. The proteolyzed, active enzymes have primary structures different from the gene product and are not active upon renaturation. In addition, zinc is a cofactor required for carboxypeptidase A activity.

11. The structure of the transition-state analog is complementary to the structure of the active site. The analog thus binds tightly to the active site.

13. (a) Substitution of Asp for Lys 86 markedly decreases activity and several explanations are possible. The lysine is a critical residue either at or near the active site or is essential for maintaining a catalytically competent conformation of the enzyme. (b) Lysines 21 and 101 are probably outside the catalytic site and may not be evolutionarily conserved. Their replacement with aspartate yielded no great change in enzymatic activity. (c) Lysine 86 is essential for enzymatic activity and would be conserved.

Chapter 9

1. The Ile in chymotrypsinogen is the 16th amino acid. During the conversion to chymotrypsin a 15 amino acid peptide is cleaved from chymotrypsinogen. The original amino acid number system was retained in chymotrypsin, making the N-terminus amino acid number 16.

3. Covalent modification requires an enzyme to control an enzyme. Allosterism requires only a binding site on an enzyme that interacts (via an equilibrium) with a particular small molecule.

5. Phosphorylation of serine or threonine (and possibly tyrosine) on the target protein may be largely influenced by the amino acid sequence around these residues. These amino acid sequences may define a specific motif or recognition site for the protein kinase.

7. The substrate concentration required for half-maximal activity ($S_{0.5}$) of an allosterically regulated enzyme will depend on the cumulative effects of allosteric activators and/or inhibitors also present. Hence $S_{0.5}$ may be decreased with allosteric activators and may be increased with allosteric inhibitors.

9. Allosteric regulation of an enzyme having a binding site for a regulatory molecule and an active site on the same subunit is not uncommon. Regulatory molecules bind at sites separate from the active site and induce a conformational change that affects substrate binding to the active site on the same as well as adjacent subunits.

11. The methylene group prevents the elimination of a phosphate, which is required to convert the postulated intermediate into carbamoyl aspartate. The suggested oxygen analog might eliminate phosphate (i.e., be a substrate for aspartate carbamoyltransferase).

Chapter 10

1. The following coenzymes contain the AMP moiety: NAD^+, NADH, $NADP^+$, NADPH, FAD, $FADH_2$, and CoASH.

3. (a) Refer to figure 10.5b in the text for the structures of each intermediate step.
   Steps 1. & 2. Form Schiff base
   3. Remove proton α-C
   4. Protonate C-4′
   5. & 6. Hydrolyze Schiff base
   Release α-keto acid and pyridoxamine phosphate
   Transfer of amino group from pyridoxamine phosphate to pyruvate, forming alanine occurs by reversal of these steps. Other transaminases use other α-amino acids and α-keto acids.
   (b) Figure 10.5b in the text gives the structures of intermediates.
   Steps   1. Decarboxylation of α-COO^-
   2. Protonate α-Carbon
   3. & 4. Hydrolyze Schiff base
   (c) β-Decarboxylation. Form Schiff base steps 1–3, solution 3a.

5. Refer to figure 10.2 in the text.
   (a) Steps 1. Deprotonation of TPP to ylid form, nucleophilic addition of ylid to α-ketogroup

2. Decarboxylation
3. Resonance stabilization
4. Protonation, elimination of TPP
   (b) Steps 1. Deprotonation to ylid form, nucleophilic addition of ylid to fructose-6-phosphate
   2. Oxidation of C-3, release of erythrose-4-phosphate
   3. Resonance stabilization of intermediate
   4. Elimination of $-OH$ from C-2
   5. Transfer of acyl group to phosphate

7. (a) Thiamine pyrophosphate, lipoic acid, FAD, NAD^+, CoASH (an α-ketoacid dehydrogenase); (b) biotin; (c) FAD.

9. (a) Reduced flavins in solution rapidly reduce O_2 to superoxide and H_2O_2, metabolites that are toxic to the cell. Enzyme-bound flavins are usually shielded from rapid oxidation by O_2. (b) Tightly bound NAD(P) is an advantage to enzymes catalyzing rapid $H:^-$ removal and readdition in a stereospecific fashion. Freely diffusing NADH is an advantage in transferring reducing equivalents among various enzyme-catalyzed reactions.

11. (a) Redox agents: FAD, FMN, NAD^+, $NADP^+$, and lipoyl; (b) acyl carriers: CoASH, lipoyl, and thiamine pyrophosphate; (c) both acyl carriers and redox agents: lipoyl.

13. The hydroxyl hydrogen on C-3 is the most acidic because the resulting anion is resonance stabilized.

Chapter 11

1. The transition from left to right involves dehydration reactions; the transition from right to left involves hydrolysis reactions.

3. Catabolism involves pathways composed of enzymes and chemical intermediates that are involved primarily in the breakdown of large molecules into small molecules, often by oxidation processes. Anabolism is the collection of the enzymes and chemical intermediates involved in the biological synthesis of larger molecules from smaller molecules. Anabolism often involves reduction processes.

5. The advantage of subcellular compartments is that a specific pathway can be isolated from comparable pathways that might use similar or identical chemical intermediates. Compartmentalization simplifies regulation of the various processes.

7. (a) The concentration of most metabolites measured under steady-state conditions in the cell usually does not exceed the K_m value. For an enzyme whose reaction can be described by simple Michaelis-Menten kinetics, the observed velocity, v, is $0.5V_{max}$ if substrate concentration equals the

K_m value. The velocity of most enzymes is likely significantly less than V_{max} *in vivo*.

(b) End-product inhibition usually occurs at the committed step in a metabolic pathway or at a branch point in the pathway.

(c) Catabolic pathways tend to be convergent rather than divergent. Metabolic convergence of precursors into common intermediates of a metabolic pathway provides an efficient route for the metabolism of a variety of metabolites by a limited number of enzymes.

(d) Enzymes that are regulated in metabolic pathways most frequently exhibit cooperative kinetic responses rather than hyperbolic responses with respect to substrate concentration and are frequently responsive to allosteric regulation by products, energy charge, or concentration ratio of $NAD(P)H/NAD(P)^+$. The rate of enzymatic activity over a narrow range of substrate concentration can be changed dramatically by allosteric activators or inhibitors. Such dramatic changes in velocity in response to small changes in substrate concentration is not observed with enzymes exhibiting Michaelis-Menten kinetics.

(e) Low energy charge signals the cell that a need for ATP formation exists, and pathways (glycolysis and Krebs cycle) leading to ATP formation are activated. Anabolic pathways that demand high ATP concentrations are inhibited at low energy charge. In the latter case, the ATP required to drive biosynthesis is in low supply. Conversely, increased energy charge inhibits pathways leading to ATP formation and activates anabolic pathways.

(f) Formation of multienzyme complexes is a strategy frequently used for efficient catalysis and control of metabolic pathways. Substrates diffuse shorter distances between active sites in multienzyme complexes than if the enzymes were not organized. Frequently, intermediates covalently bound to cofactors (e.g., lipoamide, biocytin) are moved among active sites within the complex, effectively trapping intermediates in the complex and increasing the concentration of substrates at the enzyme active site.

(g) Separation of catabolic and anabolic pathways diminishes the likelihood of futile cycling of metabolites. Enzymes catalyzing the β-oxidation of fatty acids are located in the mitochondrial matrix, whereas enzymes catalyzing the synthesis of palmitate are located in the cytosol.

9. The "committed step" in a reaction sequence steers the metabolite to a sequence of reactions whose intermediates have no other function in the cell. Control of the committed step prevents wasteful accumulation of these single-purpose intermediates and obviates the necessity of controlling each enzyme in a pathway.

Chapter 12

1. The number of sugars is 2^n where n is the number of chiral centers but does not include the chiral carbon involved in producing the D series of sugars. Because C-2 of the ketoses is a carbonyl group, only the geometry of C-3 and C-4 remain to produce the four D-ketohexoses ($2^2 = 4$). The carbonyl group on C-1 of the aldoses allow C-2, C-3, and C-4 to produce eight D-aldohexoses ($2^3 = 8$).

3. (a) Trehalose is two α-D-glucoses linked through C-1 of each sugar. However, the correct answer to this question would also include (b) two β-D-glucoses linked through C-1 and (c) one α-D-glucose linked through C-1 to C-1 of a β-D-glucose.

5. Maltose, lactose, and cellobiose are reducing sugars; sucrose and trehalose (problem 3) are nonreducing sugars.

7. Glucose C-3 and C-4 are lost as CO_2 during ethanol production.

9. By assessing the rate of $^{14}CO_2$ production from ^{14}C-1-glucose versus ^{14}C-6-glucose, we can determine the relative significance of the different pathways in a tissue.

11. Glycerate-2,3-bisphosphate was encountered as an allosteric effector of hemoglobin.

13. In the first example the actual chemistry is identical: an aldose-ketose interconversion. Note that the geometry of the alcohol on C-2 is identical. The second example has essentially the same chemistry. Logically, to do the same chemistry, nature could have stumbled upon the same mechanism twice. More likely, however, the mechanism evolved once and, following gene duplication, the segments of the gene controlling the substrate specificity changed on one copy of the gene.

15. The carboxyl group of pyruvate is lost as CO_2 during the pyruvate decarboxylase–catalyzed formation of acetaldehyde. The pyruvate carboxyl group is formed by the oxidation of glyceraldehyde-3-phosphate. The aldehyde (C-1) carbon is derived directly from the aldolase-dependent cleavage of the fructose-1,6-bisphosphate. In this cleavage, C-4 of glucose becomes C-1 of glyceraldehyde-3-P_i and C-3 of glucose becomes C-3 of dihydroxyacetone phosphate. DHAP is isomerized to Ga3P$_i$. In this isomerization, C-3 of DHAP (originally C-3 of glucose) becomes C-1 of Ga3P. Thus, labeling either C-3 or C-4 of

glucose will ensure that label is released as CO_2 upon fermentation to ethanol.

17. (a) Triose phosphate isomerase deficiency would inhibit conversion of DHAP to Ga3P and would cause accumulation of DHAP, preventing half of the glucose molecule (C1-C3) from being metabolized through the remainder of the glycolytic pathway. There would be a recovery of only 2 of the possible 4 moles of ATP from glucose, resulting in no net formation of ATP. In addition, DHAP, a product of the aldolase reaction, would likely reverse the aldolase (reaction) and eventually inhibit glycolysis. Either result would be lethal to a cell whose only energy source was glycolysis.

   (b) The small amount of TPI activity would likely allow glycolysis to proceed slowly, but low energy (ATP) level will limit the growth rate under anaerobiosis. However, the yield of ATP is significantly greater when the pyruvate, formed during glycolysis, is oxidized to CO_2 and H_2O. Hence, the growth rate of the mutant should be correspondingly greater under aerobic growth conditions but not as great as the wild type.

   (c) Cells expressing DHAP phosphatase would likely not grow anaerobically if glycolysis of glucose to lactate were the only pathway for ATP formation. The combined activities of TPI and DHAP phosphatase would be predicted to deplete the pool of triosephosphate and the yield of ATP per glucose would likely be less than 1.

19. (a) $K'_{eq} = 30$

   (b) Hexoses brought into the glycolytic pathway must be phosphorylated to provide the appropriate substrate for the glycolytic enzymes and to trap the sugar within the cell. Phosphorylation at the expense of ATP or group translocation at the expense of PEP are common methods to activate the sugar molecules. Sucrose phosphorylase uses the exergonic lysis of the glycosidic bond between the hemiacetyl OH group of glucose and the hemiketal OH group of fructose to drive the endergonic phosphorylation of the hemiacetal C-1 OH group of glucose. Transfer of the phosphate to C-6, catalyzed by phosphoglucomutase, provides substrate for entry into the glycolytic pathway without addition of ATP. The net ATP yield will therefore be 3 rather than 2 moles of ATP per mole of glucose derived from sucrose. The fructose can be phosphorylated by ATP and used in the glycolytic pathway.

21. The free energy available from a reaction depends on the energies of the products compared with the substrates. Dehydration of 2-phosphoglycerate "traps" phospho-(enol)pyruvate in the enolate form. The hydrolysis products of PEP are phosphate and the enol form of pyruvate, but (enol) pyruvate is significantly less stable than (keto) pyruvate and rapidly tautomerizes to the more stable keto form. The tautomerization drives the reaction strongly toward products, resulting in a larger free energy difference between substrate and product.

23. Given the ratio of ATP/ADP and the K'_{eq} of 10^6, the equilibrium ratio of [Pyry]/[PEP] would be about 10^5. This calculation supports the metabolic irreversibility of the pyruvate kinase reaction.

25. 6-phosphogluconate + $NADP^+$ $\longrightarrow$
$$NADPH + H^+ + CO_2 + \text{Ribulose-5-phosphate}$$

27. (a) Unregulated hepatic PK theoretically could become part of a futile cycle. Net: GTP $\longrightarrow$ GDP + P_i plus formation of cytosolic NADH at the expense of mitochondrial NADH.

   (b) Activation by fructose-1,6-bisphosphate decreases $S_{0.5}$ for PEP and increases PK activity at a given PEP concentration. During gluconeogenesis, the fructose-1,6-bisphosphate concentration should diminish, owing to the hydrolytic activity of the FBPase-1. The low FBP concentration, coupled with the elevated ATP levels, could inhibit the hepatic pyruvate kinase.

Chapter 13

1. The oxidation occurs first, creating a β-keto-carboxylate, which readily loses CO_2 via a resonance-stabilized carbanion.

3. Citrate is a tertiary alcohol, which does not oxidize readily.

5. The following sequence will do the task: α-ketoglutarate, the tricarboxylic acid cycle to oxaloacetate, to phosphoenolpyruvate, to pyruvate, to acetyl-CoA, into the tricarboxylic acid cycle.

7. Both of the oxidative-decarboxylation steps in the tricarboxylic acid cycle are bypassed by the glyoxylate cycle.

9. (a) Increased concentrations of acetyl-CoA slow the activity of pyruvate dehydrogenase.

   (b) The removal of succinyl-CoA for heme synthesis removes any control it has over lowering the activity of α-ketoglutarate dehydrogenase and citrate synthase.

11. (a) False. Lipoamide transacetylase catalyzes reduction of the disulfide on lipoamide concomitantly with oxidation and transfer of the hydroxyethyl group from thiamine pyrophosphate. Dihydrolipoamide dehydrogenase catalyzes oxidation of dihydrolipoamide and reduction of NAD^+. (b) False. Hydrolysis of

acetyl-CoA thioester should yield as much free energy as succinyl-CoA hydrolysis, a process coupled to ADP phosphorylation (succinate thiokinase).
(c) False. The methyl group of acetyl-CoA could be derived from pyruvate, from β oxidation of long-chain fatty acids, or from amino acid metabolism.
(d) False. If aconitase failed to discriminate between the ($—CH_2—COO^-$) groups, half the CO_2 would arise from oxaloacetate and half from the acetate carboxylate. The two CO_2 molecules released by oxidative decarboxylation of isocitrate and α-ketoglutarate are derived from the carboxyl groups of oxaloacetate with which the acetyl-CoA was condensed. (e) False. Malate can easily be dehydrated to fumarate by reversal of the fumarase reaction.

13. (a) Without malate synthase, the yeast would be unable to grow on 2-carbon precursors as the sole carbon source because TCA cycle intermediates could not be synthesized.

   (b) TCA cycle activity would markedly diminish if the cycle intermediates were being used in biosynthetic pathways. In addition, the biosynthetic pathways (lipids, amino acids, and carbohydrates) dependent on TCA cycle intermediates would also be inhibited by the lack of metabolites.

   (c) If the PDH were inhibited more strongly than usual by acetyl-CoA, we might suspect that acetyl-CoA concentration in the mitochondrial matrix would markedly decrease, in turn limiting activity of citrate synthase and diminishing TCA cycle activity. Hence growth of the organism may be limited by lowered energy production and by diminished concentrations of biosynthetic precursors supplied by the TCA cycle.

15. (a) Fumarate + NADH + H^+ $\longrightarrow$ succinate + NAD^+.
   (b) Phosphoenolpyruvate + CO_2 $\longrightarrow$ oxaloacetate + P_i
   (PEP carboxylase)
   Oxaloacetate + NADH + H^+ $\longrightarrow$ L-malate + NAD^+
   (malate dehydrogenase)
   L-Malate $\longrightarrow$ fumarate + HOH
   (fumarase)
   Fumarate + NADH + H^+ $\longrightarrow$ succinate + NAD^+
   (fumarate reductase)

   (c) In the reactions shown in part (b), four reducing equivalents (two hydride groups) are transferred to carbon acceptors. Reduction of oxaloacetate to malate by malate dehydrogenase (MDH) and reduction of fumarate to succinate by fumarate reductase each requires hydride (or equivalent) transfer from NADH to the organic substrate. In the reduction of pyruvate to lactate via LDH

only one hydride is used. Thus, two equivalents of NAD^+ are resupplied to glycolysis by the activities of MDH and fumarate reductase, whereas only one equivalent of NAD^+ is regenerated by LDH. Fumarate is one of the terminal electron acceptors used during anaerobic respiration in *E. coli*. However, each mole of PEP carboxylated is at the expense of 1 mole equivalent of ATP that could have been formed as a product of pyruvate kinase.

17. The committed step in a metabolic pathway is usually under metabolic control. Inhibition of the committed step in a metabolic sequence or pathway prevents the accumulation of unneeded intermediates and effectively precludes activity of the enzymes using those intermediates as substrates. The decarboxylation of pyruvate and the oxidative transfer of the hydroxyethyl group by pyruvate dehydrogenase constitutes the committed step in the pyruvate dehydrogenase catalytic sequence and is a logical control point.

Chapter 14

1. In hemoglobin and myoglobin the heme serves as a carrier with oxygen–heme iron (Fe^{2+}) interacting in a ligand-metal coordination relationship. In the cytochromes heme plays a redox role, with the iron interconverting between Fe^{2+} and Fe^{3+}.

3. (a) 12.5
   (b) 16
   (c) 10

5. Both iron and copper have two common stable ions; the transition between Fe^{2+} and Fe^{3+} or Cu^{1+} and Cu^{2+} are both redox reactions.

7. (a) Heme of the mitochondrial *b*-type cytochrome interacts hydrophobically with adjacent hydrophobic residues from the membrane-spanning α helices. The heme iron is fully coordinated through two imidazole groups from histidines in the protein. Heme in the *c*-type cytochromes is covalently bound to the protein through thioethers formed by the addition of cysteinyl sulfhydryl groups to the vinyl substituents on the heme ring. The iron is also fully liganded to a nitrogen from the imidazole group of histidine and a sulfur from the thioether linkage of methionine providing the fifth and sixth ligands. Heme *a* differs from the protoheme (heme) of the *b*- and *c*-type cytochromes by substitution of a formyl group at ring position 8 and a 17-carbon isoprenoid chain at position 2.

(b) Neither CO nor CN^- at low concentration interact with the cytochrome b or c heme iron because no open ligand position is available to the iron in these heme proteins.

9. (a) In biological systems, the iron-sulfur centers are obligatorily single electron donors/acceptors, regardless of the number of Fe atoms in the center or their initial oxidation state.

   (b) The 4Fe-4S cluster is found in iron-sulfur proteins that transfer electrons at low and at high potential. The reduction potential is a measure of the ease of addition of an electron to the couple, compared to the standard hydrogen electrode. Thus the protein component of the iron-sulfur protein affects the reduction potential.

11. $E°$ (pH 6) = +170 mV; $E°$ (pH 8) = 50 mV.

13. (a) The large amount of UQ is necessary to ensure efficient transfer of electrons from the mitochondrial dehydrogenases to complex III.

   (b) (i) If UQ (ubiquinone) were limiting, the rate of oxidation of NADH by the mitochondrial electron-transfer system would also be limited. NADH concentration would increase, the NAD^+ supply would decrease, and the NAD^+-dependent dehydrogenases would be inhibited. Subsequently, the rate of oxidation of NADH-producing substrates would decrease.

       (ii) Electrons from succinate oxidation are also transferred to ubiquinone from the succinate dehydrogenase, so a deficiency of the quinone would limit succinate oxidation.

       (iii) Ascorbate plus a redox mediator reduces cytochrome c but does not transfer electrons to ubiquinone. Limiting amounts of the quinone should not affect ascorbate-dependent reduction of cytochrome c.

   (c) The deficiency of UQ will decrease the rate of extramitochondrial NADH oxidation resulting in an increase in the amount of pyruvate reduced to lactate. The lactate content would be expected to increase rapidly upon mild exercise.

15. (a) The uncoupler 2,4-dinitrophenol circumvents respiratory control in the mitochondria by short-circuiting the proton gradient. The lipophilic weak acid transports H^+ across the membrane, bypassing the F_1-F_0 complex. Substrates will be oxidized independently of ADP or ATP concentrations, and O_2 reduction will be more rapid than in state 3 respiration.

   (b) Uncoupled mitochondria oxidize NADH and succinate but fail to phosphorylate ADP. Cellular processes will continue to utilize ATP, causing an accumulation of ADP and AMP. Increased levels of NAD^+ and ADP or AMP activate both the TCA cycle and glycolysis. The rate of oxidation of carbohydrates and fatty acids would markedly increase.

   (c) The uncoupled mitochondria use little, if any, energy to phosphorylate ADP. The energy is dissipated as heat, leading to elevated body temperature (hyperthermia) and profuse perspiration in an effort to decrease body temperature.

   (d) 2,4-Dinitrophenol is a lipid soluble weakly acidic compound thought to allow equilibration of protons across the inner mitochondrial membrane. Although protons are translocated from the matrix across the inner membrane during electron transfer, the proton gradient would immediately be depleted without passing through the F_0-F_1-dependent ADP phosphorylation system. Respiratory (ADP) control would be lost.

17. The transport process is electrogenic if the export of one molecule coupled with the import of another molecule yields a net charge difference across the membrane. In general terms, transfer of A^{3-} from the matrix and A^{3-} into the matrix yields a net negative charge on the cytoplasmic side of the membrane. Electrogenic processes are driven by the membrane potential ($\Delta\Psi$).

   Neutral transport processes exchange molecules of net identical charge (sign and magnitude) in the opposite vectorial direction or oppositely charged molecules in the same vectorial direction. Processes coupled to the export of OH^- are equivalent to H^+ import, that is, to an energetically favorable decrease in the proton gradient.

Chapter 15

1. The word fixed refers to converting a gas into a liquid or solid.

3. Chlorophyll molecules are flat, completely conjugated ring compounds that obviously have a resonance form. Chlorophylls have a ring containing 9π bonds or 18π electrons, which fits the $n = 4$ situation in the Huckel ($4n + 2$) rule. The (cis, trans, trans)$_3$ double bond pattern of the chlorophylls also fits that of [18]annulene, which is aromatic.

5. Upon illumination, the chromatophore P870 is activated by absorption of a photon of light. The absorbance at 870 nm decreases because the π-cation radical of the oxidized chromatophore has a lower absorbance at that wavelength. Thus the trace monitoring 870 nm decreases upon illumination of the

chromatophore. The *c*-type cytochrome is added initially in the reduced form (cyt c^{2+}) and absorbs at 550 nm. Oxidized cytochrome *c*(cyt c^{3+}) has only a small absorbance at 550 nm. Electron transfer from reduced cytochrome *c* to the π-cation radical ($P870^+$) regenerates the ground state P870 and forms oxidized cytochrome *c*. The absorbance of the P870 increases to the initial level and the absorbance at 550 nm of the cytochrome *c* pool decreases.

7. (a) The absorbance at 275 nm decreases because the concentration of ubiquinone is diminished and because the semiquinone radical does not absorb 275 nm light. The absorbance increase at 450 nm is consistent with the formation of semiquinone radical that absorbs 450 nm light. The second flash activates transfer of a second electron from the photocenter to reduce the bound semiquinone to dihydroquinone. The dihydroquinone absorbs at neither 275 nm nor 450 nm. Reduction of the semiquinone form abolishes the absorbance at 450 nm. The decrease in absorbance at 275 nm is consistent with the decrease in oxidized (ubiquinone) concentration.

   (b) In this experiment, reduced cytochrome *c* is the electron donor to $P870^+$. Were the reductant omitted, the $P870^+$ might oxidize the reduced quinone, or the activated reaction center P870* might return to ground state by emission of fluorescence.

9. Singlet-state oxygen causes cumulative oxidative damage to the chloroplast. Carotenoids compete with the oxygen for the triplet-state chlorophyll and inhibit singlet oxygen production. Plants deficient in carotenoids risk photooxidative damage because of an increased flux of singlet oxygen.

11. The mechanisms are different, but the energetic outcome is the same.

13. The enolate intermediate (figure 15.26) reacts with oxygen to produce a cycloperoxide intermediate that decomposes into glycerate-3-phosphate and glycolate-2-phosphate (figure 15.27).

15. Some of the many differences include the involvement of erythrose-4-phosphate and dihydroxyacetone phosphate in the production of sedoheptulose-1,7-bisphosphate, phosphorylation of ribulose-5-phosphate to ribulose-1,5-bisphosphate, the addition of CO_2 to ribulose-1,5-bisphosphate, and the production of two glyceraldehyde-3-phosphates.

Chapter 16

1. The sequence, bond geometry, and linkage type of monosaccharides in an oligosaccharide or polysaccharide is determined by the specificity of the glycosyltransferases involved.

3. In general, amino groups are added to biological molecules in locations once occupied by keto groups. Because ketoses have C-2 as the carbonyl group the majority of the amino sugars in nature are 2-amino sugars.

5. Sugars have a large number of hydroxyl groups that form glycosidic bonds with the anomeric carbons of other sugars; these anomeric hydroxyl groups can be either α or β conformation. More than eighty different glycosidic linkages have been identified.

7. Patients with I-cell disease do not phosphorylate the mannose residues on the glycoproteins that are lysosome-bound. These "lysosomal hydrolases" are therefore secreted.

9. Gluconic acid cannot form a cyclic hemiacetal and cannot form the glucosidic bonds required for participation in oligo- or polysaccharides.

11. The undecaprenol phosphate functions as a carrier and is acted upon by the series of reactions.

13. The major problem in the synthesis of complex carbohydrates outside the cell is the lack of an external energy source such as ATP. The biosynthesis of a bacterial cell wall such as the peptidoglycan occurs mainly inside the cell, the final step being the cross-linking of the peptidoglycan strands outside the cell. This reaction is a transpeptidation, which does not require any energy source.

Chapter 17

1. The lecithin serves as an emulsifying agent that allows the aqueous and lipid phases to be dispersed in each other and increases the time required for phase separation.

3. The fatty acyl groups are placed in these two different positions by different enzymes.

5. The hydrophobic amino acid side chains on the exterior of the integral membrane protein interact with the hydrophobic lipid of the membrane exterior and are stable in the nonaqueous environment. These residues pack in the interior, hydrophobic environment of globular proteins.

7. Peripheral proteins are bound to the inner or outer aspects of the membrane through weak ionic interactions that include association with phospholipid head groups, by electrostatic or ionic interaction with a hydrophilic region of an integral membrane protein or through divalent metal ion bridging to the membrane surface. Peripherally bound proteins may be released without disrupting the membrane. Thus in-

creased salt concentration shields ionic charges and weakens the charge–charge interactions between the peripheral protein and the membrane components.

   Integral proteins are dissolved into the lipid bilayer of the membrane through interactions of the hydrophobic amino acid side chains and fatty acyl groups of phospholipids. In order to remove integral membrane proteins, the membrane must be disrupted by addition of detergents or other chaotropic reagents to solubilize the protein and to prevent aggregation and precipitation of the hydrophobic proteins upon their removal from the membrane.

9. In principle, placing a hydrophilic residue in a non-aqueous environment is energetically unfavorable. In integral proteins with multiple α helices that span the membrane, hydrophilic side chains from different helical segments may interact and in some cases form a channel through which ions may diffuse. Portions of the helical segments exposed to the lipid will contain primarily hydrophobic amino acid residues.

11. Visualizing a protein actually revolving in a controlled manner within the membrane is difficult.

13. (a) $\Delta\Psi = +18$ mV; (b) from side 1 to side 2; (c) Side 1: 64.3 mM K^+, 50 mM Na^+, 144.3 mM Cl^-. Side 2: 85.7 mM K^+, 85.7 mM Cl^-. $\Delta\Psi = 7.5$ mV.

15. Sucrose uptake should be inhibited by a proton ionophore if uptake is by a proton symport. If a protein-binding system was operational, membrane vesicles or cells subjected to osmotic shock would be defective in uptake. If a Na^+ symport was involved, uptake would be dependent on extracellular Na^+. If a PTS was operational, sucrose phosphorylation would be dependent on PEP and not ATP in a crude cell extract.

Chapter 18

1. Arachidic acid produces 134 moles of ATP/mole, whereas arachidonic acid gives 126 moles of ATP/mole of fatty acid. This difference in ATP quantities is rather minor. Therefore any biological difference caused by dietary saturated versus unsaturated fats being due to their ATP yields is unlikely.

3. (a) 29.5 or 30 moles ATP
   (b) 64 moles ATP
   (c) 33% and 32%, respectively
   (d) 2.25 versus 2.13

5. (a) Oxidation of 1 mole of glucose yields 32 moles of ATP. 350 kcal of energy are stored as ATP or 1.9 kcal/g. (b) Oxidation of 1 mole of palmitate yields 106 moles of ATP. 1200 kcal of energy are stored as

ATP, or 4.7 kcal/g. (c) Lipids are more highly reduced than are carbohydrates and supply more reducing equivalents to the electron-transport system than do carbohydrates. (d) Lipids have approximately 2.5 times greater energy storage capacity per gram than do carbohydrates and are stored as compact, hydrophobic globules. Storage of an equivalent energy as carbohydrate would require at least 2.5 times the mass, not considering the water of hydration that would accompany the carbohydrate.

7. The availability of citrate has no relationship to the flow of metabolites through the tricarboxylic acid cycle. Increased citrate concentrations result in increased cytoplasmic acetyl-CoA concentrations, which in turn increases fatty acid biosynthesis.

9. From a reaction sequence viewpoint it is a cyclic process. From a substrate viewpoint it is a decreasing spiral because the substrate's length decreases with each turn around the spiral.

11. Many mechanisms are conceivable. One possibility is the approach of a phosphate on succinyl-CoA to produce succinyl phosphate (a mixed anhydride).

13. (a) Ketone body formation in liver supplies an easily transported, water-soluble, energy-rich metabolite that can be used in lieu of glucose in many nonhepatic tissues. (b) β-hydroxybutyrate supplies an additional hydride (2 reducing equivalents) compared with acetoacetate. (c) Consider the reactions catalyzed by β-hydroxybutyrate dehydrogenase, 3-ketoacyl-CoA transferase, and thiolase.

15. (a) The ^{14}C-labeled methyl group of acetyl-CoA will be C-16 of palmitate. (b) Only one deuterium atom from each labeled malonyl-CoA will remain in the reduced lipid chain. Carbons 2, 4, 6, 8, 10, 12, and 14 of palmitic acid will each have one deuterium label. (c) The ^{14}C label will be lost by decarboxylation and no label will remain in the palmitate.

17. Glucose-6-phosphate dehydrogenase, 6-phosphogluconate dehydrogenase, and the $NADP^+$-malic enzyme are sources of the 14 moles of NADPH required for biosynthesis of palmitate.

19. The thioesterase activity of the fatty acid synthase prefers the palmitoyl acyl carrier protein thioester as substrate.

Chapter 19

1. Phosphatidylserine decarboxylase is a pyridoxal phosphate enzyme.

3. The synthesis of phosphatidylcholine is thermodynamically feasible because of the ATP used to phosphorylate choline and the CTP used to form CDP-choline.

5. One possibility is to remove the fatty acyl groups that have been damaged by oxidation.

7. Carriers of a defective gene for the hexosaminidase A enzyme still have a functional copy of the gene.

9. The activated component is different. In the first case (figure 19.4) the minor component is activated, whereas in the production of phosphatidylinositol (figure 19.6) the diacylglycerol is activated.

11. Arachidonic acid is stored in membranes as phospholipids with C_{20} polyunsaturated fatty acids in the SN-2 position. Phospholipase A_2 releases arachidonic acid, which is then used to synthesize prostaglandins, which induce inflammation.

Chapter 20

1. No

3. These frequencies are consistent if one assumes that the gene has only two alleles: $A + B = 1$ and $A^2 + 2AB + B^2 = 1$ describes the population where A^2 are "normal," $2AB$ are heterozygous and B^2 are homozygous for familial hypercholesterolemia. $2AB = 1/500$, $AB = 0.001$, $A = 0.999$ and $B = 0.001$. Therefore, $B^2 = (0.001)^2 = 0.000001$, which is the one in a million indicated in the text. These frequencies would be inconsistent if the gene has more than two alleles.

5. If a normal person is given an inhibitor for HMG-CoA reductase, cholesterol synthesis is inhibited in the liver. Lower levels of cholesterol then signal the synthesis of increased levels of LDL receptors. This increases the uptake of LDL into the liver and reduces serum LDL. In a patient with FH, this has little effect because there are no LDL receptors. The only effect is that the liver does not make as much cholesterol and does not contribute as much to serum LDL levels. The new liver will make normal amounts of LDL receptors and have normal uptake of LDL from the blood. This result will dramatically lower serum LDL levels and prevent the new heart from developing coronary artery disease. If the liver transplant had not been done, the heart transplant would have been to no avail.

7. The ^{14}C in HMG-CoA derived from 2-[^{14}-C]-acetate is marked in the structure.

$$^* = {}^{14}C$$

9. The evolution of membranes, organelles, and organs combined with the evolutionary development of new metabolic sequences from previously developed enzymatic mechanisms probably best explains this phenomenon.

Chapter 21

1. The oxide ion is a poor leaving group.

3. The two sequences differ by an ATP.

5. Presumably the availability of valine inhibited an enzyme, such as acetohydroxy acid synthase, in the early stages of valine, isoleucine and leucine biosynthesis.

7. Only when the environmental tryptophan is depleted will material flow through the pathway, or in this case, because of the auxotroph, only part way through the pathway.

9. In this two-step pathway the acetyl group is added in the first reaction and an acetate leaves, and an HS^- is added in the second step. The acetyl group facilitates the removal of the serine oxygen. The OH group is a poor leaving group; acetate is better.

11. Glutamine, the product of glutamine synthase, is the source of nitrogen required in the synthesis of a number of diverse, structurally unrelated compounds synthesized by different pathways. Total inhibition of the glutamine synthase by a single product would in turn inhibit the synthesis of all compounds requiring glutamine.

13. Pyruvate, from glycolysis of glucose, is carboxylated to oxaloacetate or oxidized to acetyl-CoA. These metabolites enter the Krebs cycle, are metabolized to α-ketoglutarate and oxaloacetate, then transaminated to aspartate or glutamate. Asn, Gln, and Pro are synthesized from Asp or Glu. The cycle replenishes intermediates via the anaplerotic reactions (e.g., carboxylation of pyruvate to form oxaloacetate).

15. Hydroxyproline is formed by a posttranslational modification of proline residues in the protein. The ^{14}C-labeled hydroxyproline is not incorporated directly into the collagen because there is no genetic codon to specify the incorporation of hydroxyproline.

Chapter 22

1. The *N*-acetyl groups in the *de novo* pathway prevent the spontaneous cyclization of the semialdehyde intermediate.

3. One of the nitrogens enters the urea cycle via carbamoyl phosphate that comes from ammonia. The second urea nitrogen enters the urea cycle as part of aspartic acid that can come from transamination of oxaloacetate. Glutamate is the source of the amino group in the transamination.

5. Because of the importance of the urea cycle, the capacity to convert ornithine into arginine is obvious. Complete loss of the ability to produce ornithine (a catalyst or carrier in the urea cycle) would limit the organism's control over production of its nitrogen waste product.

7. Based on figure 22.11 Thr, Ala, Ser, Gly, Cys, Asn, Asp, Gln, Glu, His, Arg, Pro, Val, and Met are glucogenic; Lys, Trp, and Leu are ketogenic; and Phe, Tyr, and Ile are both ketogenic and glucogenic.

9. Because of its critical role in ATP production, a homozygotic defect in a gene for a protein involved in glycolysis, the citric acid cycle, or the electron transport chain probably leads to the death of the cell(s) soon after fertilization.

11. Pyridoxal phosphate forms a Schiff base (imine) with the glycine. A carbon-bound hydrogen is labile, and the resulting carbanion stabilized by resonance back into the pyridoxal phosphate. The carbanion approaches the carbonyl carbon of the succinyl-CoA. Following the elimination of the CoASH, the intermediate shown in figure 22.13 is formed. The intermediate then loses a CO_2, forming a carbanion that is resonance stabilized back into the pyridoxal phosphate.

13. We would predict an increased arginase activity in the liver of the untreated diabetic animal. The untreated diabetic animal synthesizes glucose primarily by hepatic gluconeogenesis utilizing amino acids derived from protein catabolism as the carbon source. The urea cycle activity must increase to accommodate the increased flux of amino groups removed from the amino acids. Arginase catalyzes the hydrolysis of arginine yielding urea plus ornithine and is the rate-limiting step in the urea cycle.

15. (a) L-glutathione is synthesized in successive steps, catalyzed by γ-glutamyl cysteine synthase and glutathione synthase. Glutathione synthesis is directed by the substrate specificity of these enzymes. (b) Decreased glutathione synthesis would increase the probability of oxidative damage to the cell.

Chapter 23

1. Carbamoyl phosphate synthase contributes to two processes: (a) the initial enzyme in the biosynthesis of pyrimidines and (b) a component in the synthesis of arginine biosynthesis or the urea cycle. In bacteria both of these processes occur within the same compartment. In human beings the carbamoyl phosphate synthase involved in the urea cycle is contained in the mitochondria; it is isolated from the cytosol counterpart that is involved in the biosynthesis of pyrmidines. Because the two carbamoyl phosphate synthases are in separate cellular compartments in human beings, control of the cytosol carbamoyl phosphate synthase by pyrimidine pathway products has no impact on the urea cycle.

3. Typically, the mechanism for amine addition involves a nucleophilic approach by a nitrogen, requiring a lone pair of electrons on the nitrogen. Ammonium ions are protonated at physiological pH and do not have a lone pair. The amide of glutamine is not protonated and carries a lone pair of electrons.

5. The phosphorylation step utilizing ATP converts the oxygen into a much better leaving group.

7. This oxygen becomes incorporated into an H_2O molecule.

9. Because Lesch-Nyhan patients lack the enzyme hypoxanthine-guanine phosphoribosyltransferase, they accumulate high levels of PRPP, which stimulates purine biosynthesis to high levels, leading to production of large amounts of uric acid. The brain may not have high levels of *de novo* purine biosynthesis and probably relies on the salvage pathway enzymes for its purine nucleotides.

11. The product would be 3-amino-2-methylpropanoic acid.

Chapter 24

1. In a liver cell, fatty acid synthesis takes place in the cytosol. It uses acetyl-CoA carboxylase and a large, multifunctional polypeptide fatty acid synthase. The first committed step is acetyl-CoA carboxylase, which is highly regulated by hormonal control; this results in the phosphorylation of the enzyme. Citrate is transported from the mitochondria and is used to generate acetyl-CoA and reducing power in the form of NADPH (NADPH is used in biosynthetic reactions instead of NADH). Citrate activates the carboxylase, but the end product, palmitate, inhibits the reaction. Fatty acid degradation occurs in the matrix of the mitochondria. A key point of regulation is on the uptake of the fatty acid into the matrix of the mitochondria. Malonyl-CoA, the product of the acetyl-CoA carboxylase, inhibits uptake and prevents the newly made palmitic acid from being degraded.

Gluconeogenesis utilizes many of the glycolytic enzymes, yet three of these enzymes in glycolysis have large negative free energy changes in the direction of pyruvate formation. These reactions must be replaced in gluconeogenesis to make glucose formation thermodynamically favorable. Replacement allows glycolysis and gluconeogenesis to be thermodynamically favorable and at the same time permits the pathways to be independently regulated to avoid a futile cycle.

3. A hormone receptor must do two things if it is to function properly:
   (1) Distinguish the hormone from all other surrounding chemical signals and bind it with a very high affinity (K_d ranges from 1×10^{-7} M to 1×10^{-12} M).
   (2) Upon binding the hormone, undergo a conformational change into an active form that can then interact with other molecules that initiate the molecular events leading to the hormone's elicited response.

   Proteins are the only macromolecules that can exhibit this kind of behavior (specific binding and conformation change).

5. (1) A polypeptide signal sequence must be present if the protein is to be transported into the endoplasmic reticulum and subsequently secreted.
   (2) Additional polypeptide sequences are necessary for proper peptide chain folding (e.g., C peptide of insulin).
   (3) Cleavage allows control of hormones from inactive to active form (e.g., thyroxine).
   (4) Production of a number of different hormones from the same precursor allows coordinate production of several hormones. Specific cleavage by the cell allows control of which peptides are produced (e.g., cleavage of prepro-opiocortin to corticotropin, β-lipotropin, γ-lipotropin, α-MSH, β-MSH, γ-MSH, endorphin, and enkephalin).
   (5) A large precursor of the hormone can serve as a storage form (e.g., thyroglobulin).

7. The same amino acid, named as 5-oxoproline, is an intermediate in the γ-glutamyl cycle. The name pyroglutamate suggests formation involving dehydration via heat (fire) from glutamate.

9. Vitamin D can be considered both a hormone and a vitamin. Its mode of action is like that of other steroid hormones, and it is synthesized in the body. It can be given in the diet (e.g., in supplemented milk) and would then be called a vitamin.

Chapter 25

1. Avery, working with two different strains of pneumococcus, was able to show that a fraction isolated from the pathogenic S strain that transformed the nonpathogenic R strain was DNA. This transforming activity was not affected by RNase, proteases, or enzymes that degrade capsular polysaccharides but was destroyed by treatment with DNase. Purified DNA from S cells was able to transform R cells into S cells *in vitro*.

3. dpCpGpTpA or, in abbreviated form, CGTA.

5. (a) In the major groove of B-form DNA:
   Adenine N^7; N^6
   Cytosine N^4
   Guanine N^7, O^6
   Thymine O^4
   In the minor groove of B-form DNA:
   Adenine N^3
   Cytosine O^2
   Guanine N^3; N^2
   Thymine O^2
   (b) Although the numbers of electronegative atoms capable of forming hydrogen bonds in the major and minor groove are similar, access to those in the minor groove is hindered by ribose moiety.

7. The melting temperature (T_m) of DNA is affected by the base composition, with the G-C-rich DNA having a higher T_m than the A-T-rich DNA. The DNA samples could be heated in a spectrophotometer and the increase in absorbance of ultraviolet light (hyperchromism) could be plotted against temperature. The A-T-rich DNA would have a lower T_m than the G-C-rich DNA.

9. A simple approach would be to take small aliquots of the sample and treat them with the enzymes ribonuclease (RNase) or deoxyribonuclease (DNase). Digestion of the sample by one of these enzymes would indicate whether the sample is RNA or DNA. Another method is to treat a small sample with alkali, which degrades RNA to mononucleotides, but only denatures DNA to the single-stranded form. One way to detect whether these treatments had any effect on the nucleic acid is to subject it to electrophoresis on an agarose gel. Free nucleotides, or even small oligonucleotides, are not visible on agarose gels, whereas the original viral nucleic acid should yield one (or more) discrete high molecular weight bands.

   To determine whether the nucleic acid is single-stranded or double-stranded, it could be heated. A sharp increase in absorbance at 260 nm would indicate a double-stranded RNA or DNA; a broader melting curve would suggest a single-stranded nucleic acid. We could also analyze the nucleotide composition of the nucleic acid. Equivalence between A and T and between G and C would strongly suggest that the nucleic acid is double-stranded.

11. The 2′-OH group found on the ribose in RNA sterically prevents the B duplex from forming.

13. The differences in Ethidium (Et) binding capacities between linear duplex DNA and covalently closed circular DNA (cccDNA) can be understood in terms of the differences in topological constraints imposed on the two molecules. By intercalating between two adjacent base pairs, Et unwinds the double helix, which results in an increase in the length of the helix (pitch). For cccDNA, the conformational stress introduced by unwinding is compensated for by a change in tertiary structure of the molecule (i.e., supercoiling).

At some point, the torsional stress caused by the positive supercoils will become energetically unfavorable and the tendency of the molecule will be toward winding, thus preventing further binding of Et. Linear duplex DNA does not undergo this torsional stress because it is not covalently closed and would be expected to have a greater binding capacity for Et.

15. The ratio of the histones in chromatin supports the model proposed for nucleosome structure. The core of the nucleosome is made up of an octamer of two molecules each of H2A, H2B, H3, and H4. The H1 histone seals off the nucleosome (i.e., only one H1 per nucleosome).

Chapter 26

1. If DNA replication were dispersive, each strand of each daughter molecule would have had an intermediate density in the experiment. The fact that after one generation, one strand of the DNA still had the same density as the parental (^{15}N) DNA served to further corroborate the conclusion that DNA replication is semiconservative.

3. As has been seen in many other biochemical reactions in the cell, the generation of pyrophosphate coupled with its hydrolysis by pyrophosphatase is the major driving force for DNA synthesis. The Mg^{2+} can bind to the transition state intermediate (trigonal bipyramidal intermediate) and stabilize it, lowering the energy of activation. The metal ion can also promote the reaction by charge shielding; the Mg^{2+}NTP complex is the actual substrate, with the metal reducing the negative charge on the phosphate groups so

as not to repel the electron pair of the attacking nucleophile.

5. The use of a primer increases fidelity in DNA synthesis by providing a more extensive stacked double helix to which the first few bases of DNA are added, allowing the proofreading exonuclease activity of the polymerase to evaluate more accurately the stability of the newly synthesized DNA. The reason that the primer is made of RNA and not DNA may be because RNA, even in a double-stranded nucleic acid, is readily recognized as being different from DNA (RNA–DNA duplexes adopt a different conformation because of the presence of the 2′-OH on the ribonucleotides.) A proofreading DNA polymerase recognizes the differences in the RNA primer and replaces it with DNA.

7. About 33 min would be required to replicate the chromosome because you have a bidirectional mode of replication with two replication forks. If you had multifork replication (one round of replication starting before the other finished), a 20-min division time would be possible.

9. 2.9×10^9 bp $\times$ 1 sec/60 bp $\times$ 1 h/3600 sec = 13,400 h
The human genome would need at least 13,400 replication origins to be completely replicated in one hour.

11. DNA in eukaryotic chromosomes is complexed with histone proteins in complexes called nucleosomes. These DNA-protein complexes are disassembled directly in front of the replication fork. The nucleosome disassembly may be rate-limiting for the migration of the replication forks, as the rate of migration is slower in eukaryotes than prokaryotes. The length of Okazaki fragments is also similar to the size of the DNA between nucleosomes (about 200 bp). One model that would allow the synthesis of new eukaryotic DNA and nucleosome formation would be the disassembly of the histones in front of the replication fork and then the reassembly of the histones on the two duplex strands. Histone synthesis is closely coupled to DNA replication.

13. The SOS response is reversed when the protease activity of recA can no longer be activated because most or all of the damaged DNA has been repaired or eliminated and intact lexA protein levels begin to rise. Then lexA can act as a repressor, binding to the gene control regions of all of the genes it regulates, including its own and that of recA.

15. The recA protein is very important in DNA repair. An insult to the DNA leads to the activation of the

protease function of recA, which then cleaves lexA protein, turning on the genes in the SOS response. Once the DNA is repaired, the recA protease is inactivated and new lexA protein is made, repressing the DNA repair genes again. RecA mutations are totally deficient in homologous recombination, demonstrating the important role of the recA protein in this process. The purified recA protein will catalyze the exchange between duplex and single-stranded DNAs with the hydrolysis of ATP. Also, recA protein will form a complex between two circular helices if one helix is gapped on one strand. These functions of recA protein would place it at the hub of activities in recombination. RecA mutants would be very useful in genetic research because mutants generated would not be repaired, and in recombinant DNA cloning recombinational events between vector recombinant DNA and host genomic DNA would not occur.

Chapter 27

1. Spaces are introduced every 10 nucleotides for clarity: 5'-CAAAAAACGG ACCGGGTGTA CAACTTTTAC TATGGCGTGA CACCTAAATT ATAGGCAGAA ATAAGTACAT GACTATTGGG AGGAGCAGGA ACAAGTAGG-3'.

3. The frequency of occurrence of the sequence -CTGCAG- (a PstI site) in DNA that is 80% G + C is $0.4 \times 0.1 \times 0.4 \times 0.4 \times 0.1 \times 0.4$, or $(0.4)^4(0.1)^2 = $ once every 3906 bp. An AAGCTT (HindIII) site should occur $(0.4)^2(0.1)^4 = $ once every 62,500 bp.

   The *E. coli* genome composition is approximately 26% G, 26% C, 24% A, and 24% T. PstI would cleave this genome $(0.26)^4(0.24)^2 = $ once every 3,800 bp, and HindIII, $(0.26)^2(0.24)^4 = $ once every 4,500 bp.

5. There is no difference between the single-stranded ends generated by BamHI and by MboI. These ends are complementary, and they can anneal and be ligated together by DNA ligase. The resulting sequence contains an MboI site and can be cleaved by that enzyme. It has only a 25% probability of restoring a BamHI site, however, depending on the nucleotide located adjacent to the MboI site.

7. In order to isolate a 15-kb gene from a genome containing 3×10^9 bp, it would be necessary to isolate $3 \times 10^9/15,000$ or 200,000 fragments per genome, or 200,000 clones. We recommend that a library 3 to 10 times the minimum size should be prepared, to ensure a high probability that a given fragment will be represented at least once. In the case of the gene

specified, the library should therefore contain between 6×10^5 and 2×10^6 clones.

9. (a) 2, 4
   (b) 1, 5
   (c) 1, 3

11. To reduce the hybridization stringency, researchers choose conditions that stabilize double-stranded DNA, allowing sequences that share only partial complementarity to form base pairs. Examples of such conditions include reducing the hybridization temperature and increasing the salt concentration in solution.

13. Chromosome jumping is a useful procedure to traverse long distances and skip troublesome regions of the genome (repetitious sequences, etc.). One approach with this technique is to digest the genome with rare cutting restriction enzymes (*Not*I recognizes an 8-base sequence and generates an average fragment size of 500 kb of DNA). The fragments are circularized with a small marker DNA between the ends. The marker DNA contains sequences necessary for cloning in λ. These circular DNA fragments are then cleaved with another restriction enzyme that produces fragments small enough to clone in λ. With this procedure only clones that contain the marker DNA (i.e., the ends of the original fragment) are isolated. This method would permit only one jump; thus another method used with this technique is called a linking library. The same sample of DNA is digested with a restriction enzyme that gives smaller fragments. These smaller fragments are then circularized with the same marker DNA used in the jumping library. The circular DNA is then digested with *Not*I to linearlize the fragments for insertion into the vector. The linking library carries sequences on both sides of the *Not*I restriction sites while the jumping library carries sequences from one side of two adjacent *Not*I sites.

Chapter 28

1. RNA and DNA polymerases catalyze the same reaction mechanistically, involving hydrolysis of a nucleotide triphosphate to release pyrophosphate and form a phosphodiester bond. In both cases, the order of nucleotide addition is specified by the template, and synthesis of the growing nucleic acid chain is in a 5' to 3' direction (the enzymes move in a 3' to 5' direction along the template strand). In addition to the obvious difference in substrates (RNA polymerase utilizes ribonucleotides, whereas DNA polymerase utilizes deoxyribonucleotides), these two enzymes differ in their requirements for initiating synthesis:

RNA polymerase initiates *de novo,* and does not require a primer, unlike DNA polymerase. Finally, RNA polymerase does not perform any proofreading activity, unlike DNA polymerases, which generally have a 3′ to 5′ exonuclease activity to remove misincorporated nucleotides.

3. mRNA has a short half-life (a couple of minutes); tRNA and rRNA are very stable (half-life measured in hours) and accumulate to make up most of the RNA in the cell (95%).

5. The bases within the base-paired region of each arm of the tRNA cloverleaf stack in a manner similar to the base stacking described in Chapter 25 for DNA. In addition to the base stacking within base-paired regions, there is also stacking of one helix on top of another in the tRNA molecule. In particular, the acceptor stem stacks with the TΨC stem and loop to form one nearly continuous stacked double helix. The anticodon stem and the D stem also stack on top of one another.

7. The D and T loops in tRNA interact with each other to form the tertiary structure, leaving only the anticodon with a single-stranded loop able to be cleaved by RNase.

9. The RNA polymerase binds to the same side of the duplex at the -10 and -35 regions with about two turns of the duplex helix between these boxes. This binding site can be determined by a variety of "footprint" experiments. DNA that is 5′-labeled can be mixed with the polymerase, and regions that are protected from digestion with an enzyme like DNase I can be determined on a sequencing gel. Sequences with tight contact with the RNA polymerase are observed as a series of blank spots in the sequencing ladder that look like footprints.

11. The 5′ terminus is capped (G^m pppX—), generating a guanosine nucleoside with 2′ and 3′-OH groups and a 5′-5′ pyrophosphate linkage. The other end of the mRNA has poly(A) added, so the 3′ end is adenosine. Therefore most of the mRNA in the cell has a guanosine and an adenosine 2′ and 3′-OH and no typical 5′-triphosphate ending (pppN—).

13. By flattening and widening the DNA minor groove, TATA binding protein may assist other factors in forming an open complexlike structure with separated strands. The bending induced by TATA binding protein would bring the DNA upstream and downstream of the TATA box closer together, promoting interactions between proteins bound to upstream elements and the start site of transcription.

15. If an intron or part of an intron containing a stop signal for translation was not removed, this mRNA would be longer but would yield a shorter polypeptide. Alternative splicing would remove the intron or use an alternative splice site, generating a shorter mRNA and a longer polypeptide.

Chapter 29

1. The difference in the mechanism of translation initiation in prokaryotes compared to eukaryotes has profound consequences for the strategy used to coordinate the expression of a set of genes in the two systems. In prokaryotes this coordination is achieved by organizing genes into transcription units that are transcribed to give polycistronic mRNAs. Each cistron within a polycistronic mRNA begins with a Shine-Dalgarno sequence and initiation AUG codon. In contrast, in eukaryotes, in which translation begins almost invariably (and exclusively) at the 5′-most AUG codon, monocistronic mRNAs are necessarily the order of the day. Genes that must be coordinately expressed are consequently not organized into transcriptional units, and coordination must be achieved in some other way.

3. This serine tRNA has inosine at the 5′ position of its anticodon, which can pair with U, C, or A. Thus the anticodon of this tRNA is most likely IGA (given in the correct 5′ to 3′ direction).

5. A number of variations have been found in the universal genetic code in genes located in mitochondria and chloroplast. These variations in the meaning of some of the code words represent divergences from the standard genetic code and not an independent origin of another genetic code. These divergences probably arose in these organelles because of the limited number of genes coded and requirements for the synthesis of ribosomes and tRNA. Clearly, something was unusual when only 24 types of tRNAs were found in mitochondria. Mitochondria do not use all 61 codons.

7. Met Val Glu Ile Arg Asp Thr His Leu
Lys Lys Gln Ile Ala Phe Ter Ter

9. (5′)AAY TGG GCN CAR TGY AAY CC(3′). R is the abbreviation for a mixture of A or G (R = puRine), Y is the abbreviation for a mixture of C and T (Y = pYrimidine), and N represents a mixture of all four Nucleotides.

11. The x-ray crystal structure of EF-Tu bound to GDP and GTP is known. Based on this structure, many of the amino acids shared among IF-2, EF-Tu, EF-G, and RF-3 apparently are somehow involved in binding of GTP and GDP. Thus these proteins (and many others, including the proto-oncogene, Ras) share a similar GTP-binding domain.

13. Import into the endoplasmic reticulum requires an N-terminal signal sequence that contains a long stretch of hydrophobic amino acids. The mitochondrial transit peptide is a hydrophilic sequence rich in serine and threonine, with regularly spaced basic amino acids. Import into the ER requires the signal recognition particle and its receptor, but mitochondrial import does not require the SRP and presumably uses a different receptor. Import into mitochondria requires a membrane potential, but import into the ER does not.

15. GAA encodes a glutamic acid residue. If this glu is essential for the catalytic activity of the protein, any mutation (except GAA to GAG) will be harmful to the activity of the protein. If the glu residue is not essential for catalysis or proper folding, the possible mutations, in order of increasing potential severity, are:

GAA changes to	Glu changes to
GAG	Glu
GAT or GAC	Asp
CAA	Gln
GCA	Ala
AAA	Lys
GTA	Val
GGA	Gly
TAA	Terminator

Chapter 30

1. The existence of a repressor in *lac* operon regulation was first suggested by the results of studies of merodiploids. In merodiploids of the type i^+z^-/Fi^-z^+, Jacob and Monod were able to demonstrate that the i^+ (inducible) allele is dominant to the i^- (constitutive) allele when on the same chromosome (*cis*) or on a different chromosome (*trans*) with respect to the z^+ allele.

3. (a) Cells with the genotype $i^so^+z^+$ have a "super-repressor" i^s mutation, which causes the repressor to be insensitive to an inducer. Thus even though the operator and β-galactosidase loci are wild-type, no β-galactosidase will be produced in this mutant. On media containing X-gal, with or without IPTG, colonies will be white.

(b) Cells with the genotype $i^so^cz^+$ still have the superrepressor, but the β-galactosidase gene is under the control of a constitutive operator, o^c. This mutation interferes with the ability of the repressor to bind and repress transcription. Therefore, β-galactosidase probably would be produced continuously. On X-gal, with or without IPTG, colonies will be blue.

(c) Merodiploids with the genotype $i^so^cz^-/i^+o^+z^+$ would behave like the mutant described in part (a).

(d) Merodiploid cells with the genotype $i^+o^cz^-/i^-o^+z^+$ would exhibit β-galactosidase regulation essentially identical to that of wild-type cells; on X-gal alone, colonies would be white. In the presence of an inducer, the β-galactosidase gene would be induced; on X-gal and IPTG, colonies would be blue.

5.

	Uninduced		Induced	
Strains	Enz A	Enz B	Enz A	Enz B
Haploid				
(1) $R^+O^+A^+B^+$	1	1	100	100
(2) $R^+O^cA^+B^+$	1–100	1–100	100	100
(3) $R^-O^+A^+B^+$	100	100	100	100
Diploid				
(4) $R^+O^+A^+B^+/$ $R^+O^+A^+B^+$	2	2	200	200
(5) $R^+O^cA^+B^+/$ $R^+O^+A^+B^+$	2–101	2–101	200	200
(6) $R^+O^+A^-B^+/$ $R^+O^+A^+B^+$	1	2	100	200
(7) $R^-O^+A^+B^+/$ $R^+O^+A^+B^+$	2	2	200	200

7. The explanation for the less than perfect match of most promoters to the consensus sequence is to be found in the need to regulate transcription. Transcriptional regulation is achieved in many instances by the selective improvement of the affinity of specific promoters for RNA polymerase. Such selective improvement is well illustrated in the case of regulation of the *lac* operon.

9. One possibility is the binding of the repressor non-specifically to the DNA and then searching in one dimension (binding to the DNA and then sliding along until the promoter is reached). Also, a long section of DNA would not be randomly distributed but would form a loose ball of DNA that would define a domain much smaller than the solution in the test tube. When the repressor was released from the DNA it could more quickly find another strand of DNA to bind (effectively giving a much higher concentration of DNA).

11. The synthesis of ribosomal proteins is regulated in *E. coli* by translational regulation (i.e., free ribosomal proteins inhibit the translation of their own

mRNA). As long as rRNA is being made, these proteins bind to the rRNA and the translation of the ribosomal proteins continues. The genes for the ribosomal proteins are clustered in a number of operons that produce polycistronic mRNAs. One of the simplest operons is P_{L11}, which codes for proteins L1 and L11. L1 is the regulatory protein and can bind to the 23S rRNA or to the 5' end of its own polycistronic mRNA. If the levels of L1 increase, it binds to its own mRNA and inhibits translation of both L1 and L11 proteins. This mechanism keeps the levels of L1 and L11 in register with the amount of rRNA. The other ribosomal proteins are regulated in a similar manner.

13. Translation is an amplification process, in that each molecule of ribosomal protein mRNA can yield many copies of the corresponding protein.

15. The genes encoded by bacteriophage lambda DNA are organized so that gene products required for the same function or process are clustered. Clustering facilitates regulation because all the gene products required at the same time can be induced simultaneously. Clustering also leads to more efficient organization and tighter packing of the bacteriophage genome.

17. The final proof of the two-site model came with the cocrystallization of the regulatory protein and its DNA-binding site. The protein binds on one side of the DNA in two adjacent major grooves. The hydrogen-bonding groups exposed in the major groove present many possibilities for interactions with the amino acid side chains of the protein.

Chapter 31

1. One definition is: *the DNA (or RNA) sequences necessary to produce a peptide (or RNA).* Some viruses have RNA as their genetic material and some genes do not code for a protein but make a functional RNA such as tRNA or rRNA.

3. No, this construct would not be activated by binding of *lexA* because *lexA* does not have a eukaryotic activation domain.

5. A deletion of *HMLE* would remove the element repressing transcription of HML_α at the "storage" location, and result in the constitutive expression of HML_α genes from this locus. The consequences of such constitutivity would depend on the mating type of the mutant. If HML_α were at the *MAT* locus (MAT_α), the arrangement characteristic of the α mating type, the cell exhibits the behavior expected of the α mating type. However, if HMR_a were at the

MAT locus (MAT_a), the arrangement giving rise to the a mating type, the cell would resemble the diploid, expressing both a and α genes simultaneously, and thus be sterile.

Mutants that began as α-type would become sterile at a rate conditioned by the frequency of transposition of the HMR_a to the *MAT* locus. In homothallic strains this frequency is very high, approximately once per cell division, while in heterothallic strains it is considerably lower. Deletion of the gene encoding $\alpha 2$ from MAT_α would result in a failure to inhibit expression of the a-specific genes. Haploid mutants that contained such a deletion would therefore resemble the diploid, and be sterile. If the $\alpha 2$ gene contained by the HML_α locus were similarly mutated, mutants of this type would be unable to switch to the α-type. However, transposition of HMR_a to the *MAT* locus would allow for expression of the a mating type. Again the frequency of transposition would condition the rate of switching between sterile and a-type, the frequency being far greater in homothallic strains than in heterothallic strains.

7. The results of the nuclear transplantation experiments of Briggs and King indicated that nuclei, to a certain stage in embryonic development, remain totipotent, in the sense that they can support the normal development of enucleated differentiating embryos. They found that blastula nuclei, although already "committed" to differentiated pathways, were capable of being "reprogrammed" by a proper environment to allow for some degree of dedifferentiation. Gastrula nuclei, in contrast, were found to support normal development only at a much reduced efficiency.

9. Histones bind to DNA by electrostatic interaction of basic amino acids (arginine and lysine) with the negative charge on the phosphate. These electrostatic (negative-positive charge) interactions are weakened by high salt, which will allow the histone proteins to disassociate from the DNA. Many regulatory proteins may initially interact in a nonspecific fashion with DNA until they locate their high-affinity binding sites. Once these proteins bind at their specific recognition sites, the major interaction is through hydrogen bonding and hydrophobic interactions. These hydrophobic interactions are stabilized by high salt.

11. MspI would cleave at every CCGG sequence in or near the globin genes, regardless of the cell type from which the DNA had been isolated because this enzyme is insensitive to methylation. Globin genes isolated from erythroblast cells would not differ from those of other tissues in their cleavage pattern with MspI.

HpaII does not recognize CCGG sequences when the second C is methylated. In tissues in which genes are *not* transcribed, CpG sequences tend to be methylated more often than in tissues where the same genes are transcribed. Thus HpaII is expected to recognize and cleave more sites in globin genes in erythroblast DNA than in DNA from other cell types (where the globin genes are not transcribed and the corresponding DNA is more highly methylated).

13. Most transcription factors have two-domain structures, one for DNA binding and another for protein binding. The DNA-binding domains involve various structural motifs that interact with the DNA in the major groove. The type of structure previously discussed in prokaryotes is the helix-turn-helix motif, which is also found in some eukaryotic transcription factors. A more common motif in eukaryotes is the zinc finger (zinc forms a tetrahedral complex with histidines and cysteines). Another type of structure found is the leucine zipper (a highly conserved stretch of amino acids with net basic charge followed by a region of four leucine residues at intervals of seven amino acids). All of these motifs found in these transcription factors involve interactions in the major groove of the DNA with either an α-helix segment (more common) or a two-stranded antiparallel β sheet.

15. One strategy employed to generate sufficient amounts of a product required at a specific stage of development is gene amplification. An alternative strategy used to provide large amounts of required products early in development is the accumulation of mRNAs or proteins prior to need.

17. Although the precise regulatory relationships between homeotic genes remain unclear, it is apparent that a hierarchy of interactions exists among gap, pair-rule, and segment polarity genes. Each gene class is influenced by the action of earlier acting genes that control larger units of pattern, and by some members of the same class. Thus gap genes that are expressed early influence the pattern of expression of pair-rule, (e.g., fushi terazu, *ftz*), segment polarity (e.g., engrailed, *en*) and homeotic genes (e.g., ultrabithorax, *ubx*), which are expressed later in a hierarchical fashion. Although segment polarity genes do not affect the expression of gap or pair-rule genes that occupy a higher position in the hierarchy, some gap genes, and presumably genes at other levels in the hierarchy, have been found to be mutually negative regulators of each other.

Glossary

A

A form. A duplex DNA structure with right-handed twisting in which the planes of the base pairs are tilted about 70° with respect to the helix axis.

Acetal. The product formed by the successive condensation of two alcohols with a single aldehyde. It contains two ether-linked oxygens attached to a central carbon atom.

Acetyl-CoA. Acetyl-coenzyme A, a high-energy ester of acetic acid that is important both in the tricarboxylic acid cycle and in fatty acid biosynthesis.

Actin. A protein found in combination with myosin in muscle and also found as filaments constituting an important part of the cytoskeleton in many eukaryotic cells.

Actinomycin D. An antibiotic that binds to DNA and inhibits RNA chain elongation.

Activated complex. The highest free energy state of a complex in going from reactants to products.

Active site. The region of an enzyme molecule that contains the substrate-binding site and the catalytic site for converting the substrate(s) into product(s).

Active transport. The energy-dependent transport of a substance across a membrane.

Adenine. A purine base found in DNA or RNA.

Adenosine. A purine nucleoside found in DNA, RNA, and many cofactors.

Adenosine diphosphate (ADP). The nucleotide formed by adding a pyrophosphate group to the 5′-OH group of adenosine.

Adenosine triphosphate (ATP). The nucleotide formed by adding yet another phosphate group to the pyrophosphate group on ADP.

Adenylate cyclase. The enzyme that catalyzes the formation of cyclic 3′,5′-adenosine monophosphate (cAMP) from ATP.

Adipocyte. A specialized cell that functions as a storage depot for lipid.

Aerobe. An organism that utilizes oxygen for growth.

Affinity chromatography. A column chromatographic technique that employs attached functional groups that have a specific affinity for sites on particular proteins.

Alcohol. A molecule with a hydroxyl group attached to a carbon atom.

Aldehyde. A molecule containing a doubly bonded oxygen and a hydrogen attached to the same carbon atom.

Alleles. Alternative forms of a gene.

Allosteric enzyme. An enzyme the active site of which can be altered by the binding of a small molecule at a nonoverlapping site.

Allosteric site. Location on an allosteric enzyme where the allosteric effector binds.

Aminotransferase. An enzyme that catalyzes the transfer of an amino group from an α-amino acid to an α-keto acid.

Amphibolic pathway. A metabolic pathway that functions in both catabolism and anabolism.

Anabolism. Metabolism that involves biosynthesis.

Anaerobe. An organism that does not require oxygen for maintenance or growth.

Anaplerotic reaction. An enzyme-catalyzed reaction that replenishes the intermediates in a cyclic pathway.

Angstrom (Å). A unit of length equal to 10^{-8} cm.

Anomers. The sugar isomers that differ in configuration about the carbonyl carbon atom. This carbon atom is called the anomeric carbon atom of the sugar.

Antibiotic. A natural product that inhibits bacterial growth (is bacteriostatic) and sometimes results in bacterial death (is bacteriocidal).

Antibody. A specific protein that interacts with a foreign substance (antigen) in a specific way.

Anticodon. A sequence of three bases on the transfer RNA that pair with the bases in the corresponding codon on the messenger RNA.

Antigen. A foreign substance that triggers antibody formation and is bound by the corresponding antibody.

Antiparallel β-pleated sheet (β sheet). A hydrogen-bonded secondary structure formed between two or more extended polypeptide chains.

Antiport. A protein system that can transport different molecules in opposite directions.

Apoactivator. A regulatory protein that stimulates transcription from one or more genes in the presence of a coactivator molecule.

Asexual reproduction. Growth and cell duplication that does not involve the union of nuclei from cells of opposite mating types.

Asymmetrical carbon. A carbon that is covalently bonded to four different groups.

Attenuator. A provisional transcription stop signal.

Autoradiography. The technique of exposing film in the presence of disintegrating radioactive particles. Used to obtain information on the distribution of radioactivity in a gel or a thin cell section.

Autoregulation. The process by which a gene regulates its own expression.

Autotroph. An organism that can form its organic constituents from CO_2.

Auxin. A plant growth hormone usually concentrated in the apical bud.

Auxotroph. A mutant that cannot grow on the minimal medium on which a wild-type member of the same species can grow.

Avogadro's number. The number of molecules in a gram molecular weight of any compound (6.022×10^{23}).

B

β bend. A characteristic way of turning an extended polypeptide chain in a different direction, involving the minimum number of residues.

β oxidation. Oxidative degradation of fatty acids that occurs by the successive oxidation of the β-carbon atom.

β sheet. A sheetlike structure formed by the interaction between two or more extended polypeptide chains.

B cell. One of the major types of cells in the immune system. B cells can differentiate to form memory cells or antibody-forming cells.

B form. The most common form of duplex DNA, containing a right-handed helix and about 10 (10.5 exactly) bp per turn of the helix axis.

Base analog. A compound, usually a purine or a pyrimidine, that differs somewhat from a normal nucleic acid base.

Base stacking. The close packing of the planes of base pairs, commonly found in DNA and RNA structures.

Bidirectional replication. Replication in both directions away from the origin, as opposed to replication in one direction only (unidirectional replication).

Bilayer. A double layer of lipid molecules with the hydrophilic ends oriented outward, in contact with water, and the hydrophobic parts oriented inward toward each other.

Bile salts. Derivatives of cholesterol with detergent properties that aid in the solubilization of lipid molecules in the digestive tract.

Biochemical pathway. A series of enzyme-catalyzed reactions that results in the conversion of a precursor molecule into a product molecule.

Bioluminescence. The production of light by a biochemical system.

Blastoderm. The stage in embryogenesis when a unicellular layer at the surface surrounds the yolk mass.

Bond energy. The energy required to break a bond.

Branchpoint. An intermediate in a biochemical pathway that can follow more than one route in subsequent steps.

Buffer. A conjugate acid–base pair that is capable of resisting changes in pH when acid or base is added to the system. This tendency is maximal when the conjugate forms are present in equal amounts.

C

cAMP. 3′,5′ cyclic adenosine monophosphate. The cAMP molecule plays a key role in metabolic regulation.

CAP. The catabolite gene activator protein, sometimes incorrectly referred to as the CRP protein. The latter term, in small letters (*crp*), should be used to refer to the gene but not to the protein.

Capping. Covalent modification involving the addition of a modified guanine group in a 5′-5″ linkage. It occurs only in eukaryotes, primarily on mRNA molecules.

Carbanion. A negatively charged carbon atom.

Carbohydrate. A polyhydroxy aldehyde or ketone.

Carboxylic acid. A molecule containing a carbon atom attached to a hydroxyl group and to an oxygen atom by a double bond.

Carcinogen. A chemical that can cause cancer.

Carotenoids. Lipid-soluble pigments that are made from isoprene units.

Catabolism. That part of metabolism concerned with degradation reactions.

Catabolite repression. The general repression of transcription of genes associated with catabolism that is seen in the presence of glucose.

Catecholamines. Hormones that are amino derivatives of catechol, for example, epinephrine or norepinephrine.

Catalyst. A compound that lowers the activation energy of a reaction without itself being consumed.

Catalytic site. The site of the enzyme involved in the catalytic process.

Catenane. An interlocked pair of circular structures, such as covalently closed DNA molecules.

Catenation. The linking of molecules without any direct covalent bonding between them, as when two circular DNA molecules interlock like the links in a chain.

cDNA. Complementary DNA, made *in vitro* from the mRNA by the enzyme, reverse transcriptase, and deoxyribonucleotide triphosphates.

Cell commitment. That stage in a cell's life when it becomes committed to a certain line of development.

Cell cycle. All of those stages that a cell passes through from one cell generation to the next.

Cell line. An established clone originally derived from a whole organism through a long process of cultivation.

Cell lineage. The pedigree of cells resulting from binary fission.

Cell wall. A tough outer coating found in many plant, fungal, and bacterial cells that accounts for their ability to withstand mechanical stress or abrupt changes in osmotic pressure. Cell walls always contain a carbohydrate component and frequently also a peptide and a lipid component.

Channeling. The direct transfer of a reaction intermediate from one enzyme to the next.

Chelate. A molecule that contains more than one binding site and frequently binds to another molecule through more than one binding site at the same time.

Chemiosmotic coupling. The coupling of ATP synthesis to an electrochemical potential gradient across a membrane.

Chimeric DNA. Recombinant DNA the components of which originate from two or more different sources.

Chiral compound. A compound that can exist in two forms that are nonsuperimposable images of one another.

Chlorophyll. A green photosynthetic pigment that is made of a magnesium dihydroporphyrin complex.

Chloroplast. A chlorophyll-containing photosynthetic organelle, found in eukaryotic cells, that can harness light energy.

Chromatin. The nucleoprotein fibers of eukaryotic chromosomes.

Chromatography. A procedure for separating chemically similar molecules. Segregation is usually carried out on paper or in glass or metal columns with the help of different solvents. The paper or glass columns contain porous solids with functional groups that have limited affinities for the molecules being separated.

Chromosome. A threadlike structure, visible in the cell nucleus during metaphase, that carries the hereditary information.

Chromosome puff. A swollen region of a giant chromosome; the swelling reflects a high degree of transcription activity.

Cis dominance. Property of a sequence or a gene that exerts a dominant effect on a gene to which it is linked.

Cistron. A genetic unit that encodes a single polypeptide chain.

Citric acid cycle. *See* tricarboxylic acid (TCA) cycle.

Clone. One of a group of genetically identical cells or organisms derived from a common ancestor.

Cloning vector. A self-replicating entity to which foreign DNA can be covalently attached for purposes of amplification in host cells.

Closed system. A system that exchanges neither matter nor energy with its surroundings.

Coactivator. A molecule that functions in conjunction with a protein apoactivator. For example, cAMP is a coactivator of the CAP protein.

Codon. In a messenger RNA molecule, a sequence of three bases that represents a particular amino acid.

Coenzyme. An organic molecule that associates with enzymes and affects their activity.

Cofactor. A small molecule required for enzyme activity. It could be organic, like a coenzyme, or inorganic, like a metallic cation.

Competitive inhibition. An inhibitor that competes with the substrate for binding to the enzyme.

Complementary base sequence. For a given sequence of nucleic acids, the nucleic acids that are related to them by the rules of base pairing.

Configuration. The spatial arrangement of atoms in a molecule that can only be changed by breaking and reforming covalent bonds.

Conformation. The spatial arrangement of groups in a molecule that can be changed without breaking covalent bonds. Often molecules with the same configuration can have more than one conformation.

Consensus sequence. In nucleic acids, the "average" sequence that signals a certain type of action by a specific protein. The sequences actually observed usually vary around this average.

Constitutive enzymes. Enzymes synthesized in fixed amounts, regardless of growth conditions.

Cooperative binding. A situation in which the binding of one substituent to a macromolecule favors the binding of another. For example, DNA cooperatively binds histone molecules, and hemoglobin cooperatively binds oxygen molecules.

Coordinate induction. The simultaneous expression of two or more genes.

Cosmid. A DNA molecule with *cos* ends from λ bacteriophage that can be packaged *in vitro* into a virus for infection purposes.

Cot curve. A curve that indicates the rate of DNA–DNA annealing as a function of DNA concentration and time.

Covalent bond. A chemical bond that involves sharing electron pairs.

Cytidine. A pyrimidine nucleoside found in DNA and RNA.

Cytochromes. Heme-containing proteins that function as electron carriers in oxidative phosphorylation and photosynthesis.

Cytokinin. A plant hormone produced in root tissue.

Cytoplasm. The contents enclosed by the plasma membrane, excluding the nucleus.

Cytosine. A pyrimidine base found in DNA and RNA.

Cytoskeleton. The filamentous skeleton, formed in the cytoplasm, that is largely responsible for controlling cell shape.

Cytosol. The liquid portion of the cytoplasm, including the macromolecules but not including the larger structures, such as subcellular organelles or cytoskeleton.

D

D loop. An extended loop of single-stranded DNA displaced from a duplex structure by an oligonucleotide.

Dalton. A unit of mass equivalent to the mass of a hydrogen atom (1.66×10^{-24} g).

Dark reactions. Reactions that can occur in the dark, in a process that is usually associated with light, such as the dark reactions of photosynthesis.

De novo **pathway.** A biochemical pathway that starts from elementary substrates and ends in the synthesis of a biochemical.

Deamination. The enzymatic removal of an amine group, as in the deamination of an amino acid to an α-keto acid.

Dehydrogenase. An enzyme that catalyzes the removal of a pair of electrons (and usually one or two protons) from a substrate molecule.

Deletion mutation. A mutation in which one or more nucleotides is removed from a region of the gene.

Denaturation. The disruption of the native folded structure of a nucleic acid or protein molecule; may be due to heat, chemical treatment, or change in pH.

Density-gradient centrifugation. The separation, by centrifugation, of molecules according to their density, in a gradient varying in solute concentration.

Dialysis. Removal of small molecules from a macromolecule preparation by allowing them to pass across a semipermeable membrane.

Diauxic growth. Growth on a mixture of two carbon sources in which one carbon source is used up before the other one is mobilized. For example, in the presence of glucose and lactose, *E. coli* utilizes the glucose before the lactose.

Difference spectra. Display comparing the absorption spectra of a molecule or an assembly of molecules in different states, for example, those of mitochondria under oxidizing or reducing conditions.

Differential centrifugation. Separation of molecules or organelles by sedimentation rate.

Differentiation. A change in the form and pattern of a cell and the genes it expresses as a result of growth and replication, usually during development of a multicellular organism. Also occurs in microorganisms (e.g., in sporulation).

Diploid cell. A cell that contains two chromosomes ($2N$) of each type.

Dipole. A separation of charge within a single molecule.

Directed mutagenesis. In a DNA sequence, an intentional alteration that can be genetically inherited.

Dissociation constant. An equilibrium constant for the dissociation of a molecule into two parts (e.g., dissociation of acetic acid into acetate anion and proton).

Disulfide bridge. A covalent linkage formed by oxidation between two SH groups either in the same polypeptide chain or in different polypeptide chains.

DNA. Deoxyribonucleic acid. A polydeoxyribonucleotide in which the sugar is deoxyribose; the main repository of genetic information in all cells and most viruses.

DNA cloning. The propagation of individual segments of DNA as clones.

DNA library. A mixture of clones, each containing a cloning vector and a segment of DNA from a source of interest.

DNA polymerase. An enzyme that catalyzes the formation of $3'-5'$ phosphodiester bonds from deoxyribonucleotide triphosphates.

DNA supercoiling. The coiling of double helix DNA upon itself.

Domain. A segment of a folded protein structure showing conformational integrity. A domain can comprise the entire protein or just a fraction of the protein. Some proteins, such as antibodies, contain many structural domains.

Dominant. Describing an allele the phenotype of which is expressed regardless of whether the organism is homozygous or heterozygous for that allele.

Double helix. A structure in which two helically twisted polynucleotide strands are held together by hydrogen bonding and base stacking.

Duplex. Synonymous with double helix.

Dyad symmetry. Property of a structure that can be rotated by $180°$ to produce the same structure.

E

Ecdysone. A hormone that stimulates the molting process in insects.

Edman degradation. A systematic method of sequencing proteins, proceeding by stepwise removal of single amino acids from the amino terminal of a polypeptide chain.

Eicosanoid. Any fatty acid with 20 carbons.

Electron carrier. A protein or coenzyme that can reversibly gain and lose electrons and that serves the function of carrying electrons from one site to another.

Electron donor. A substance that donates electrons in an oxidation–reduction reaction.

Electrophoresis. The movement of particles in an electrical field. A commonly used technique for analysis of mixtures of molecules in solution according to their electrophoretic mobilities.

Elongation factors. Protein factors uniquely required during the elongation phase of protein synthesis. Elongation factor G (EF-G) brings about the movement of the peptidyl-tRNA from the A site to the P site of the ribosome.

Eluate. The effluent from a chromatographic column.

Embryo. Plant or animal at an early stage of development.

Enantiomers. Isomers that are mirror images of each other.

Endergonic reaction. A reaction with a positive free energy change.

Endocrine glands. Specialized tissues the function of which is to synthesize and secrete hormones.

Endonuclease. An enzyme that breaks a phosphodiester linkage at some point within a polynucleotide chain.

Endopeptidase. An enzyme that breaks a polypeptide chain at an internal peptide linkage.

Endoplasmic reticulum. A system of double membranes in the cytoplasm that is involved in the synthesis of transported proteins. The rough endoplasmic reticulum has ribosomes associated with it. The smooth endoplasmic reticulum does not.

End-product (feedback) inhibition. The inhibition of the first enzyme in a pathway by the end product of that pathway.

Energy charge. The fractional degree to which the AMP–ADP–ATP system is filled with high-energy phosphates (phosphoryl groups).

Enhancer. A DNA sequence that can bind protein factors that stimulate transcription at an appreciable distance from the site where it is located. It acts in either orientation and either upstream or downstream from the promoter.

Entropy. A thermodynamic measure of the randomness of a system.

Enzyme. A protein that contains a catalytic site for a biochemical reaction.

Epimers. Two stereoisomers with more than one chiral center that differ in configuration at one of their chiral centers.

Equilibrium. In chemistry the point at which the concentrations of two compounds are such that the interconversion of one compound into the other compound does not result in any change in free energy.

***Escherichia coli* (*E. coli*).** A gram negative bacterium commonly found in the vertebrate intestine. It is the bacterium most frequently used in the study of biochemistry and genetics.

Established cell line. A group of cultured cells derived from a single origin and capable of stable growth for many generations.

Ether. A molecule containing two carbons linked by an oxygen atom.

Eukaryote. A cell or organism that has a membrane-bound nucleus.

Excision repair. DNA repair in which a damaged region is replaced.

Excited state. An energy-rich state of an atom or a molecule, produced by the absorption of radiant energy.

Exergonic reaction. A chemical reaction that takes place with a negative change in free energy.

Exon. A segment within a gene that carries part of the coding information for a protein.

Exonuclease. An enzyme that breaks a phosphodiester linkage at one or the other end of a polynucleotide chain so as to release a single nucleotides or small oligonucleotides.

F

F factor. A large bacterial plasmid, known as the sex-factor plasmid because it permits mating between F^+ and F^- bacteria.

Facilitated diffusion. Diffusion of a substance across a membrane through a protein transporter.

Facultative aerobe. An organism that can use molecular oxygen in its metabolism but that also can live anaerobically.

Fatty acid. A long-chain hydrocarbon containing a carboxyl group at one end. Saturated fatty acids have completely saturated hydrocarbon chains. Unsaturated fatty acids have one or more carbon–carbon double bonds in their hydrocarbon chains.

Feedback inhibition. *See* end-product inhibition.

Fermentation. The energy-generating breakdown of glucose or related molecules by a process that does not require molecular oxygen.

Fingerprinting. The characteristic two-dimensional paper chromatogram obtained from the partial hydrolysis of a protein or a nucleic acid.

Fluorescence. The emission of light by an excited molecule in the process of making the transition from the excited state to the ground state.

Footprinting. A technique that results in a DNA sequence ladder in which part of the ladder is missing due to the binding of protein to the DNA before processing.

Frameshift mutations. Insertions or deletions of genetic material that lead to a shift in the translation of the reading frame. The mutation usually leads to nonfunctional proteins.

Free energy. That part of the energy of a system that is available to do useful work.

Furanose. A sugar that contains a five-member ring as a result of intramolecular hemiacetal formation.

Futile cycle. *See* pseudocycle.

G

G_1 phase. That period of the cell cycle in which preparations are being made for chromosome duplication, which takes place in the S phase.

G_2 phase. That period of the cell cycle between S phase and mitosis (M phase).

Gametes. The ova and the sperm, haploid cells that unite during fertilization to generate a diploid zygote.

Gel exclusion chromatography. A technique that makes use of certain polymers that can form porous beads with varying pore sizes. In columns made from such beads, it is possible to separate molecules, which cannot penetrate beads of a given pore size, from smaller molecules that can.

Gene. A segment of the genome that codes for a functional product.

Gene amplification. The duplication of a particular gene within a chromosome two or more times.

Gene splicing. The cutting and rejoining of DNA sequences.

General recombination. Recombination that occurs between homologous chromosomes at homologous sites.

Generation time. The time it takes for a cell to double its mass under specified conditions.

Genetic map. The arrangement of genes or other identifiable sequences on a chromosome.

Genome. The total genetic content of a cell or a virus.

Genotype. The genetic characteristics of an organism (distinguished from its observable characteristics, or phenotype).

Globular protein. A folded protein that adopts an approximately globular shape.

Gluconeogenesis. The production of sugars from nonsugar precursors, such as lactate or amino acids. Applies more specifically to the production of free glucose by vertebrate livers.

Glycogen. A polymer of glucose residues in 1,4 linkage and 1,6 linkage at branchpoints.

Glycogenic. Describing amino acids the metabolism of which may lead to gluconeogenesis.

Glycolipid. A lipid containing a carbohydrate group.

Glycolysis. The catabolic conversion of glucose to pyruvate with the production of ATP.

Glycoprotein. A protein linked to an oligosaccharide or a polysaccharide.

Glycosaminoglycans. Long, unbranched polysaccharide chains composed of repeating disaccharide subunits in which one of the two sugars is either *N*-acetylglucosamine or *N*-acetylgalactosamine.

Glycosidic bond. The bond between a sugar and an alcohol. Also the bond that links two sugars in disaccharides, oligosaccharides, and polysaccharides.

Glyoxylate cycle. A pathway that uses acetyl-CoA and two auxiliary enzymes to convert acetate into succinate and carbohydrates.

Glyoxysome. An organelle containing key enzymes of the glyoxylate cycle.

Goldman equation. An equation expressing the quantitative relationship between the concentrations of charged species on either side of a membrane and the resting transmembrane potential.

Golgi apparatus. A complex series of double-membrane structures that interact with the endoplasmic reticulum and that serve as a transfer point for proteins destined for other organelles, the plasma membrane, or extracellular transport.

Gram molecular weight. For a given compound, the weight in grams that is numerically equal to its molecular weight.

Ground state. The lowest electronic energy state of an atom or a molecule.

Growth factor. A substance that must be present in the growth medium to permit cell proliferation.

Growth fork. The region on a DNA duplex molecule where synthesis is taking place. It resembles a fork in shape because it consists of a region of duplex DNA connected to a region of unwound single strands.

Guanine. A purine base found in DNA or RNA.

Guanosine. A purine nucleoside found in DNA and RNA.

H

Hairpin loop. A single-stranded complementary region that folds back on itself and base-pairs into a double helix.

Half-life. The time required for the disappearance of one half of a substance.

Haploid cell. A cell containing only one chromosome of each type.

Heavy isotopes. Forms of atoms that contain greater numbers of neutrons (e.g., ^{15}N, ^{13}C).

Helix. A spiral structure with a repeating pattern.

Heme. An iron–porphyrin complex found in hemoglobin and cytochromes.

Hemiacetal. The product formed by the condensation of an aldehyde with an alcohol; it contains one oxygen linked to a central carbon in a hydroxyl fashion and one oxygen linked to the same central carbon by an ether linkage.

Henderson-Hasselbach equation. An equation that relates the pK_a to the pH and the ratio of the proton acceptor (A^-) and the proton donor (HA) species of a conjugate acid–base pair.

Heterochromatin. Highly condensed regions of chromosomes that are not usually transcriptionally active.

Heteroduplex. An annealed duplex structure between two DNA strands that do not show perfect complementarity. Can arise by mutation, recombination, or the annealing of complementary single-stranded DNAs.

Heteropolymer. A polymer containing more than one type of monomeric unit.

Heterotroph. An organism that requires preformed organic compounds for growth.

Heterozygous. Describing an organism (a heterozygote) that carries two different alleles for a given gene.

Hexose. A sugar with a six-carbon backbone.

High-energy compound. A compound that undergoes hydrolysis with a high negative standard free energy change.

Histones. The family of basic proteins that is normally associated with DNA in most cells of eukaryotic organisms.

Holoenzyme. An intact enzyme containing all of its subunits with full enzymatic activity.

Homeobox. A conserved sequence of 180 bp encoding a protein domain found in many eukaryotic regulatory proteins.

Homologous chromosomes. Chromosomes that carry the same pattern of genes but not necessarily the same alleles.

Homopolymer. A polymer composed of only one type of monomeric building block.

Homozygous. Describing an organism (a homozygote) that carries two identical alleles for a given gene.

Hormone. A chemical substance made in one cell and secreted so as to influence the metabolic activity of a select group of cells located elsewhere in the organism.

Hormone receptor. A protein that is located on the cell membrane or inside the responsive cell and that interacts specifically with the hormone.

Host cell. A cell used for growth and reproduction of a virus.

Hybrid (or chimeric) plasmid. A plasmid that contains DNA from two different organisms.

Hydrogen bond. A weak attractive force between one electronegative atom and a hydrogen atom that is covalently linked to a second electronegative atom.

Hydrolysis. The cleavage of a molecule by the addition of water.

Hydrophilic. Preferring to be in contact with water.

Hydrophobic. Preferring not to be in contact with water, as is the case with the hydrocarbon portion of a fatty acid or phospholipid chain.

I

Ion-exchange resin. A polymeric resinous substance, usually in bead form, that contains fixed groups with positive or negative charge. A cation exchange resin has negatively charged groups and is therefore useful in exchanging the cationic groups in a test sample. The resin is usually used in the form of a column, as in other column chromatographic systems.

Isoelectric pH. The pH at which a protein has no net charge.

Isomerase. An enzyme that catalyzes an intramolecular rearrangement.

Isomerization. Rearrangement of atomic groups within the same molecule without any loss or gain of atoms.

Isoprene. The hydrocarbon 2-methyl-1,3-butadiene, which in some form serves as the precursor for many lipid molecules.

Isozymes. Multiple forms of an enzyme that differ from one another in one or more properties.

K

K_m. *See* Michaelis constant.

Ketogenic. Describing amino acids that are metabolized to acetoacetate and acetate.

Ketone. A functional group of an organic compound in which a carbon atom is double-bonded to an oxygen. Neither of the other substituents attached to the carbon is a hydrogen. Otherwise the group would be called an aldehyde.

Ketone bodies. Refers to acetoacetate, acetone, and β-hydroxybutyrate made from acetyl-CoA in the liver and used for energy in nonhepatic tissues.

Ketosis. A condition in which the concentration of ketone bodies in the blood or urine is unusually high.

Kilobase. One thousand bases in a DNA molecule.

Kinase. An enzyme catalyzing phosphorylation of an acceptor molecule, usually with ATP serving as the phosphate (phosphoryl) donor.

Kinetochore. A structure that attaches laterally to the centromere of a chromosome; it is the site of chromosome tubule attachment.

Krebs cycle. *See* tricarboxylic acid (TCA) cycle.

L

Lampbrush chromosome. Giant diplotene chromosome found in the oocyte nucleus. The loops that are observed are the sites of extensive gene expression.

Law of mass action. The finding that the rate of a chemical reaction is a function of the product of the concentrations of the reacting species.

Leader region. The region of an mRNA between the 5′ end and the initiation codon for translation of the first polypeptide chain.

Lectins. Agglutinating proteins usually extracted from plants.

Ligase. An enzyme that catalyzes the joining of two molecules together. In DNA it joins 3′-OH to 5′ phosphates.

Linkers. Short oligonucleotides that can be ligated to larger DNA fragments, then cleaved to yield overlapping cohesive ends, suitable for ligation to other DNAs that contain comparable cohesive ends.

Linking number. The net number of times one polynucleotide chain crosses over another polynucleotide chain. By convention, right-handed crossovers are given a plus designation.

Lipid. A biological molecule that is soluble in organic solvents. Lipids include steroids, fatty acids, prostaglandins, terpenes, and waxes.

Lipid bilayer (*see* Bilayer). Model for the structure of the cell membrane based on the hydrophobic interaction between phospholipids.

Lipopolysaccharide. Usually refers to a unique glycolipid found in Gram negative bacteria.

Lyase. An enzyme that catalyzes the removal of a group to form a double bond, or the reverse reaction.

Lysogenic virus. A virus that can adopt an inactive (lysogenic) state, in which it maintains its genome within a cell instead of entering the lytic cycle. The circumstances that determine whether a lysogenic (temperate) virus adopts an inactive state or an active lytic state are often subtle and depend on the physiological state of the infected cell.

Lysosome. An organelle that contains hydrolytic enzymes designed to break down proteins that are targeted to that organelle.

Lytic infection. A virus infection that leads to the lysis of the host cell, yielding progeny virus particles.

M

M phase. That period of the cell cycle when mitosis takes place.

Meiosis. Process in which diploid cells undergo division to form haploid sex cells.

Membrane transport. The facilitated transport of a molecule across a membrane.

Merodiploid. An organism that is diploid for some but not all of its genes.

Mesosome. An invagination of the bacterial cell membrane.

Messenger RNA (mRNA). The template RNA carrying the message for protein synthesis.

Metabolic turnover. A measure of the rate at which already existing molecules of the given species are replaced by newly synthesized molecules of the same type. Usually isotopic labeling is required to measure turnover.

Metabolism. The sum total of the enzyme-catalyzed reactions that occur in a living organism.

Metamorphosis. A change of form, especially the conversion of a larval form to an adult form.

Metaphase. That stage in mitosis or meiosis when all of the chromosomes are lined up on the equator (i.e., an imaginary line that bisects the cell).

Micelle. An aggregate of lipids in which the polar head groups face outward and the hydrophobic tails face inward; no solvent is trapped in the center.

Michaelis constant (K_m). The substrate concentration at which an enzyme-catalyzed reaction proceeds at one-half maximum velocity.

Michaelis-Menten equation (also known as the Henri-Michaelis-Menten equation). An equation relating the reaction velocity to the substrate concentration of an enzyme.

Microtubules. Thin tubules, made from globular proteins, that serve multiple purposes in eukaryotic cells.

Mismatch repair. The replacement of a base in a heteroduplex structure by one that forms a Watson-Crick base pair.

Missense mutation. A change in which a codon for one amino acid is replaced by a codon for another amino acid.

Mitochondrion. An organelle, found in eukaryotic cells, in which oxidative phosphorylation takes place. It contains its own genome and unique ribosomes to carry out protein synthesis of only a fraction of the proteins located in this organelle.

Mitosis. The process whereby replicated chromosomes segregate equally toward opposite poles prior to cell division.

Mixed-function oxidases. Enzymes that use molecular oxygen to oxidize two different molecules simultaneously, usually a substrate and a coenzyme.

Mobile genetic element. A segment of the genome that can move as a unit from one location on the genome to another, without any requirement for sequence homology.

Molecularity of a reaction. The number of molecules involved in a specific reaction step.

Monolayer. A single layer of oriented lipid molecules.

Mutagen. An agent that can bring about a heritable change (mutation) in an organism.

Mutagenesis. A process that leads to a change in the genetic material that is inherited in subsequent generations.

Mutant. An organism that carries an altered gene or change in its genome.

Mutarotation. The change in optical rotation of a sugar that is observed immediately after it is dissolved in aqueous solution, as the result of the slow approach of equilibrium of a pyranose or a furanose in its α and β forms.

Mutation. The genetically inheritable alteration of a gene or group of genes.

Myofibril. A unit of thick and thin filaments in a muscle fiber.

Myosin. The main protein of the thick filaments in a muscle myofibril. It is composed of two coiled subunits (M_r about 220,000) that can aggregate to form a thick filament that is globular at each end.

N

Nascent RNA. The initial transcripts of RNA, before any modification or processing.

Negative control. Regulation of the activity by an inhibitory mechanism.

Negative feedback. Regulation of a reaction or a pathway by the end product.

Nernst equation. An equation that relates the redox potential to the standard redox potential and the concentrations of the oxidized and reduced form of the couple.

Nitrogen cycle. The passage of nitrogen through various valence states, as the result of reactions carried out by a wide variety of different organisms.

Nitrogen fixation. Conversion of atmospheric nitrogen into a form that can be converted by biochemical reactions to an organic form. This reaction is carried out by a very limited number of microorganisms.

Nitrogenous base. An aromatic nitrogen-containing molecule with basic properties. Such bases include purines and pyrimidines.

Noncompetitive inhibitor. An inhibitor of enzyme activity the effect of which is not reversed by increasing the concentration of substrate molecule.

Nonsense mutation. A change in the base sequence that converts a sense codon (one that specifies an amino acid) to one that specifies a stop (a nonsense codon). There are three nonsense codons.

Northern blotting. *See* Southern blotting.

Nuclease. An enzyme that cleaves phosphodiester bonds of nucleic acids.

Nucleic acids. Polymers of the ribonucleotides or deoxyribonucleotides.

Nucleohistone. A complex of DNA and histone.

Nucleolus. A spherical structure visible in the nucleus during interphase. The nucleolus is associated with a site on the chromosome that is involved in ribosomal RNA synthesis.

Nucleophilic group. An electron-rich group that tends to attack an electron-deficient nucleus.

Nucleosome. A complex of DNA and an octamer of histone proteins in which a small stretch of the duplex is wrapped around a molecular bead of histone.

Nucleotide. An organic molecule containing a purine or pyrimidine base, a five-carbon sugar (ribose or deoxyribose), and one or more phosphate groups.

Nucleus. In eukaryotic cells, the centrally located organelle that encloses most of the chromosomes. Minor amounts of chromosomal substance are found in some other organelles, most notably the mitochondria and the chloroplasts.

O

Okazaki fragment. A short segment of single-stranded DNA that is an intermediate in DNA synthesis. In bacteria, Okazaki fragments are 1,000–2,000 bases in length; in eukaryotes, 100–200 bases in length.

Oligonucleotide. A polynucleotide containing a small number of nucleotides. The linkages are the same as in a polynucleotide; the only distinguishing feature is the small size.

Oligosaccharide. A molecule containing a small number of sugar residues joined in a linear or a branched structure by glycosidic bonds.

Oncogene. A gene of cellular or viral origin that is responsible for rapid, unruly growth of animal cells.

Operon. A group of contiguous genes that are coordinately regulated by two *cis*-acting elements: A promoter and an operator. Found only in prokaryotic cells.

Optical activity. The property of a molecule that leads to rotation of the plane of polarization of plane-polarized light when the latter is transmitted through the substance. Chirality is a necessary and sufficient property for optical activity.

Organelle. A subcellular membrane-bound body with a well-defined function.

Osmotic pressure. The pressure generated by the mass flow of water to that side of a membrane-bound structure that contains the higher concentration of solute molecules. A stable osmotic pressure is seen in systems in which the membrane is not permeable to some of the solute molecules.

Oxidation. The loss of electrons from a compound.

Oxidative phosphorylation. The formation of ATP as the result of the transfer of electrons to oxygen.

Oxido-reductase. An enzyme that catalyzes oxidation–reduction reactions.

P

Palindrome. A sequence of bases that reads the same in both directions on opposite strands of the DNA duplex (e.g., GAATTC).

Pentose. A sugar with five carbon atoms.

Pentose phosphate pathway. The pathway involving the oxidation of glucose-6-phosphate to pentose phosphates and further reactions of pentose phosphates.

Peptide. An organic molecule in which a covalent amide bond is formed between the α-amino group of one amino acid and the α-carboxyl group of another amino acid, with the elimination of a water molecule.

Peptide mapping. Same as fingerprinting.

Peptidoglycan. The main component of the bacterial cell wall, consisting of a two-dimensional network of heteropolysaccharides running in one direction, cross-linked with polypeptides running in the perpendicular direction.

Periplasm. The space between the inner and outer membranes of a bacterium.

Permease. A protein that catalyzes the transport of a specific small molecule across a membrane.

Peroxisomes. Subcellular organelles that contain flavin-requiring oxidases and that regenerate oxidized flavin by reaction with oxygen.

Phenotype. The observable trait(s) that result from the genotype in cooperation with the environment.

Phenylketonuria. A human disease caused by a genetic deficiency in the enzyme that converts phenylalanine to tyrosine. The immediate cause of the disease is an excess of phenylalanine. The condition can be alleviated by a diet low in phenylalanine.

Pheromone. A hormonelike substance associated with insects that acts as an attractant.

Phosphodiester. A molecule containing two alcohols esterified to a single molecule of phosphate. For example, the backbone of nucleic acids is connected by 5′-3′ phosphodiester linkages between the adjacent individual nucleotide residues.

Phosphogluconate pathway. Another name for the pentose phosphate pathway. This name derives from the fact that 6-phosphogluconate is an intermediate in the formation of pentoses from glucose.

Phospholipid. A lipid containing charged hydrophilic phosphate groups; a component of cell membranes.

Phosphorolysis. Phosphate-induced cleavage of a molecule. In the process the phosphate becomes covalently linked to one of the degradation products.

Phosphorylation. The formation of a phosphate derivative of a biomolecule.

Photoreactivation. DNA repair in which the damaged region is repaired with the help of light and an enzyme. The lesion is repaired without excision from the DNA.

Photosynthesis. The biosynthesis that directly harnesses the chemical energy resulting from the absorption of light. Frequently used to refer to the formation of carbohydrates from CO_2 that occurs in the chloroplasts of plants or the plastids of photosynthetic microorganisms.

Pitch length (or pitch). The number of base pairs per turn of a duplex helix.

pK. The negative logarithm of the equilibrium constant.

Plaque. A circular clearing on a lawn of bacterial or cultured cells, resulting from cell lysis and production of phage or animal virus progeny.

Plasma membrane. The membrane that surrounds the cytoplasm.

Plasmid. A circular DNA duplex that replicates autonomously in bacteria. Plasmids that integrate into the host genome are called episomes. Plasmids differ from viruses in that they never form infectious nucleoprotein particles.

Polar group. A hydrophilic (water-loving) group.

Polar mutation. A mutation in one gene that reduces the expression of a gene or genes distal to the promoter in the same operon.

Polarimeter. An instrument for determining the rotation of polarization of light as the light passes through a solution containing an optically active substance.

Polyamine. A hydrocarbon containing more than two amino groups.

Polycistronic messenger RNA. In prokaryotes, an RNA that contains two or more cistrons; note that only in prokaryotic mRNAs can more than one cistron be utilized by the translation system to generate individual proteins.

Polymerase. An enzyme that catalyzes the synthesis of a polymer from monomers.

Polynucleotide. A chain structure containing nucleotides linked together by phosphodiester (5′-3′) bonds. The polynucleotide chain has a directional sense, with a 5′ and a 3′ end.

Polynucleotide phosphorylase. An enzyme that polymerizes ribonucleotide diphosphates. No template is required.

Polypeptide. A linear polymer of amino acids held together by peptide linkages. The polypeptide has a directional sense, with an amino- and a carboxyl-terminal end.

Polyribosome (polysome). A complex of an mRNA and two or more ribosomes actively engaged in protein synthesis.

Polysaccharide. A linear or branched-chain structure containing many sugar molecules linked by glycosidic bonds.

Porphyrin. A complex planar structure containing four substituted pyrroles covalently joined in a ring and frequently containing a central metal atom. For example, heme is a porphyrin with a central iron atom.

Positive control. A system that is turned on by the presence of a regulatory protein.

Posttranslational modification. The covalent bond changes that occur in a polypeptide chain after it leaves the ribosome and before it becomes a mature protein.

Primary structure. In a polymer, the sequence of monomers and the covalent bonds.

Primer. A structure that serves as a growing point for polymerization.

Primosome. A multiprotein complex that catalyzes synthesis of RNA primer at various points along the DNA template.

Prochiral molecule. A nonchiral molecule that may react with an enzyme so that two groups that have a mirror image relationship to each other are treated differently.

Prokaryote. A unicellular organism that contains a single chromosome, no nucleus, no membrane-bound organelles, and has characteristic ribosomes and biochemistry.

Promoter. The region of the gene that signals RNA polymerase binding and the initiation of transcription.

Prophage. The silent phage genome. Some prophages integrate into the host genome; others replicate autonomously. The prophage state is maintained by a phage-encoded repressor.

Prophase. The stage in meiosis or mitosis when chromosomes condense and become visible as refractile bodies.

Proprotein. A protein that is made in an inactive form, so that it requires processing to become functional.

Prostaglandin. An oxygenated eicosanoid that has a hormonal function. Prostaglandins are unusual hormones in that they usually have effects only in that region of the organism where they are synthesized.

Prosthetic group. Synonymous with coenzyme except that a prosthetic group is usually more firmly attached to the enzyme it serves.

Protamines. Highly basic, arginine-rich proteins found complexed to DNA in the sperm of many invertebrates and fish.

Protein subunit. One of the components of a complex multicomponent protein.

Protein targeting. The process whereby proteins following synthesis are directed to specific locations.

Proteoglycan. A protein-linked heteropolysaccharide in which the heteropolysaccharide is usually the major component.

Protist. A relatively undifferentiated organism that can survive as a single cell.

Proton acceptor. A functional group capable of accepting a proton from a proton donor molecule.

Proton motive force (Δp). The thermodynamic driving force for proton translocation. Expressed quantitatively as $\Delta G_{H^+}/F$ in units of volts.

Protooncogene. A cellular gene that can undergo modification to a cancer-causing gene (oncogene).

Pseudocycle. A sequence of reactions that can be arranged in a cycle but that usually do not function simultaneously in both directions. Also called a futile cycle because the net result of simultaneous functioning in both directions would be the expenditure of energy without accomplishing any useful work.

Pulse-chase. An experiment in which a short labeling period is followed by the addition of an excess of the same, unlabeled compound to dilute out the labeled material.

Purine. A heterocyclic ring structure with varying functional groups. The purines adenine and guanine are found in both DNA and RNA.

Puromycin. An antibiotic that inhibits polypeptide synthesis by competing with aminoacyl-tRNA for the ribosomal binding site A.

Pyranose. A simple sugar containing the six-member pyran ring.

Pyrimidine. A heterocyclic six-member ring structure. Cytosine and uracil are the main pyrimidines found in RNA, and cytosine and thymine are the main pyrimidines found in DNA.

Pyrophosphate. A molecule formed by two phosphates in anhydride linkage.

Q

Quaternary structure. In a protein, the way in which the different folded subunits interact to form the multisubunit protein.

R

R group. The distinctive side chain of an amino acid.

R loop. A triple-stranded structure in which RNA displaces a DNA strand by DNA–RNA hybrid formation in a region of the DNA.

Rapid-start complex. The complex that RNA polymerase forms at the promoter site just before initiation.

Recombination. The transfer to offspring of genes not found together in either of the parents.

Redox couple. An electron donor and its corresponding oxidized form.

Redox potential (E). The relative tendency of a pair of molecules to release or accept an electron. The standard redox potential ($E°$) is the redox potential of a solution containing the oxidant and reductant of the couple at standard concentrations.

Regulatory enzyme. An enzyme in which the active site is subject to regulation by factors other than the enzyme substrate. The enzyme frequently contains a nonoverlapping site for binding the regulatory factor that affects the activity of the active site.

Regulatory gene. A gene the principal product of which is a protein designed to regulate the synthesis of other genes.

Renaturation. The process of returning a denatured structure to its original native structure, as when two single strands of DNA are reunited to form a regular duplex, or the process by which an unfolded polypeptide chain is returned to its normal folded three-dimensional structure.

Repair synthesis. DNA synthesis following excision of damaged DNA.

Repetitive DNA. A DNA sequence that is present in many copies per genome.

Replica plating. A technique in which an impression of a culture is taken from a master plate and transferred to a fresh plate. The impression can be of bacterial clones or phage plaques.

Replication fork. The Y-shaped region of DNA at the site of DNA synthesis; also called a growth fork.

Replicon. A genetic element that behaves as an autonomous replicating unit. It can be a plasmid, phage, or bacterial chromosome.

Repressor. A regulatory protein that inhibits transcription from one or more genes. It can combine with an inducer (resulting in specific enzyme induction) or with an operator element (resulting in repression).

Resonance hybrid. A molecular structure that is a hybrid of two structures that differ in the locations of some of the electrons. For example, the benzene ring can be drawn in two ways, with double bonds in different positions. The actual structure of benzene is a blending of these two equivalent structures.

Restriction-modification system. A pair of enzymes found in most bacteria (but not eukaryotic cells). The restriction enzyme recognizes a certain sequence in duplex DNA and makes one cut in each unmodified DNA strand at or near the recognition sequence. The modification enzyme methylates (or modifies) the recognition sequence, thus protecting it from the action of the restriction enzyme.

Reverse transcriptase. An enzyme that synthesizes DNA from an RNA template, using deoxyribonucleotide triphosphates.

Rho factor. A protein involved in the termination of transcription of some messenger RNAs.

Ribose. The five-carbon sugar found in RNA.

Ribosomal RNA (rRNA). The RNA parts of the ribosome.

Ribosomes. Small cellular particles made up of ribosomal RNA and protein. They are the site, together with mRNA, of protein synthesis.

RNA (ribonucleic acid). A polynucleotide in which the sugar is ribose.

RNA polymerase. An enzyme that catalyzes the formation of RNA from ribonucleotide triphosphates, using DNA as a template.

RNA splicing. The excision of a segment of RNA, followed by a rejoining of the remaining fragments.

Rolling-circle replication. A mechanism for the replication of circular DNA. A nick in one strand allows the 3′ end to be extended, displacing the strand with the 5′ end, which is also replicated, to generate a double-stranded tail that can become larger than the unit size of the circular DNA.

S

S phase. The period during the cell cycle when the chromosome is replicated.

Salting in. The increase in solubility that is displayed by typical globular proteins on the addition of small amounts of certain salts such as ammonium sulfate.

Salting out. The decrease in protein solubility that occurs when salts such as ammonium sulfate are present at high concentrations.

Salvage pathway. A family of reactions that permits nucleosides or purine and pyrimidine bases resulting from the partial breakdown of nucleic acids, to be reutilized in nucleic acid synthesis.

Satellite DNA. A DNA fraction the base composition of which differs from that of the main component of DNA, as revealed by the fact that it bands at a different density in a CsCl gradient. Usually repetitive DNA or organelle DNA.

Scissile. Capable of being cut smoothly or split easily.

Second messenger. A diffusible small molecule, such as cAMP, that is formed at the inner surface of the plasma membrane in response to a hormonal signal.

Secondary structure. In a protein or a nucleic acid, any repetitive folded pattern that results from the interaction of the corresponding polymeric chains.

Semiconservative replication. Duplication of DNA in which the daughter duplex carries one old strand and one new strand.

Sigma factor. A subunit of bacterial RNA polymerase that recognizes specific sites on DNA for initiation of RNA synthesis.

Signal sequence. A sequence in a protein that serves as a signal to guide the protein to a specific location.

Signal transduction. The process by which an extracellular signal is converted into a cellular response.

Single-copy DNA. A region of the genome the sequence of which is present only once per haploid complement.

Somatic cell. Any cell of an organism that cannot contribute its genes to a subsequent generation.

SOS system. A set of DNA repair enzymes and regulatory proteins that regulate their synthesis so that maximum synthesis occurs when the DNA is damaged.

Southern blotting. A method for detecting a specific DNA restriction fragment, developed by Edward Southern. DNA from a gel electrophoresis pattern is blotted onto nitrocellulose paper; then the DNA is denatured and fixed on the paper. Subsequently, the pattern of specific sequences in the Southern blot can be determined by hybridization to a suitable probe and autoradiography. A northern blot is similar, except that RNA is blotted instead onto the nitrocellulose paper.

Splicing. *See* RNA splicing.

Sporulation. Formation from vegetative cells of metabolically inactive cells that can resist extreme environmental conditions.

Stacking energy. The energy of interaction that favors the face-to-face packing of purine and pyrimidine base pairs.

Steady-state. In enzyme kinetic analysis, the time interval when the rate of reaction is approximately constant with time. The term is also used to describe the state of a living cell in which the concentrations of many molecules are approximately constant because of a balancing between their rates of synthesis and breakdown.

Stem cell. A cell from which other cells stem or arise by differentiation.

Stereoisomers. Isomers that are nonsuperimposable mirror images of each other.

Steroids. Compounds that are derivatives of a tetracyclic structure composed of a cyclopentane ring fused to a substituted phenanthrene nucleus.

Structural domain. An element of protein tertiary structure that recurs in many structures.

Structural gene. A gene encoding the amino acid sequence of a polypeptide chain.

Structural protein. A protein that serves a structural function.

Subunit. Individual polypeptide chains in a protein.

Supercoiled DNA. Supertwisted, covalently closed duplex DNA.

Suppressor gene. A gene that can reverse the phenotype of a mutation in another gene.

Suppressor mutation. A mutation that restores a function lost by an initial mutation and that is located at a site different from the initial mutation.

Svedberg unit (S). The unit used to express the sedimentation constant s: $1\,S = 10^{-13}$ s. The sedimentation constant s is proportional to the rate of sedimentation of a molecule in a given centrifugal field and is related to the size and shape of the molecule.

Synapse. The chemical connection for communication between two nerve cells or between a nerve cell and a target cell such as a muscle cell.

Synapsis. The pairing of homologous chromosomes, seen during the first meiotic prophase.

T

Tandem duplication. A duplication in which the repeated regions are immediately adjacent to one another.

TCA cycle. *See* tricarboxylic acid cycle.

Template. A polynucleotide chain that serves as a surface for the absorption of monomers of a growing polymer and thereby dictates the sequence of the monomers in the growing chain.

Termination factors. Proteins that are exclusively involved in the termination reactions of protein synthesis on the ribosome.

Terpenes. A diverse group of lipids made from isoprene precursors.

Tertiary structure. In a protein or nucleic acid, the final folded form of the polymer chain.

Tetramer. Structure resulting from the association of four subunits.

Thioester. An ester of a carboxylic acid with a thiol or mercaptan.

Thymidine. One of the four nucleosides found in DNA.

Thymine. A pyrimidine base found in DNA.

Topoisomerase. An enzyme that changes the extent of supercoiling of a DNA duplex.

Transamination. Enzymatic transfer of an amino group from an α-amino acid to an α-keto acid.

Transcription. RNA synthesis that occurs on a DNA template.

Transduction. Genetic exchange in bacteria that is mediated via phage.

Transfection. An artificial process of infecting cells with naked viral DNA.

Transfer RNA (tRNA). Any of a family of low-molecular-weight RNAs that transfer amino acids from the cytoplasm to the template for protein synthesis on the ribosome.

Transferase. An enzyme that catalyzes the transfer of a molecular group from one molecule to another.

Transformation. Genetic exchange in bacteria that is mediated via purified DNA. In somatic cell genetics the term is also used to indicate the conversion of a normal cell to one that grows like a cancer cell.

Transgenic. Describing an organism that contains transfected DNA in the germ line.

Transition state. The activated state in which a molecule is best suited to undergoing a chemical reaction.

Translation. The process of reading a messenger RNA sequence for the specified amino acid sequence it contains.

Transport protein. A protein the primary function of which is to transport a substance from one part of the cell to another, from one cell to another, or from one tissue to another.

Tricarboxylic acid (TCA) cycle. The cyclical process whereby acetate is completely oxidized to carbon dioxide and water, and electrons are transferred to NAD^+ and FAD. The TCA cycle is localized to the mitochondria in eukaryotic cells and to the plasma membrane in prokaryotic cells. Also called the Krebs cycle.

Trypsin. A proteolytic enzyme that cleaves peptide chains next to the basic amino acids arginine and lysine.

Tryptic peptide mapping. The technique of generating a chromatographic profile characteristic of the fragments resulting from trypsin enzyme cleavage of the protein.

Tumorigenesis. The mechanism of tumor formation.

Turnover number. The maximum number of molecules of substrate that can be converted to product per active site per unit time.

U

Ultracentrifuge. A high-speed centrifuge that can attain speeds up to 60,000 rpm and centrifugal fields of 500,000 times gravity. Useful for characterizing and separating macromolecules.

Uncoupler. A substance that uncouples phosphorylation of ADP from electron transfer; for example, 2,4-dinitrophenol.

Unidirectional replication. *See* bidirectional replication.

Unwinding proteins. Proteins that help to unwind double-stranded DNA during DNA replication.

Urea cycle. A metabolic pathway in the liver that leads to the synthesis of urea from amino groups and CO_2. The function of the pathway is to convert the ammonia resulting from catabolism to a nontoxic form, which is subsequently secreted.

UV irradiation. Electromagnetic radiation with a wavelength shorter than that of visible light (200–390 nm). Causes damage to DNA (mainly pyrimidine dimers).

V

van der Waals forces. Refers to two types of interactions, one attractive and one repulsive. The attractive forces are due to favorable interactions among the induced instantaneous dipole moments that arise from fluctuations in the electron charge densities of neighboring nonbonded atoms. Repulsive forces arise when noncovalently bonded atoms come too close together.

Viroids. Pathogenic agents, mostly of plants, that consist of short (usually circular) RNA molecules.

Virus. A nucleic acid–protein complex that can infect and replicate inside a specific host cell to make more virus particles.

Vitamin. A trace organic substance required in the diet of some species. Many vitamins are precursors of coenzymes.

W

Watson-Crick base pairs. The type of hydrogen-bonded base pairs found in DNA, or comparable base pairs found in RNA. The base pairs are A-T, G-C, and A-U.

Wild-type gene. The form of a gene (allele) normally found in nature.

Wobble. A proposed explanation for base-pairing that is not of the Watson-Crick type and that often occurs between the 3′ base in the codon and the 5′ base in the anticodon.

X

X-ray crystallography. A technique for determining the structure of molecules from the x-ray diffraction patterns that are produced by crystalline arrays of the molecules.

Y

Ylid. A compound in which adjacent, covalently bonded atoms, both having an electronic octet, have opposite charges.

Z

Z form. A duplex DNA structure in which the usual type of hydrogen bonding occurs between the base pairs but in which the helix formed by the two polynucleotide chains is left-handed rather than right-handed.

Zwitterion. A dipolar ion with spatially separated positive and negative charges. For example, most amino acids are zwitterions, having a positive charge on the α-amino group and a negative charge on the α-carboxyl group but no net charge on the overall molecule.

Zygote. A cell that results from the union of haploid male and female sex cells. Zygotes are diploid.

Zymogen. An inactive precursor of an enzyme. For example, trypsin exists in the inactive form trypsinogen before it is converted to its active form, trypsin.

Credits

Portions of this text have been adapted from Geoffrey Zubay, *Genetics*. To be published.

Molecular Modeling

Molecular Graphics The molecular graphics photos listed below were developed by Michael Pique at the SCRIPPS Research Institute using software by Yng Chen, Michael Connolly, Michael Carson, Alex Shah, and AVS, Inc. Images copyright 1994 by the SCRIPPS Research Institute. Visualization advice was provided by Holly Miller, Wake Forest University Medical Center.

Atomic Coordinates All atomic coordinates for the molecular graphics images were obtained from the Protein Data Bank at Brookhaven National Laboratory, Upton, N.Y.

4.21: Phosphoglycerate kinase: entry 3PGK of July 1992. T. N. Bryant, et al. (To be published). **7.7**: Thermolysin: entry 1TLP of June 1987. D. E. Tronrud, et al., *European Journal of Biochemistry,* 157:261, 1986. **8.6a:** Trypsin: entry 3PTB of Sept. 1982. M. Marquart, et al., *ACTA Crystallography,* Section B, 39:480, 1983: Chymotrypsin: entry 4CHA of Nov. 1984. H. Tsukada & D. M. Blow, *Journal of Molecular Biology,* 184:703, 1985; Elastase: entry 2EST of Mar. 1986. D. L. Hughes, et al., *Journal of Molecular Biology,* 162:645, 1982. **8.6b:** Trypsin: entry 3PTB of Sept. 1982. M. Marquart, et al., *ACTA Crystallography,* Section B, 39:480, 1983; Subtilisin: entry 1SBC of May 1988. D. J. Neidhart & G. A. Petsko, (To be published). **8.7a-b:** Trypsin: entry 3PTB of Sept. 1982. M. Marquart, et al., *ACTA Crystallography,* Section B, 39:480, 1983. **8.19a-b:** Ribonuclease A: entry 6RSA of Feb. 1986. B. Borah, et al., *Biochemistry,* 24:2058, 1985. **8.20a-b:** Triose phosphate isomerase: entry 1TIM of Sept. 1976. D. W. Banner, et al., *Biochemical & Biophysical Research Communications,* 72:146, 1976. **9.9:** Phosphofructokinase: entry 4PFK of Jan. 1988. P. R. Evans, et al., *Philosophical Transactions of the Royal Society of London,* 293:53, 1981. **21.4a-b:** Glutamine synthetase: entry 2GLS of May 1989. M. M. Yamashita, et al., *Journal of Biological Chemistry,* 264:17681, 1989.

Chapter 1
1.1: From Sylvia S. Mader, *Inquiry into Life,* 7th edition. Copyright © 1994 Wm. C. Brown Communications, Inc., Dubuque, Iowa. All Rights Reserved. Reprinted by permission.
1.10: From R. E. Dickerson and I. Geis, *The Structure and Action of Proteins,* Benjamin/Cummings, Menlo Park, Calif., 1969. Illustration copyright by Irving Geis. Reprinted by permission.
1.15: From R. E. Dickerson and I. Geis, *The Structure and Action of Proteins,* Benjamin/Cummings, Menlo Park, Calif., 1969. Coordinates courtesy of D. C. Phillips, Oxford. Illustration copyright by Irving Geis. Reprinted by permission.

Chapter 3
3.6: From R. H. Haschenmeyer and A. E. V. Haschenmeyer, *A Guide to Study by Physical and Chemical Methods,* John Wiley & Sons, New York, 1973.

Chapter 4
4.2: From R. E. Dickerson and I. Geis, *The Structure and Action of Proteins,* Benjamin/Cummings, Menlo Park, Calif., 1969. Illustration copyright by Irving Geis. Reprinted by permission.
4.5: From R. E. Dickerson and I. Geis, *The Structure and Action of Proteins,* Benjamin/Cummings, Menlo Park, Calif., 1969. Illustration copyright by Irving Geis. Reprinted by permission.
4.24a: From Lansing M. Prescott, John P. Harley, and Donald A. Klein, *Microbiology,* 2d edition. Copyright © 1993 Wm. C. Brown Communications, Inc., Dubuque, Iowa. All Rights Reserved. Reprinted by permission.
4.25: From Lansing M. Prescott, John P. Harley, and Donald A. Klein, *Microbiology,* 2d edition. Copyright © 1993 Wm. C. Brown Communications, Inc., Dubuque, Iowa. All Rights Reserved. Reprinted by permission.
4a: From C. H. Bamford et al., *Synthetic Polypeptides,* Academic Press, Orlando, Fla., 1956.

Chapter 5
5.15: From John W. Hole, Jr., *Human Anatomy and Physiology,* 5th edition. Copyright © 1990 Wm. C. Brown Communications, Inc., Dubuque, Iowa. All Rights Reserved. Reprinted by permission.

Chapter 6
6.2: From E. J. Cohn and J. T. Edsall, *Proteins, Amino Acids, and Peptides as Ions and Dipolar Ions,* Reinhold, New York, 1942.

Chapter 7
7.5: From *Enzyme Structure and Mechanism,* 2d edition, by Alan Ferscht. Copyright © 1985 by W. H. Freeman and Company. Reprinted with permission.
7.3: From *Enzyme Structure and Mechanism* by Alan Ferscht. Copyright © 1985 by W. H. Freeman and Company. Reprinted with permission.

Chapter 9
9.20: From N. B. Madsen in *The Enzymes,* 3d edition, vol. XVII, ed. by P. D. Boyer and E. G. Krebs, Academic Press, New York, 1986.

Chapter 16
16.5: Reproduced with permission from B. Alberts, D. Bray, J. Lewis, M. Raff, K. Roberts, and J. D. Watson, *Molecular Biology of the Cell* (New York: Garland Publishing, 1989).

Chapter 17
17.3: Adapted from M. K. Jain and R. C. Wagner, *Introduction to Biological Membranes,* John Wiley & Sons, New York, 1980.
17.18: From J. T. Segrest and L. D. Kohn, Protein-lipid interactions of the membrane penetrating MN-glycoprotein from the human erythrocyte, *Protides of the Biological Fluids,* 21st colloquium, ed. by J. Peeters, Pergamon Press, New York, 1973.
17.4: Adapted from M. K. Jain and R. C. Wagner, *Introduction to Biological Membranes,* John Wiley & Sons, New York, 1980.
17.23: From B. Alberts et al., *Molecular Biology of the Cell,* 2d edition, Garland Publishing, New York, 1989.

Supplement 1
S1.3: Adapted from S. W. Kuffler and J. G. Nicholls, *From Neuron to Brain,* Sinauer Associates, Sunderland, Mass., 1976.

Chapter 25
25.16: From Geoffrey Zubay, *Genetics,* Benjamin/Cummings, Menlo Park, Calif., 1987. Reprinted by permission of the author.
25.18b: From Robert F. Weaver and Philip W. Hedrick, *Basic Genetics.* Copyright © 1991 Wm. C. Brown Communications, Inc., Dubuque, Iowa. All Rights Reserved. Reprinted by permission.

Chapter 26
26.16: From Robert F. Weaver and Philip W. Hedrick, *Basic Genetics.* Copyright © 1991 Wm. C. Brown Communications, Inc., Dubuque, Iowa. All Rights Reserved. Reprinted by permission.
26.21: From Geoffrey Zubay, *Genetics,* Benjamin/Cummings, Menlo Park. Calif., 1987. Reprinted by permission of the author.
26.22: From Geoffrey Zubay, *Genetics,* Benjamin/Cummings, Menlo Park, Calif., 1987. Reprinted by permission of the author.

Chapter 27
27.4: From Robert F. Weaver and Philip W. Hedrick, *Basic Genetics.* Copyright © 1991 Wm. C. Brown Communications, Inc., Dubuque, Iowa. All Rights Reserved. Reprinted by permission.

Chapter 28
28.2: From Geoffrey Zubay, *Genetics,* Benjamin/Cummings, Menlo Park, Calif., 1987. Reprinted by permission of the author.

28.5: From Robert F. Weaver and Philip W. Hedrick, *Genetics,* 2d edition. Copyright © 1992 Wm. C. Brown Communications, Inc., Dubuque, Iowa. All Rights Reserved. Reprinted by permission.
28.6: From Robert F. Weaver and Philip W. Hedrick, *Genetics.* Copyright © 1989 Wm. C. Brown Communications, Inc., Dubuque, Iowa. All Rights Reserved. Reprinted by permission.
28.8: From Geoffrey Zubay, *Genetics,* Benjamin/Cummings, Menlo Park, Calif., 1987. Reprinted by permission of the author.

Chapter 29
29.1: From Robert F. Weaver and Philip W. Hedrick, *Basic Genetics.* Copyright © 1991 Wm. C. Brown Communications, Inc., Dubuque, Iowa. All Rights Reserved. Reprinted by permission.
29.8: From Robert F. Weaver and Philip W. Hedrick, *Genetics,* 2d edition. Copyright © 1992 Wm. C. Brown Communications, Inc., Dubuque, Iowa. All Rights Reserved. Reprinted by permission.
29.23: From *Molecular Cell Biology,* 2d edition, by Darnell, Lodish, and Baltimore. Copyright © 1990 by Scientific American Books, Inc. Reprinted with permission of W. H. Freeman and Company.

Chapter 30
30.25: From C. Branden and J. Tooze, *Introduction to Protein Structure,* Garland Publishing, 1991, p. 102. Copyright 1991 by Garland Publishing, New York. Reprinted by permission.
30.28: From C. Branden and J. Tooze, *Introduction to Protein Structure,* Garland Publishing, New York, 1991, p. 109. Adapted from unpublished diagrams, courtesy of S. Phillips. Reprinted by permission.
30.29: From C. Branden and J. Tooze, *Introduction to Protein Structure,* Garland Publishing, New York, 1991, p. 105. Adapted from R.-g. Zhang et al., The crystal structure of trp aporepressor at 1.8 Å shows how binding tryptophan enhances DNA affinity, *Nature* 327:591, 1987. Reprinted by permission.

Chapter 31
31.10: From Geoffrey Zubay, *Genetics,* Benjamin/Cummings, Menlo Park, Calif., 1987. Reprinted by permission of the author.
31.13: From Geoffrey Zubay, *Genetics,* Benjamin/Cummings, Menlo Park, Calif., 1987. Reprinted by permission of the author.
31.21: From C. Branden and J. Tooze, *Introduction to Protein Structure,* Garland Publishing, New York, 1991, p. 126. Adapted from C. R. Vinson, P. B. Sigler, and S. L. McKnight, Scissors-grip model for DNA recognition by a family of leucine zipper proteins, *Science* 246:911, 1989. Copyright 1989 by the AAAS. Reprinted by permission.
31.24: Adapted from A. P. Bird, Gene reiteration and gene amplification, *Cell Biology,* vol. 3, *Gene Expression: The Production of RNAs,* ed. by L. Goldstein and D. M. Prescott, Academic Press, New York, 1980.

Supplement 3
S3.6: From Geoffrey Zubay, *Genetics,* Benjamin/Cummings, Menlo Park, Calif., 1987. Reprinted by permission of the author.
S3.7: From Geoffrey Zubay, *Genetics,* Benjamin/Cummings, Menlo Park, Calif., 1987. Reprinted by permission of the author.
S3.2: From Geoffrey Zubay, *Genetics,* Benjamin/Cummings, Menlo Park, Calif., 1987. Reprinted by permission of the author.
S3.10: From Geoffrey Zubay, *Genetics,* Benjamin/Cummings, Menlo Park, Calif., 1987. Reprinted by permission of the author.
S3.16: From Geoffrey Zubay, *Genetics,* Benjamin/Cummings, Menlo Park, Calif., 1987. Reprinted by permission of the author.

Supplement 4
S4.0: © Scientific American, Inc., George V. Kelvin. Reprinted by permission.
S4.1: Reproduced with permission from B. Alberts, D. Bray, J. Lewis, M. Raff, K. Roberts, and J. D. Watson, *Molecular Biology of the Cell* (New York: Garland Publishing, 1989).
S4.5: From Robert F. Weaver and Philip W. Hedrick, *Genetics,* 2d edition. Copyright © 1992 Wm. C. Brown Communications, Inc., Dubuque, Iowa. All Rights Reserved. Reprinted by permission.
S4.6: From Geoffrey Zubay, *Genetics,* Benjamin/Cummings, Menlo Park, Calif., 1987. Reprinted by permission of the author.

Index

Note: Page numbers followed by F indicate illustrations; page numbers followed by T indicate tables.

A band, 110, 111F
abl gene, 857T
ABO blood groups, 368, 368T, 369F
Abscisic acid, 592, 593F, 594
Acceptor arm, 735
Acceptor stem, 733, 734F, 735
Accessory pigments, in antenna systems, 342
Acetaldehyde, 201F
Acetals, 245, 246F
Acetate, 201F
  standard free energy of formation of, 36T
Acetic anhydride, formation of, 155
Acetoacetate, synthesis of, 524F
Acetoacetic acid decarboxylase, 157, 157F
Acetoacetyl-CoA thiolase, reaction catalyzed by, 418
Acetohydroxy acid isomeroreductase, reaction catalyzed by, 497, 498F
Acetohydroxy acid synthase, reaction catalyzed by, 497, 498F
Acetoin, 201F
Acetolactate, 201F
Acetylcholine
  hydrolysis of, 609
  receptor for, 610, 611F, 612F
  structure of, 609F
  synaptic transmission for, 609
Acetylcholinesterase
  specificity constant of, 145T
  turnover number of, 144T
Acetyl-CoA
  formation of, 287–289, 288F
  metabolic fates of, 418
  pyruvate dehydrogenase and pyruvate carboxylase regulation by, 298F, 299
  supply for biosynthetic sequences in cytosol, 300F, 301
Acetyl-CoA carboxylase
  reactions catalyzed by, 420F, 420–421, 421F
  regulation of, 431–432, 432F
Acetyl-CoA transacylase, 421, 422F
N-Acetyl-D-galactosamine, structure of, 357F
N-Acetyl-D-glucosamine, structure of, 357F
β-*N*-Acetylhexosaminidase A, deficiency of, 452
N-Acetylmuramic acid, structure of, 357F
N-Acetylneuraminic acid, structure of, 357F
Acetylphosphate, 201F
Acetyl phosphate, hydrolysis of, standard free energy of, 42F
O-Acetylserine sulfhydrylase, reaction catalyzed by, 495, 496F
Acetyl-thiamine pyrophosphate, 201F
Acid anhydride bonds, 537, 537F
Acid dissociation constant, 52
Acidemia
  isovaleric, 525T
  methylmalonic, 525T
Acivicin, 550T, 551
Aconitase, reaction catalyzed by, 289F, 290
Acquired immunodeficiency syndrome
  pharmacotherapy for, 552, 553T
  virion structure in, 851, 852F

Acromegaly, 589
ACTH, 571, 572T
Actin, in skeletal muscle, 112T, 113
Actin-myosin cross-bridges, 113, 114F
Actinomycin C₁, 503T
Actinomycin D, 724F, 725
Action potentials, 602, 604–608
  mediation by voltage-gated ion channels, 605–608, 606–608F
  recording, 604F
Activated transcription, 713
Activation domains, of transcription factors, 815–817
Activation free energy, 138, 139F
  catalysts and, 139F, 139–140
Activators, 769
Active site, of enzymes, 140, 141F, 150, 150F, 154
  serine proteases, 160–162, 162F, 163, 164–165
Active transport, against electrochemical potential gradient, energy
  requirement for, 401F, 401–402, 402T
Acyclovir, 552, 553T
Acyl carrier protein, 421, 423F
Acyl-CoA, reactions catalyzed by, 429–430, 430F
Acyl-CoA:cholesterol acyltransferase, 471
  reaction catalyzed by, 469, 470F
Acyl-CoA dehydrogenases
  reaction catalyzed by, 414
  reactions involving, 209T
Acyl-CoA ligase, reaction catalyzed by, 414, 429
Acyl-CoA synthase, in eukaryotes, 425
Acyl-enzyme ester intermediate, 157
1-Acylglycerol-3-phosphate acyltransferase, reaction catalyzed by, 438
Acyl-group transfers
  linked to oxidation-reduction, α-lipoic acid and, 212F, 212–213
  phosphopantetheine coenzymes and, 210–212, 211F
N-Acyl-sphingenine, 447
Addison's disease, 590
Adenine, 535
  structure of, 12F
  tautomeric forms of, 537, 537F
Adenine nucleotide translocator, 325
Adenine phosphoribosyltransferase, reaction catalyzed by, 548
Adenomas, benign and malignant, 849F
Adenosine, in tRNA, structure of, 704F
Adenosine diphosphate. *See* ADP
Adenosine kinase, reaction catalyzed by, 548
Adenosine monophosphate. *See* AMP
Adenosine triphosphate. *See* ATP
Adenoviruses, oncogenes in, 854–855, 855T
Adenylate cyclase
  hormones affecting, 580T
  reaction catalyzed by, 567
Adenylate cyclase pathway, triggering of, 580, 581F, 582, 582F
Adenylation, enzyme regulation by, 178–179, 179F, 179T
Adenylosuccinate lyase, 543
Adenylylsulfate pyrophosphorylase, reaction catalyzed by, 497
Adenylyltransferase, 493
Adipocytes, as source of fatty acids, 413–414

Adipose tissue, 413
    cells of, 6F
    energy-related reactions in, 568F
    fatty acid release from, regulation of, 427, 428F, 429, 429F
    fuel reserves in, 563T, 564T
    triacylglycerol stores of, 563, 563T, 566, 567F
ADP
    interconversion with ATP, 19, 20F, 266
        bidirectionality of, 232–233, 233F
    oxidative phosphorylation to ATP, 283, 284F
    phosphate groups of, resonance forms of, 40, 42F
    phosphorylation of, 259, 260F, 260T
    relative concentration as fuction of adenylate energy charge, 236F
    uptake of, coupling of ATP export to, 324–325, 325F
ADP-glucose, in glycogen synthesis, 266
ADP-glucose synthase, reaction catalyzed by, 265
Adrenal cortical hormones, 573T
Adrenalin. See Epinephrine
Adrenal medullary hormones, 573T
Aerobic pathway, for fatty acid synthesis, 424, 425–426, 426F
Affinity chromatography, 121
Agrobacterium, 594, 595F
Agrobacter tumefaciens, as cloning system, 689
AIDS. See Acquired immunodeficiency syndrome
Alanine, 205F, 513T
    enantiomers for, 56, 56F
    pK value of, 54T
    structure of, 51F
    titration curve of, 53, 53F
D-Alanine, 503, 503T
Albinism, 525T
Albumin
    diffusion constant of, 124T
    fatty acid binding to, 429
    isoelectric point of, 124T
    molecular weight of, 124T
    sedimentation constant of, 124T
Alcohol dehydrogenase, zinc in, 158
Aldehyde hydrate, 245, 246F
Aldimine, 202–203, 203F
Aldolase
    reaction catalyzed by, 260T
    reactions involving, 257, 257F
Aldoses, 253
D-Aldoses, 244F
Aldosterone, 576
    structure and function of, 479T
    synthesis of, 576–577
Algae, 26F
    photosynthetic properties of, 333T
Alimentary tract. See also specific organs
    hormones of, 573T
Alkaptonuria, 523, 525T
Alleles, 666, 773
Allelic exclusion, 836
Allolactose, 771F
    synthesis of, 772F
Allopurinol, 555
    structure of, 555F
D-Allose, 244F
Allosteric effectors, 176
    enzyme sensitivity to, 178
    regulation of energy metabolism by, 267, 268T
Allosteric enzymes, 176, 180F, 180–193, 181F, 183–194F
    sigmoidal dependence on substrate concentration, 180F, 180–182, 181F
Allosteric proteins, 101–110. See also Hemoglobin
    function of, models of, 109–110, 110F
Allostery, 103
Allotypes, 833
All-trans-retinal, structure of, 617F
α,β Eliminations, 202

Alpha helix, 75T, 75–77, 76F, 77F
α pattern, of fibrous proteins, 74–75
Altman, Sid, 723
D-Altrose, 244F
Amanita phalloides, 725
α-Amanitin, 724F, 725
Ames, Bruce, 850
Ames tests, 850
Amethopterin. See Methotrexate
Amide(s), hydrolysis of, 157
Amide plane, 56–57
Amino acid(s), 9, 11F, 50–56, 511–530. See also specific amino acids
    acidic and basic properties of, 52T, 52–54, 53–56F, 54T
    activated, ordering on mRNA template, 731, 733, 734F, 735
    aminoacyl-tRNA synthase recognition of, 743–744, 744F, 745F
    aromatic, 499–502
        light absorption by, 55, 56F
        negative regulation of synthesis and, 502
        tryptophan synthesis and, 499, 500F. 501T, 501–502
    aspartate family of, 497–499, 498F
        isoleucine and valine synthesis and, 497, 498F, 499
        safe herbicides and, 499, 499F
    asymmetry of, 56, 56F
    attachment to tRNAs, 742, 743F
    catabolism of
        genetic diseases and, 523, 524F, 525T
        as source of carbon skeletons and energy, 521, 522F, 523
    classification of, 50, 51T
    contact between base pairs and, DNA-protein cocrystal revelation of, 790–791, 792–794F
    deamination of, detoxification of ammonia resulting from, 516–521, 517F
    essential, 513T, 513–514, 514F, 515T
        sources of, 514–515, 515F
    families of, 488–489
    fates of, 19, 19F
    free, 63F
    function of, 487–488
    glutamate family of, 489–495
        glutamate amidation in synthesis of, 491F, 491–493, 492F
        α-ketoglutarate amination in synthesis of, 490F, 491
        nitrogen cycle and, 493, 493F, 495
    glycogenic, 523
    ketogenic, 521
    as negative regulators of their own synthesis, 502
    nonessential, 513T, 513–514, 514F, 515T
        synthesis of, 515T
    nonprotein, 502–503
    oxidative deamination of, 515–516, 517F
    polypeptide chain formed by, 11F
    as precursors for compounds other than proteins, 526–530
    protein composition of, determination of, 58–60, 59F, 60F
    pyruvate family of, 497–499, 498F
        isoleucine and valine synthesis and, 497, 498F, 499
        safe herbicides and, 499, 499F
    R groups of, 9, 11F, 53–54, 54T
        classification according to, 50, 51T
    sequence of
        in proteins, determination of, 61, 62–66F, 65–66
        in serine proteases, 159, 160, 160F, 161F
    serine family of, 495–497
        cysteine synthesis and, 495, 496F, 497
        sulfate reduction to sulfide and, 497, 497F
    structure of, 50, 50F, 51T
    synthesis of, 20–21, 21F, 488F, 488–506
        auxotrophs and, 489, 489F
        de novo, 512F, 512–513, 514, 514F, 515T
        number of proteins participating in, 489
        regulation of enzymes catalyzing, 777–780, 778F
    titration curve of, 53F, 53–54, 54F
    transamination of, 515, 516F

tRNA as carrier of, for protein synthesis, 703F, 703–704, 704F
zwitterion form of, 11F, 50, 50F, 51T
α-Amino acid(s), reactions of, pyridoxal-5'-phosphate and, 200, 201F, 202–203, 203F, 204F
D-Amino acid(s), in microbes, 503, 503T
Amino acid analyzer, 60, 60F
Amino acid oxidases, reactions involving, 209T
Aminoacyl-tRNA, synthesis of, 742, 743F
Aminoacyl-tRNA binding factor, 748–749
Aminoacyl-tRNA synthases, 703
    amino acid attachment to tRNAs and, 742, 743F
    correction of acylation errors by, 744–745
    recognition of amino acids and regions on cognate tRNA by, 743–744, 744F, 745F
p-Aminobenzoic acid, 199T, 552F
γ-Aminobutyrate, structure of, 609F
Amino group, 17F
δ-Aminolevulinate, 526
Ammonia
    amino acids derived from. *See* Amino acid(s), glutamate family of
    detoxification of, 516–521, 517F
        carriers for transporting ammonia to liver and, 520–521, 521F
        Krebs bicycle and, 520, 520F
        urea formation as mode of, 517, 518F, 519F, 519–520
Ammonium ion, bacterial production of, 495
Ammonotelic animals, 516
Amoeba, 26F
AMP
    as allosteric effector, 267, 268T
    phosphorylases *a* and *b* and, 191, 191F, 192
    relative concentration as fuction of adenylate energy charge, 236F
    synthesis of, 542F, 543
Amphibians, gene products required during embryonic development and, 819–820
Amphibolic pathway, tricarboxylic acid cycle as, 295, 296F
Amphipathic molecules, 9
Ampholytes, 122
Amylose, complex with iodine, structure of, 249, 251F
Anabolic reactions, 19, 20–21, 21F, 231, 233
    end-product inhibition of, 235, 235F
    regulation by energy status of cell, 235–236, 236F, 237F
    starting materials for
        differences among organisms and, 228T, 228–229
        requirement for, 227–228
Anaerobic metabolism, catabolic reactions in, 282–283, 283F
Anaerobic pathway, for fatty acid synthesis, 424, 425, 425F
Anaplerotic reactions, 295
Anderson, W., 451
Androgens, 573T. *See also* Testosterone
    gene regulation by, 586, 587T
Anemia
    megaloblastic, 545
    sickle cell, 109
Anfinsen, Christian, 82
Angiosperms, 26F
Angiotensin, 573T
Ankyrin, 396
Annealing temperature, 640
Anomers, 245, 246F
Antennapedia mutation, 822, 823F
Anthranilate phosphoribosylisomerase, 501T
Anthranilate phosphoribosyltransferase, 501, 501T
Anthranilate synthases, 501, 501T
Antibiotics
    D-amino acids in, 503, 503T
    binding to sites on ribosome by, 756, 756F
    inhibition of transpeptidation reaction by, 374
    structure of, 756F
Antibodies. *See* Immunoglobulin(s)
Antibody response. *See* Humoral response

Anticodon(s), interactions with codons
    species-specificity of rules regarding, 741–742, 742T
    wobble and, 738–739, 740F, 740T
Anticodon arm, 735
Anticodon loop, 733, 734F, 735
Anticooperative binding, 103
Antidiuretic hormone, 571, 572T
Antigen(s), 831
    antigen-specific reactions involving T cells and, 845, 846F
Antigen-antibody complexes, removal of, 841, 841F
Antiparallel chains, 633, 633F
Antiparallel β-pleated sheet, in proteins, 77–78, 78F, 88F
Antiport, 400
apo B100, 471, 471F, 472F
Apoenzymes, 137, 207
Apolar groups, 13–14, 14F
Apoprotein(s), 467
    functions of, 467, 469
    properties of, 469, 469T
    synthesis of, 469
Apoprotein C-II, 470–471
D-Arabinose, 244F
Arachidonic acid, 384T, 426, 447, 585
    release by phospholipid A$_2$, 453
    synthesis of, 426, 427F
Archebacteria, 26F, 27F
Arginine, 513T
    equilibrium between charged and uncharged forms of, 54F
    p*K* value of, 54T
    structure of, 51F
    synthesis of, 515T
Argininemia, 525T
Arnold, William, 341, 342
Aromatic amino acids. *See* Amino acid(s), aromatic
Arreguin, Barbara, 461
Arrestin, 622
Arthropods, 26F
Artificial transformation, 628
Ascorbic acid, 199T
    reactions involving, 216
    structure of, 216, 216F
Asparagine, 513T
    p*K* value of, 54T
    structure of, 51F
Aspartae, 205F
Aspartate, 204F, 513T
    at active site, 160, 161–162, 162F, 163, 164
Aspartate amino acid family, 497–499, 498F
    isoleucine and valine synthesis and, 497, 498F, 499
    safe herbicides and, 499, 499F
Aspartate aminotransferase, Michaelis constants for, 143T
Aspartate carbamoyltransferase
    catalytic activity of, 187–189, 187–191F, 191
    cytidine triphosphate inhibition by, 187–188, 188F, 234–235
    molecular weight of, 91T
    R and T conformations of, 189, 190F, 191, 191F
    subunit composition of, 91T
Aspartate β-decarboxylase, pyridoxal phosphate action in, 205F
Aspartic acid
    equilibrium between charged and uncharged forms of, 54F
    p*K* value of, 54T
    structure of, 51F
Aspirin
    anti-inflammatory effect of, 453, 454F
    prostaglandin synthesis inhibition by, 453
Asymmetry, of amino acids, 56, 56F
Atherosclerosis, 472
    heart disease and, 474–475
    prostaglandins in, 454
Atmosphere, composition of, 23–24

ATP
  as allosteric effector, 267, 268T
  consumption in gluconeogenesis, 263
  export of, coupling to ADP uptake, 324–325, 325F
  fatty acid oxidation as source of, 414, 415T, 416
  functional coupling and, 231
  hydrolysis of, 40–43
    active transport driven by, 401, 401F, 402T
    alternative sites of, 40, 41F
    standard free energy, 36, 37T, 42F
  interconversion with ADP, 19, 20F, 266
    bidirectionality of, 232–233, 233F
  muscle contraction and, 113
  oxidative phosphorylation of ADP to, 283, 284F
  phosphate groups of, resonance forms of, 40, 42F
  phosphorylase a and, 191, 191F
  protein degradation and, 764
  regeneration of, 258
  relative concentration as fuction of adenylate energy charge, 236F
  structure of, 20F
  synthesis of, 257–258, 258F, 259, 260F, 260T. See also Oxidative
        phosphorylation
    by chloroplasts in dark, 347, 348F
    by complete oxidation of glucose, 325–326, 326T, 327F
    cyclic electron-transport chain and, 339–340, 340F
    electron transfer coupling to, 316F, 316–317, 317F
    flow of protons back into matrix driving, 321–322, 322F
    localization in cell, 20
    oxidative phosphorylation and, 318, 319F
    in photoautotrophs, 231
    in photoheterotrophs, 231
    by photophosphorylation, 347–348
ATP-synthase
  components of, 322–324, 323F, 324F
  proton return to stroma through, 346–348, 348F
Atrial natriuretic peptide, 573T
AUG codon, 746
Autotrophs, 228
Auxins, 592, 593F
Auxotrophs, 489, 489F
Avery, Oswald T., 628
Avian sarcoma virus 17, 861
Axon, 602
3'-Azido-3'-deoxythymidine, 552, 553T
Azidothymidine. See 3'-Azido-3'-deoxythymidine
AZT. See 3'-Azido-3'-deoxythymidine

Bacillus, 5F
Bacillus stearothermophilus, phosphofructokinase of, 184, 184F
Bacillus subtilis, as cloning system, 689
Bacitracin, 372
Bacitracin A, 503T
Bacteria, 4, 5F, 7, 26F. See also Cyanobacteria; Escherichia coli; Purple
        bacteria; specific bacteria
  ammonium ion production by, 495
  cell membranes of, 385
  cell wall synthesis in, 369F, 370–374
    activated monomer synthesis and, 370–371, 372F
    cross-linking of linear polymers and, 372, 374F
    linear polymer formation and, 371–372, 373F
    penicillin inhibition of transpeptidation reaction and, 374, 375F
  chemical composition of, 7T
  chromosomes of, structure of, 641–642, 642F
  glycogen synthesis in, 266
  protein transport during translation in, 760
  reverse transcriptases of, 673
  RNA polymerase of, subunits of, 706–707
  toxins of, G protein cycle as target for, 583, 583F
  tryptophan synthesis in, 501T

Bacterial viruses, gene expression in. See Prokaryotic gene expression,
        regulation in bacterial viruses
Bacteriochlorophyll(s)
  absorption bands of, 336–337, 337–338, 339F
  as antenna system, 340–342, 341F, 342F
  molecular weight of, 91T
  subunit composition of, 91T
Bacteriochlorophyll a
  absorption band of, 333, 335F
  structure of, 334F
Bacteriochlorophyll b, structure of, 334F
Bacteriophage gamma, gene expression in. See Prokaryotic gene
        expression, regulation in bacterial viruses
Bacteriophage gamma vectors, in DNA cloning, 685F, 685–686
Bacteriopheophytin(s), 333, 338
Bacteriopheophytin a, structure of, 334F
Bacteriorhodopsin, 322, 324F
  structure of, 390, 391F
Ball-and-stick models, 75, 76F
Baltimore, David, 854
Band-3 protein, 396
Band-4.1 protein, 396
Banting, Frederick G., 570
Barr body, 810, 810F
β Barrel, in proteins, 88F
Basal body, 4F
Basal transcription, 713
Base(s)
  in nucleotides, 12F
  Schiff's, 202–203, 203F
Base exchange reaction, 443, 445F
Base pairs, standard and wobble, 739, 739F
Basic region, 860
Bathorhodopsin, 616, 618
  synthesis of, 619, 620F
bcd gene, 823, 823F, 825
B cells, 831, 831F
  antigenic stimulation of B-cell clones and, 839, 839F
  helper T-cell triggering of division and differentiation of, 839–840,
        840F
  interaction with T cells, antibody formation and, 838F, 838–840
B DNA, 636F
Beadle, George, 523
Beckwith, Jonathan, 776
Beer-Lambert law, 70
Bendall, F., 342
Bends, in DNA, 636
Benzamidine, trypsin inhibition by, 147, 148F
Bergström, Sune, 453
Berzelius, Jöns Jakob, 49, 136
Best, Charles, 438, 570
β Oxidation, of fatty acids, 414, 415F
β Pattern, of fibrous proteins, 74
β-sheet motif, regulatory proteins using, 791, 795F
Bicarbonate, 36
  standard free energy of formation of, 36T
Bicoid gene, 823, 823F
Big-T oncoprotein, 855–856
Bile acids, 473, 473F, 475, 476–478F
  functions of, 473
Bimolecular reactions, kinetics of, 138
Bingham, P., 817
Biochemical pathways, 20F, 20–21, 21F. See also specific pathways
Biochemical reactions
  anabolic. See Anabolic reactions
  anaplerotic, 295
  catabolic. See Catabolic reactions
  on catalytic surfaces of enzymes, 18F, 18–19, 19F
  energy required by, 19, 19F, 20F
  in glycolytic pathway, 260T
  localization in cell, 20

organization into pathways, 20F, 20–21, 21F
  regulation of, 21
  as subset of ordinary chemical reactions, 16–18, 17F, 17T
Biochemical systems, evolution of, 22–26, 25F, 25T
  common, 25–26, 26F, 27F
Biological systems, types of work performed by, 38
Biosphere, flow of energy in, 19F
Biosynthesis. *See* Anabolic reactions
Biosynthetic pathways, inhibition by end products of pathway, 180
Biotin, 199T, 420
  reactions involving, 213–214
  structure of, 213, 213F
Biotin carboxylase, 420
Biotin carboxyl carrier protein, 420, 420F
Bipolar cell, 615
Bithorax mutation, 822, 823F
Blobel, Gunter, 757
Bloch, K., 461
Bloch, Konrad, 419
Blood cells, 6F
Blood coagulation cascade, 177, 177F
Blood groups, ABO system of, 368, 368T, 369F
Bohr effect, 103–104, 104F
Boltzmann's constant, 32
Bomb calorimeter, 31
Bond(s)
  acid anhydride (pyrophosphate), 537, 537F
  C-C and C-X, cleavage of, thiamine pyrophosphate and, 199–200, 200F, 201F
  covalent, 16, 17T
  disulfide. *See* Disulfide bonds
  double, entropy and, 32
  glycosidic, 9, 10F
  hydrogen. *See* Hydrogen bonds
  peptide. *See* Peptide bonds
  phosphodiester, 12F
    formation of, 709, 710F
Bond energies, 86, 86F
Bone, cells of, 6F
Bonner, James, 461
Bordetella pertussis, 583
Boyer, Paul, 323
Brady, Roscoe, 452
Bragg's law, 96
Brain
  energy demands of, 563–564
  energy-related reactions in, 568F
  fuel reserves in, 563T
Branched-chain ketoaciduria, 525T
Branchpoints
  in biochemical pathways, 20–21
  of monosaccharides, 10F
Breast cancer, incidence of, 850, 851T
Breslow, Ronald, 200
Briggs, G.E., 142
Briggs, R., 807
Briggs-Haldane equation, 142
Brown, Michael, 471, 472
Brown algae, 26F
Buchner, Eduard, 136
Buffered solutions, 53
Bundle sheath cells, 351, 352F
Bungarus, 610
Burkitt's lymphoma, 851, 860
  genome of cells of, 851, 853F

c-abl oncogene, 852
Cadmium ion, standard enthalpy and entropy changes on forming complexes between methylamine or ethylenediamine and, 35T
Cairns, John, 652, 658

Calcitonin, 572T
Calcitonin gene-related peptide, 572T
Calcium
  in earth's crust, 25T
  gap junctions and, 407
  in human body, 25T
  inositol triphosphate pathway and, 584F, 584–585, 585F
  muscle contraction and, 113, 115F
  in ocean, 25T
Calcium channels, voltage-gated, 609
Calmodulin, 584
Calvin, Melvin, 348
Calvin cycle. *See* Reductive pentose cycle
cAMP. *See* Cyclic AMP
Cancer, 848–862, 849F. *See also* Oncogenes; Tumor(s)
  arising by mutational events in cellular protooncogenes, 852–853
  arising from genetically recessive mutations, 852, 854T
  genetic abnormality of cancerous cells and, 851–852, 853F
  incidence of, environmental factors influencing, 850–851, 851T, 852F
  p53 gene and, 855–856
  similarity of transformed cells to cancer cells and, 850
  tumor-causing viruses and, 853–855
    association of oncogenes with, 853–854
    role in transformation, 854–855, 855T
CAP formation, 840, 840F
Capping, 719, 720F
CAP protein, 776–777, 777F, 790, 791F, 795
  CAP-binding region in *lac* promoter and, 777F
  stimulation of *lac* operon, 776
Carbamoyl aspartate, formation of, 187–189, 187–191F, 191
Carbamoyl phosphate, synthesis of, 519, 519F
Carbamoyl phosphate synthase, reaction catalyzed by, 544F, 544–545
Carbohydrates, 7, 9, 227–240. *See also* Electron transport; Gluconeogenesis; Glycolysis; Oxidative phosphorylation; Pentose phosphate pathway; Photosynthesis; Tricarboxylic acid cycle; *specific carbohydrates and types of carbohydrates*
  ATP-ADP system and, 232–233, 233F
  clustering of sequentially related enzymes and, 229–231, 230F
  conversion into lipids, in liver, 430, 431F
  differing sources of energy, reducing power, and starting materials for biosynthesis and, 228T, 228–229
  in energy metabolism, 243–249
  enzyme regulation and, 234–236
  functional coupling and, 231F, 231–232, 232F
  kinetic regulation of conversions and, 233–234
  monomers and polymers of, 9, 10F
  pathways and, 229, 229F, 230F
  as recognition markers, 359, 361
  starting materials and energy required by living cells and, 227–228
  strategies for pathway analysis and, 236–239
Carbon
  amino acid catabolism as source of carbon skeletons and, 521, 522F, 523
  in atmosphere, 24
  in earth's crust, 25T
  fixation of, via reductive pentose cycle, 348, 349F, 350
  in human body, 25T
  interdependence of organisms and, 21–22
  in ocean, 25T
Carbon dioxide
  hemoglobin's role in disposing of, 104, 104F
  oxygen binding to hemoglobin and, 103–104, 104F
  standard free energy of formation of, 36T
  transport in circulatory system, 103–104, 104F
Carbonic anhydrase, 104
  molecular weight of, 91T
  specificity constant of, 145T
  subunit composition of, 91T
  turnover number of, 144T
  zinc in, 158, 220, 220F
Carbonylcyanide-*p*-trifluoromethoxyphenylhydrazone, structure of, 317F

Carbonyl group, 17F
$N^{1'}$-Carboxybiotin, 213F
N'-Carboxybiotin, structure of, 420F
Carboxylation, biotin and, 213F, 213–214
Carboxyl group, 17F
Carboxyltransferase, 420
Carboxymethyl cellulose (CMC), 121T
Carboxypeptidase A, zinc in, 158
Carcinogenesis, multistep nature of, 861–862, 862F
Cardiac muscle
    energy demands of, 564T, 564–565
    energy-related reactions in, 568F
    fatty acid utilization by, 429
    lactate dehydrogenase of, 566
Cardiolipin. See Diphosphatidylglycerol
Cardiovascular disease, cholesterol and, 463, 472
Carnitine, 199T
Carnitine acyltransferase I, reactions catalyzed by, 430, 430F
Carnitine acyltransferase II, 430
β-Carotene, structure of, 342F, 617F
Carotenoids, 342, 342F
Cartilage, cells of, 6F
Cashel, M., 781
Castanospermine, 370T
Catabolic reactions, 19, 231, 233
    under anaerobic conditions, 282–283, 283F
    regulation by energy status of cell, 235–236, 236F, 237F
Catabolite repression, 769, 776
Catalase
    diffusion constant of, 124T
    isoelectric point of, 124T
    Michaelis constant for, 143T
    molecular weight of, 124T
    sedimentation constant of, 124T
    specificity constant of, 145T
    turnover number of, 144T
Catalysts, 19. See also Enzymes; specific catalysts
    coupling of favorable and unfavorable reactions and, 39
Catenation, 659, 660F
Cathepsins, 763
CCA enzyme, 716–717
cca gene, 716t
C-4 cycle, 351
cDNA libraries, 686, 687–688, 688F, 696–697
cDNA probe, β-globin characterization using, 694
CDP-choline:1,2-diacylglycerol phosphocholinetransferase, reaction catalyzed by, 441
CDP-diacylglycerol, 438
Cech, Tom, 722
Cell(s), 3–9
    bacterial. See Bacteria; specific bacteria
    composition of, 7, 7T, 8F, 9, 10–12F
    elements in, 16
    as fundamental unit of life, 4–6F, 7
    hydrolysis in, hydrolysis in intestine compared with, 251, 253
    localization of biochemical reactions in, 20
    specialized types of, 6F, 7
Cell adhesion proteins, 849
    required to maintain immune response, 844–845, 846F
Cell division, 22, 23F
Cell line, 850
Cell-mediated response, 831–832, 841–845
    cell adhesion proteins required for mediation of immune response and, 844–845, 846F
    evolutionary relationship among immune recognition molecules and, 845
    interleukins and, 842
    major histocompatibility molecules and, 843–844
        class I and class II, 843–844, 845F
        graft rejection and, 843
    T-cell receptors and, 844
    T-cell recognition of combination of self and nonself and, 843, 844F
    tolerance and, 842–843, 843F
Cell membrane. See Plasma membrane
Cellobiose, 248
    structure of, 247F
Cellulose(s), 248–249, 249F, 358
    in column procedures, 121
    configuration of, 249, 250F, 251F
    degradation of, 249
Cell wall, 5F, 7
    bacterial, synthesis of. See Bacteria, cell wall synthesis in
Central vacuole, 5F
Centriole, 4F
Ceramide
    glycosphingolipid synthesis from, 448, 450, 450F
    sphingomyelin synthesis from, 447–448, 449F
Ceruloplasmin
    molecular weight of, 91T
    subunit composition of, 91T
Chaikoff, Irving, 438
Chance, Britton, 309
Changeaux, Jean-Pierre, 180
Changeaux, Pierre, 109
Chargaff, Erwin, 631
Charge-charge interactions, in polypeptide chains, 87–88, 88T
Charge-dipole interactions, in polypeptide chains, 87–88, 88T
Chelation effect, 34–35, 35T
Chemical reactions
    bimolecular, kinetics of, 138
    biochemical. See Biochemical reactions
    energy changes in. See Free energy; Thermodynamics
    kinetics of. See Kinetics
    reversible, kinetics of, 137–138
Chemiosmotic theory, 318, 319F
Chemoautotrophs, 228, 228T
Chirality, of amino acids, 56, 56F
Chitin, structure of, 358–359, 359F
Chloramphenicol, structure of, 756F
Chlorella pyrenoidosa
    O_2 evolution as function of excitation wavelengths in, 343, 344F
    reaction centers versus antenna in, 341F
Chlorine
    in earth's crust, 25T
    in human body, 25T
    in ocean, 25T
Chloroacetone phosphate, 169
Chlorophyll
    as antenna system, 340–342, 341F, 342F
    electron release from, 336, 337F
    photochemical reactivity of, 332–333, 334F, 335F
Chlorophyll a, 333
    absorption band of, 333, 335F
    structure of, 334F
Chlorophyll b, 333
    structure of, 334F
Chloroplasts, 5F, 7, 332, 332F
    ATP formation by, in dark, 347, 348F
    photosystem linkage in, 342–344, 343F–345F
Cholecalciferol, 220–221, 577, 579F
Cholecystokinin, 573T
Cholera, 583, 583F
Cholesterol, 459–482, 460F, 477, 480–481F
    cardiovascular disease and, 463
    conversion to bile acids, 473, 473F, 475, 476–478F
    diseases resulting from, 472
        cardiac, 472–473, 474–475
    low-density lipoproteins uptake and, 471
    melting transition of phospholipid bilayer and, 396
    plasma concentration of, 467T
    reduction of deposits by low-density lipoproteins, 472–473, 473F, 474–475

steroid hormones derived from, 574, 576–577, 577–579F
structure of, 383, 386F
synthesis of, 460F, 461F, 461–465, 464, 467F
  lanosterol conversion into cholesterol in, 464, 467F
  lanosterol formation and, 463–464, 464–466F
  mevalonate in, 461–463, 462F, 463F
transport of, 470–471
  reverse, 472
Cholesteryl esters, synthesis of, 469, 470F
Cholesteryl ester transfer protein, 472
Cholic acid, 475
synthesis of, 475, 476F
Choline, 199T, 384T
dietary requirement for, 441
Choline kinase, reaction catalyzed by, 441
Chondroitin sulfate, structure of, 360F
Choriogonadotropin, 573T
Chorismate, 299F
Chromatin, 4F, 5F, 642, 642F
active versus inactive, 810–812
  DNA methylation and, 810–811, 811F
  DNA susceptibility to DNase degradation and, 810
  enhancers and, 811–812, 812F, 813F
Chromatography. *See also* High-performance liquid chromatography
affinity, 121
gel-exclusion, 120, 121F
ion-exchange, 59, 59F, 120–121, 121T
Chromium
in earth's crust, 25T
in human body, 25T
in ocean, 25T
Chromosome(s), 22
eukaryotic, DNA of, 662F, 662–663
heterochromatic, 809–810, 810F
homologous, 666
Philadelphia, 851, 853F
polytene, 809, 809F
replication of
  in *Escherichia coli,* 660–661, 662F
  in eukaryotes, 673, 673F
structure of, 641–644
  in bacteria, 641–642, 642F
  in eukaryotes, 642F, 642–644, 643T, 644F, 645F
  variation with gene activity, 809
translocations of, in cancerous cells, 851
X, 810
Chromosome jumping, mapping of cystic fibrosis gene using, 694–697, 695F, 696F
Chromosome walking
identification and isolation of regions around adult β-globin genes by, 694, 695F
mapping of cystic fibrosis gene using, 694–697, 695F, 696F
Chylomicrons, 413, 414F, 465
cholesterol and triacylglycerol transport by, 470–471
composition and density of, 468T
structure of, 468F
Chymotrypsin
catalytic activity of, 159–164F, 159–165
Michaelis constants for, 143T, 163F
molecular weight of, 91T
site of action in polypeptide chain cleavage, 64F
structure of, 159–163
  amino acid sequence and, 159, 160F
  crystal, 159–160, 161F, 162–163, 176, 176F
subunit composition of, 91T
turnover number of, 144T, 163F
Chymotrypsinogen
diffusion constant of, 124T
isoelectric point of, 124T
molecular weight of, 124T
sedimentation constant of, 124T

structure of, crystal, 176, 176F
cII protein, lysogeny and, 786, 788F, 789
Cilia, 4F, 6F
Ciliates, 26F
cI protein, prevention of buildup during lysis, 785, 788F
Circular dichroism, protein structure analysis using, 100T
Circularly polarized light, structural analysis methods using, 281, 281F
Circulin A, 503T
gamma cI repressor, 790, 793F
cis orientation, 773
Citrate
activation of hepatic acetyl-CoA carboxylase by, 431
binding of, 293, 293F
metabolism of, stereochemical relationships in, 292F
regulatory effects of, 301
synthesis of, 289F, 290, 295, 297F
  stereochemical relationships in, 292F
Citrate cycle. *See* Tricarboxylic acid cycle
Citrate synthase
regulation of, 300
in tricarboxylic cycle, 289F, 290
Class switching, antibody diversity and, 836, 837F
Clathrate structure, 14, 14F
Cloning, of DNA. *See* DNA cloning; DNA libraries
Closed promoter complexes, 708, 709F
Clostridium pasteurianum, nitrogen fixation in, 494F, 495
c-myc oncogene, 852
Coagulation, prostaglandins in, 454, 456F
Coagulation cascade, 177, 177F
Coated regions (pits), 471, 471F, 472F
Cobalt
in earth's crust, 25T
in human body, 25T
in ocean, 25T
Coding ratio, 736
Codons, 22, 731, 736
interactions with anticodons
  species-specificity of rules regarding, 741–742, 742T
  wobble and, 738–739, 740F, 740T
Codon usage, 737
Coelenterates, 26F
Coenzyme(s), 9, 136
of water-soluble vitamins, 199–220
Coenzyme A esters, binding of, 212
Coenzyme Q, 199T
Cofactors, 136–137
Cognate tRNAs, 733
aminoacyl-tRNA synthase recognition of, 743–744, 744F, 745F
Cohen, P.P., 519
Collagen
processing of, 760, 761F
structure of, 79–80, 80F, 81F
Collagen pattern, of fibrous proteins, 74
Colon cancer, incidence of, 850, 851T
Colony hybridization method, 688, 689F
Columnar epithelium, 6F
Column procedures, for protein purification, 120–122, 121F, 121T
Competitive inhibition, of enzymes, 147F, 147–149, 148F
Complementation analysis, 237–238, 238F
Complementation group, 237
Complement system, 841
activation of, 841, 841F
Concanavalin A
molecular weight of, 91T
subunit composition of, 91T
Concentration work, 38
Conditional mutants, 655
Cone cells, 615, 615F
electrophysiological response to light, 620–621, 621F
Conformation, 72
Consensus sequences, 709

Constitutive heterochromatin, 809
Constitutive mutation, 772
Controlling elements, 771
Conversions
  bidirectional, 232–233
  kinetic regulation of, 233–234
  sequences distinguished from, 232
Cooperative binding, 103
Copper
  in earth's crust, 25T
  in human body, 25T
  in ocean, 25T
Cordycepin, 724F
Core polymerase, 707
Corey, Robert, 73–80
Cori, Carl, 191
Cori, Gerty Radnitz, 191
Cori cycle, 565, 565F
Coronary artery disease, high-density lipoproteins and, 472
Corticosterone, 576
Corticotropin, 571, 572T
Corticotropin-releasing factor, 572T
Cortisol, 576
  structure and function of, 479T
Corynebacterium diphtheriae, toxin produced by, mechanism of damage
    produced by, 752–753, 753F
Cosmids, in DNA cloning, 686
Cot curve, 640
Cotranslational transport, 760
Covalent bonds, 16, 17T
Crick, Francis, 633, 649, 651, 739
Cristae, 307, 307F
Cro protein, 785, 788F
Crotonase, specificity constant of, 145T
Crown gall tumors, 594, 595F
Cruciforms, in DNA, 636
Crystallins, 615
c-src protooncogene, 855
CTP, synthesis of, 545, 545F
CTP:phosphocholine cytidylyltransferase, reaction catalyzed by, 441, 446,
    447F
Cuboidal epithelium, 6F
Cushing's syndrome, 589
Cyanobacteria, 26F
  photosynthetic properties of, 333T
Cyanogen bromide, 61, 64F
Cyclic AMP
  function of, 567, 569, 582
  mediation of glucagon effects by, 268, 269F, 270
  synthesis of, 268, 269F, 427, 429F
Cyclic AMP phosphodiesterase, reaction catalyzed by, 268
$3',5'$-Cyclic GMP, guanine, 621–622, 622F
γ-Cystathionase, reaction catalyzed by, 497
Cystathionine, synthesis of, 495, 496F, 497
Cysteic acid, 63F
Cysteine, 513T
  equilibrium between charged and uncharged forms of, 54F
  pK value of, 54T
  structure of, 51F
  synthesis of, 495, 496F, 497, 515T
Cystic fibrosis, mapping of gene for, chromosome walking and jumping
    in, 694–697, 695F, 696F
Cystine, disulfide reduction of, 178–179, 179F
Cytidine, in tRNA, structure of, 704F
Cytidine triphosphate, aspartate carbamoyl transferase inhibition by,
    187–188, 188F, 234–235
Cytochrome(s), 307–308, 308F
Cytochrome b_5, 425–426, 426F
Cytochrome bc_1 complex, 312, 313T, 314F, 314–315
Cytochrome b_6f complex, 342
Cytochrome b_5 reductase, reaction catalyzed by, 425–426, 426F

Cytochrome c, 307
  absorption spectra of, 307, 308F
  diffusion constant of, 124T
  isoelectric point of, 124T
  molecular weight of, 91T, 124T
  sedimentation constant of, 124T
  structure of, 15F
  subunit composition of, 91T
Cytochrome c_1, targeting of, 758, 759F
Cytochrome oxidase, 308
Cytochrome oxidase complex, 312, 313T, 315, 315F
Cytochrome P450, 218–220
  catalytic activity of, 219–220, 220F
  synthesis of, 219, 219F
Cytokinins, 592, 593F
Cytomegalovirus infections, pharmacotherapy for, 552, 553T
Cytoplasm, 7
  influence on nuclear expression, 807–808, 808F
Cytosine, 535
  structure of, 12F
Cytosine arabinoside, 552, 553T
Cytoskeleton, 7
  connection of membrane proteins to, 396, 397F, 398F
  structure of, 396, 397F
Cytosol, 4F, 7
  molecules in, 9

Dam, Henrik, 221
Danielli, J.F., 388
Dansyl chloride method, 71
Daughter cells, 22, 23F
Davern, Ric, 652
Davison, H., 388
Dawson, 628
D-Biotibyl-protein, 213F
Deamination, of amino acids, detoxification of ammonia resulting from,
    516–521, 517F
Death, rigor and, 113, 115F
β-Decarboxylases, 202, 203F
Decatenation, 659
DeDuve, Christian, 763
Degradative reactions. *See* Catabolic reactions
Dehydroascorbic acid, structure of, 216, 216F
7-Dehydrocholesterol, 590
17-Dehydroxylase, deficiency in, 576
21-Dehydroxylase, deficiency in, 576–577
Deisenhofer, Johann, 337
De Lucia, 658
Denaturation, of DNA, 638F, 638–639, 639F
Dendrites, 602
Deoxyadenosine, 548–549
Deoxyadenosine-5'-phosphate, structure of, 536F
5'-Deoxyadenosylcobalamin
  reactions involving, 216–217
  structure of, 216, 217F
Deoxycytidine kinase, reaction catalyzed by, 548
Deoxycytidine-5'-phosphate, structure of, 536F
Deoxyguanosine-5'-phosphate, structure of, 536F
Deoxyhemoglobin, 104–105, 105F, 106F
  iron-porphyrin complex in, 107F
Deoxymannojirimycin, 370T
Deoxyribonucleic acid. *See* DNA
Deoxyribonucleotide, synthesis of, regulation of, 558F, 559, 559F
Deoxyribonucleotides, synthesis of, 545–547, 546F
2-Deoxyriboses, 535
Dermatan sulfate, structure of, 360F
Desensitization, 587
Desmolase, reaction catalyzed by, 475, 478F
Detergents, phospholipid bilayer disruption by, 389, 389F

Developmental processes, gene expression regulation and, in eukaryotes. *See* Eukaryotic gene expression

α-D-Glucose-6-phosphate, 272F

Diabetes mellitus, juvenile-onset, 590

Diacylglycerol, 443, 447, 584
in phospholipid synthesis, 441, 442F

Diacylglycerol acyltransferase, reaction catalyzed by, 441

Diatoms, 26F

2′,3′-Dideoxyinosine, 552, 553T

2,4-Dienoyl-CoA reductase, reaction catalyzed by, 416, 417F, 418

Diet
choline requirement in, 441
in genetic diseases associated with errors in amino acid catabolism, 523
heart disease and, 474
long-term changes in, enzyme levels and, 432
as source of fatty acids, 412, 413F

Diethylaminoethyl cellulose (DEAE), 121T

Differential centrifugation, of proteins, 119, 119T

Differential precipitation, of proteins, 119–120

Differential scanning calorimetry, of phospholipids, 395, 395F

Diffusion
across membranes, asymmetrical, 393–394, 394F
facilitated, 400

Digestion
proteolytic, 514–515
of starch and glycogen, intestinal, 251, 253

Digestive enzymes, 61, 64F. *See also specific enzymes*
catalytic activity of, 159–164F, 159–165

Dihydroacetone, formation of, 243F

Dihydrolipoyl dehydrogenase, reaction catalyzed by, 287

Dihydrolipoyltransacetylase, reaction catalyzed by, 287

Dihydrouridine, in tRNA, structure of, 704F

Dihydroxyacetone phosphate, in glycolysis, 256–257, 257F

Dihydroxy acid dehydrase, reaction catalyzed by, 497, 498F

5α-Dihydroxytestosterone, synthesis of, 577, 577F

1,25-Dihydroxyvitamin D₃, 221, 573T
gene regulation by, 586, 587T
synthesis of, 577, 579F
underproduction of, 590

Diisopropylfluorophosphate, enzyme inhibition by, 150, 150F

Dimethylallyl pyrophosphate, 465

N^2,N^2-Dimethylguanosine, in tRNA, structure of, 704F

Dineflagelates, 26F

2,4-Dinitrophenol, structure of, 317F

Dinitrophenyl-amino acid, 63F

Dinucleotides, formation of, 12F

Dipalmitoylphosphatidylcholine, 383, 441, 447
synthesis of, 441, 443F

Dipeptides
basic dimensions of, 74F
synthesis of, 66, 67F

Diphosphatidylglycerol, 384T
synthesis of, 438, 440F

Diphtheria toxin, mechanism of damage produced by, 752–753, 753F

Diplococcus pneumoniae, transformation of, 628, 629F

Diploids, 773
partial, 772–773, 774T

Dipole-dipole interactions, in polypeptide chains, 87–88, 88T

Disaccharides
glycosidic bonds in monosaccharides in, 245–248, 247F
structure of, 247F

Disks, in cone cells, 615

Disulfide bonds
cleavage of, 61, 63F
covalent, 57, 58F

Disulfide reduction, enzyme regulation by, 178–179, 179F

D loop, 733, 734F, 735

DNA, 9, 628, 630F, 630–641. *See also* Transcription
amount per cell, 630, 630F
of bacteriophage gamma, 787F
base composition of, 632T

B form of, 636F
conformation in rapid-start complex, 769, 770F
denaturation of, 638F, 638–639, 639F
double-helix structure of, 631, 632T, 632–634F, 633, 635–636, 651, 651F
conformational variants of, 635–636, 636F
stabilization by hydrogen bonds and stacking forces, 633, 635, 635F
duplex
structure of, supercoils formed by, 636–638, 637F, 638F
synthesis of, 671, 672F
enzymes acting on, 673
interaction with DNA-binding proteins, 789–795
prokaryotic gene expression and. *See* Prokaryotic gene expression, DNA-regulatory protein interaction and
mediation of transformation by, 628, 629F
methylation of, correlation with inactive chromatin, 810–811, 811F
polynucleotide chain of, 630–631, 631F
protein synthesis and, 22, 23F, 24F
recombinant DNA techniques and. *See* Recombinant DNA techniques
recombination of, 666–671, 667–669F, 668F, 669F
enzymes mediating in *Escherichia coli,* 668–671, 670F
homologous, 667–668
nonspecific, 671
site-specific, 671
relaxed, 637
renaturation of, 639–641, 640F, 641F
repair of, 664F, 664–666, 665F
regulation of synthesis of repair proteins and, 665–666, 666F
replication of
in bacteria. *See Escherichia coli,* DNA replication in
in eukaryotes, 661–663
semiconservative, 651–652, 652F
RNA hybrid duplexes with, 701–702, 702F
RNA inhibitors binding to, 725
sequencing, 66
structure of, 16, 16F
supercoiled, 636–638, 637F, 638F
susceptibility to DNase degradation, in active chromatin, 810
synthesis of, localization in cell, 20
template and growing strands of, 656, 657F
x-ray diffraction of, 649, 649F
Z form of, 635–636

DNA amplification, 679–680, 681F
elevation of rRNA in frog eggs by, 819–820, 820F

DNA cloning, 682–686
bacteriophage gamma vectors in, 685F, 685–686
cosmids in, 686
libraries and. *See* DNA libraries
plasmids in, 683–685, 684F, 685F
restriction enzymes in, 682F, 682–683, 683T
shuttle vectors in, 686

dnaG gene, 716t

DNA gyrase, 637

DNA libraries, 686–689
approaches for picking clones from, 688, 689F
cDNA, 687–688, 688F, 696–697
cloning systems for, 688–689
genomic, 686–687, 687T
jumping and walking, 696, 696F

DNA ligase, 659, 659F

DNA linkers, 687–688, 688F

DNA polymerase δ, 663, 664F

DNA polymerase α, 663

DNA polymerase, RNA-directed. *See* Reverse transcriptases

DNA polymerase I
in *Escherichia coli*
role of, 658
in vitro, 656, 657F
turnover number of, 144T

DNA polymerase II
in *Escherichia coli,* 656–657, 658F

role of, 658
Escherichia coli RNA polymerase compared with, 710–711, 712T
DNA polymorphisms, 690, 692F, 692–694, 693F
DNA sequencing, 679, 680F
Docking protein, 759
Docosahexaenoic acid, 426
Doisy, Edward, 221
Dolichol phosphate, in oligosaccharide biosynthesis, 362–363, 364F
Domains, as functional units of protein tertiary structure, 88–90, 88–90F
Dominant gene interactions, 773
Dopamine, 610
    structure of, 609F
Doty, Paul, 639
Double bonds, entropy and, 32
Double-helix structure of DNA. *See* DNA, double-helix structure of
Double-reciprocal plot, 143, 143F
Down-regulation, 587, 589F
Dozy, 692
Drosophila
    homeodomain in, 813–814, 814F
    potassium channel in, 608
    splicing in, 817
    Z DNA of, 636
Drosophila melanogaster
    analysis of genes controlling early events of embryogenesis of, 823–825
        maternal-effect gene products for oocytes and, 823F, 823–824
        segmentation genes and, 824–825, 825F
    coding regions of, 644
    early development of segmented structure preserved to adulthood in, 820–823, 821F, 822F
        cascade of regulatory events and, 822
        regulatory genes involved in, 822–823, 823F, 823T
    visualization of active genes in, 809, 809F
Dulbecco, Renato, 854
Duysens, Louis, 336, 344
Dwarfism, 590

Earth
    atmosphere of, 23–24, 25T
    crust of, 23, 25T
    water on surface of, 23, 25T
Ecdysone, gene regulation by, 586, 587T
β-Ecdysone, structure of, 589F
Echinoderms, 26F
Edidin, M., 393
Editing, of RNA, 721, 722F
Edman degradation, 65, 65F
EF-1, 749, 749F
EF-2, 749
    inactivation of, 752, 753F
    protein kinases, 817–818, 819F
EF-G, 749
EF-T, 749, 749F
EF-Tu, 749, 749F
Ehrlich's ascites tumor, purification of uridine 5'-monophosphate synthase from, 125–127, 126T, 127F
Eicosanoids, 447, 452F, 452–454
    local action of, 454, 456F
    oxygenated, 452–453
    synthesis of, 453–454, 453–455F
eIF-1, protein kinases, phosphorylating and inactivating, 817–818, 819F
Einstein, 335
Elastase
    catalytic activity of, 159–164F, 159–165
    domains in, 89, 89F
    structure of, 159–163
        amino acid sequence and, 159, 160F
        crystal, 159–160, 161F, 163
Electrical work, 38

Electrochemical potential gradients
    diffusion of solutes down, 400
    electron transport and, 318–319, 320F, 321, 322F
    energy required for active transport against, 401F, 401–402, 402T
Electronic entropy, 32T
Electron microscopy, freeze-fracture, 390, 391F
Electron paramagnetic resonance, protein structure analysis using, 100T
Electron transfer
    coupling to ATP formation, 316F, 316–317, 317F
    decay of excited molecules via, 336
    flavin coenzymes and, 209
    flavins and, 207F, 207–209, 208F, 209T, 210F
    α-lipoic acid and, 212F, 212–213
    proton pumps driven by, 402
    pumping of protons across membrane and, 321, 321F
Electron transport, 306, 306F, 307F, 307–316
    complexes in, components of, 313T
    cyclic electron-transport chain and, ATP formation and, 339–340, 340F
    electrochemical potential gradient and, 318–319, 320F
    electron carriers in, 307–310, 308F, 309F
        complexes of, 312F, 312–316, 313T
            mediation of electron transfer between complexes and, 316
        sequence of, 309–310
    redox potentials and, 310, 311F, 311T, 312, 312F
    release from phosphorylation, 317F, 317–318, 318F
Electron-transport chain, 283, 284F
Electron volts, 335
Electrophiles, 155
Electrophoresis, gel, of proteins, 122F, 122–123, 123F
Electrostatic effects, formation of transitional state and, 156–157
Electrostatic forces, in polypeptide chains, 87–88, 88T
Elements. *See also specific elements*
    in atmosphere, 23–24, 25T
    in earth's crust, 23, 25T
    in living cells, 16
    in oceans, 23, 25T
    trace, 16
Elongation
    GTP in, 750, 751F
    repetition of, 748–749, 749–751F, 752–753, 753F
    site-specific variation in, 739
    in transcription, 709–710
Embden, Gustav, 250
Embden-Meyerhof pathway. *See* Glycolysis
Emerson, Robert, 341, 342
Enantiomers, 56, 56F
Endocrine glands, 570F. *See also* Hormones; *specific hormones*
    diseases associated with, 589–591
Endocytic vesicle, 4F
Endocytosis, 763
Endopeptidases, 61, 64F, 159. *See also specific enzymes*
Endoplasmic reticulum, 7
    phospholipid synthesis on cytosolic surface of, 445, 446F
    protein synthesis in, in eukaryotes, 758–760
    rough, 4F, 5F
        mammalian, 731, 733F
        protein synthesis in, in eukaryotes, 758–760
    smooth, 4F, 5F
Endoribonucleases, posttranscriptional tRNA processing and modification and, 717
Endorphin, 571
End-product inhibition, 234F, 234–235, 235F
Energy, 30
    amino acid catabolism as source of, 521, 522F, 523
    carbohydrates important in metabolism of, 243–249
    flow in biosphere, 19F
    of formation, 31, 36T, 36–38, 37T
    free. *See* Free energy
    ketone bodies as source of, 418–419, 419F
    metabolism of, regulation of, 267–270, 268T
    needed for reactants to reach transition state, 138, 139F

of photons, 335
required by biochemical reactions, 19, 19F, 20F
required by living cells, 227–228
required for active transport against electrochemical potential gradients, 401F, 401–402, 402T
sources of, differences among organisms and, 228T, 228–229
tissue demands and contributions to pool of, 563T, 563–569
    adipose tissue and, 566, 567F
    of brain tissue, 563–564
    of heart muscle, 566
    liver function and, 566–567, 567F, 568T
    pancreatic hormones and, 567, 569, 569F
    of skeletal muscle, 565F, 565–566
tissue storage of, 563, 563T, 564F
transfer to reaction centers, antenna system and, 340–342, 341F, 342F
Energy charge, 236, 236F, 237F
  citrate synthase regulation by, 300
  isocitrate dehydrogenase regulation by, 300F, 300–301
Enhancer binding proteins, 802F
Enhancers, 715
  as promoter elements, 811–812, 812F, 813F
Enhansons, 812
Enkephalin, 571
Enolase
  reaction catalyzed by, 260T
  reactions involving, 259, 259F
Enolization, 210–212
2,3-*trans*-Enoyl-ACP reductase, 421
Enoyl-CoA hydrase, reaction catalyzed by, 414
Enoyl-CoA isomerase, reaction catalyzed by, 416
Enthalpy, 30, 31, 33, 35
Entropy, 30, 31–35, 32T, 33F, 34F, 35T
  proximity effect in enzyme catalysis and, 155
Environmental factors, cancers associated with, 850–851, 851T, 852F
Enzymatic photoreactivation, DNA repair and, 664F, 664–665
Enzymes, 135–151. *See also specific enzymes*
  acting on DNA, 673
  active site of, 140, 141F, 150, 150F, 154
    of serine proteases, 160–162, 162F, 163, 164–165
  allosteric, 176, 180F, 180–193, 181F, 183–194F
  for amino acid synthesis, *de novo*, 514, 515T
  binding of substrate or inhibitor to, entropy and, 34F, 34–35, 35T
  blood clotting and, 177, 177F
  catalytic activity of, 136, 136T, 154–173
    activation free energy and, 139F, 139–140
    detailed mechanisms of, 159–172
    electrostatic interactions and, 156–157
    functional groups and, 157F, 157–158
    general-base and general-acid catalysis and, 155–156, 156F
    proximity effect and, 155
    specificity of, 158–159
    structural flexibility and, 158F, 158–159
  catalytic surfaces of, as site of biochemical reactions, 18F, 18–19, 19F
  cellulose degradation by, 249
  classes of, 136, 136T
  controlling fluxes in glycolysis-gluconeogenesis pathways, 267, 268T
  digestive, 61, 64F
    catalytic activity of, 159–164F, 159–165
  discovery of, 136
  fatty acid biosynthesis and, 426
  fatty acid oxidation and, 414, 416, 416F, 417F, 418
  free, enzyme-substrate complex in equilibrium with, 140–141
  of *GAL* system, regulation of, 801, 803F, 804
  in glycolytic pathway, 260T
  inhibition of, 146–150
    competitive, 147F, 147–149, 148F
    entropy and, 34, 34F, 35T
    irreversible, 149T, 149–150, 150F
    noncompetitive, 147F, 149, 149F
    uncompetitive, 147F, 149
  kinetics of, 139–146

energy needed for reactants to reach transition state and, 138, 139F
  enzymes with two identical subunits and, 181F, 181–182
  Henri-Michaelis-Menten treatment and, 140–141
  lowering of free energy of activation and, 139F, 139–140
  measuring initial reaction velocity as function of substrate concentration and, 140, 140F, 141F
  pH and, 146, 147F
  reactions involving two substrates and, 144–146, 145F, 146F
  steady-state kinetic analysis and, 141–144, 142F, 143F
  temperature and, 146, 146F
long-term dietary changes and, 432
lysosomal, targeting of, 365, 367
mediating DNA recombination in *Escherichia coli*, 668–671, 670F
Michaelis constants for, 142, 143T, 163T
pathway regulation by, 234–236
  of anabolic and catabolic pathways, by energy status of cell, 235–236, 236F, 237F
  cooperative behavior of regulatory enzymes and, 235, 236F
  position of regulatory enzymes and, 234F, 234–235, 235F
  regulation of enzyme activity by regulatory factors and, 234
posttranscriptional tRNA processing and modification and, 717–719, 718F
prochiral compounds and, 293, 293F
regulation of, 175–195
  by adenylation, 178–179, 179F, 179T
  allosteric, 180F, 180–193, 181F, 183–194F
  by disulfide reduction, 178–179, 179F
  by partial proteolysis, 176F, 176–177, 177F, 177T
  by phosphorylation, 177–178, 178F, 178T, 191–193, 191–194F
  symmetry model of, 182–183, 183F
regulatory. *See* Enzymes, pathway regulation by
restriction, in DNA cloning, 682F, 682–683, 683T
RNAs functioning like, 722–723
RNA-synthesizing, 715–716, 716T
self-splicing, 722, 723F
sequentially related, clustering of, 229–231, 230F
specificity constants of, 144, 145T
subunits of, multiple, allosteric regulation of, 181F, 181–182
terminology for, 136F, 136–137
turnover numbers of, 143–144, 144T, 163F
with two catalytic activities, purification of, 125–127, 126T, 127F
Enzyme-substrate complex
  concentration at steady state condition, 142, 142F
  in equilibrium with free enzyme and substrate, 140–141
Epidermal growth factor, 591, 591T, 858
Epinephrine, 573T
  acetyl-CoA carboxylase regulation by, 431–432, 432F
  glycolysis stimulation by, 270
  interconversion of phosphorylases *a* and *b* and, 191
  structure of, 609F
  synthesis of, 574, 576F
Epstein-Barr virus, Burkitt's lymphoma and, 851
Equilibrium constant, standard free energy related to, 36–38, 37T
erbB gene, 857T, 858–859
Erythrocytes. *See* Red blood cells
Erythromycin, structure of, 756F
Erythropoietin, 591T
D-Erythrose, 244F
D-Erythrulose, 245F
Escherichia coli, 5F
  acetyl-CoA carboxylase of, 420F, 420–421, 421F
  adenylation in, 178
  amino acid synthesis in, 489
  aminoacyl-tRNA synthases of, 743–744
  chromosomes of, 643
    structure of, 641–642, 642F
  cysteine synthesis in, 495, 496F, 497
  deoxyribonucleotide synthesis in, regulation of, 558F, 559
  DNA cloning of, 683–685, 684F, 685F
  DNA of, 640
  DNA recombination in, enzymes mediating, 668–671, 670F

DNA repair in, regulation of synthesis of repair proteins and, 665–666, 666F
DNA replication in, 651, 652–661
  bidirectional growth during, 652–653, 653F
  chromosome replication and, 660, 661F
  discontinuity of growth at replication forks and, 653–654, 654F
  DNA polymerase I and, 656, 657F, 658
  DNA polymerase II and, 656–658, 658F
  initiation and termination of chromosomal replication and, 660–661, 662F
  proteins involved in, 654–656, 655T
fatty acid synthase of, 230, 424, 424F
fatty acid synthesis in, 424, 425, 425F
β-galactosidase synthesis in, augmentation by small-molecule inducer, 771F, 771–772, 772F
genes for ribosomes in, regulation of, 780–783
genetic code of, 736
genetic map of, 642
glutamine synthesis in, 491F, 491–493, 492F
infection by bacteriophage gamma, lysis or lysogeny following, 784–785, 786F
initiation factors of, 747, 747F
introns in, 721
lysogeny of, parallel with transformation tumor-causing viruses, 854
phosphofructokinase of, 184, 184F
phospholipid synthesis in, 438, 439F, 440F
polysomes of, 733F
processing of ribosomal precursor in, 719, 719F
protein degradation in, 763
pyruvate dehydrogenase complex in, 287, 289F
release reaction in, 754, 754F
ribosomes of, composition of, 705, 705F, 706F
RNA in, 701–702, 702F
RNA polymerase of
  comparison with DNA polymerase I and II, 710–711, 712T
  subunits of, 706–707
transcription in
  binding at promoters and, 708–709, 709F
  control of, as regulatory mode, 769
  initiation of, 709, 711T
  termination of, 710, 711F
translational control of ribosomal protein synthesis in, 783, 783F
translational frameshifting in, 755
transport system synthesis in, 399
tRNA in, posttranscriptional processing and modification of, 718F, 718–719
trp operon of, 777–780, 778F
tryptophan synthesis in, 237, 238F, 499, 500F, 501T, 501–502, 509–510
without glutathione, 526
Esophageal cancer, incidence of, 851T
Essential fatty acids, 426
Esters
  cholesteryl, synthesis of, 469, 470F
  coenzyme A, binding of, 212
  hydrolysis of, 155, 156F, 157
  oxygen, enolization of, 212
  thiol, enolization of, 211–212
17β-Estradiol, 576
  synthesis of, 577, 578F
Estradiol, structure and function of, 479T
Estrogens, 573T. See also specific estrogens
  gene regulation by, 586, 587T
Ethanol
  in glycolysis, 252F
  yeast production of, 261
Ethanolamine, 384T
  metabolic fate of, 445
  synthesis of, 443, 445
Ethanolamine kinase. See Choline kinase
Ethidium bromide, 724F

Ethylene, 592, 593F, 594
  synthesis of, 594, 594F
Ethylenediamine, standard enthalpy and entropy changes on forming complexes between cadmium and, 35T
Ethylenediaminetetraacetate, in protein purification, 125
Eubacteria, 27F
Euchromatin, 809
Eukaryotes, 7, 27F. See also specific eukaryotes
  cell life cycle in, 559F
  chromosome replication in, 673, 673F
  chromosomes of, 22, 23F, 24F
  connection of membrane proteins to cytoskeleton in, 396, 397F, 398F
  DNA of, 642F, 642–643, 643T, 644F, 645F
  DNA replication in, 661–663
  evolution of, 25, 26, 26F, 27F
  fatty acid biosynthesis in, 424, 425–426, 426F, 427F
  glucose breakdown to carbon dioxide and water in, 20F
  half-lives of proteins of, 762, 762T
  initiation factors of, 747–748, 748F
  internal membranes of, 382
  multicellular, 26F
  organelles of, 382, 382F
  organization of genes within chromosomes of, 643–644
  phospholipid synthesis on, 438, 441–443
    CDP-diacylglycerol in, 441, 443, 444F
    diacylglycerol in, 441, 442F
    replaceability of fatty acid substituents at SN-1 ans SN-2 positions and, 441, 443F
  phosphorylation in, 178, 178T
  protein synthesis in, 731, 734F
  protein targeting in, 758–760
  single-cell, 26F
  transcription in, 711–715
    promoter elements and, 715, 715F
    RNA polymerases and, 712–713, 713T, 714F, 715
    TATA-binding protein and, 713, 715
Eukaryotic gene expression, 800–826, 802F
  asymmetry of DNA-binding proteins regulating transcription and, 812–817
    helix-loop-helix motif and, 815
    homeodomain and, 813–814, 814F
    leucine zipper and, 815, 816F
    transcription activation domains of transcription factors and, 815–817
    zinc finger and, 814–815, 815F, 816F
  developmental processes and, 819
  in Drosophila, 820–825, 821F, 822F
    analysis of genes controlling early events of embryogenesis and, 823–825
    cascade of regulatory events during early development and, 822
    regulatory genes involved in, 822–823, 823F, 823T
  during embryonic development, 819–820
    elevation of ribosomal RNA in frog eggs by DNA amplification and, 819–820, 820F
    regulatory protein required for 5S rRNA synthesis in frog eggs and, 820
  at levels of translation and polypeptide processing, 817–818, 819F
  mRNA splicing modes and, 817, 818F
  multicellular, 807–812
    active versus inactive chromatin and, 810–812
    direct visualization of active genes and, 809, 809F
    heterochromatic chromosomes and, 809–810, 810F
    timing of nuclear differentiation and, 807F, 807–809, 808F
    variation of chromosome structure with gene activity and, 809
  in yeasts, 801, 803F, 804–807
    determination of mating type by transposable elements and, 804–807, 805–807F
    galactose metabolism and, 801, 803F, 804, 804T
    separation of GAL4 protein into domains with different functions and, 804, 805F
Evans, Philip, 184
Evolution, of biochemical systems, 22–26, 25F, 25T

common, 25–26, 26F, 27F
Exons, 720
Exonucleases, posttranscriptional tRNA processing and modification and, 717
exoV enzymes, DNA unwinding by, 670, 670F
Extensive properties, 30
Extinction coefficient, 70
Extracellular matrix, 6F
Eyes. *See* Vision

Fabry, J., 451
Fabry's disease, 451–452
Facilitated diffusion, 400
Facultative heterochromatin, 809
Familial hypercholesterolemia, 472, 474
Faraday constant, 310, 322
Farnesyl pyrophosphate, 464
Fat(s), emulsification of, 475, 477F
Fat-soluble vitamins, 199T, 220–222, 221F
Fatty acid(s), 411–433, 412F
    degradation of, 411–419, 412F, 428F
        ATP yielded by, 414, 415T, 416
        in blocks of two carbons, 414
        enzymes required for, 416, 416F, 417F, 418
        hepatic ketone bodies and, 418F, 418–419, 419F
        limitation of simultaneous biosynthesis with, 432
        in liver, 427, 428F
        in mitochondria, 414, 415F
        sources of fatty acids and, 412–414, 413F, 414F
    essential, 426
    in membrane phospholipids, 383, 384T
    monounsaturated, synthesis of, 424–426, 425F, 426F
    polyunsaturated
        oxidation of, 416, 417F, 418
        synthesis of, 426, 427F
    regulation of metabolism of, 427–432
        by first step in synthetic pathway, 430–432, 432F
        limitation of simultaneous synthesis and breakdown and, 432
        long-term dietary changes and enzyme levels and, 432
        release from adipose tissue and, 427, 428F, 429, 429F
        substrate supply and, 430, 431F
        transport into mitochondria and, 429–430, 430F
    replaceability of substituents at sN-1 and sN-2 positions, 441, 443F
    during starvation, 567, 567F
    structure of, 8F
    synthesis of, 411–412, 412F, 419–427, 428F
        acetyl-CoA carboxylase in, 420F, 420–421, 421F
        in *Escherichia coli,* 424, 425, 425F
        in eukaryotes, 424, 425–426, 426F, 427F
        fatty acid synthase in, 421–424, 422–424F
        limitation of simultaneous breakdown with, 432
        in liver, 427, 428F
        regulation of, 301
    unsaturated, oxidation of, 416, 416F
Fatty acid synthase
    in *Escherichia coli,* 230, 424, 424F
    reactions catalyzed by, 421–424, 422F, 423F
Female pseudohermaphroditism, 576–577
Ferns, 26F
Ferredoxin, 345
Ferredoxin-NADP oxireductase, 345
Ferritin
    molecular weight of, 91T
    subunit composition of, 91T
F helix, 107
Fibrinogen
    diffusion constant of, 124T
    isoelectric point of, 124T
    molecular weight of, 124T
    sedimentation constant of, 124T

Fibroblast growth factor, 591T
Fibronectin, 396, 398F
Filmer, 109
First-order kinetics, 137F, 137–138
Fischer, Edmund, 191
Fischer, Emil, 56
Fischer projections, 245, 246F
Flavin adenine dinucleotide
    oxidation states of, 207–208, 208F
    reactions involving, 208–209, 209T, 210F
    reduced, reoxidation of, 306. *See also* Electron transport
    structure of, 207F
Flavin mononucleotide
    oxidation states of, 207–208, 208F
    reactions involving, 208–209, 209T, 210F
    structure of, 207F
Flavoproteins, 207, 308
    reactions catalyzed by, 209T
Fleming, Alexander, 374
Fletterick, Robert, 192
Flipases, 445
Fluid-mosaic model, 391, 392F, 393, 393F
Fluorescence, 335
    protein structure analysis using, 100T
Fluorescence polarization, protein structure analysis using, 100T
Fluorine
    in earth's crust, 25T
    in human body, 25T
    in ocean, 25T
5-Fluoro-2'-deoxyuridine-5'-phosphate, 551F
Fluorodinitrobenzene, 61, 63F
5-Fluorouracil, 550T, 551
Folate, coenzymes of, 214F, 215F, 215–216
Folic acid, 199T, 552F
Folkers, Karl, 461
Follicle-stimulating hormone, 572T
Footprinting technique, 729, 729F
Formaldehyde, 215, 216
Formate, 215–216
N^{10}-Formyltetrahydrofolate
    reactions involving, 215, 215F, 216
    structure of, 214F, 215
fos gene, 861
Four-helix cluster, in proteins, 88F
fps gene, 857T
Frameshifting, translational, 755, 756, 756F
Frameshift mutation, 736
Free energy
    activation, 138, 139F
        catalysts and, 139F, 139–140
    applications of free energy function, 35–40
    ATP as carrier of, 40–43, 41F, 42F
    change in reactions, relation to logarithm constant, 36–38, 37T
    coupling of favorable and unfavorable reactions and, 38–40, 39F
    as criterion for spontaneity, 35
    of formation, standard, 36T, 36–38, 37T
    functional coupling and, 233
    in glycolytic pathway, 260T
    as maximum energy available for useful work, 38
    work performed by biological systems and, 38
Free radicals, cationic, generated by photooxidation of chlorophyll, 336–337, 337F
Freeze-fracture electron microscopy, 390, 391F
Frog eggs
    rRNA in, elevation by DNA amplification, 819–820, 820F
    5S rRNA synthesis in, 820
D-Fructose, 245F
Fructose-1,6-biphosphatase, 178–179
Fructose-1,6-biphosphate
    in glycolysis, 252F, 256F, 256–257, 257F
    interconversion into fructose-6-phosphate, 233, 233F, 234

reaction catalyzed by, 270
synthesis of, 264
Fructose biphosphate phosphatase, 268T
    allosteric effector regulation of, 267, 268T
    reaction catalyzed by, 264
Fructose-6-phosphate
    in first metabolic pool of glycolysis, 251, 253F
    in glycolysis, 252F, 254, 255F, 256
    interconversion into fructose-1,6-biphosphate, 233, 233F, 234
    synthesis of, 264
Frye, 393
Fumarase
    Michaelis constants for, 143T
    reaction catalyzed by, 292
    specificity constant of, 145T
Fumarate, synthesis of, 291–292, 524F
Functional groups
    in biomolecules, 17F, 17–18
    of esters, 157f, 157–158
Fungi, tryptophan synthesis in, 501T
Fungisporin, 503T
Furanoses, 245
Futile cycles, 234

D-Galactose, 244F
    structure of, 357F
Galactose, regulation of, in yeasts, 801, 803F, 804, 804T
Galactosemia, 358
β-Galactosidase
    expression of, as function of genotype, 773, 774T
    synthesis of, augmentation by small-molecule inducer, 771F, 771–772, 772F
α-Galactosidase A, 452
β-Galactoside, 771F
    as bacterial transporter
        α-helices of, 403–404, 404F, 405F
        study of, 403
Galactosylceramide, 383
Gallant, J., 781
GAL system
    enzymes of, regulation of, 801, 803F, 804
    separation of GAL4 protein into domains with different functions and, 804, 805F
Gancyclovir, 552, 553T
Ganglion cell, 621
Gap genes, 824–825, 825F
    influence on maternal-effect genes, 825, 825T
Gap junctions, 407, 407F
Garrod, 523
Gastric inhibitory peptide, 573T
Gastrin, 573T
Gating current, 607–608, 608F
Gehring, Walter, 813
Gel electrophoresis, of proteins, 122F, 122–123, 123F
Gel-exclusion chromatography, 120, 121F
Gene expression. *See* Eukaryotic gene expression; Prokaryotic gene expression
Gene mutations. *See* Mutations
General-acid catalysis, 156, 156F
General-base catalysis, 155–156, 156F
Genetic code, 22, 736–742, 738T
    degenerative nature of, 737, 738T
    lack of universality of, 740, 741T
    nondegenerate, 736
    species-specificity of codon-anticodon pairing rules and, 741–742, 742T
    synthetic messengers and, 736–737, 737F
    wobble and, 738–739, 740F, 740T
Genetic concepts, 773
Genetic map, of *Escherichia coli*, 642
Genetic notation, 773

Genome
    of Burkitt's lymphoma cells, 851, 853F
    human, 644
    size of, 630, 630F
Genomic DNA libraries, 686–687, 687T
Geranylgeraniol, 333
Geranylgeranyl side chain, 334F
Geranyl pyrophosphate, synthesis of, 464, 465F
Germ cells, 7
Giantism, 589
Gibberellins, 592, 593F
Gibbs, Josiah, 35
Gilbert, Walter, 775
β-Globin, characterization using cDNA probe, 694
Globin chain, 102, 102F
Globin gene family, characterization by recombinant DNA techniques. *See* Recombinant DNA techniques, characterization of globin gene family by
β-Globin genes, identification and isolation using chromosome walking, 694, 695F
Gluatamate, 204F
Glucagon, 573T
    acetyl-CoA carboxylase regulation by, 431–432, 432F
    function of, 567, 569, 569F
    interconversion of phosphorylases *a* and *b* and, 191
    molecular weight of, 91T
    regulation of energy metabolism by, 268, 269F, 270
    subunit composition of, 91T
Glucocorticoids, 573T
    gene regulation by, 586, 587T
Glucokinase, reactions involving, 254
Gluconeogenesis, 262–266
    ATP consumption in, 263
    fructose-1,6-biphosphate production in, 264
    fructose-6-phosphate production in, 264
    phosphoenolpyruvate production in, 263F, 263–264
    regulation of, 266–271
        cyclic AMP in, 268, 269F, 270
        hormonal, 267–270
        intracellular signals and, 267, 268T
    relationship to glycolysis, 262F
    storage polysaccharide production in, 264–266, 265F
Glucose
    ATP yielded by complete oxidation of, 325–326, 326T, 327F
    breakdown of, 229, 229F, 230F
    breakdown to carbon dioxide and water in eukaryotic cells, 20F
    catabolism of, 305–306
    phosphorylase inhibition by, 192
    pyruvate from, 229, 230F. *See also* Glycolysis
    structure of, 10F
    tissue stores of, 563T
D-Glucose, 244F
    configuration of, 249, 250F
    structure of, 357F
Glucose-alanine cycle, 521, 521F
Glucose monophosphate, synthesis of, 542F, 543
Glucose-6-phosphatase
    deficiency of, 270
    regulation of, 270
Glucose-1-phosphate
    in glycolysis, 252F, 254, 255F
        first metabolic pool of, 251, 253F
    hydrolysis of, standard free energy of, 42F
    synthesis of, 251, 253F
Glucose-6-phosphate
    in glycolysis, 252F, 254, 255F, 256
        first metabolic pool of, 251, 253F
    hydrolysis of, standard free energy of, 42F
    phosphorylase inhibition by, 192
    synthesis of, 39, 39F
Glucose-6-phosphate dehydrogenase

deficiency of, 272
  genetic coding of, 810
  reaction catalyzed by, 272–273
Glucosyl-ceramide, 448
Glucuronic acid, steroid conjugation to, 475, 479F
D-Glucuronic acid, structure of, 357F
Glutamate, 513T
  amidation of, 491F, 491–493, 492F
  deamination of, 515–516, 517F
  synthesis of, 301
D-Glutamate, 503, 503T
Glutamate amino acid family. *See* Amino acid(s), glutamate family of
Glutamate dehydrogenase, reaction catalyzed by, 490F, 491, 515–516, 517F
Glutamate-oxaloacetate transaminase, pyridoxal phosphate action in, 204F
Glutamate synthase, reaction catalyzed by, 491F, 491–493, 492F
Glutamic acid
  equilibrium between charged and uncharged forms of, 54F
  pK value of, 54T
  structure of, 51F
  titration curve of, 53, 54F
Glutamine, 513T
  pK value of, 54T
  structure of, 51F
  synthesis of, 491F, 491–493, 492F
Glutamine phosphoribosylpyrophosphate amidotransferase, reaction catalyzed by, 538, 540
Glutamine synthase
  molecular weight of, 91T
  regulation of, 492–493
  structure of, 491F
  subunit composition of, 91T
Glutaminyl-tRNA synthase, structure of, 744, 745F
γ-Glutamyl cycle, 528, 529F, 530
γ-Glutamylcysteinylglycine, synthesis of, 526, 528F, 529, 530
Glutathione, synthesis of, 526, 528F, 529, 530
Glutathione reductase, reactions involving, 209T, 210F
Glyceraldehyde, synthesis of, 243F, 243–244
D-Glyceraldehyde, 244F
Glyceraldehyde-3-phosphate, 251
  in glycolysis, 252F, 256–258, 257F, 258F
  synthesis of, 273, 273F, 274F
Glycerate-1,3-biphosphate
  in glycolysis, 251, 252F
  hydrolysis of, standard free energy of, 42F
Glycerate-2,3-biphosphonate, oxygen binding to hemoglobin and, 103, 103F, 105, 107F
Glycerate-2-phosphate, in glycolysis, 252F, 259, 259F
Glycerate-3-phosphate, in glycolysis, 252F, 257–258, 258F, 259, 259F
Glycerol, 384T
  conversion to glyceraldehyde and dihydroacetone, 243F, 243–244
  in protein purification, 125
  standard free energy of formation of, 36T
Glycerol-3-phosphate, hydrolysis of, standard free energy of, 42, 42F
Glycerol-3-phosphate acyltransferase, reaction catalyzed by, 438
Glycerol-3-phosphate dehydrogenase, reaction catalyzed by, 309
Glycine, 513T
  at active site, 163, 164
  pK value of, 54T
  porphyrin synthesis from, 526, 527F, 528F
  structure of, 51F
Glycocholate, synthesis of, 475, 478F
Glycoconjugates, oligosaccharides in. *See* Oligosaccharides
Glycogen
  configuration of, 249, 250F, 251F
  functions of, 266
  intestinal digestion of, 251, 253
  structure of, 10F, 248, 248F
  synthesis of. *See* Glycolysis
  tissue stores of, 563T
Glycogenic amino acids, 523

Glycogen phosphorylase, 268T
  catalytic activity of, 191–193, 191–194F
  conversion from phosphorylase *b* to phosphorylase *a*, 268
  molecular weight of, 91T
  reactions involving, 251, 253F
  subunit composition of, 91T
Glycogen synthase, 268T
  phosphorylation of, 268
Glycolipids. *See* Glycosphingolipids
Glycolysis, 249–261, 252F, 264–266, 265F
  aldose cleavage of fructose-1,6-biphosphate in, 257, 257F
  fructose-1,6-biphosphate formation and, 256, 256F
  hexokinase and glucokinase conversion of free sugars to hexose phosphates in, 253–254, 254F
  metabolic pools in, 251
    first, 251, 253F
    flux in, 263
    interconnection of, 262F, 262–263
    second, 256–257
    third, 252F, 259, 259F
  NAD$^+$ regeneration in, 259, 261, 261F
  pathway of, 252F
  phosphoenolpyruvate conversion to pyruvate in, 259, 260F, 260T
  phosphoglucomutase interconversion of glucose-1-phosphate and glucose-6-phosphate in, 254, 255F
  phosphohexoisomerase interconversion of glucose-6-phosphate and fructose-6-phosphate in, 254, 255F, 256
  phosphorylase conversion of storage carbohydrates to hexose phosphates in, 251, 253, 253F
  reactions, enzymes, and standard free energies for steps in, 260F
  regulation of, 266–271
    cyclic AMP in, 268, 269F, 270
    hormonal, 267–270
    intracellular signals and, 267, 268T
  relationship to gluconeogenesis, 262F
  triose phosphate conversion to phosphoglycerates in, 257–258, 258F
  triose phosphate isomerase interconversion of trioses in, 257
Glycophorin
  hydropathy index for, 390F
  topography in mammalian erythrocyte membrane, 394, 394F
Glycoproteins
  functions of, inhibitors and mutants in study of, 368–369, 370T, 371T
  glycosidic bonds in, 361F, 361–362
Glycosaminoglycans, structure of, 359, 360T
Glycosides, synthesis of, 245–248, 247F
Glycosidic bonds, 9, 10F
  in glycoproteins, 361F, 361–362
  in monosaccharides in disaccharides, 245–248, 247F
Glycosphingolipids
  degradation of, defects in, 450–452, 451F
  functions of, 450
  synthesis of, 448, 450, 450F
Glycosylation mutants, 369, 370T
Glycosyltransferases
  in glycosphingolipid synthesis, 448
  oligosaccharide synthesis by, 362–364F
Glyoxylate cycle, 295, 297F
Glyphosphate, 499
Goiter, 590
Goldstein, Joseph, 471, 472
Golgi complex, 4F, 5F, 7, 758
  glycosylation of proteins passing through, 760
  oligosaccharide biosynthesis in, 362, 364F, 365, 366F, 367F, 367–368
Gonadal hormones, 573T
Gonadotropin-releasing factor, 572T
Gorter, E., 388
Gout, 555
G protein(s), 579
  interacting with GTP, 750
  receptors and effectors for, 582, 582T
G protein cycle, as target for bacterial toxins, 583, 583F

Graft rejection, 843
Gramicidin S, 503T
Gram positive pacteria, 26F
Grana, 332, 332F
Granulocyte colony-stimulating factor, 591T
Granulocyte-macrophage colony stimulating factor, 591T
Graves' disease, 590
Green, David, 312
Green algae, 26F
Grendel, F., 388
Griffith, Fred, 628
Grisolia, S., 519
GroEl proteins, 86
Ground state, 335, 336F
Group translocation, 402
Growth factors, 591T, 591–592. *See also specific growth factors*
    phosphorylation and, 178
Growth fork, 652, 653F
Growth hormone, 572T, 590
    cascade of, 588F
Growth hormone-release inhibiting factor, 571, 572T
Growth hormone-releasing factor, 572T
Grunberg-Managgo, Marianne, 701
Guanine, 535
    structure of, 12F
    tautomeric forms of, 537, 537F
Guanine nucleotides, mediation of effect of light by, 621–622, 622F, 623F
Guanosine
    in tRNA, structure of, 704F
    uncharged and pronated forms of, 538F
Guanosine-5'-monophosphate, structure of, 536F
Guanosine tetraphosphate
    concentration of, protein synthesis and, 781F, 781–782
    synthesis of, 782F, 782–783
Guanosine triphosphate
    in elongation, 750, 751F
    synthesis of, 291
Guanylate cyclase pathway, 583–584
Guanylyl methylene diphosphonate, structure of, 751F
Guarente, Leonard, 713
Guide RNA, 721
D-Gulose, 244F
Gurson, J., 807
Gymnosperms, 26F

Hair, α-helices of, 75–77, 76F, 77F
Haldane, J.B.S., 142
Halobacterium halobium
    bacteriorhodopsin from, 322, 324F
    bacteriorhodopsin of, 390, 391F
Halophiles, 27F
Handedness, of amino acids, 56, 56F
Haploids, 773
Harden, Arthur, 136
Hatefi, Yousef, 312
Haworth projections, 245, 246F
hb gene, 824–825, 825F
Heart. *See also* Cardiac muscle
    hormone of, 573T
Heart disease, cholesterol and, 472–473, 474–475
Heat, transfer of, 31
Heat shock protein(s), 763
Heat shock protein 70, 86
Heat shock response, 763
Helicase, 659, 660
α-Helices
    transmembrane, of integral membrane proteins, 389–390, 390F, 391F
    of transport proteins, 403–404, 404F, 405F
Heliozoans, 26F

Helix-loop-helix motif, 815, 860
Helix-turn-helix motif, 789–790, 790F, 791F
Helper T cells, triggering of B-cell division and differentiation by, 839–840, 840F
Hemagglutinin, 359
Hematin, 217
Heme, 101–102, 102F, 307–308, 308F
    oxygen binding to, hemoglobin conformation and, 107–109, 107–109F
    reactions involving, 217, 218
Hemiacetals
    formation of, 245, 246F
    stereoisomeric, 245, 246F
Hemoglobin, 101–110, 102F
    allostery of, 109, 110F
    amino acid sequences of, 108, 108F
    conformations of, 104–105, 105–107F
        changes in, initiation by oxygen binding, 107–109, 107–109F
    defective genes for, detection of, 690, 692F, 692–694, 693F
    diffusion constant of, 124T
    dimers of, 105, 106F
    folding of polypeptide chains of, 102, 102F
    function of, theories of, 109–110, 110F
    isoelectric point of, 124T
    molecular weight of, 91T, 124T
    negative effects on oxygen binding of, 103F, 103–104, 104F
    pathological mutations in, 108–109, 109F
    physiological responses of, 102
    sedimentation constant of, 124T
    solubility of, in different salt solutions, 119, 120F
    subunit composition of, 91T
Henderson-Hasselbach equation, 52–53
Henri, Victor, 140–141
Henri-Michaelis-Menten equation, 140–141, 181
Heparan sulfate, structure of, 360F
Heparin, structure of, 360F
Hepatitis B virus
    liver cancer and, 851
    replication of, 671
Herbicides, safe, 499, 499F
Herpes virus infections, pharmacotherapy for, 552, 553T
Heterochromatic chromosomes, 809–810, 810F
Heterochromatin, 809
    constitutive, 809
    facultative, 809
Heterotrophs, 228, 228T, 229, 330
    aerobic cells of, 231, 231F
    photosynthesis in. *See* Photosynthesis
Hexokinase
    domains in, 89, 90F
    Michaelis constants for, 143T
    molecular weight of, 91T
    reaction catalyzed by, 260T
    reactions involving, 253–254, 254F
    structural changes in, 158, 158F, 159
    subunit composition of, 91T
Hexose(s), synthesis of, 357–358, 358F
Hexose monophosphate pool, 251, 253F
Hexose phosphates, free sugar conversion to, 253–254, 254F
High-density lipoproteins, 465
    composition and density of, 468T
    heart disease and, 474
    plasma level of, coronary artery disease and, 472
    reduction of cholesterol deposits by, 472–473, 473F, 474–475
    structure of, 468F
    synthesis of, 470, 470F
High-performance liquid chromatography, 121–122
High pressure liquid chromatography, 65
Hill, R., 342
Hill, Robin, 336, 342
Hill coefficient, 182
Hill equation, 182

Hirsutism, 475, 479F
Histamine, structure of, 609F
Histidine(s), 502, 504–505F, 513T
    at active site, 160–162, 162F, 163, 164–165
    equilibrium between charged and uncharged forms of, 54F
    oxygen binding to hemoglobin and, 105, 107
    pK value of, 54T
    structure of, 51F
    synthesis of, 180, 502, 504–505F
    titration curve of, 53, 54F
Histidinemia, 525T
Histones, DNA complexed with, 642F, 642–643, 643T, 644F, 645F
HML locus, determination by transposable elements, 507F, 804–807
Hodgkin, A.L., 604
Hodgkin, Dorothy, 216
Holliday, R., 668
Holoenzyme, 137
Holoenzymes, 207, 706
Homeobox sequence, 813, 814F
Homeodomain, 813–814, 814F
Homeotic genes, 822
Homocystinuria, 525T
Homologous chromosomes, 666
Homologous recombination, 667–668
    models of, 668, 668F, 669F
Hormones, 572–573T. *See also* Eicosanoids; *specific hormones*
    circulating, regulation of, 578
    intercellular communication via, 570, 570F, 572–573T
    interconversion of phosphorylases *a* and *b* and, 191
    mediation of actions by receptors for, 578–589, 580F
        adenylate cyclase pathway triggering and, 580, 581F, 582, 582T
        calcium and inositol triphosphatase pathway and, 584F, 584–585, 585F
        G protein cycle and, 583, 583F
        guanylate cyclase pathway and, 583–584
        hierarchical organization of hormones and, 586–587, 588F, 589, 589F
        intercellular signals and, 580, 580T
        multicomponent hormonal systems and, 583
        rate of transcription and, 586, 586F, 587T
    overproduction of, 589–590, 590F
    pancreatic, blood glucose level and, 567, 569, 569F
    phosphorylation and, 178
    of plants, 592, 593–595F, 594
    polypeptide, storage of, 570–571, 571F, 574F
    regulation of energy metabolism by, 267–270
    steroid. *See* Steroid hormones
    synthesis of, 570–571, 574–577
    target-cell insensitivity to, 590–591
    thyroid, synthesis of, 574, 575F
    underproduction of, 590
H-ras gene, 857T
Hubbard, Ruth, 616
Huber, Robert, 337
Human body, composition of, 22–23, 25T
Human cytomegalovirus infections, pharmacotherapy for, 552, 553T
Human immunodeficiency virus infection, pharmacotherapy for, 552, 553T
Human T-cell leukemia virus, 851
Humoral response, 831–841
    augmentation of antibody diversity by genetic mechanisms and, 834–838
    complement system facilitation of removal of microorganisms and antigen-antibody complexes and, 841, 841F
    requirement for interaction of B cells and T cells for antibody formation and, 838F, 838–840
    specificities of immunoglobulins and, 832F, 832–834, 833F, 834T
Hunchback gene, 824–825, 825F
Huxley, A.F., 604
Huxley, Hugh, 110
Hyaluronic acid, structure of, 359, 359F, 360F

Hyaluronic acid-proteoglycan complex, 359, 361F
Hydrides, transfers of, nicotinamide coenzymes and, 203, 205F, 206, 206F
Hydrogen
    in atmosphere, 24
    dinitrogen reduction and, 495
    in earth's crust, 25T
    in human body, 25T
    in ocean, 25T
    standard free energy of formation of, 36T
Hydrogen bonds, 13, 13F
    in DNA, 632F, 633, 633F, 635, 635F
    donor and acceptor groups in proteins, 73–74, 74F
    in phosphofructokinase, at interface between subunits A and D, 185, 186F
    in polypeptide chains, 86–87
    of ribonuclease A, 166, 167F, 168
Hydrolases, 136, 136T
Hydrolysis, in cells versus intestine, 251, 253
Hydropathy, 389–390, 390F
Hydrophilic molecules, 9
Hydrophobic effects, in polypeptide chains, 87
Hydrophobic heme pocket, 102, 102F
Hydrophobic molecules, 9
Hydroxide ion catalysis, 156F
3-Hydroxyacyl-ACP dehydrase, 421
Hydroxyacyl-CoA dehydrogenase, reaction catalyzed by, 414
β-Hydroxybutyrate dehydrogenase, reaction catalyzed by, 418
7-Hydroxycholesterol
    conversion to cholic acid, 475, 476F
    synthesis of, 473F
Hydroxyethyl thiamine pyrophosphate, 287
7α-Hydroxylase, reaction catalyzed by, 473, 473F
21-Hydroxylase deficiency, 475, 479F
Hydroxyl group, 17F
β-Hydroxy-β-methylglutaryl-CoA, 418
    in cholesterol synthesis, 461–462
β-Hydroxy-β-methylglutaryl-CoA lyase, 462
    reaction catalyzed by, 418
β-Hydroxy-β-methylglutaryl-CoA reductase
    degradation of, 463
    reaction catalyzed by, 461
    regulation of, 463
    structure of, 462–463, 463F
β-Hydroxy-β-methylglutaryl-CoA synthase, reaction catalyzed by, 418, 461
Hydroxyproline, 216, 216F
5-Hydroxytryptamine, structure of, 609F
Hydroxyurea, 550T, 551
25-Hydroxyvitamin D$_3$, 221
Hyperammonemia, 525T
Hypercholesterolemia, familial, 472, 474
Hyperglycinemia, 525T
Hyperlysinemia, 525T
Hyperphenylalaninemia, 525T
Hyperprolinemia, type I, 525T
Hypervariable regions, 833, 833F
Hypothalamic hormones, 572T
Hypoxanthine, 535
Hypoxanthine-guanine phosphoribosyltransferase, reaction catalyzed by, 548
H zone, 110, 111F

I band, 110, 111F
I-cell disease, 367
Icosahedral viruses, structure of, 92, 92F
Ideal solutions, mixing of, 33
Idiotype, 833–834
D-Idose, 244F
L-Iduronic acid, structure of, 357F
Imaginal disks, 821, 822F

Imazaquin, structure of, 499F
Imidazolinones, 499
Imino group, 17F
Immunobiology, 830–847, 831F. *See also* Cell-mediated response; Humoral response
Immunogenic immunoglobulins, 833
Immunoglobulin(s), 831
    augmentation of diversity by genetic mechanisms, 834–838
        DNA splicing and, 834–837, 835–837F, 838, 838F
        RNA splicing and, 837
        somatic mutation and, 837T, 837–838
    formation of, 838F, 838–840
        antigen stimulation of formation of B-cell clones and, 839, 839F
        helper T-cell triggering of B-cell division and differentiation and, 839–840, 840F
    heavy and light chains for, 832–833
    IgA, 834T
    IgD, 834T
    IgE, 834T
    IgG, 833, 834T
        structure of, 832, 832F
    IgM, 834T
    immunogenic, 833
    specificities of, 832F, 832–834, 833F, 834T
    structure of, 832F, 832–833
Inactivation, of enzymes, 125
Inactive complex, 148
Indole, tryptophan synthesis and, 509–510
Indole-glycerol-P synthase, 501T
Induced fit, 158
Inflammation, eicosanoids as mediators of, 454
Infrared absorption spectroscopy, protein structure analysis using, 100T
Inhibins, 573T
Inhibition
    end-product, 234F, 234–235, 235F
    of enzymes. *See* Enzymes, inhibition of
Initiation context, 731
Initiation factors, 747F, 747–748, 748F
Inosine, in tRNA, structure of, 704F
Inosine-5′-monophosphate, structure of, 536F
Inosine monophosphate, synthesis of, 538, 540, 541F, 542F, 543
Inositol, 199T
myo-Inositol, 384T
Inositol-1,4,5-P_3, 443, 447
Inositol triphosphate pathway, calcium and, 584F, 584–585, 585F
Insulin, 573T
    function of, 567, 569, 569F
    molecular weight of, 91T
    processing of, 757, 758F
    sequence determination of B chain of, 62F
    storage of, 570–571
    subunit composition of, 91T
Insulin-like growth factor, 590
Insulin-like growth factor-1, 591, 591T
Insulin-like growth factor-2, 591T
Intensive properties, 30
Intercalation, 725
Intercellular space, 5F
Interleukin(s), 841, 842
Interleukin-2, 591T
Interleukin-3, 591T
Intermediary metabolism. *See* Carbohydrates; Electron transport; Gluconeogenesis; Glycolysis; Oxidative phosphorylation; Pentose phosphate pathway; Photosynthesis; Tricarboxylic acid cycle; *specific carbohydrates and types of carbohydrates*
Intermediate-density lipoproteins, 467, 471
    composition and density of, 468T
Intermembrane space, 307, 307F
Intestinal cancer, incidence of, 850, 851T
Intestine
    fat emulsification in, 475, 477F

hydrolysis in, hydrolysis in cells compared with, 251, 253
    lipoprotein synthesis in, 469–470, 470F
Intracellular signals
    for glycolysis-gluconeogenesis pathways, 267, 268T
    hormone regulation and, 580, 580T
Introns, 694, 719, 720–721
In vitro studies, for pathway analysis, 239
In vivo studies, for pathway analysis, 239
Iodine
    in earth's crust, 25T
    in human body, 25T
    in ocean, 25T
Iodoacetamide, enzyme inhibition by, 150, 150F
Iodoacetate, ribonuclease A inhibition by, 165–166, 166F
Ion channels, voltage-gated, mediation of action potentials by, 605–608, 606–608F
Ion-exchange chromatography, 59, 120–121, 121T
Ionization reaction, entropy and, 33, 34F
Ionophores, 319, 320F
Iron
    in earth's crust, 25T
    in human body, 25T
    in ocean, 25T
Iron-porphyrin complex, in deoxyhemoglobin, 107F
Iron-sulfide proteins, NADP$^+$ reduction by, 345
Iron-sulfur clusters, 218, 219F
Iron-sulfur proteins, 308, 309
Isoacceptor tRNAs, 733
Isocitrate, synthesis of, 289F, 290
Isocitrate dehydrogenase
    reaction catalyzed by, 289, 291F
    regulation of, 300F, 300–301
Isoelectric focusing, 122–123
Isoelectric point, protein solubility at, 119, 120F
Isoleucine, 513T
    pK value of, 54T
    structure of, 51F
    synthesis of, 21, 21F, 497, 498F, 499
Isomerases, 136, 136T
N^6-Isopentenyladenosine, in tRNA, structure of, 704F
Isopentenyl pyrophosphate, 464–465
Isopropyl-β-D-thiogalactoside, 771F
Isopropyl-β-D-thiogalactoside isopropylthiogalactopyranoside, 775
Isothermal process, 33
Isotope incorporation studies, 509
Isotypes, in humans, 834T
Isotypic exclusion, 836–837
Isovaleric acidemia, 525T

Jacob, François, 774
Jagendorf, André, 347
Johnson, Louise, 192
Joliot, Pierre, 345
Jones, Mary Ellen, 125
Joyce, Gerald, 723, 725
Jumping, translational, 755
jun gene, 861
Juvenile hormone, structure of, 589F

Kalckar, Herman, 316
Kan, 692
Keilin, David, 307–308
Kendrew, John, 82
Kennedy, Eugene, 307, 316, 438
Keratan sulfate, structure of, 360F
α-Keratins, structure of, determination of, 75T, 75–77, 76F, 77F
β-Keratins, sheetlike structures formed by, 77–78, 78F, 79F
Ketoaciduria, branched-chain, 525T
3-Ketoacyl-ACP reductase, 421

3-Ketoacyl-ACP synthase, 421
Ketogenic amino acids, 521
α-Ketoglutarate, 204F
    amination of, 490F, 491
α-Ketoglutarate dehydrogenase
    reaction catalyzed by, 290–291
    regulation of, 301
Ketone bodies
    as energy source, 418–419, 419F
    during starvation, 567, 567F
    synthesis of, 418F, 418–419, 462
3-Keto-6-phospho-Dgluconate, 272F
Ketoses, 253
D-Ketoses, 245F
Khoury, George, 811
Kidney, hormone of, 573T
Kilocalorie, 31
Kilojoule, 31
Kinetics, 137F, 137–140
    of enzymes. See Enzymes, kinetics of
    first-order, 137F, 137–138
        reaction velocity as function of substrate concentration and, 235, 236F
    fourth-order, reaction velocity as function of substrate concentration and, 235, 236F
    regulation of conversions and, 233–234
    second-order, 138
King, T.J., 807
Klenow fragment, 657, 658F
Klionsky, Bernard, 452
kni gene, 824–825, 825F
Knirps gene, 824–825, 825F
Knowles, Jeremy, 171
Kok, Bessel, 336, 345
Kornberg, Arthur, 656, 658
Koshland, 109
K-ras gene, 857T
Krebs, Edwin, 191
Krebs, Hans, 239, 284–285, 517
Krebs bicycle, 520, 520F
Krebs cycle. See Tricarboxylic acid cycle
Kr gene, 824–825, 825F
Krüppel gene, 824–825, 825F

lac operon, 773
    activator protein augmenting expression of, 775–777, 777F
    CAP-binding region in promoter and, 776–777, 777F
    inducers of, 771F, 771–772
    locus required for repressor action and, 774
    mutations in genes associated with, 772–774, 774T
    regulation of, 770, 770F
β-Lactamase, specificity constant of, 145T
Lactate
    in glycolysis, 252F
    pyruvate reduction to, 259, 261, 261F
    standard free energy of formation of, 36T
Lactate dehydrogenase
    of heart muscle, 566
    molecular weight of, 91T
    reactions involving, 259, 261, 261F
    subunit composition of, 91T
    turnover number of, 144T
D-Lactate dehydrogenase, reactions involving, 209T
Lactate oxidase, reactions involving, 209T
β-Lactoglobulin
    diffusion constant of, 124T
    isoelectric point of, 124T
    molecular weight of, 124T
    sedimentation constant of, 124T
    solubility of, as function of pH and ionic strength, 119, 120F

titration curve for, 54, 56F
Lactose, 248
    structure of, 247F
Lactose carrier protein, purification of, 127–128, 128T, 129F
Lagging-strand synthesis, 653, 654F
λ-Carboxyglutamic acid, 221F
    in protein, 221F
λ-Glutamyl-enzyme complex, 490F, 491
Lamella, middle, 5F
Lanosterol
    conversion to cholesterol, 464, 467F
    synthesis of, 463–464, 464–466F
Laron dwarfs, 590
Leader peptidase, 760
Leading-strand synthesis, 653, 654F
Lecithin, 383, 384T, 385F
Lecithin:cholesterol acyltransferase, reaction catalyzed by, 472, 473F
Lectins, 369
Leder, Philip, 737
Lehninger, Albert, 307, 316
Leishmania tarentolae, RNA editing in, 721, 722F
Lesch-Nyhan syndrome, 548
Leucine, 513T
    pK value of, 54T
    structure of, 51F
Leucine zipper, 815, 816F, 860
Leukemia, pharmacotherapy for, 552, 553T
lexA, DNA repair and, 665–666, 666F
Libraries, DNA. See DNA libraries
Life
    cells as fundamental unit of, 4–6F, 7
    origin of, 25, 25F
Ligaments, cells of, 6F
Ligases, 136, 136T
Light. See also Photosynthesis
    amino acid absorption of, 55, 56F
    DNA repair and, 664F, 664–665
    photons in, 333, 335F, 335–336
        interaction with electrons in molecules, 335–336, 336F, 337F
    polarized, structural analysis methods using, 281, 281F
    properties of, 333, 335, 335F
    protein absorption of, 70
    vision and. See Vision
Light microscopy, image reconstruction in, 97, 97F
Light scattering, protein structure analysis using, 100T
Linearly polarized light, structural analysis methods using, 281, 281F
Lineweaver-Burk plot, 143, 143F
Linking number, 636
Linoleic acid, 384T
Lipases, 412, 413F
Lipids, 7, 9. See also specific lipids
    carbohydrate conversion into, in liver, 430, 431F
    dietary, as source of fatty acids, 412–413, 413F
    membrane. See Eicosanoids; membrane lipids; Phospholipid(s); Phospholipid bilayer; Sphingolipid(s)
    structure of, 8F
Lipid-soluble vitamins, 199T, 220–222, 221F
Lipmann, Fritz, 210
α-Lipoic acid, 199T
    reactions involving, 212–213
    structure of, 212, 212F
Lipoprotein(s), 413, 465, 467–473. See also Cholesterol; Chylomicrons
    cholesterol and triacylglycerol transport by, 470–471
    composition and density of, 465, 467, 468T
    high-density. See High-density lipoproteins
    intermediate-density. See Intermediate-density lipoproteins
    low-density. See Low-density lipoproteins
    normal plasma ranges of, 465, 467T
    removal from plasma, 471, 471F, 472F
    synthesis of, 469–470, 470F
    types of, 465, 467–469T, 468F, 469, 469F

very-low-density. *See* Very-low-density lipoproteins
Lipoprotein lipase, 413, 470–471
   deficiency of, 471
Liposome, 387, 387F
Lipotropin, 572T
β-Lipotropin, 571
Lipscomb, William, 189
Liver
   ammonia transport from muscle to, 520–521, 521F
   carbohydrate conversion into lipid in, 430, 431F
   energy-related reactions in, 568F
   fatty acid synthesis and degradation in, 427, 428F
   fuel reserves in, 563T, 564F
   function of, 566–567, 567F, 568T
   glycolysis stimulation in, 270
   hormonal regulation of energy metabolism in, 267–268
   hormone of, 573T
   ketone body synthesis in, 418F, 418–419
   lipoprotein synthesis in, 469–470, 470F
   regulation of lipid synthesis in, 445–446, 447F
Liver cancer, incidence of, 850–851, 851T
Liverworts, 26F
Lovastatin
   in heart disease, 474
   inhibition of β-hydroxy-β-methylglutaryl-CoA by, 463
Lovastatin acid, structure of, 463F
Low-density lipoproteins, 467
   composition and density of, 468T
   defective receptor for, 472
   heart disease and, 474
   removal from plasma, 471, 471F, 472F
   structure of, 468F
   synthesis of, 470, 470F
Lowry, Oliver, 267
Lumirhodopsin, 616, 618
Lung cancer, incidence of, 850, 851T
Lung surfactant, 441
Luteinizing hormone, 572T
Lyases, 136, 136T
Lynen, Feodor, 419
Lysine, 513T
   equilibrium between charged and uncharged forms of, 54F
   pK value of, 54T
   structure of, 51F
   synthesis of, 20–21, 21F
D-Lysine, 503, 503T
Lysis
   infection of *Escherichia coli* by bacteriophage gamma and, 784–785, 786F
   prevention of buildup of *cI* protein by *Cro* protein during, 785, 788F
Lysogeny
   buildup of *cII* regulatory protein and, 786, 788F, 789
   of *Escherichia coli,* parallel with transformation tumor-causing viruses, 854
   infection of *Escherichia coli* by bacteriophage gamma and, 784–785, 786F
Lysosomal storage diseases, 367, 763
Lysosomes, 4F, 7
   glycosphingolipid degradation in, 450–452, 451F
   mammalian, proteolytic hydrolysis in, 763
Lysozyme
   complex formed with substrate, structure of, 18F
   molecular weight of, 91T
   subunit composition of, 91T
   turnover number of, 144T
D-Lyxose, 244F

McKnight, Steve, 815
Macromolecules, 7, 8F, 9, 10–12F, 13–16
   three-dimensional folding of, 9, 13–16
   water and, 9, 13–14, 13–16F, 16
Macrophage colony-stimulating factor, 591, 591T
Magnesium
   in earth's crust, 25T
   in human body, 25T
   in ocean, 25T
Major histocompatibility complex, 843–844
   class I and class II proteins and, 843–844, 845F
   graft rejection and, 843
Malate, synthesis of, 292
Malate dehydrogenase, 40
   reaction catalyzed by, 292
Malformin A$_1$, 503T
Malformin C, 503T
Malonyl-CoA, 419, 430
Malonyl-CoA transacylase, 421, 422F
Maltose, structure of, 247F
Manganese
   in earth's crust, 25T
   in human body, 25T
   in ocean, 25T
D-Mannose, 244F
   structure of, 357F
Mannose-6-phosphate, protein targeting and, 760
Man-6-P receptor, 367
Maple syrup urine disease, 523, 525T
Marmur, Julius, 639
Masculinization, 475, 479F, 576–577
Maternal-effect genes, 822, 823–824, 824F
   influence on pair-rule genes, 825, 825T
Mating types, of yeasts, determination by transposable elements, 804–807, 805–807F
MAT locus, determination by transposable elements, 507F, 804–807
Matrix, in mitochondria, 307, 307F
Mattaei, Heinrich, 736
Mechanical work, 38
Meelson, Matthew, 651
Megaloblastic anemia, 545
Meiosis, 666, 667F
   in yeasts, regulation of, 806, 807F
Melanocyte-stimulating hormones, 571, 572T
Melanotonin, 572T
Melting curve, of DNA, 638F, 638–639, 639F
Membrane(s). *See also specific membranes*
   of organelles, 382
   pore size in, 406F, 406–407, 407F
   structure of, 383–396
      asymmetrical, 393–394, 394F
      complex lipids in, 383, 384T, 385, 385F, 386F, 386T
      integral and peripheral proteins in, 388–390, 388–391F
      phospholipid bilayer in, 386–388, 387F, 388F
      protein and lipid movement in membranes and, 390–391, 392F, 393, 393F
      proteins connected to cytoskeleton and, 396, 397F, 398F
      sensitivity of fluidity to temperature and lipid composition and, 395F, 395–396, 396T
   transport across, 398–407
      active. *See* Active transport
      electrochemical potential gradient and, 400
      interactions mediated by membrane proteins and, 407
      molecular models of, 403–404, 404F, 405F
      Na$^+$-K$^+$ pump and, 404–406, 406F
      size of pores and, 406F, 406–407, 407F
      study of, 402–403, 403F
      transport proteins and, 398–399, 399F
Membrane lipids, 383, 384T, 385, 385F, 386F, 386T, 436–457. *See also* Eicosanoids; Phospholipid(s); Phospholipid bilayer; Sphingolipid(s)
   asymmetry of, 394
   membrane fluidity and, 396
   movement of, 390–391, 392F, 393, 393F
   synthesis of, 437F

Memory cells, 839

Menten, Maude, 140–141

Mercaptoethanol, in protein purification, 125

6-Mercaptopurine, 549, 550T, 551

Merodiploids, 772–773, 774T

Merrifield, Robert, 66

Merrifield process, 66, 67F

Meselson, Matthew, 651, 667–668

Mesophyll cells, 351, 352F

Messenger RNA. *See* mRNA

Metabolic pools, in glycolysis. *See* Glycolysis, metabolic pools in

Metabolic reactions. *See* Biochemical reactions

Metal cofactors, 220, 220F

Metallothionein-growth hormone, 590, 590F

Metals, bound, in enzymatic catalysis, 158

Metarhodopsin I, 618

Metarhodopsin II, 618, 622

    synthesis of, 619

Methanogens, 27F

N^5,N^{10}-Methenyltetrahydrofolate

    reactions involving, 215, 215F

    structure of, 214F, 215

Methionine, 513T, 731

    pK value of, 54T

    in protein synthesis, 745–746

    structure of, 51F

    synthesis of, 21, 21F, 515T

Methotrexate, 550T, 551–552

1-Methyladenosine, in tRNA, structure of, 704F

Methylamine, standard enthalpy and entropy changes on forming complexes between cadmium and, 35T

5-Methylcytidine, in tRNA, structure of, 704F

N^5,N^{10}-Methylenetetrahydrofolate

    reactions involving, 215, 215F

    structure of, 214F, 215

Methylgalactosides, 245, 247F

Methylglucosides, 245, 247F

Methylglycosides, 245, 247F

1-Methylinosine, in tRNA, structure of, 704F

Methylmalonic acidemia, 525T

N^5-Methyltetrahydrofolate

    reactions involving, 215, 215F

    structure of, 214F, 215

Mevalonate, in cholesterol synthesis, 461–463

    as key intermediate, 461–462, 462F

    rate of synthesis and, 462–463, 463F

Meyerhof, Otto, 250

Micelles, 387, 387F

Michaelis, Leonor, 140–141

Michaelis constants, 142, 143T, 163T

Michaelis-Menten equation, 142, 180–181

Michel, Hartmut, 337

Microbody, 4F, 5F

Microfibrils, in cellulose, 249, 249F

Microfilaments, 4F, 5F, 396

Microtubules, 4F, 5F, 396

Microvillus, 4F

Middle lamella, 5F

Mineralocorticoids, 573T

    cell specificity of action of, 577

Mitchell, Peter, 318, 319

Mitochondria, 4F, 5F, 7, 307, 307F

    electron-transport chain in. *See* Electron transport

    fatty acid oxidation in, 414, 415F

    fatty acid transport into, regulation of, 429–430, 430F

    proteins of, posttranslational transport of, 757–758, 759F

    pumping of protons out of, during respiration, 318, 321, 321F

    transport of substrates into and out of, 324–325

        ATP yielded by complete oxidation of glucose and, 325–326, 326T, 327F

        coupling of ATP export to ADP uptake and, 324–325, 325F

        coupling of uptake of P_i and oxidizable substrates to release of other compounds and, 324, 325F

        shuttle systems in import of electrons from cytosolic NADH and, 325

    of yeasts, genetic code of, 740, 741T

Mitogen(s). *See* Growth factors

Mitogenesis, in normal and transformed cells, 858F

Mitosis, 22, 23F

Mixed-function oxidases, 473

M line, 110, 111F

Molecular models, of protein structure, 75T, 75–77, 76F, 77F

Molecules

    energy of formation of, 31

    functional groups in, 17F, 17–18

Mollusks, 26F

Molybdenum

    in earth's crust, 25T

    in human body, 25T

    in ocean, 25T

Monoamine oxidase, reactions involving, 209T

Monocistrionic mRNAs, 731

Monod, Jacques, 109, 180, 184, 774

Monomers, 9, 10F

Mononucleotides, in polynucleotide chain, 630–631, 631F

Monosaccharides, 243–245, 244F, 245F

    glycosidic bonds in, 245–248, 247F

    hemiacetal formation from, 245, 246F

    integration into polysaccharides, 357F, 357–359

    interconversions of, 357–358, 358F

    isomers of, 244–245

    structure of, 10F

    synthesis of, 243F, 243–244

Moore, Stanford, 165

Mosses, 26F

Motilin, 573T

Movement, between domains, 89–90, 90F

Moyle, Jennifer, 318, 319

mRNA, 22

    binding to ribosome, 746, 746T

    as carrier of information for polypeptide synthesis, 702–703

    coding ratio and, 736

    hormone synthesis from, 571

    monocistrionic, 731

    polycistrionic, 731

    splicing of, as mechanism for posttranscriptional regulation, 817, 818F

    as template for protein synthesis, 731, 734F

    tRNA ordering of activated amino acids in, 731, 733, 734F, 735

Mucins, 368

Mulder, Gerardus, 49

Muller-Hill, Benno, 775

Multicolony-stimulating factor, 591T

Muscle

    ammonia transport to liver from, 520–521, 521F

    cardiac. *See* Cardiac muscle

    cells of, 6F

    fuel reserves in, 563T, 564F

    glycolysis stimulation in, 270

    interconversion of phosphorylases *a* and *b* in, 191

    pentose phosphate pathway in, 272

    rigor mortis and, 113, 115F

    skeletal. *See* Skeletal muscle

Muscle contraction, 110, 111F, 112F, 112T, 113, 114F, 115, 115F

    sliding-filament model of, 110, 112F

Muscle fibers, 110

Mutagenesis, site-directed, 689, 690F

Mutants, conditional, 655

Mutarotation, 245

Mutations

    in cellular protooncogenes, cancers arising by, 852–853

    constitutive, 772

    in *Drosophila*, 822, 823F

errors in amino acid catabolism and, 523, 525T
frameshift, 736
point, in biosynthetic pathways, immediate consequences of, 489, 489F
recessive, cancers arising from, 852, 854T
somatic, 834, 837T, 837–838
myc gene, 857T, 860
Mycobacillin, 503T
Mycoplasma, 5F
Myelin, 602
Myofibrils, 110, 111F
Myoglobin, 101
  diffusion constant of, 124T
  folding of polypeptide chains of, 102, 102F
  isoelectric point of, 124T
  molecular weight of, 91T, 124T
  physiological responses of, 102
  sedimentation constant of, 124T
  subunit composition of, 91T
Myohemerythrin, 88F
Myosin
  diffusion constant of, 124T
  isoelectric point of, 124T
  molecular weight of, 124T
  sedimentation constant of, 124T
  in skeletal muscle, 112F, 112T, 113
  structure of, 112F, 113
Myristic acid, 384T
Myxobacteria, 26F

NAD$^+$
  isocitrate dehydrogenase regulation by, 300F, 300–301
  reactions of, 203, 206, 206F
  regeneration of, 259, 261, 261F
  structure of, 203, 205F
NADH
  citrate synthase regulation by, 300
  isocitrate dehydrogenase regulation by, 300F, 300–301
  α-ketoglutarate dehydrogenase regulation by, 301
  reoxidation of, 306. *See also* Electron transport
  shuttle systems in import of electrons from, 325
  structure of, 205F
NADH dehydrogenase, 308–309
NADH dehydrogenase complex, 312F, 312–313, 313F
NADP$^+$
  reactions of, 203, 206, 206F
  reduction by iron-sulfide proteins, 345
  structure of, 203, 205F
NADPH
  formation in pentose phosphate pathway, 272F, 272–273
  functional coupling and, 231
  structure of, 205F
NADP-malate dehydrogenase, 178–179
Nalidixic acid, 724F
Negative cooperativity, 103
Nematodes, 26F
Nemethy, 109
Nerve cells, 6F
  sodium and potassium channels of, 400
Nerve growth factor, 591T
Neurospora
  pyrimidines in, 543
  tryptophan synthesis in, 501–502, 509–510
Neurotransmission, 602–613, 603F
  action potentials and, 604F, 604–608
    mediation by voltage-gated ion channels, 605–608, 606–608F
  electric potential difference across plasma membrane created by pumping and diffusion of ions and, 603
  mediation of synaptic transmission by ligand-gated ion channels and, 609–610, 609–612F
Newton, 31

Niacin, 199T
Nickel
  in earth's crust, 25T
  in human body, 25T
  in ocean, 25T
Nicolson, Garth, 390–391
Nicotinamide, 205F
  coenzyme forms of, 203, 204F, 206, 206F
  structure of, 205F
Nicotinamide adenine dinucleotide
  oxidized. *See* NAD$^+$
  reduced. *See* NADH
Nicotinamide-adenine dinucleotide phosphate. *See* NADP
Nicotinamide adenine dinucleotide phosphate, oxidized. *See* NADP$^+$
Nicotinic acid, 199T
Ninhydrin reaction, 59–60, 60F
Nirenberg, Marshall, 736, 737
Nitrate reduction, function of, 495
Nitrogen
  in atmosphere, 24
  in earth's crust, 25T
  in human body, 25T
  interdependence of organisms and, 21, 22
  in ocean, 25T
  valence states of, 16–17
Nitrogenase, reaction catalyzed by, 494F, 495
Nitrogen cycle, 493, 493F, 495
p-Nitrophenol, formation of, 161, 162F, 163F
o-Nitrophenyl-β-galactoside, 403
Nomura, M., 783
Noncompetitive inhibition, 147F, 149, 149F
Nonspecific recombination, 671
Noradrenalin. *See* Norepinephrine
Norepinephrine, 573T
  structure of, 609F
  synthesis of, 574, 576F
Northrop, John, 136
nos gene, 823, 823F, 824, 825
Novobiocin, 724F
N protein, as antiterminator, 785, 787F
N-ras gene, 857T
N-terminal rule, 762, 762T
Nuclear envelope, 4F, 5F
Nuclear magnetic resonance spectrometry, protein structure analysis using, 100T
Nuclear pore, 5F
Nuclear transplantation, 807, 807F
Nucleation, 640
Nucleic acids, 7, 9, 535
  genetic significance of, 628
  sequences of, evolution and, 25–26, 27F
  structure of, 9, 12F
Nucleolus, 4F, 5F
Nucleophiles, 155, 157–158
Nucleoprotein complexes, 22
Nucleoside(s), nucleotide formation from, 548–549
Nucleoside diphosphate kinase, reaction catalyzed by, 264
Nucleoside monophosphates, conversion to triphosphates, 549
Nucleoside triphosphates, synthesis of, 549
Nucleosomes, 642–643
  structure of, 643, 644F, 645F
Nucleotide(s), 533–560. *See also specific nucleotides*
  absorption maxima of, 538, 539F
  catabolism of, 553–556, 554F
  coding ratio and, 736
  components of, 535, 535T, 536–539F, 537T, 537–538
  dinucleotide formation from, 12F
  guanine, mediation of effect of light by, 621–622, 622F, 623F
  intracellular concentrations of, 559–560
  pronation and depronation of, 537F, 537T, 537–538, 538F
  purine, *de novo* synthesis of, 538–543, 539F, 540F

pyrimidine, *de novo* synthesis of, 543F, 543–545
   regulation of, 556–560
   structure of, 9, 12F
   synthesis of, 533, 534F, 535
      from bases and nucleosides, 548–549
      inhibitors of, 549, 550T, 551F, 551–552, 552F, 553T
   in tRNA, structure of, 704, 704F
   utilization of, 534F, 535
Nucleotide bases (nucleobases), 535, 535T, 536F
   tautomeric forms of, 537, 537F
Nucleotide complexity, of DNA, 640
Nucleus, 4F, 7

Oceans, composition of, 23, 25T
Ochoa, Severo, 701
Ocytocin, 571, 572T
Ogston, 293
Oleic acid, 384T
   structure of, 8F
Oleoyl-CoA, reaction catalyzed by, 416, 416F
Oligomycin, as uncoupler, 318, 318F
Oligopeptides, 57
Oligosaccharides, 359, 361F, 361–368, 592, 593F, 594
   blood types and, 368, 368T, 369F
   protein targeting of, 365, 367
   synthesis of, 362–364F, 362–365
      in *cis*-Golgi, 367–368
      lipid carrier in, 362–363, 364–367F, 365
Olins, A.L., 642
Olins, D.E., 642
Oncogenes, 592, 851, 853
   association with tumor-causing viruses, 853–854
   retroviral-associated, involved in growth regulation, 856–861, 857F, 857T
   transition from protooncogenes to, 861–862
   in tumor-causing viruses, 854–855, 855T
Open promoter complexes, 708, 709F
Operon hypothesis, 774–775, 775F, 776F
Opsin, 614, 615–616
Optical absorption spectroscopy, protein structure analysis using, 100T
Optical rotatory dispersion, protein structure analysis using, 100T
Ordered pathway, 144, 145–146, 146F
Organelles, 7, 7T, 8F, 9, 10–12F, 382, 382F
Organisms. *See also specific organisms*
   biochemical interdependence of, 21–22
   differences in sources of energy, reducing power, and starting materials for biosynthesis among, 228T, 228–229
   metabolic classification of, 228T
   starting materials and energy required by, 227–228
OriC segment, 661
D-Ornithine, 503, 503T
Ornithinemia, 525T
Orotate 5′-monophosphate decarboxylase, 125, 127
Orotate phosphoribosyltransferase, 125, 127, 545
Orotic acid, 543
   structure of, 543F
Orotidine phosphate decarboxylase, 545
Ouabain, 404
Oust, 499
Ovarian cancer, incidence of, 851T
Overwound, 637
Oxaloacetate, 204F
   formation of, 39–40, 264, 292
   standard free energy of formation of, 36T
Oxaloacetate amino acid family
   isoleucine and valine synthesis and, 497, 498F, 499
   safe herbicides and, 499, 499F
Oxidases, mixed-function, 473
Oxidative deamination, of amino acids, 515–516, 517F

Oxidative phosphorylation, 306, 306F, 316–324
   of ADP to ATP, 283, 284F
   ATP formation in, 321–322, 322F
   ATP-synthase in, 322–324, 323F, 324F
   chemiosmotic theory of, 318, 319F
   coupling with respiration, 317, 317F
   electrochemical potential gradient in, 318–319, 320F
   electron transfer coupling to ATP in, 316F, 316–317, 317F
   electron transport release from, 317F, 317–318, 318F
   proton pumping across membranes in, 321, 321F
Oxidoreductases, 136, 136T
β-Oxoacid-CoA transferase, reaction catalyzed by, 419, 419F
Oxygen
   in atmosphere, 24
   binding to hemoglobin
      hemoglobin conformation and, 107–109, 107–109F
      negative influences on, 103F, 103–104, 104F
   in earth's crust, 23, 25T
   evolution of, photosystem II and, 345–346, 346F
   hemoglobin affinity for, 102, 102F
   in human body, 25T
   interdependence of organisms and, 21
   myoglobin affinity for, 102, 102F
   in ocean, 25T
Oxygen esters, enolization of, 212
Oxygen transport, 103–104, 104F
Oxyhemoglobin, 104–105, 105F, 106F
Oxytocin, 571, 572T

P680, 336, 337, 340, 346
   excitation of, 342
P700, 336, 337F, 340
   excitation of, 342
P870, 336, 340
   reduction of, 339–340
Pair-rule genes, influence on maternal-effect genes, 825, 825T
Palmitic acid, 384T
   synthesis of, 419
Palmitoleic acid, 384T
Palmitoyl-CoA, oxidation of, 414, 415T, 416
1-Palmitoyl-2-oleoylphosphatidylcholine, 383
Pancreatic hormones, 573T
Pancreatic polypeptide, 573T
Pantothenic acid, 199T
   coenzymes of, 210–212, 211F
Papain
   domains in, 89, 89F
   site of action in polypeptide chain cleavage, 64F
Parallel β-pleated sheet, 77, 78, 78F
Parathyroid hormone, 572T
Partial diploids, 772–773, 774T
Partial proteolysis, 176F, 176–177, 177F, 177T
Pastan, I., 776
Pasteur, Louis, 136
Pathways, 20F, 20–21, 21F. *See also specific pathways*
   analysis of, 236–239
      of multistep pathways, 237–238, 238F, 239F
      radiolabeled compounds for, 238–239
      of single-step pathways, 237
      in vitro and *in vivo* studies for, 239
   biosynthetic, inhibition by end products of pathway, 180
   Embden-Meyerhof. *See* Glycolysis
   enzymatic regulation of. *See* Enzymes, pathway regulation by
   functional coupling of, 231F, 231–232, 232F
   glycolytic. *See* Glycolysis
   ordered, 144, 145–146, 146F
   organization of reactions into, 229, 229F, 230F
   random-order, 144
   study of, 489, 489F
Pauling, Linus, 73–80

pBR322, in DNA cloning, 683–685, 684F, 685F
Pelleted fraction, 119
Penefsky, Harvey, 323
Penicillin, inhibition of transpeptidation reaction by, 374, 375F
Penicillinase, turnover number of, 144T
Pentose(s), 535
Pentose phosphate pathway, 272–276
  interconversion of phosphorylated sugars in, 273, 273F, 274F
  NADPH generation by, 272F, 272–273
  ribose-5-phosphate and xylulose-5-phosphate production in, 274, 275F, 276
Pepsin, site of action in polypeptide chain cleavage, 64F
Peptide(s), 56–57. *See also* Polypeptide chain(s)
  synthesis of, 66, 67F
Peptide bonds, 9, 11F, 18
  cleavage of, 58–59, 59F
  covalent, 56–57, 57F, 58F
  formation of, 750F
    rRNA and, 723
  planar structure of, 73, 73F
  resonance and, 73, 73F
Peptidoglycans
  cross-linking of, 372, 374F
  synthesis of, 371–372, 373F
Peptidyl transferase, 749
Perlman, R., 776
Permeases. *See* Transporters
Peroxidase
  molecular weight of, 91T
  subunit composition of, 91T
Pertussis, 583
Perutz, Max, 82
PEST sequences, 762
Pettijohn, D.E., 641
p53 gene, 855–856
pH, 52, 52T, 53F
  enzymatic activity and, 146, 147F
  gradient created by respiration, 319
  oxygen binding to hemoglobin and, 105
  ribonuclease A activity and, 166, 166F
Phage repressor, 790, 792F
Phagocytosis, 763
Phenotypes, 666
Phenylalanine, 513T
  abnormal metabolism of, 523
  conversion to fumarate and acetoacetate, 524F
  light absorption by, 55, 56F
  pK value of, 54T
  structure of, 51F
  synthesis of, 515T
Phenylisothiocyanate, 65F
Phenylketonuria, 525T
Phenylthiocarbamoyl tetrapeptide, 65F
Pheophytin(s), 333
Pheophytin *a*, structure of, 334F
Philadelphia chromosome, 851, 853F
Phosphatase, dephosphorylation by, 177, 178F
Phosphate group, 17F
Phosphatidic acid, structure of, 383, 385F
Phosphatidylcholine, 383, 384T, 385, 385F
  acylation of, 441
  synthesis of, regulation of, 445–446, 447F
  transition temperatures for aqueous suspensions of, 396T
Phosphatidylethanolamine, 384T, 385
  metabolism of, 443, 445, 445F
  synthesis of, 438, 440F
    regulation of, 445–446
  transition temperatures for aqueous suspensions of, 396T
Phosphatidylglycerol, 384T
  synthesis of, 438, 440F
  transition temperatures for aqueous suspensions of, 396T

Phosphatidylinositol, 384T
  synthesis of, 441, 443, 444F
Phosphatidylinositol-4,5-biphosphate, 584, 584F
  synthesis of, 441, 443, 444F
Phosphatidylinositol cycle, 584, 585F, 586
Phosphatidylinositol-4-phosphate, 441, 443
Phosphatidylserine, 384T
  metabolism of, 443, 445, 445F
3′-Phosphoadenosine-5′-phosphosulfate, synthesis of, 497, 497F
Phosphoarginine, hydrolysis of, standard free energy of, 42F
Phosphocellulose (PC), 121T
6-Phospho-D-gluconate, 272F
6-Phospho-D-gluconolactone, 272F
Phosphodiesterase. *See* Phospholipase C
Phosphodiester bonds, 12F
  formation of, 709, 710F
Phosphoenolpyruvate
  in glycolysis, 252F, 259, 259F, 260F, 260T
  hydrolysis of, standard free energy of, 42F
  synthesis of, 263F, 263–264
Phosphoenolpyruvate carboxykinase, reaction catalyzed by, 264
Phosphoenolpyruvate carboxylase, reaction catalyzed by, 295, 297
Phosphofructokinase
  allosteric effector regulation of, 267, 268T
  catalytic activity of, 180F, 181, 183–186, 184–186F
  hydrogen bonds at interface between subunits A and D of, 185, 186F
  molecular weight of, 91T
  reaction catalyzed by, 260T
  reactions involving, 256, 256F
  structure of
    crystal, 184, 184F
    T and R conformations and, 184–186, 185F, 186F
  subunit composition of, 91T
Phosphoglucomutase, reactions involving, 254, 255F
Phosphoglyceraldehyde dehydrogenase, reaction catalyzed by, 260T
3-Phosphoglyceraldehyde dehydrogenase, reactions involving, 257–258, 258F
3-Phosphoglycerate kinase
  reaction catalyzed by, 260T
  reactions involving, 258
Phosphoglycerate kinase, domains in, 89, 90F
Phosphoglycerides, 383, 384T
  in membranes, 383, 384T, 385F
Phosphoglyceromutase, reaction catalyzed by, 260T
Phosphohexoisomerase, reactions involving, 254, 255F, 256
Phosphohexose isomerase, reaction catalyzed by, 260T
Phospholipase(s), 412, 413F
  phospholipid degradation by, 447, 448F
Phospholipase A$_1$, phospholipid degradation by, 447, 448F
Phospholipase A$_2$, 441
  arachidonic acid release by, 453
  phospholipid degradation by, 447, 448F
Phospholipase C, 443, 584
  phospholipid degradation by, 447, 448F
Phospholipase D, phospholipid degradation by, 447, 448F
Phospholipid(s), 438–447. *See also* Phospholipid bilayer
  aggregates of, 14, 14F
  amphipathic characteristic of, 386
  degradation of, 447, 448F
  differential scanning calorimetry of, 395, 395F
  in membranes, 383, 384T, 385F
    fatty acids in, 383, 384T
  plasma concentration of, 467T
  structure of, 8F
  synthesis of, 438–446
    CDP-diacylglycerol in, 441, 443, 444F
    on cytosolic surface of endoplasmic reticulum, 445, 446F
    diacylglycerol in, 441, 442F
    in *Escherichia coli,* 438, 439F, 440F
    in eukaryotes, 438, 441–443

linkage of phosphatidylserine and phosphatidylethanolamine metabolism and, 443, 445, 445F
regulation of, 445–446, 447F
replaceability of fatty acid substituents at sN-1 and sN-2 positions and, 441, 443F
transition temperatures for aqueous suspensions of, 395–396, 396T
in water, structures formed by, 386–388, 387F, 388F
Phospholipid bilayer, 387, 387F, 388, 388F, 438
detergent disruption of, 389, 389F
phase transition of, 395, 395F
N-Phosphonacetyl-L-aspartate, 188–189, 189F, 191F, 550T, 551
Phosphopantetheine, 421, 423F
4′-Phosphopantetheine coenzymes, 210–212, 211F
Phosphoribosyl anthranilate isomerase, 501T
Phosphoribosyl anthranilate transferase, 501, 501T
Phosphoribosylpyrophosphate, synthesis of, 538, 539F
Phosphoribulokinase, 178
Phosphorus
in earth's crust, 25T
in human body, 25T
in ocean, 25T
Phosphorylase a, 191, 191F
interconversion with phosphorylase b, 191, 193, 194F
phosphorylase b conversion to, 268
structure of, 192, 193F
Phosphorylase b, 191, 191F
conversion to phosphorylase a, 268
interconversion with phosphorylase a, 191, 193, 194F
structure of, T and R conformations and, 192–193, 193F, 194F
Phosphorylation
enzyme regulation by, 177–178, 178F, 178T
oxidative. See Oxidative phosphorylation
Phosphoryl groups, 535
Photoautotrophs, 228, 228T
ATP production in, 231
Photoheterotrophs, 228T, 228–229
ATP production in, 231
Photons, 333, 335, 335F, 335–336
conductivity changes resulting from absorption of, 619–621, 621F
energy of, 335
interaction with electrons in molecules, 335–336, 336F, 337F
Photophosphorylation, 347–348
Photoreactivation, enzymatic, DNA repair and, 664F, 664–665
Photorespiration, 350–351, 351F, 352F
Photosynthesis, 330–353, 331F
antenna system in, 340–342, 341F, 342F
carbon fixation utilizing reductive pentose cycle in, 348, 349F, 350
chlorophyll and, 332–333, 334F, 335F
binding to proteins in reaction centers, 337–338, 338F, 339F
cationic free radical generated by photooxidation of, 336–337, 337F
cyclic electron-transport chain in, 339–340, 340F
enzyme activation in, 178, 179T
interdependence of organisms and, 21–22
light and. See Light
photosystems
linking in chloroplasts, 342–344, 343–345F
NADP+ reduction by iron-sulfur proteins by, 345
oxidizing equivalent accumulation in reaction centers of, 345–346, 346F
proton transport in, 346–348, 347F, 348F
reaction centers in
chlorophyll binding to proteins in, 337–338, 338F, 339F
crystal structure of, 337–338, 339F
electron movement in, 338, 340F
energy transfer to, 340–342, 341F, 342F
oxidizing equivalent accumulation in, 345–346, 346F
ribulose-biphosphate carboxylase-oxygenase in, 350–351, 350–352F
site of, 332, 332F, 333F, 333T
Photosystem I, 336
electron carriers in, 344, 345F
NADP+ reduction by, 345

reaction center of, 338
Photosystem II, 336
electron carriers in, 344, 345F
O_2 evolution in, 345–346, 346F
reaction centers of, 338
Phsophofructokinase, 268T
Phytol, 333
Phytyl side chain, 334F
Pigments
accessory, in antenna systems, 342
visual, 614–615, 615F, 616F
Pineal hormone, 572T
Ping-pong mechanism, 144–145, 145F
Pinocytosis, 763
Pituitary hormones, 572T
control of synthesis and secretion of, 588F
Placental hormones, 573T
Placental lactogen, 573T
Planck's constant, 335
Plants. See also Photosynthesis
cell membranes of, 385
glycogen synthesis in, 266
hormones of, 592, 593–595F, 594
photosynthetic properties of, 333T
tumors of, 594, 595F
Plasma membrane, 4, 4F, 5F, 7, 381, 382F
action potential and, 604–605
electric potential difference across, created by pumping and diffusion of ions, 603
Plasmids, in DNA cloning, 683–685, 684F, 685F
Plastocyanin, 342
Platelet(s), 6F
aggregation of, 475
Platelet-derived growth factor, 591, 591T, 856, 858, 858F
pnp gene, 716t
Point mutations, in biosynthetic pathways, immediate consequences of, 489, 489F
Polarized light, structural analysis methods using, 281, 281F
Polyalcohols, 253
Polycistrionic mRNAs, 731
Polydeoxyribonucleotide, structure of, 630
Polydextran column, 121F
Polyhydroxyaldehydes, 253
Polyhydroxyketones, 253
Polymerase chain reaction, 679–680, 681F
site-directed mutagenesis and, 689, 690F
Polymers, 9, 10F
Polymorphic genes, 690
Polymyxin B_1, 503T
Polynucleotide(s), structure of, 9, 12F
Polynucleotide chain, mononucleotides in, 630–631, 631F
Polynucleotide ligase, 659, 659F
Polynucleotide phosphorylase, 701
Polyols, 253
Polyoma virus, oncogenes in, 854–855, 855T
Polypeptide(s)
eukaryotic gene expression regulation at level of processing of, 817–818, 819F
structure of, 9, 11F
synthesis of, 66, 67F
mRNA as carrier of information for, 702–703
Polypeptide chain(s), 14, 15F
amino acid reaction to form, 11F
end groups of, determination of, 61, 63F, 71
folding of, 72
fragmentation of, 61, 64F
secondary valence forces in, 86T, 86–88
electrostatic forces and, 87–88, 88T
hydrogen bonds and, 86–87
hydrophobic effects and, 87
van der Walls forces and, 87

synthesis of, 57, 57F, 58F
Polypeptide-chain-binding proteins, 757
Poly(A) polymerase, 717
Polysaccharides, 9, 10F, 248–249
    configurations of, 249, 250F, 251F
    energy storage by, 248, 248F
    monosaccharide integration into, 357F, 357–359
    storage, synthesis of, 264–266, 265F
    structural, 358–359, 359F, 360T, 361F
    structure of, 10F
Polysome, 4F, 5F, 731, 733F
Poly(A) tail, 731
Polytene chromosomes, 809, 809F
Porins, 406F, 407
Porphyridium cruentum, oxidation of cytochrome *f* in, 344, 344F
Porphyrin, synthesis of, 526, 527F, 528F
Positive cooperativity, 103
Posttranslational transport, of mitochondrial proteins, 757–758, 759F
Potassium
    in earth's crust, 25T
    electric potential difference across plasma membrane created by
        pumping and diffusion of, 603
    in human body, 25T
    in ocean, 25T
    plasma membrane conductivity to, 604–605
Potassium channels
    inactivation of, 608, 608F
    of nerve cells, 400
    voltage-gated, mediation of action potentials by, 606–608,
        606–608F
Power stroke, 113
Precipitation, differential, of proteins, 119–120
Pregnenolone, 576
    synthesis of, 475, 478F
Pre-mRNA, processing of, 719–721, 720F
Preproinsulin, 571
Preproopiomelanocortin, 571, 571F
Preproprotein, 757
Preprotein, 757
Primase, 660
Primer, 653
Prochiral compounds, 293, 293F
Progesterone, 475, 576
    structure and function of, 479T
    synthesis of, 478F
Progestins, 573T
    gene regulation by, 586, 587T
Prokaryotes, 7, 26F. *See also specific prokaryotes*
    chromosomes of, 22
    evolution of, 25, 26, 26F, 27F
    functional sites on ribosomes of, 735F, 735–736
    photosynthesis in, 332
    protein synthesis in, 731, 734F
Prokaryotic gene expression, 768–797
    control of transcription as mode of, in *Escherichia coli,* 769
    coordinate regulation of genes for ribosomes and, 780–783
        rel gene control of rRNA and tRNA synthesis and, 781F,
            781–783, 782F
        translational control of ribosomal protein synthesis and, 783, 783F
    DNA-regulatory protein interaction and, 789–795
        helix-turn-helix motif and, 789–790, 790F, 791F
        recognition of specific regions in DNA complex and, 789
        β-sheet motif and, 791, 795F
        small molecule involvement in, 791, 795, 795F
        specific contacts between base pairs and amino acid side chains and,
            790–791, 792–794F
    enzyme regulation at transcription level, *trp* operon regulation after
        initial point for transcription and, 777, 779F, 779–780, 780F
    enzyme regulation at transcription level and, 777–780, 778F
    initiation point for transcription and major site for regulation of, 769,
        770F

lac operon regulation at transcription level and, 770F, 770–777
    activator protein augmenting operon expression and, 775–777, 777F
    augmentation of β-galactosidase synthesis by small-molecule inducer
        and, 771F, 771–772, 772F
    gene leading to repression of synthesis in absence of inducer and,
        772–774, 774T
    locus required for repressor action and, 774
    operon hypothesis and, 774–775, 775F, 776F
regulation in bacterial viruses, 783–789
    buildup of *cII* regulatory protein and, 786, 788F, 789
    Cro protein prevention of buildup of *cI* protein during lytic cycle
        and, 785, 788F
    lysis or lysogeny following infection of *Escherichia coli* by
        bacteriophage gamma and, 784–785, 786F
    maintenance of dormant prophage state of gamma by phage-encoded
        repressor and, 784, 785F
    N protein as autoterminator and, 785, 787F
    Q protein as autoterminator and, 785
    regulatory proteins directing metabolism and, 784, 784T
Prolactin, 572T
Prolactin-releasing factor, 572T
Proline, 513T
    p*K* value of, 54T
    structure of, 51F
    synthesis of, 515T
Proline hydrolase, 216
Promoter(s), 706
    binding and, 708–709, 709F
    eukaryotic, location of, 715, 715F
    initiation of transcription at, 709, 710F, 711T
Promoter locus, 771
Proofreading hydrolysis reactions, 744–745
Properties, intensive and extensive, 30
Proprotein, 757
Prostaglandin(s), 452F, 452–453
    in blood clotting, 454, 456F
    inhibition of, aspirin inhibition of, 453
    synthesis of, 455F
Prostaglandin E₂, structure of, 452F
Prostaglandin endoperoxide synthase, reaction catalyzed by, 453, 453F
Prostate cancer, incidence of, 851T
Prosthetic groups, 199
Proteases, 514–515, 515F
Protein(s), 7, 9, 118–130. *See also specific proteins*
    allosteric, 101–110. *See also* Hemoglobin
        function of, models of, 109–110, 110F
    amino acid composition of, determination of, 58–60, 59F, 60F
    amino acid sequence of. *See* Protein(s), primary structure of
    cell adhesion, 849
        required to maintain immune response, 844–845, 846F
    characterization of. *See* Protein(s), fractionation and characterization of
    diffusion constants of, 124T
    DNA-binding
        Footprinting technique to determine binding site of, 729, 729F
        interaction with DNA, prokaryotic gene expression and. *See*
            Prokaryotic gene expression, DNA-regulatory protein interaction
            and
        regulating transcription in eukaryotes, 812–817
    in DNA replication, 654–656, 655T
    docking, 759
    fibrous, 72
        structure of, 73F, 73–80, 74F
        x-ray diffraction patterns, 74–75
    fractionation and characterization of, 119–124
        column procedures for, 120–122, 121F, 121T
        differential centrifugation for, 119, 119T
        electrophoresis for, 122F, 122–123, 123F
        sedimentation and diffusion for size and shape determination and,
            123F, 123–124, 124T
        solubility differences as basis of, 119–120
    functions of, 101–115

globular, 72
  folding of, 82–92, 83F
heat shock, 763
isoelectric points of, 124T
light absorption by, 70
membrane-bound, purification of, 127–128, 128T, 129F
in membranes, 388F, 388–390, 389F
  connection to cytoskeleton in eukaryotes, 396, 397F, 398F
  α-helices of, 389–390, 390F, 391F
  movement of, 390–391, 392F, 393, 393F
  orientations of, 394, 394F
mitochondrial, posttranslational transport of, 757–758, 759F
molecular weights of, 124T
mRNA and. *See* Translation
muscle contraction and, 110, 111F, 112F, 112T, 113, 114F, 115, 115F
oligosaccharide attachment to, 362, 363F
primary structure of, 61, 82, 83F
  determination of, 61, 62–66F, 65–66
  determination of tertiary structure by, 82, 86, 86F
purification of, 124–128
  of enzyme with two catalytic activities, 125–127, 126T, 127F
  of membrane-bound proteins, 127–128, 128T, 129F
quaternary structure of, 82, 83F, 91T, 91–92, 92F
regulatory, gamma metabolism control by, 784, 784T
repair, synthesis of, regulation of, 665–666, 666F
secondary structure of, 82, 83F
sedimentation constants of, 124T
in skeletal muscle, 112T, 113
structure of, 11F, 72–93
  of fibrous proteins, 73F, 73–80, 74F
  of globular proteins, 80, 82–92, 83F
  molecular models of, 75T, 75–77, 76F, 77F, 82, 84–85F
  radiation techniques for analysis of, 99–100, 100F
  x-ray diffraction for analysis of, 97–99, 97–99F
subunits of, 91T, 91–92, 92F
synthesis of, 731, 732F, 733F, 733–736. *See also* Translation
  DNA and, 22, 23F, 24F
  localization in cell, 20
  mRNA as template for, 731, 734F
  ordering of activated amino acids in mRNA template by tRNAs in, 731, 733, 734F, 735
  ribosomes as site of, 735F, 735–736
  tRNA as carrier of amino acids to template for, 703F, 703–704, 704F
targeting and posttranslational modification of, 757–760
  of bacterial proteins, 760
  collagen and, 760, 761F
  glycosylation of proteins passing through Golgi apparatus and, 760
  of mitochondrial proteins, 757–758, 759F
  signal sequences and, 757, 758F
  synthesis of eukaryotic proteins in endoplasmic reticulum and, 758–760
tertiary structure of, 82, 83F
  determination by primary structure, 82, 86, 86F
  domains as functional units of, 88–90, 88–90F
  predicting, 90F, 90–91
tissue stores of, 563T
transporter. *See* Transporters
turnover of, 760–764
  ATP in, 764
  differing lifetimes of proteins and, 761–762, 762T
  partial, 176F, 176–177, 177F, 177T
  proteolytic hydrolysis in mammalian lysosomes and, 763
  selective degradation of abnormal proteins and, 762–763
  ubiquitin tagging of proteins for proteolysis and, 763, 764F
Protein kinase(s)
  phosphorylating and inactivating eIF-1 and EF-s, 817–818, 819F
  phosphorylation by, 177, 178F
Protein kinase C, 443, 584, 586
Protein targeting, 365, 367
Proteolysis. *See* Protein(s), turnover of

Proton(s)
  flow back into matrix, 321–322, 322F
  pumping out of mitochondria during respiration, 318, 321, 321F
  return to stroma through ATP-synthase, 346–348, 348F
  transport into thylakoid lumen, 346, 347F
Proton gradient, 283, 284F
Proton pumps, driven by electron-transfer reactions, 402
Protooncogenes
  transition to oncogenes from, 861–862
  tumors arising by mutational events in, 852–853
Protoporphyrin IX
  reactions involving, 217–218
  structure of, 218F, 334F
Proximity effect, in enzyme catalysis, 155
Pseudocycles, 234
Pseudohermaphroditism, female, 576–577
Pseudomonas aeruginosa, toxin produced by, 752
Pseudouridine, in tRNA, structure of, 704F
D-Psicose, 245F
Ptashne's domain-swap experiment, 805F
Pteroylglutamic acid. *See* Folic acid
Puffs, 809, 809F
Purine(s), 535
  catabolism of, 554–556F, 555
  conversion to nucleotides, 548–549, 549F
  synthesis of, regulation of, 556, 558, 558F
Purine bases, nucleotide formation from, 548–549
Purine nucleoside(s), 548
Purine nucleoside phosphorylase
  deficiency of, 555
  reaction catalyzed by, 555
Purine ribonucleotides, synthesis of, *de novo*, 538–543, 539F, 540F
Purple bacteria, 26F
  photosynthetic properties of, 333T
  reaction centers of, 337, 338F
    electron movement in, 338, 340F
Pyranoses, 245
Pyridoxal-5′-phosphate
  reactions involving, 200, 202–203, 203F, 204F
  structure of, 200, 201F
Pyridoxal phosphate, covalent bonding to phosphorylase, 192, 193F
Pyridoxamine, structure of, 201F
Pyridoxine (vitamin B_6), 199T
  coenzyme form of, 200, 201F, 202–203, 203F, 204F
  structure of, 201F
Pyrimidine(s), 535
  catabolism of, 555–556, 557F
  synthesis of, regulation of, 558
Pyrimidine bases, DNA repair and, 664, 664F, 665, 665F
Pyrimidine ribonucleotides, synthesis of, *de novo*, 543F, 543–545
Pyrophosphate, 265F, 265–266
Pyrophosphate bonds, 537, 537F
Pyrophosphate group, 17F
Pyruvate
  decarboxylation of, 287–289, 288F
  from glucose, 229, 230F. *See also* Glycolysis
  partitioning between acetyl-CoA and oxaloacetate, 299–300
  reduction to lactate, 259, 261, 261F
  standard free energy of formation of, 36T
Pyruvate amino acid family, 497–499, 498F
  isoleucine and valine synthesis and, 497, 498F, 499
  safe herbicides and, 499, 499F
Pyruvate carboxylase
  acetyl-CoA regulation of, 298F, 299
  reaction catalyzed by, 263F, 263–264, 295
Pyruvate dehydrogenase
  acetyl-CoA regulation of, 298F, 299
  molecular weight of, 91T
  subunit composition of, 91T
Pyruvate dehydrogenase complex, 287–289, 288F
  nature of, 287, 289, 289F

Pyruvate kinase, reaction catalyzed by, 260T
Pyruvate-phosphoenolpyruvate pseudocycle, 263F, 263–264
Pyruvate-thiamine adduct, 201F

Q cycle, 314F, 314–315
Q protein, 785
Quantum requirement, 345
Quantum yield, 338

Rabbit reticulocyte protein synthesis, 819F
Racemase, 202
Racker, Efraim, 317, 322
Radiation techniques, for protein structure analysis, 99–100, 100F
Radiolabeled compounds, for pathway analysis, 238–239
Ramachandran, 80
Raman scattering, protein structure analysis using, 100T
Random-order pathway, 144
Range, in DNA library, 687
ras gene, 859F, 859–860
ras sequences, 811
Rate constant, 137
Ratner, S., 519
Reaction centers, 337. *See also* Photosynthesis, reaction centers in
Reading frame, 731
recA enzyme
    DNA repair and, 665–666, 666F
    reaction catalyzed by, 670F, 670–671
Recessive gene interactions, 773
Recognition helix, 789
Recognition markers, carbohydrates as, 359, 361
Recombinant DNA techniques, characterization of globin gene family by, 689–697, 691F
    chromosome walking and jumping and, 694–697, 695F, 696F
    DNA sequence differences and, 690, 692F, 692–694, 693F
    β-globin DNA probe and, 694
Rectal cancer, incidence of, 851T
Red algae, 26F
Red blood cells, 6F. *See also* Hemoglobin
    anion transporter of, 400
    glycophorin topography in, 394, 394F
    pentose phosphate pathway in, 272
Redox couples, $E°$ values of, 310, 311T
Redox potentials, 310, 311F, 311T, 312, 312T
Redox reactions
    flavin coenzymes and, 209
    iron-containing coenzymes and, 217–220, 218–220F
Reducing power, sources of, differences among organisms and, 228T, 228–229
Reductive pentose cycle, 348, 349F, 350
Relaxed DNA, 637
Relaxin, 573T
Release factors, termination of translation and, 754F, 754–755, 755F
rel gene, rRNA and tRNA synthesis controlled by, 781F, 781–783, 782F
Renaturation, of DNA, 639–641, 640F, 641F
Replication eye, 652, 653F
Replication forks
    discontinuous growth at, 653–654, 654F
    in *Escherichia coli,* 661F
Repressors, 769
Resonance energy transfer, 336
Respiration, 307
    coupling with phosphorylation, 317, 317F
    electron transport and. *See* Electron transport
    photorespiration, 350–351, 351F, 352F
    pumping of protons out of mitochondria during, 318, 321, 321F
Respiratory metabolism, 283, 284F
Restriction enzymes, in DNA cloning, 682F, 682–683, 683T
Restriction fragment length polymorphisms, 692, 693F, 693–694
    linked to cystic fibrosis gene, 694, 695F

Reticulocyte, protein synthesis in, regulation of, 819F
Retina, 615
11-*cis*-Retinal, 614, 616
    absorption spectrum of, 617F
    isomerization of, 616
    structure of, 617F
Retinol, 615
trans-Retinol, 199T, 221–222
Retroviruses
    replication of, 671, 672F
    RNA polymerases of, 716
Reverse cholesterol transport, 472
Reverse genetics, 695–696
Reverse transcriptases, 671–673
    bacterial, 673
    hepatitis B virus replication and, 671
    retrovirus replication and, 671, 672F
    telomerase and, 673, 673F
    transposable genetic elements and, 671
Reversible reactions, 33
    kinetics of, 137–138
R groups, of amino acids, 9, 11F, 53–54, 54T
    classification according to, 50, 51T
Rhodanese
    domains in, 89, 89F
    molecular weight of, 91T
    subunit composition of, 91T
Rhodobacter capsulatus, poring from outer membrane of, 406F, 407
Rhodobacter sphaeroides, reaction centers in, 337
Rhodopseudomonas viridis, reaction centers of, 337–338, 339F
Rhodopsin, 614, 615–619, 617F
    absorption spectrum of, 616, 617F, 618, 618F
    isomerization of retinal of, 616
        structural changes caused by, 618–619, 619F, 620F
    photochemical transformations of, 616, 618, 618F
Rhodospirillum rubrum, cell of, 333F
rho protein, termination of transcription and, 710
Riboflavin (vitamin B$_2$), 199T
Ribonuclease(s)
    as RNAs, 722–723
    unfolding of, 82, 86, 86F
Ribonuclease A
    catalytic activity of, 165–166, 165–169F, 168–169
    complex with uridine vanadate, crystal structure of, 169, 169F
    inhibition of, 165–166, 166F
    molecular weight of, 91T
    structure of, crystal, 166, 167F
    subunit composition of, 91T
Ribonucleic acid. *See* RNA
Ribonucleotide(s), structure of, 536F
Ribonucleotide reductase, reaction catalyzed by, 545–546, 546F
D-Ribose, 244F
Ribose(s), 535
    structure of, 10F
Ribose-5-phosphate, synthesis of, 274, 275F, 276
Ribosomal RNA. *See* rRNA
Ribosome(s), 4F, 5F
    antibiotic binding to sites on, 756, 756F
    find structure of, 705, 706F, 707F
    mRNA binding to, 746, 746T
    processing of ribosomal precursor and, 719, 719F
    prokaryotic, functional sites on, 735F, 735–736
    protein synthesis in, translational control of, 783, 783F
    protein synthesis on, 746–747
        dissociable protein factors and, 746–747
    as protein synthesis site, 735F, 735–736
    reading frame change by, during translation, 755, 756, 756F
    regulation of genes for, 780–783
        rel gene and, 781F, 781–783, 782F
        translational control of protein synthesis and, 783, 783F
    rRNA as integral part of, 705, 705F

Ribosome-binding site, 731
Ribosome entry site, 731
Ribosome releasing factor, 754–755
Ribothymidine, in tRNA, structure of, 704F
D-Ribulose, 245F
Ribulose biphosphate carboxylase, reaction catalyzed by, 348, 349F, 350, 350F
Ribulose biphosphate oxygenase, reaction catalyzed by, 350–351, 351F
D-Ribulose-5-phosphate, 272F, 275F
Rich, Alex, 635
Richards, Frederick, 165
Rickets, 590
Rickettsia, 5F
Rifampicin, 724F
Rifamycin, 724F
Rigor mortis, 113, 115F
Rings, entropy and, 32
Rittenberg, David, 419, 461
RNA, 9. *See also* mRNA; rRNA; Transcription; Translation; tRNA
    catalytic, evolutionary significance of, 723, 725
    classes of, 702T, 702–705. *See also* mRNA; rRNA; tRNA
    cleavage by pancreatic ribonuclease, 165, 165F
    DNA polymerases directed by. *See* Reverse transcriptases
    DNA sequences carried by, 701–702, 702F
    fine structure of ribosome and, 705, 706F, 707F
    functioning like enzymes, 722–723
        ribonucleases and, 722–723
        rRNA catalysis of peptide bond formation and, 723
        self-splicing and, 722, 723F
    guide, 721
    inhibitors of metabolism of, 724F, 725
    leader, secondary structures in, 779–780, 780F
    messenger. *See* mRNA
    posttranscriptional alterations of transcripts and, 717F, 717–721
        enzymes required for, 717–719, 718F
        eukaryotic pre-mRNA processing and, 719–721, 720F
        ribosomal precursor processing and, 719, 719F
    ribonucleases as, 722–723
    ribosomal. *See* rRNA
    RNA inhibitors incorporated into growing chain of, 725
    splicing of, antibody diversity and, 837
    synthesis of, 701, 701F. *See also* Transcription
        localization in cell, 20
        types of, 716–717
    transfer. *See* tRNA
RNA editing, 721, 722F
RNA polymerases, 701
    bacterial, subunits of, 706–707
    DNA-dependent, reaction catalyzed by, 705–706
    eukaryotic, 712–715
        functionality of, 712–713, 714F
        number of, 712, 713T
        TATA-binding protein requirement of, 713, 715
    reaction catalyzed by, 660
    RNA-dependent, of viruses, 715–716, 716T
    RNA inhibitors binding to, 725
    of viruses, 715–716
        encoding of, 715–716
        RNA-dependent, 715–716, 716T
Robert, Ric, 720
Robinson, Robert, 461
Rod cells, 615, 615F, 616F
    electrophysiological response to light, 620–621, 621F
Rodlike viruses, structure of, 92, 92F
Rose, W.C., 513
Rotation, entropy associated with, 32
Rotational entropy, 32T
Rough endoplasmic reticulum, 4F, 5F
    mammalian, 731, 733F
    protein synthesis in, in eukaryotes, 758–760
Roundup, 499

Rous, Peyton, 853
Rous sarcoma virus, 853
rpoB gene, 716t
rpoC gene, 716t
rpoD gene, 716t
rpsA gene, 716t
rRNA
    catalysis of peptide bond formation by, 723
    in frog eggs, elevation by DNA amplification, 819–820, 820F
    as integral part of ribosome, 705, 705F
    5S, synthesis in frogs, 820
    synthesis of, control by *rel* gene, 781F, 781–783, 782F

Saccharomyces cerevisiae
    as cloning system, 689
    gene expression in. *See* Eukaryotic gene expression, in yeasts
    introns in, 721
    ras genes of, 859
Salmonella typhimurium, glutamine synthase of, 491F
Salt bridges, 643
Salting in, 119–120, 120F
Salting out, 120, 120F
Samuelsson, Bengt, 453
Sanger, Fred, 61, 570
Sanger method
    of amino acid sequence determination, 61, 63F
    of DNA sequencing, 679, 680F
Sarcolemma, 110, 111F
Sarcomere, 110, 111F
Sarcoplasmic reticulum, 110F
Satellite tobacco necrosis virus
    molecular weight of, 91T
    subunit composition of, 91T
Saxitoxin, 605
Sceptor, 499
Schachman, Howard, 187
Schiff's base, 157, 157F, 202–203, 203F
    pronated, 618–619, 618–620F
    retinal binding to, 616
    structure of, 617F
Schoenheimer, Rudolf, 461
Scurvy, 216
Second-messenger roles, 584, 585F, 586
Second-order kinetics, 138
Secretin, 573T
Sediment, protein fractionation and, 119
Sedimentation and diffusion, for protein size and shape determination, 123F, 123–124, 124T
Sedoheptulose-7-phosphate, synthesis of, 273, 273F, 274F
Segmentation genes, 822, 824–825, 825F
Segment-polarity genes, influence on maternal-effect genes, 825, 825T
Seitz, Tom, 657
Selenium
    in earth's crust, 25T
    in human body, 25T
    in ocean, 25T
Sequenators, 65
Sequences
    conversions distinguished from, 232
    coupling of favorable and unfavorable reactions and, 39–40
    organization into oppositely directed pairs, 233, 233F
Sequential model, of hemoglobin function, 109–110, 110F
Serine, 384T, 513T
    at active site, 160, 161, 162, 162F, 163, 164
    equilibrium between charged and uncharged forms of, 54F
    pK value of, 54T
    structure of, 51F
Serine amino acid family, 495–497
    cysteine synthesis and, 495, 496F, 497
    sulfate reduction to sulfide and, 497, 497F

Serine proteases. *See also specific enzymes*
 catalytic activity of, 159–164F, 159–165
Serine transacetylase, reaction catalyzed by, 495, 496F
Serotonin. *See* 5-Hydroxytryptamine
Sharp, Phil, 713, 720
Shemin, David, 526
Shine-Delgarno sequence, 731, 734F, 746, 746T, 788F, 789
Shingles, pharmacotherapy for, 552, 553T
Shuttle systems, in import of electrons from NADH, 325
Shuttle vectors, in DNA cloning, 686
Sia, 628
Sialic acid, structure of, 357F
Sickle cell anemia, 109
Sickle-cell disease, inheritance pattern of restriction fragment length
   polymorphism associated with, 693F, 693–694
Signal hypothesis, 757
Signal recognition particle, 362, 759
Signal sequences, 362
 targeting or proteins to destination by, 757, 758F
Silicon
 in earth's crust, 25T
 in human body, 25T
 in ocean, 25T
Silks, β-keratin structure in, 78, 79F
Singer, Jon, 390–391
Single-stranded binding protein, 659–660
sis gene, 856, 857T, 858, 858F
Site-directed mutagenesis, 689, 690F
Site-specific recombination, 671
Site-specific variation, in translation elongation, 739
Skeletal muscle
 aerobic and anaerobic function of, 565F, 565–566
 cells of, 6F
 contraction of, 110, 111F, 112F, 112T, 113, 114F, 115, 115F
 energy-related reactions in, 568F
 proteins of, 112T, 113
 rigor mortis and, 113, 115F
Skin cancer, incidence of, 851T
Slime molds, 26F
Smooth endoplasmic reticulum, 4F, 5F
Snake venoms
 neurotoxins of, 610
 phospholipase A$_2$ in, 447
snRNAs, 719–720
snRNPs, 719–720
Sodium
 in earth's crust, 25T
 electric potential difference across plasma membrane created by
    pumping and diffusion of, 603
 in human body, 25T
 in ocean, 25T
 plasma membrane conductivity to, 604–605
Sodium channels
 of nerve cells, 400
 voltage-gated, mediation of action potentials by, 605, 606F
Sodium dodecyl sulfate gel electrophoresis, 122, 123F
Sodium-potassium pump
 catalytic cycle of, 404–406, 406F
 electrophysiological response to light and, 620–621, 621F
Solubility, of proteins, at isoelectric point, 119, 120F
Solutions
 buffered, 53
 entropy of, 33, 33F, 34F
 ultraviolet light absorption in, measurement of, 70
Solvation, 33, 33F, 34F
Somatic mutation, 834, 837T, 837–838
Somatic recombination, 834
Somatostatin, 571, 572T, 573T
Somatotropin, 572T
D-Sorbose, 245F
Southern blotting, 692F, 692–693, 693F

Space-filling models, 75, 76F
Specific activity, 125
Specificity constant, of enzymes, 144, 145T
Spectrin, structure and location of, 396, 397F
Spectrometry, protein structure analysis using, 100T
Spectrophotometer, 70
Spectrophotometric assays, for study of transport, 403
Spheroidene, structure of, 342F
Sphinganine, synthesis of, 447, 449F
Sphingolipid(s), 447–452
 degradation of, defects in, 450–452, 451F
 functions of, 450
 structure of, 383, 385F
 synthesis of, 447–448, 449F, 450, 450F
Sphingolipidoses, 451–452
Sphingomyelin
 structure of, 383, 385F
 synthesis of, 447–448, 449F
Spiegelman, Sol, 701
Spirillum, 5F
Spliceosome, 721
Splicing, 719–721
 of DNA, antibody diversity and, 834–837, 835–837F, 838, 838F
 of mRNA, as mechanism for posttranscriptional regulation, 817, 818F
 removal of internal sequences and, 720–721, 721F
 of RNA, antibody diversity and, 837
 self-splicing RNAs and, 722, 723F
 steps in, 721, 722F
Sponges, 26F
Spontaneity, free energy as criterion for, 35
Spontaneous processes, entropy and, 31–35, 32T, 33F, 34F, 35T
Squalene, synthesis of, 464, 466F
Squalene-2,3-oxide, synthesis of, 464
Squamous epithelium, 6F
src gene, 856, 857T
SRP receptor, 759
Stahl, Frank, 651
Standard free energy of formation, 36T, 36–38, 37T
Standard hydrogen half-cell, 310
Standard redox potential, 310, 311F, 311T, 312, 312T
Staphylococcus, 5F
Starch
 intestinal digestion of, 251, 253
 structure of, 248, 248F
 synthesis of, 264–266, 265F
   elongation step in, 265F
Starch grain, 5F
Start codons, 731
Starvation, fatty acids and ketone bodies during, 567, 567F
States, 30
 entropy and, 32–33
 ground, of molecules, 335, 336F
 standard, of protons, 38
Steady state, 228
 enzymatic reactions and, 141–144, 142F, 143F
Stearic acid, 384T
 structure of, 8F
Stearoyl-CoA desaturase, reaction catalyzed by, 425–426, 426F
Stein, William, 165
Steitz, Tom, 657, 744
Stem cells, pluripotent, 838F, 839
Stereochemistry, of tricarboxylic acid cycle, 292F, 293, 293F
Stereoisomeric pairs, 56, 56F
Steroid(s), 459. *See also specific steroids*
 biological roles of, 459, 461
Steroid hormones, 459, 475, 478F, 479F, 479T
 activation of, 586, 586F
 defective synthesis of, 475, 479F
 modulation of rate of transcription by, 586, 586F, 587T
 receptors for, 579
 synthesis of, 574, 576–577, 577–579F

Sterols, in membranes, 383, 385
Stoeckenius, Walter, 322
Stoichiometry, of tricarboxylic acid cycle, 293–294
Stomach cancer, incidence of, 850, 851T
Stop codons, 731
Stopped-flow device, 140, 140F
Streptococcus pneumoniae, transformation of, 628, 629F
Streptolydigin, 724F
Streptomyces, as cloning system, 689
Streptomyces antibioticus, 725
Streptomycin, structure of, 756F
Streptomycin sulfate, nucleic acid aggregation by, 126
Stringent response, 781
Stroke, 475
Stroma, 332, 332F
    proton return to, 346–348, 348F
Structural analysis methods, 281, 281F
Substrates, 18F, 19, 135. *See also* Enzyme-substrate complex
    binding to enzymes, entropy and, 34F, 34–35, 35T
    concentration of, kinetic parameter determination by measuring initial
        reaction velocity as function of, 140, 140F, 141F
    enfolding of, 158, 158F
    enzyme inhibition and. *See* Enzymes, inhibition of
    fatty acid biosynthesis limitation by supply of, 430, 431F
    free, enzyme-substrate complex in equilibrium with, 140–141
    kinetics enzymatic reactions involving two substrates and, 144–146,
        145F, 146F
    radioactively labeled, for study of transport, 402
    reaction velocity as function of, 235, 236F
    sigmoidal dependence of allosteric enzymes on concentration of, 180F,
        180–182, 181F
Subtilisin, amino acid sequence of, 160, 161F
Succinate, standard free energy of formation of, 36T
Succinate dehydrogenase, 309
    reaction catalyzed by, 291–292, 292
    reactions involving, 209T
Succinate dehydrogenase complex, 312, 312F, 313, 313F, 313T
Succinate thiokinase, reaction catalyzed by, 291
Succinyl anhydride, synthesis of, 155
Succinyl-CoA
    in porphyrin synthesis, 526, 527F, 528F
    synthesis of, 290–291
Sucrose, 248
    structure of, 247F
Sugar(s). *See also* Monosaccharides; *specific sugars*
    free, conversion to hexose phosphates, 253–254, 254F
    synthesis and breakdown of, 249
Sugar monomers, 9, 10F
Sugar polymers, 9, 10F
Sulfometuron methyl, structure of, 499F
Sulfonamides, 550T, 551–552, 552F
Sulfonylureas, 499
Sulfur
    in earth's crust, 25T
    in human body, 25T
    in ocean, 25T
Sumner, James, 136
Supercoiled DNA, 636–638, 637F, 638F
Supernatant fraction, 119
Surroundings
    first law of thermodynamics and, 30–31, 31F
    second law of thermodynamics and, 31–35, 32T, 33F, 34F, 35T
Sutherland, Earl, 191, 268, 776
SV40
    cleavage map of, 682, 682F
    enhancer in, 811–812, 812F
    oncogenes in, 854–856, 855T
    replication of, 663, 664F
Swainsonine, 370T
Sweeley, Charles, 452
Symmetry model, of hemoglobin function, 109

Synapses, 602
Synaptic transmission, mediation by ligand-gated ion channels, 609–610,
    609–612F
Synaptic vesicles, 609
Synthases. *See* Ligases
Synthetic reactions. *See* Anabolic reactions
Synthetic work, 38–40
    coupling of favorable and unfavorable reactions to perform, 38–40, 39F
Systems
    first law of thermodynamics and, 30–31, 31F
    second law of thermodynamics and, 31–35, 32T, 33F, 34F, 35T
Szent-Györgyi, Albert, 284
Szostak, Jack, 723, 725

D-Tagatose, 245F
D-Talose, 244F
TATA body, 713
    TATA-binding protein requirement of eukaryotes and, 713, 715
TATA box, 811, 812F
Tatum, E.L., 523
Tay, Warren, 452
Tay-Sachs disease, 452
T cell(s), 831, 831F, 841–845. *See also* Cell-mediated response
    helper, 831
        triggering of B-cell division and differentiation by, 839–840, 840F
    interaction with B cells, antibody formation and, 838F, 838–840
    suppressor, 831
T-cell receptors, 844
Telomerase, eukaryotic chromosome replication and, 673, 673F
Telomeres, 673
Temin, Howard, 854
Temperature
    annealing, 640
    DNA mutants sensitive to, 655–656
    enzymatic activity and, 146, 146F
    membrane fluidity and, 395F, 395–396, 396T
Tendons, cells of, 6F
Teratocarcinoma, 851–852
Teratoma, cytoplasmic influence on nuclear expression and, 808
Testicular feminization, 590
Testosterone, 576
    conversion to 5α-dihydroxytestosterone, 577, 577F
    metabolic conversion by target cells, 577, 578F
    structure and function of, 479T
Tetapeptide, 65F
Tetracycline, structure of, 756F
Tetrahydrofolate
    reactions of, 215
    structure of, 214F, 215
Tetrapyrrole
    linear, polymerization in, 526, 528F
    synthesis of, 526, 527F
Tetrodotoxin, 605
TFIIIA protein, 5S rRNA synthesis in frogs and, 820
T7 *gene1,* 716t
Thermoacidophiles, 27F
Thermodynamics, 29–33. *See also* Energy; Free energy
    first law of, 30–31, 31F
    second law of, 31–35, 32T, 33F, 34F, 35T
    of tricarboxylic acid cycle, 294, 294T
Thermolysin, 141F
    zinc in, 158
Thiamine (vitamin B$_1$), 199T
Thiamine, coenzyme forms of, 199–200, 200F, 201F
Thiamine pyrophosphate (TTP)
    reactions involving, 199–200, 201F
    structure of, 199, 200F
Thiamine pyrophosphate, ylid form of, 287, 288F
Thick filaments, 110, 111F, 112F, 113, 114F
Thin filaments, 110, 111F, 112F, 112T, 113, 114F

Thin-layer chromatography, 65, 66F
2-Thiocytidine, in tRNA, structure of, 704F
Thiolase, reaction catalyzed by, 414, 461, 462F
Thiol esters, enolization of, 211–212
Thiol group, 17F
Thiopurines, 549, 550T, 551, 551F
Thioredoxin, 178, 179F, 545–546, 546F
6-Thiosine-5′-monophosphate, 551, 551F
4-Thiouridine, in tRNA, structure of, 704F
Threonine, 513T
    equilibrium between charged and uncharged forms of, 54F
    pK value of, 54T
    structure of, 51F
    synthesis of, 20–21, 21F
D-Threose, 244F
Thrombosis, prostaglandins in, 454, 456F
Thromboxane(s), synthesis of, 455F
Thromboxane A$_2$
    structure of, 452F
    synthesis of, 454
Thromboxanes, 452, 452F
Thudichum, Johann L.W., 447
Thylakoid lumen, 332, 332F
    proton transport into, 346, 347F
Thylakoid membrane, 332, 332F
Thymidine-5′-phosphate, structure of, 536F
Thymidylate, synthesis of, 546–547, 547F
Thymidylate synthase, reaction catalyzed by, 546–547, 547F
Thymine, 535
    structure of, 12F
Thyroglobulin, 574
Thyroid hormones, 572T
    gene regulation by, 586, 587T
Thyroid-stimulating hormone, goiter and, 590
Thyrotropin, 572T
Thyrotropin-releasing factor, 572T
Thyrotropin-releasing hormone, synthesis of, 571, 574F
Thyroxine, 572T
    synthesis of, 574, 575F
    underproduction of, 590
Titration curves
    of amino acids, 53F, 53–54, 54F
    for β-lactoglobulin, 54, 56F
T killer cells, 831
Tobacco mosaic virus
    diffusion constant of, 124T
    genetic code of, 736
    isoelectric point of, 124T
    molecular weight of, 91T, 124T
    sedimentation constant of, 124T
    structure of, 92, 92F
    subunit composition of, 91T
α-Tocopherol, 199T, 221, 222F
Todd, Alexander, 630
Tolerance, 842–843, 843F
Tomato bushy stunt virus domain 3, 88F
Tonegawa, S., 834, 836
Topoisomerases, 636–637, 659, 660F
Torpedo californica, acetylcholine receptor of, 610
TΨC loop, 734F, 735
Trace elements, 16
Transaldolase, reaction catalyzed by, 273, 273F
Transamination, of amino acids, 515, 516F
Transcript, elongation of, 709–710
Transcription, 22, 24F, 705–711, 708F
    activated, 713
    basal, 713
    binding at promoters and, 708–709, 709F
    control of, as regulatory mode in Escherichia coli, 769
    elongation in, 706, 708F
        of transcript, 709–710

eukaryotic versus prokaryotic, 711–715
    initiation of, 706, 708F, 709
        alternative sigma factor triggering, 709, 711T
        at promoters, 709, 710F
        as site for gene expression regulation, 769, 770F
        lac operon regulation and, 770, 770F
        RNA polymerases and. See RNA polymerases
        steroid hormone modulation of rate of, 586, 586F, 587T
    termination of, 706, 708F, 710, 711F
Transcription factors, activation domains of, 815–817
Transcription unit, 710, 711F
Transducin, 622, 623F
Transfection, 368
Transferases, 136, 136T
Transfer RNA. See tRNA
Transformation
    artificial, 628
    DNA mediation of, 628, 629F
Transforming genes. See Oncogenes
Transition state
    energy required for reactants to reach, 138, 139F
    formation of, electrostatic interactions and, 156–157
Transketolase, reaction catalyzed by, 273, 274F
Translation, 22, 24F, 742–756
    acylation errors in, aminoacyl-tRNA synthase correction of, 744–745
    amino acid attachment to tRNAs in, 742, 743F
    control of ribosomal protein synthesis and, 783, 783F
    entropy associated with, 32
    eukaryotic gene expression regulation at level of, 817–818, 819F
    GTP in elongation and, 750, 751F
    initiation of, 746, 746T
        protein factors and, 747F, 747–748, 748F
    protein factors in
        dissociable, 746–747
        initiation and, 747F, 747–748, 748F
    repetition of elongation reactions in, 748–749, 749–751F, 752–753, 753F
    ribosomal changing of reading frame during, 755, 756, 756F
    site-specific variation in elongation and, 739
    synthase recognition of amino acids and regions in cognate tRNA and, 743–744, 744F, 745F
    termination of, 754F, 754–755, 755F
    tRNA initiating protein synthesis and, 745–746
Translational entropy, 32T
Translational frameshifting, 755, 756, 756F
Translational jumping, 755
Translation factors, interacting with GTP, 750
Translocation, 749, 751F
Translocation factor, 749
trans orientation, 773
Transpeptidation reaction, 749, 750F
    penicillin inhibition of, 374, 375F
Transporters, 398
    diffusion of solutes down electrochemical potential gradients and, 400
    inhibitors of, 399
    interactions mediated by, 407
    specificity of, 398–399, 399F
Transposable genetic elements, 671
Transverse tubules, 110F
Triacylglycerol(s), 427, 428F
    emulsification of, 475, 477F
    plasma concentration of, 467T
    structure of, 8F
    synthesis of, 412, 413F
        regulation of, 445–446
    tissue stores of, 563, 563T, 566, 567F
    transport of, 470–471
Triacylglycerol lipase, reaction catalyzed by, 427, 569
1,2,4-Triazolo-(1,5-a)-dimethyl-3-(N-sulfonyl)2-nitro-6-methyl sulfonanilide, structure of, 499F

Triazopyrimidines, 499
Tricarboxylic acid cycle, 282–302, 283F, 284F, 286F, 521, 523
  acetyl-CoA production in, 287–289, 288F
  amphibolic nature of, 295, 296F
  ATP stoichiometry of, 293–294
  citrate synthase in, 289, 290F
  discovery of, 283–285, 285F
  fumarate production in, 291–292
  glyoxalate cycle and, 295
  GTP production in, 291
  intermediates in, 295, 297, 299
  isocitrate dehydrogenase in, 289, 291F
  isocitrate production in, 289, 290F
  linkage to urea cycle, 520, 520F
  malate production in, 292
  originally proposed form of, 285, 285F
  oxaloacetate production in, 292
  regulation of, 298F, 299–301
    of citrate synthase, 300
    of isocitrate dehydrogenase, 300F, 300–301
    of α-ketoglutarate dehydrogenase, 301
    at pyruvate branchpoint, 299–300
  stereochemical aspects of reactions in, 292F, 293, 293F
  succinyl-CoA production in, 290–291
  thermodynamics of, 294, 294T
Triiodothyronine, 572T
  synthesis of, 574, 575F
  underproduction of, 590
Triose phosphate(s), in glycolysis, 256–258, 258F
Triose phosphate isomerase, 88F
  reaction catalyzed by, 260T
  reactions involving, 257
Triosephosphate isomerase
  specificity constant of, 145T
  structure of, crystal, 169, 170F
Triosephosphate isomerase, catalytic activity of, 170–172F, 171–172
Tripeptide, 63F
Triplet, 22
Trisphosphate, 584
tRNA
  amino acid attachment to, 742, 743F
  as carrier of amino acids to template for protein synthesis, 703F, 703–704, 704F
  cognate, 733
    aminoacyl-tRNA synthase recognition of, 743–744, 744F, 745F
  identity elements in, 744, 745F
  initiation of protein synthesis by, 745–746
  isoacceptor, 733
  nucleotides in, structure of, 704, 704F
  ordering of activated amino acids on mRNA template by, 731, 733, 734F, 735
  posttranscriptional processing and modification of, 717–719, 718F
  structure of, 733, 734F, 735
  synthesis of, control by rel gene, 781F, 781–783, 782F
tRNA nucleotidyltransferase, 716–717
Tropomyosin, in skeletal muscle, 112T, 113
Troponin, in skeletal muscle, 112T, 113
trpA gene, fusion with trpB gene, 502
trp operon, 777–780, 778F
  regulation of, 777, 779F, 779–780, 780F
trp repressor, 790, 794F
Trypsin
  catalytic activity of, 159–164F, 159–165
  inhibition of, by benzamidine, 147, 148F
  partial proteolysis and, 176
  site of action in polypeptide chain cleavage, 64F
  structure of, 159–163
    amino acid sequence and, 159, 160F
    crystal, 159–160, 161F, 162–163
Tryptophan, 513T
  light absorption by, 55, 56F

pK value of, 54T
structure of, 51F
synthesis of, 229, 230F, 499, 500F, 501T, 501–502
  biosynthetic pathway for, 237, 238F
  indole and, 509–510
Tryptophan synthase, 501T
tsf gene, 716t
Tubulin, 396
tuf gene, 716t
Tumor(s). See also Cancer; Oncogenes
  hormone overproduction associated with, 589–590, 590F
  of plants, 594, 595F
Tumor-causing viruses, 853–855
  oncogene association with, 853–854
  role in transformation, 854–855, 855T
Tumor growth factor-α, 591T
Tumorigenesis, multistep nature of, 861–862, 862F
Tunicamycin, 368
Turnover number, of enzymes, 143–144, 144T, 163F
Tyrocidine A, 503T
Tyrocidine B, 503T
Tyrosine, 513T
  adenylation of, 178, 179F
  conversion to fumarate and acetoacetate, 524F
  equilibrium between charged and uncharged forms of, 54F
  light absorption by, 55, 56F
  pK value of, 54T
  structure of, 51F
  synthesis of, 515T
Tyrosine kinase, growth factors and, 591

UAS sequences, 811
Ubiquinone, 308, 309, 309F
Ubiquitin, tagging of proteins for proteolysis by, 763, 764F
UDP-N-acetylmuramyl-pentapeptide, synthesis of, 370, 372F
UDPgalactose-4-epimerase, NAD+ action in, 206, 206F
UDP-glucose, in glycogen synthesis, 266
Ultracentrifugation, apparatus for, 123F
Ultraviolet absorption spectroscopy, protein structure analysis using, 100T
Ultraviolet light, absorption of, measurement in solutions, 70
Uncompetitive inhibition, 147F
Uncouplers, electron transport release from phosphorylation by, 317F, 317–318, 318F
Undecaprenol phosphate, 371, 373F
Underwound, 637
Upstream activator sequences, 715, 715F, 804
Uracil, 535
  structure of, 12F
Urea, ammonia detoxification by formation of, 517, 518F, 519F, 519–520
Urea cycle, 517, 518F, 519F, 519–520
  linkage to tricarboxylic acid cycle, 520, 520F
Urease
  diffusion constant of, 124T
  isoelectric point of, 124T
  molecular weight of, 124T
  sedimentation constant of, 124T
Uric acid, purine catabolism to, 554–556F, 555
Uridine, in tRNA, structure of, 704F
Uridine-5'-monophosphate, structure of, 536F
Uridine monophosphate, synthesis of, 544F, 544–545
Uridine 5'-monophosphate synthase, purification of, 125–127, 126T, 127F
Uridine triphosphate, aspartate carbamoyl transferase inhibition by, 187–188, 188F
Uridine vanadate, complex with ribonuclease A, crystal structure of, 169, 169F
Uridylyltransferase, 493
Urochordates, 26F
Useful work, energy available for, 38

Vacuole, 4F
Valence state, 16–17, 17T
Valine, 513T
  pK value of, 54T
  structure of, 51F
  synthesis of, 497, 498F, 499
Valinomycin, 319, 320F, 503T
Vanadium
  in earth's crust, 25T
  in human body, 25T
  in ocean, 25T
van der Waals forces, in polypeptide chains, 87
van der Waals separation, 75, 87
Vane, John, 453
van Neil, C.B., 336
Variable loop, 734F, 735
Vasoactive intestinal peptide, 573T
Vasopressin, 571, 572T
v-erbA oncogene, 858–859
v-erbB oncogene, 858–859
Vertebrates, 26F
Very-low-density lipoproteins, 467
  cholesterol and triacylglycerol transport by, 470–471
  composition and density of, 468T
  structure of, 468F
  synthesis of, 470, 470F
Vibrational entropy, 32T
Vibrio cholerae, 583, 583F
Viruses. See also specific viruses
  bacterial, gene expression in. See Prokaryotic gene expression,
      regulation in bacterial viruses
  icosahedral, structure of, 92, 92F
  RNA polymerase encoding in, 715–716
  tumor-causing, 853–855
    oncogene association with, 853–854
    role in transformation, 854–855, 855T
Vision, 614–624
  conductivity changes resulting from absorption of photons and,
      619–621, 621F
  guanine nucleotide mediation of effect of light and, 621–622, 622F,
      623F
  rhodopsin and. See Rhodopsin
  rod and cone cells and, 614–615, 615F, 616F
  structure of eye and, 614–615, 615F
Vitamin(s), 198–222
  coenzymes of, 199–220
  lipid-soluble, 199T, 220–222, 221F
  water-soluble, 199T, 199–220
Vitamin A, 199T, 221–222
Vitamin A_1, 615
Vitamin B_2, coenzymes of, 207F, 207–209, 208F, 209T, 210F
Vitamin $B_1$2, coenzymes of, 216–217, 217F, 218F
Vitamin $B_1$2, 199T
Vitamin C. See Ascorbic acid
Vitamin D, 199T
Vitamin D_3, 220–221, 577, 579F
Vitamin E, 199T, 221, 222F
Vitamin K, 199T, 221, 221F
Vitaminlike nutrients, 199T
Voltage-gated ion channels, mediation of action potentials by,
      605–608, 606–608F
von Euler, Ulf, 453
von Gierke's disease, 270

Wakil, Salih, 419
Wald, George, 616
Warburg, Otto, 250, 307, 308
Water
  in cells, 7
  dipolar properties of, 9, 13, 13F, 14F
  on earth's surface, 23, 25T
  interaction with other molecules, 9, 13, 13F, 14F
  phospholipid structures formed in, 386–388, 387F, 388F
  standard free energy of formation of, 36T
  three-dimensional folding of macromolecules and, 9, 13–14,
      13–16F, 16
  vaporization of, entropy increase on, 33
Water-soluble vitamins, 199T, 199–220
Watson, James, 633, 649, 650–651
Weiss, Sam, 701
White blood cells, 6F
Whooping cough, 583
Wire models, 75, 76F
Witt, H., 336
Wobble hypothesis, 738–739, 740F, 740T
Woodward, Robert Burns, 461
Wool, α-helices of, 75–77, 76F, 77F
Work
  concentration, 38
  electrical, 38
  mechanical, 38
  performance of, 31
  synthetic, 38
  useful, energy available for, 38
Wyman, Jeffreys, 109, 180

Xanthomas, 472
X chromosomes, 810
Xenopus laevis
  eggs of
    rRNA in, elevation by DNA amplification, 819–820, 820F
    5S rRNA synthesis in, 820
  potassium channel in, 608
X-ray crystallography, image reconstruction in, 97, 97F
X-ray diffraction, 96F, 96–99
  disadvantages of, 99
  of DNA, 649, 649F
  of fibrous proteins, 74–75
  hemoglobin conformation analysis by, 104–105, 105–107F
  protein structure analysis using, 97–99, 97–99F, 100T
D-Xylose, 244F
D-Xylulose, 245F
Xylulose-5-phosphate, synthesis of, 274, 275F, 276

Yanofsky, Charles, 499, 777
Yeasts, 26F
  alcohol fermentation by, 261
  gene expression in. See Eukaryotic gene expression, in yeasts
  genetic code of mitochondria of, 740, 741T
  haploid cells of, 801, 803F
  RNA of, structure of, 703, 703F
Yoshizawa, Toru, 616
Young, William, 136

Z DNA, 635–636
Zilversmit, Don, 438
Zinc
  in carbonic anhydrase, 220, 220F
  in earth's crust, 25T
  in enzymes, 158
  in human body, 25T
  in ocean, 25T
Zinc finger, 814–815, 815F, 816F
Zippering, 640
Z line, 110, 111F, 112F
Z scheme, 348
Z system, 342–344, 343–345F
Zubay, Geoffrey, 776
Zwitterion form, of amino acids, 11F, 50, 50F, 51T
Zymogens, 159